Innovative Renewable Energy

Series editor

Ali Sayigh
World Renewable Energy Congress, Brighton, UK

The primary objective of this book series is to highlight the best-implemented worldwide policies, projects and research dealing with renewable energy and the environment. The books will be developed in partnership with the World Renewable Energy Network (WREN). WREN is one of the most effective organizations in supporting and enhancing the utilisation and implementation of renewable energy sources that are both environmentally safe and economically sustainable. Contributors to books in this series will come from a worldwide network of agencies, laboratories, institutions, companies and individuals, all working together towards an international diffusion of renewable energy technologies and applications. With contributions from most countries in the world, books in this series will promote the communication and technical education of scientists, engineers, technicians and managers in this field and address the energy needs of both developing and developed countries.

Each book in the series will contain contributions from WREN members and will cover the most-up-to-date research developments, government policies, business models, best practices, and innovations from countries all over the globe. Additionally, the series will publish a collection of best papers presented during the annual and bi-annual World Renewable Energy Congress and Forum each year.

More information about this series at http://www.springer.com/series/15925

Ali Sayigh
Editor

Transition Towards 100% Renewable Energy

Selected Papers from the World Renewable Energy Congress WREC 2017

Springer

Editor
Ali Sayigh
World Renewable Energy Congress
Brighton, United Kingdom

ISSN 2522-8927 ISSN 2522-8935 (electronic)
Innovative Renewable Energy
ISBN 978-3-319-69843-4 ISBN 978-3-319-69844-1 (eBook)
https://doi.org/10.1007/978-3-319-69844-1

Library of Congress Control Number: 2017959613

Printed on acid-free paper

This Springer imprint is published by Springer Nature
The registered company is Springer International Publishing AG
The registered company address is: Gewerbestrasse 11, 6330 Cham, Switzerland

Preface

The World Renewable Energy Congress XVII was held from 5 to 9 February 2017 on the beautiful campus of Murdoch University, just outside of Perth, close to the coastal town of Fremantle, Western Australia. The Congress, with the theme of Transition Towards 100% Renewable Energy, featured keynote speakers and parallel technical sessions highlighting technical, policy, and investment progress towards achieving 100% renewable energy, ranging in scale from households to cities to large regions. There was a specific focus on the challenges and opportunities in transforming the global energy system.

Murdoch University organized the Congress under the chairmanship of Prof. Ali Sayigh and co-chair Prof. Martin Anda.

Over 330 attendees representing more than 33 countries participated, with nearly 200 technical oral presentations and posters.

These proceedings present more than 48 papers covering all aspects of renewable energy topics and building sustainability, following peer review organized by the technical committee and published by Springer Nature in one volume. In addition, 40 papers were selected by the Technical Committee, peer reviewed, and published in the WREC open access journal *Renewable Energy and Environmental Sustainability*, of which Prof. Sayigh is the editor in chief.

Professor Sayigh thanks HE Dr. Abdulaziz bin Othman Altwaijri, director general of ISESCO, for his support of clean energy and renewable energy, and for his contribution to help scientists from ISESCO countries to attend the Congress. He also thanks Springer for their financial donation and excellent book exhibition. The chairman thanks the sponsors as well as the three pillars of the Congress: Dr. Anda, Dr. Mathew, and Dr. Goodfield. On behalf of WREN, Prof. Sayigh thanks Murdoch University, all the student volunteers, and the registration staff for their excellent support and hard work. He would also like to thank the mayor of Fremantle, all the exhibitors, and all the contributors of this volume of proceedings for making this such a successful and enjoyable Congress.

This is the fourth World Renewable Energy Congress to be held at Murdoch University. I hope the readers of this volume will have a rewarding experience in enhancing their knowledge and updating their information on renewable energy progress.

Brighton, United Kingdom

Ali Sayigh
David Goodfield

Contents

Chapter 1
To Import Coal or Invest in Renewables? A Real Options Approach to Energy Investments in the Philippines

Casper Agaton

1.1 Introduction

In the recent years, the Philippines is making a significant stride to become energy independent by developing more sustainable sources of energy. Considering the current share of imported coal for electricity generation at 75% of total supply, the country's security of energy supply has been vulnerable to sudden changes in prices of coal. Renewable energy is seen as a promising alternative to suffice the country's energy needs. In 2015, renewable energy accounts 25% of the country's total energy generation mix, mostly from geothermal and hydropower (DOE 2016). The country is aiming to increase this capacity to 60% of the generation mix by 2030 by developing localized renewable energy resources (DOE 2012). However, investment in renewable energy is challenged by very high start-up cost, immature technology, and competitive prices of fossil fuels. It is therefore important to make a study that analyzes the attractiveness of renewable energy investments to address the country's concern on energy sufficiency and sustainability.

Investment in renewable energy, compare with other types of investments, has important characteristics. First, it is irreversible because the investment cost cannot be recovered once it is installed (Pindyck 1991). Second, investment is affected by various uncertainties including interest rates, technological progress, energy policy, and market prices (Kumbaroğlu et al. 2008). Third, investment is flexible that investors can invest immediately or delay the decision into a lesser risk and more profitable investment period (Yang et al. 2008). These characteristics are not captured by traditional investment valuation methods, including net present value (NPV), internal rate of return (IRR), and return of investment (ROI), making them

C. Agaton (✉)
Institute of Development Research and Development Policy, Ruhr University of Bochum, Bochum, Germany
e-mail: Casper.Agaton@ruhr-uni-bochum.de

A. Sayigh (ed.), *Transition Towards 100% Renewable Energy*,
Innovative Renewable Energy, https://doi.org/10.1007/978-3-319-69844-1_1

inappropriate for evaluating renewable energy investment (Dixit and Pindyck 1995). Real options approach (ROA) overcomes these limitations as it combines risk and uncertainty with flexibility of investment as a potential positive factor which gives additional value to the project (Brach 2003).

The main objective of this paper is to analyze the comparative attractiveness of either investing on renewable energy or continue importing coal for electricity generation under uncertainty in coal prices. Using ROA, this paper evaluates the investment value and optimal timing of switching technologies from coal to renewable. This also identifies the most profitable renewable energy investment among wind, solar, geothermal, and hydropower. Finally, a sensitivity analysis is conducted to investigate the dynamics of investment value and optimal timing under the changes in uncertainty of coal prices.

1.2 Methodology

1.2.1 Real Options Approach

Myers (1977) referred the term "real options approach" (ROA) to the application of option pricing theory to valuate nonfinancial or "real" assets. It is useful in project appraisal when revenues from investment contain uncertainty in the future cash flow and when there is a possibility to choose the timing of investment (Yang et al. 2008). A number of studies have been undertaken applying ROA to evaluate energy projects. Recent applications of ROA, particularly in renewable energy investments, include Kim et al. (2017) who assessed the renewable energy investment in developing countries with a case study involving a hydropower project in Indonesia; Barrera et al. (2016) who analyzed the impact of public R&D financing on renewable energy projects, specifically on concentrated solar power; Wesseh and Lin (2016) who evaluated whether the feed-in tariffs outweigh the cost of wind energy projects in China; and Loncara et al. (2016) who used a compound real options valuation method to examine a potential onshore wind farm project in Serbia.

This research uses ROA to assess renewable energy investment in the Philippines by incorporating uncertainty in coal prices. Methodology uses Matlab program codes to generate (a) transition probability matrix, (b) Monte Carlo simulation, and (c) dynamic optimization. The transition probability matrix is used to generate a matrix of random numbers that approximate stochastic prices of coal. This matrix is incorporated on the social revenue function for coal. The Monte Carlo simulation is used to estimate the expected net present values of energy from coal. The dynamic optimization is used to compute the maximized value of either continuing the use of imported coal or investing in renewable energy. Finally, spreadsheet is used to visualize the dynamics of maximized options values and evaluate the optimal timing of investment.

1.2.2 Stochastic Process

Stochastic process is a variable that changes randomly over time. Assuming coal prices following GBM, Dixit and Pindyck (1994) present the stochastic price process as

$$\frac{dP}{P} = \alpha dt + \sigma dz \tag{1.1}$$

where α and σ are parameters of drift and variance representing mean and volatility of the price process, dt is the infinitesimal time increment, and dz is the increment of the Wiener process equal to $\varepsilon_t\sqrt{dt}$ such that $\varepsilon_t \sim N(0, 1)$. Since percent changes in P, $\Delta P/P$, are normally distributed and that these are changes in ln$P$, then ΔP is *lognormally* distributed. Applying Ito's Lemma, it can be shown that P follows GBM and $F(P) = \ln P$ follows simple Brownian motion with a drift. Since $\partial F/\partial P = 1$, $\partial F/\partial t = 0$, and $\partial^2 F/\partial t^2 = -1/P^2$, then

$$dF = \alpha dt + \sigma dz - \frac{1}{2}\sigma^2 dt = \left(\alpha - \frac{1}{2}\sigma^2\right)dt + \sigma dz. \tag{1.2}$$

Therefore, the changes in ln$P$ over a finite time interval Δt are normally distributed with mean $\left(\alpha - \frac{1}{2}\sigma^2\right)\Delta t$ and variance $\sigma^2\Delta t$ (Dixit and Pindyck 1994).

Adopting Insley (2002), Eq. 1.2 is approximated in discrete time as follows:

$$p_t - p_{t-1} = \left(\alpha - \frac{1}{2}\sigma^2\right)\Delta t + \sigma\varepsilon_t\sqrt{dt} \tag{1.3}$$

where p_t is the natural logarithm of P at time t. To determine the drift and the variance rate for the price level P, the augmented Dickey-Fuller (ADF) unit root test is run using the following regression equation:

$$p_t - p_{t+1} = c(1) + c(2)p_{t-1} + \sum_{j=1}^{L}\lambda_j \Delta y_{t-j} + e_t \tag{1.4}$$

where $c(1) = \left(\alpha - \frac{1}{2}\sigma^2\right)\Delta t$ and $e_t = \sigma\varepsilon_t\sqrt{\Delta t}$. The lagged dependent variables in the Eq. 1.4 are included to account for the serial correlation with the number of lags L chosen during the experimentation. Further, the maximum likelihood estimates of the drift α and the variance rate σ are computed using the equation $\alpha = \mu + \frac{1}{2}s^2$ and $\sigma = s$, where α is the mean and s is the standard deviation of the series $p_t - p_{t-1}$ (Insley 2002).

This research used the average annual imported coal prices from 1970 to 2015 and analyzed the data using EViews software. Table 1.1 shows the ADF test result and implies that the null hypothesis that p_t has a unit root at all significant levels cannot be rejected. Hence, P conforms to a discretized GBM. The computed results

Table 1.1 Augmented Dickey-Fuller unit root test of GBM for coal prices

Test statistic and significance levels for critical values		t-Statistic	Prob*
Augmented dickey-fuller test statistic		−2.160787	0.4956
Test critical values:	1% level	−4.243644	
	5% level	−3.544284	
	10% level	−3.204699	

of $\alpha = 0.01113$ and $\sigma = 0.250153$ are then used in the Monte Carlo simulation and dynamic programming in the succeeding sections.

1.2.3 Social Revenue

In this section describes the revenue functions for coal and renewable energy at each period. First, the social benefit of continuing the use of imported coal-based energy in each period t is represented as

$$\pi_{\mathrm{c},t} = P_{\mathrm{E}}Q_{\mathrm{E}} - P_{\mathrm{c},t}Q_{\mathrm{C}} - C_{\mathrm{C}} \tag{1.5}$$

where P_{E} is the domestic electricity price, Q_{E} is the electricity consumption, $P_{\mathrm{c},t}$ is the stochastic prices of coal, Q_{C} is the amount of coal needed to generate Q_{E}, and C_{C} is the annual marginal production cost.

Second, the social revenue for the renewable energy is represented as

$$\pi_{\mathrm{R},t} = P_{\mathrm{E}}Q_{\mathrm{E}} - C_{\mathrm{R}} \tag{1.6}$$

where C_{R} is the annual marginal production cost for renewable. It should be noted that the overnight cost is not yet presented in this model. This variable will be discussed in the next section.

1.2.4 Monte Carlo Simulations

The work of Detert and Kotani (2013) is adopted to compute the expected net present value of using imported coal for Monte Carlo simulations. Applying the Bellman equation backward to solve the problem, the net present value of the coal energy is represented as

$$\mathrm{NPV_C} = \sum_{t=0}^{T_\mathrm{C}} \mathrm{PV}_{\mathrm{C},t} = \sum_{t=0}^{T_\mathrm{C}} \rho^t \pi_{\mathrm{C},t} \tag{1.7}$$

where $\rho = \frac{1}{1+r}$ is the discount factor, r is the discount rate, and T_{C} is the useful life of energy infrastructure using imported coal after the terminal period. In this model, it is assumed that the present value of this stream of revenues is discounted back to the first period of investment. Stochastic process of GBM in discrete time is used to approximate a vector of potential fuel prices from initial to terminal period using

$$P_{C,t} = P_{C,t-1} + \alpha P_{C,t-1} + \sigma P_{C,t-1}\varepsilon_{t-1} \quad (1.8)$$

This equation shows that the current fuel price is affected by the previous fuel price, the drift, and the variance of stochastic price process. From the initial fuel price $P_{C,0}$, succeeding fuel prices in each period are computed using (1.8) and the present values of revenue in each period in Eq. 1.5 as well.

To estimate the expected net present value at each initial price node i, the process is repeated in a sufficiently large number of J times and take the average as given by

$$\mathbb{E}\{\mathrm{NPV}_{C,j}|P_{C,0}\} \approx \frac{1}{J}\sum_{j=1}^{J}\mathrm{NPV}_{C,j} \approx \mathbb{E}\{\mathrm{NPV}_{C}|P_{C,0}\} \quad (1.9)$$

Finally, a series of same process is done to compute the net present value at each terminal period price node from the minimum $P_{C,0} = 0$ to the maximum $P_{C,0} = \$1,000$ at a step of \$1. The time length for Monte Carlo simulation T_C for coal is set to 40 and $J = 10{,}000$ times.

1.2.5 Dynamic Optimization

This paper follows the method described by Dixit and Pindyck (1994) on optimization of investment decision-making under uncertainty using dynamic programming. Going along with the work of Detert and Kotani (2013), dynamic programming is used to determine the optimal timing and corresponding trigger prices for each alternative energy investment. In the real options approach, we describe a model of an investor that is given a specific period to decide whether to continue importing coal for electricity generation or shift to renewable energy. After that period, the investor has no other choice but to continue using coal. The model shows an investor that maximizes the expected net present value of energy investment as

$$\max_{0\le\tau<T+1} \mathbb{E}\left\{\left[\sum_{0\le t<\tau}\rho^t\pi_{C,t} + \rho^T\mathrm{NPV}_{C,t}\left(1-\mathbb{I}_{\{\tau\le T\}}\right)\right]\middle|P_{C,0}\right\} + \mathbb{E}\{\mathrm{NPV}_R\}\left(\mathbb{I}_{\{\tau\le T\}}\right) \quad (1.10)$$

where T is the length of time where an investor has an option to switch to renewable energy, τ is the period where the switching is made, $\mathbb{I}_{\{\tau\le T\}}$ is an indicator function equal to 1 with $\tau \le T$ otherwise 0, and $\mathbb{E}\{\mathrm{NPV}_R\}$ is the expected net present value of renewable investment. In the given equation, the investor's problem is to find the optimal timing τ that maximizes the expected NPV of social revenues.

The problem is solved by backward induction using dynamic programming from the terminal period (Bertsekas 2012; Detert and Kotani 2013). From the Bellman equation, the optimal switching timing at every period is evaluated using

$$V_t(P_{C,t}) = \max\{NPV_{R,\pi_{C,t}} + \rho\mathbb{E}(V_{t+1}(P_{C,t+1})|P_{C,t})\} \tag{1.11}$$

and

$$NPV_R = \sum_{t=0}^{T_R} \rho^t \pi_R - I_R = \left(\frac{\rho - \rho^{T_R+1}}{1-\rho}\right)\pi_R - I_R \tag{1.12}$$

where NPV_R is the net present value of renewable energy (Eq. 1.12) and I_R is the investment costs for renewable energy.

Lastly, optimal trigger strategy of switching to renewable energy is represented by

$$\hat{P}_C = \inf\{P_{C,t}|V_0(P_{C,t}) = V_{T_R}(P_{C,t})\} \tag{1.13}$$

where $\hat{P}_C$ is the trigger price or the minimum coal price $P_{C,t}$, while $V_0(P_{C,0})$ and $V_{T_R}(P_{C,0})$ are the maximized values of the investment at time $t = 0$ and $t = T_R$ (Dixit and Pindyck 1994; Davis and Cairns 2012).

The dynamic optimization process T is set to a middle range of 40 years which represents a situation where the investor has 40 years to make an investment decision or not. After that period, they can no longer choose between alternatives. The choice is valued for another 40 years to represent the lifetime of renewable energy investment. Finally, T_R values are set to 25 years, which represent the number of years of electricity generation using renewable energy resources.

1.3 Results

1.3.1 Baseline Result

To determine the parameters for the baseline scenario, the data from the Philippines' Department of Energy, Philippine Statistics Authority, and US Energy Information Agency are used. Various types of renewables are then standardized such that the power generation output from each type is equal to the one generated from using imported coal (27,514 GWh/year). Investment costs and annual marginal production costs are then estimated for each type of energy source. The estimated parameters are finally substituted in the model for the dynamic optimization process.

Table 1.2 Net present values of renewable energy investments in the Philippines

Source	NPV
Geothermal	$45,400M
Wind	$28,964M
Hydroelectric	$16,685M
Solar	$573M

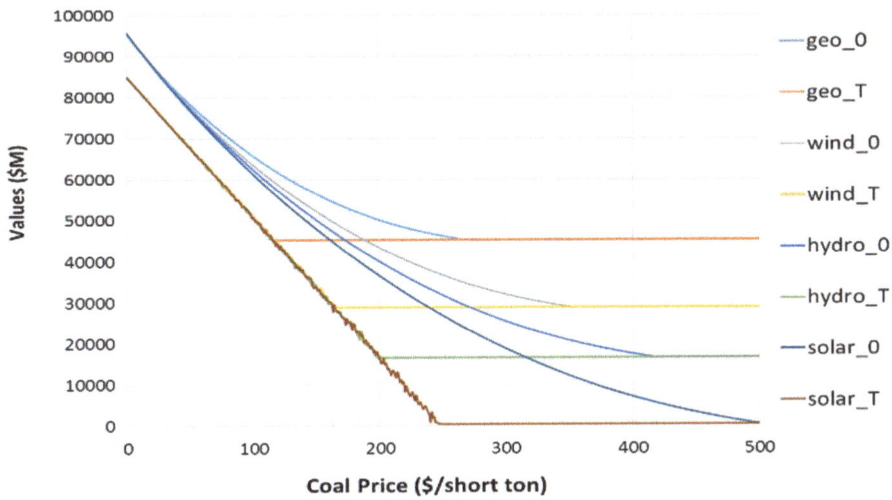

Fig. 1.1 Options values for baseline parameters

The results of dynamic optimization are illustrated in Table 1.2 and Fig. 1.1. Table 1.1 shows positive net present values for all sources of energy. This implies that investing in any of the renewable energy sources in the Philippines will be profitable. Among the available options, geothermal seems to be the most profitable, followed by wind, hydroelectric, and solar. This result is expected since majority of renewable sources in the country are coming from both geothermal and hydroelectric. Currently, the country is the world's second largest producer of geothermal energy second to the United States (Sovacool 2010). Caution must be applied, however, because NPV is not the sole determinant of investments in a real options approach. The optimal timing that maximizes the value of investment opportunity must also be accounted for (Dixit and Pindyck 1994).

Figure 1.1 shows the dynamics of the options values which is represented by the curves. Options values are the maximized value of either investing in renewable or continue importing coal for energy generation. It can be observed that the options values decrease with coal price. This implies that the value of any of the investment decreases with higher cost in coal input. The straight line on the right of the options value curve indicates the price of the renewable energy is more profitable than coal. These straight lines also confirm the result in Table 1.2 that geothermal is the most profitable source of renewable energy. The vertical distance between the options

Table 1.3 Value of waiting to invest in renewable energy at current coal price ($67/short ton)

Source	Initial period	Terminal period	Value of waiting
Geothermal	$73,929M	$62,563M	−$11,366M
Wind	$72,772M	$62,017M	−$10,755M
Hydroelectric	$72,178M	$62,252M	−$9926M
Solar	$71,603M	$62,153M	−$9450M

Table 1.4 Sensitivity of trigger prices at various uncertainties in coal price

Source	Low uncertainty*	Base**	High uncertainty***
Geothermal	$184	$268	$487
Wind	$242	$353	$643
Hydroelectric	$285	$418	$761
Solar	$343	$502	$915

*$\sigma = 0.1$, **$\sigma = 0.250153$ (base), ***$\sigma = 0.5$

value curves from initial, r_0, to terminal period, r_T, illustrates the value of waiting or delaying to invest. Therefore, the intersection of the two curves indicates the optimal timing of shifting from coal to renewable. The intersections of curves in Fig. 1.1 imply that it is better to invest earlier in more mature technology as geothermal and invest later in solar.

Table 1.3 shows the option values of various renewable energies at 2015 coal price. The difference in options values represents the value of waiting to invest. The result shows negative for all sources. This indicates that waiting or delaying the decision to invest in renewable incurs welfare loses.

1.3.2 Sensitivity Analysis

In this portion of analysis, the values of σ are varied to represent uncertainty or randomness of coal prices. Sensitivity analysis is conducted using three different uncertainty values: 0.1 (low), base, and 0.5 (high). The results show that at lower uncertainty, the trigger price of coal to shift to renewable decreases. This implies that with more deterministic prices of coal input, the better to invest earlier in renewables. On the other hand, when prices of coal input are very volatile in the market, it is better to wait and delay the investment to avoid risks (Table 1.4).

1.4 Conclusion

This paper analyzed investment environments and decision-making process for switching from imported coal energy to renewable energy sources by incorporating the option to wait to make a decision. The value of investments is evaluated by applying real options approach that determines options value and timing of investment. Under various coal price uncertainties, this paper successfully characterized when renewable energy becomes more attractive than continue importing coal for electricity generation. Among the available local renewable energy sources in the Philippines, geothermal is the most profitable followed by wind, hydroelectric, and solar. By posing several scenarios of coal price uncertainties, this paper identified that delaying or waiting to invest in renewable energy may lead to welfare loses. Higher uncertainty in coal prices lengthens the waiting to invest on renewable energy. Results suggest that early shifting from coal to renewable decreases the possibility of having welfare loses from waiting to invest. This also suggests that the Philippine government should boost its programs to reduce coal imports and invest on renewable energy sooner, especially with the current trend of volatile coal prices.

The real options approach framework introduced in this research is a good benchmark for further application, with some new modification of future research suggested. First, other investment uncertainties, particularly in the Philippines setting, may be incorporated in the model. This includes uncertainty in electricity prices, technology innovations, and natural disaster risk. Second, applying real options approach to localized or island-based or community-based project will be more practical and helpful for investors that specialize on small- and medium-scale renewable energy projects. Finally, environmental impact of renewable energy source may be incorporated in the real options approach model. This may include air and water pollution, damage to public health, wildlife and habitat loss, water use, land use, and global warming emissions. This research becomes one step for further analysis of investment of a more sustainable and greener source of energy.

References

Barrera, G. M., Ramírez, C. Z., & González, J. M. (2016). Application of real options valuation for analysing the impact of public R&D financing on renewable energy projects: A company's perspective. *Renewable and Sustainable Energy Reviews, 63*, 292–301. https://doi.org/10.1016/j.rser.2016.05.073.

Bertsekas, D. P. (2012). *Dynamic programming and optimal control* (4th ed.). Nashua: Athena Scientific.

Brach, M. A. (2003). *Real options in practice*. Hoboken: Wiley.

Davis, G. A., & Cairns, R. D. (2012). Good timing. The economics of optimal stopping. *Journal of Dynamics and Control, 36(2)*, 255–265. https://doi.org/10.1016/j.jedc.2011.09.008

Department of Energy. (2012). Philippine energy plan 2012–2030.

Department of Energy. (2016). 2015 Philippine power statistics.

Detert, N., & Kotani, K. (2013). Real options approach to renewable energy investment in Mongolia. *Energy Policy, 56*, 136–150. https://doi.org/10.1016/j.enpol.2012.12.003

Dixit, A. K., & Pindyck, R. S. (1994). *Investment under uncertainty*. Princeton: Princeton University Press.

Dixit, A. K., & Pindyck, R. S. (1995). The options approach to capital investment. *Harvard Business Review, 73*, 105–115.

Insley, M. (2002). A real options approach to the valuation of a forestry investment. *Journal of Environmental Economics and Management, 44*(3), 471–492. https://doi.org/10.1006/jeem.2001.1209

Kim, K., Park, H., & Kim, H. (2017). Real options analysis for renewable energy investment decisions in developing countries. *Renewable and Sustainable Energy Reviews, 75*, 918–926. https://doi.org/10.1016/j.rser.2016.11.073

Kumbaroğlu, G., Madlener, R., & Demirel, M. (2008). A real options evaluation model for the diffusion prospects of new renewable power generation technologies. *Energy Economics, 30(4)*, 1882–1908. https://doi.org/10.1016/j.eneco.2006.10.009

Loncar, D., Milovanovic, I., Rakic, B., & Radjenovic, T. (2017). Compound real options valuation of renewable energy projects: The case of a wind farm in Serbia. *Renewable and Sustainable Energy Reviews, 75*, 354–367.

Myers, S. C. (1977). The determinants of corporate borrowing. *Journal of Financial Economics, 5(2)*, 147–175. https://doi.org/10.1016/0304-405X(77)90015-0

Pindyck, R. (1991). Irreversibility, uncertainty, and investment. *Journal of Economic Literature, 29(3)*, 1110–1148.

Sovacool, B. K. (2010). A comparative analysis of renewable electricity support mechanisms for Southeast Asia. *Energy, 35*, 1779–1793. https://doi.org/10.1016/j.energy.2009.12.030

Wesseh, P. K., Jr., & Lin, B. (2016). A real options valuation of Chinese wind energy technologies for power generation: Do benefits from the feed-in tariffs outweigh costs? *Journal of Cleaner Production, 112*(2), 1591–1599. https://doi.org/10.1016/j.jclepro.2015.04.083

Yang, M., Blyth, W., Bradley, R., Bunn, D., Clarke, C., & Wilson, T. (2008). Evaluating the power investment options with uncertainty in climate policy. *Energy Economics, 30(4)*, 1933–1950. https://doi.org/10.1016/j.eneco.2007.06.004

Chapter 2
Geothermal Energy Barriers, Policies and Economics in East Asia

Venkatachalam Anbumozhi

2.1 Introduction

Developing new methods of low-carbon energy resource and consumption is crucial for meeting the climate goals as agreed in Paris. Against the backdrop of rising demand for sustainable energy solutions, there is a growing convergence that the role of renewables such as geothermal can play in addressing the climate change while providing access to affordable energy. Electricity production from geothermal only occurs in 24 countries worldwide, six which are in East Asia. Across these countries, installed power capacity for geothermal reached 3,743 MW between 2010 and 2011 (WWF 2012). These countries produced around 3% of the world's geothermal capacity. The Philippines has the second largest most installed capacity in the world, after the United States. In 2012, the 1,904 MW of installed capacity in the country was enough to supply around 17% of total electricity need which was equivalent to 10,324 GWh (DOE 2014). China is the world leader in direct use of geothermal energy with 12,605 GWh in 2010 (Zheng et al. 2015). There are strengths with geothermal energy in terms of steady supply, scalability and operation costs. Unlike wind and solar, electricity generated from geothermal is not intermittent, and it can be used to provide reliable base load power. Geothermal can be used for various purposes at various scales, ranging from heating for individual households to powering an entire city. Once constructed, geothermal generation can be operated cost effectively. But there are challenges to large-scale application of geothermal application in the region, which includes high upfront cost (ESMAP 2014) and limited areas for resource extraction (Ardiansyah and Putri 2013) and technologies for enhanced geothermal systems (Sanyal et al. 2014). Sanchez-Alfaro et al. (2015) found that the absence of medium- to long-term energy policies and the

V. Anbumozhi (✉)
Economic Research Institute for ASEAN and East Asia, Jakarta, Indonesia
e-mail: v.anbumozhi@eria.org

A. Sayigh (ed.), *Transition Towards 100% Renewable Energy*,
Innovative Renewable Energy, https://doi.org/10.1007/978-3-319-69844-1_2

lack of incentives for companies to overcome financial risks are perceived as main barriers in Chile. They have also identified the main perceived advantages, barriers and incentives related to geothermal development, assessing their relevance and feasibility through a survey to propose guidelines for geothermal stakeholders. Such structured analysis on a regional scale is lacking in the East Asia region. This paper investigates the economic opportunities with geothermal development in the East Asia region, analyses the barriers to achieving the targets and proposes integrated policy solutions.

2.2 Geothermal Potentials in East Asia

Currently there is no single information source for existing geothermal energy use. Several initiatives exist within the region to collect geothermal data, but these often use widely different methodologies. Table 2.1 shows the current ERIA (2015) estimates of geothermal use in China, Indonesia, Japan, Korea, the Philippines, Thailand and Vietnam.

The estimates conform to Lund and Boyd (2010) and Bertani (2015) estimates on the direct utilisation and power generation values. Nevertheless, estimating geothermal requires several distinct data set and various complex investigation and analysis (Hochstein and Crosetti 2011). Based on the available information, Sakaguchi and Anbumozhi (2015) are working to provide such information in a consolidated way. Indeed the estimation currently consists of a combination of preliminary reconnaissance studies – geological, geochemical and geophysical surveys, all of which provide data sets that can increase the rate of geothermal energy development. However, for policy decisions and investment decisions, it may be sufficient to display basic data that provide some measures of the presence and significance of the resources wherever it was found.

Table 2.2 shows development trends of geothermal energy use in each country. All the study countries have target capacity addition within 5 years. However, the development plan differs from country to country.

Some countries put geothermal development as national plans, while only the private sector or institutes have plans in other countries. National plans on geothermal development may help its promotion since countries where geothermal development is advancing, such as the Philippines and Indonesia, have national plans. Only China has a clear plan for all power production, direct use, and GSHP. No other countries have plans for direct use, while all countries show targets for power production. Long-term programmes for geothermal power generation are necessary because geothermal development takes 5–7 years. China, Japan and South Korea, which have cold seasons, have targets for GSHP.

Deployment of these geothermal technologies also presents significant opportunities for economic development and employment. Their adoption is expected to be critical in meeting the goal of clean energy access and in stimulating socio-economic development. Regionally installed geothermal capacity has doubled since 1990 to 11,765 MW in 2014 and equivalent planned capacity additions are

Table 2.1 Present status of geothermal use in each country

Country	Installed capacity			Used (produced) Energy			Reference
	PG (MW_e)	DU (MW_t)	GSHP (MW_t)	PG (GW_e-h/y)	DU (GW_t-h/y)	GSHP (GW_t-h/y)	
China	27.8	6,089	11,781	155.1	20,801	27,864	Zheng et al. (2015)
Indonesia	1341.0	2.3	–	9,332.32	11.8	–	MEMR (2013), Lund et al. (2010)
Japan	540.1	2,099.5	44.0	2,688.82	7,138.9	–	TNPES (2013), Lund et al. (2010)
Korea	–	43.7	792.2	–	164.9	580.7	Song and Lee (2015)
Philippines	1,848.0	–	–	10,230.5	–	–	Department of Energy (2014)
Thailand	0.3	–	–	–	–	–	DEDE (2012)
Viet Nam	0.0	30.7	–	–	22.36	–	Nguyen et al. (2005)

DU direct use, *GSHP* ground source heat pump, *PG* power generation

in the early stages of development or under construction (Motek 2013). By implication, investment and local employment opportunities can be assumed to have expanded considerably and will continue to grow. The largest installed geothermal power capacities are in the Philippines (1,884 MW) and Indonesia (1,333 MW), and geothermal is also heavily used in Japan, Korea and New Zealand (Sakaguchi and Anbumozhi 2015).

Countries with installed geothermal capacities derive employment benefits in construction and O&M, which are by nature more domestic; however, the production of geothermal energy may result in job creation elsewhere. The Japanese and New Zealand companies play a central role in manufacturing geothermal turbines, controlling more than half of the global market. In the Philippines, the energy development corporation (EDC) controls about 60% of the countries geothermal capacity and close to 2,500 employees. Local hires account for 75% of the company's workforce (DOE 2011). Though no robust investment figures exist, in general the investment and a number of jobs appear to expand as new markets emerge for geothermal exploration and exploitation in Southeast Asia and East Asian countries.

2.3 Technology and Management of Geothermal Energy

Geothermal resources are the thermal energy available and stored as steam or hot water in active geothermal areas. Higher-temperature water or stream resources (>180 °C) are the best for electricity generation, as the liquid can be directly used

Table 2.2 Development trends of geothermal energy use and opportunities

Country	Target capacity addition			Date
	Power generation	Direct use	GSHP	
China	100 MW_e (national plan)	3,700 MW_t (national plan)	18,200 MW_t (for residential, office buildings, school, hospital, mall, etc.) (national plan)	By 2019
Indonesia	1,160 MW_e (national plan)	NA	NA	By 2019
Japan	Several small binary (50 kW_e–1 MW_e) and a 40 MW_e (by private sector with government's support)	No specific plan	GSHP at 990 units (2011) to increase for the next 5 years (estimation by related organisation)	By 2019
Korea	Pilot plant, EGS technology (1–3 MW_e)	No significant development	>100 MW_t new installations each year (for large office buildings, greenhouse, small residential houses) (estimation by related organisation)	By 2019
Philippines	1,465 MW_e (Fronda et al. 2015) (national plan)	–	–	By 2030
Thailand	At least 5 MW_e	Spa, drying system would be supported by hot springs	No application	By 2019
Viet Nam	20 MW_e	Agricultural drying, industrial process heat, bathing, swimming	Projects to find out potential and application for office buildings and residential houses	By 2019

Source: Sakaguchi and Anbumozhi (2015) based on EPPO (2014), Honag (1998), Meier et al. (2018), Song et al. (2011) and Wang et al. (2013)

by dropping the pressure to create stream that can drive turbine. Where only medium-temperature resources are available, more expensive binary plants are required. They use a heat exchanger to create steam from a liquid with a low boiling point for subsequent use in a steam turbine. Table 2.3 shows challenges for sustainable use of geothermal energy pointed out by the countries in the region. These topics listed in order of priority are (i) monitoring and reservoir engineering, (ii) reinjection, (iii) anti-scaling and (iv) anticorrosion and anti-erosion.

In Korea, the sustainable issue of geothermal power generation is not of common interest yet. They focus on sustainability of GSHP, amongst others. Thailand and Viet Nam have yet to develop a binary system for sustainable use of geothermal energy. The study of the second year of this project was decided based on this result.

Although geothermal power generation is a mature and commercially available solution to low-cost base load capacity in areas with excellent high-temperature

Table 2.3 Countrywide challenges for sustainable use of geothermal energy in East Asia

Country	Reinjection	Monitoring and reservoir engineering	Anticorrosion and anti-erosion	Anti-scaling	Others
China	X	X			
	(a) In key cities of geothermal utilisation, the geothermal resource administration stipulates that geothermal district heating has to install reinjection				
	(b) Geothermal monitoring is popularly carried out in key cities and developing areas				
Indonesia	X	X	X	X	
Japan	X	X		X	
Korea	(e) Sustainability issue of geothermal energy is not of common interest yet, because no systematic deep geothermal utilisation is operating now. There are concerns about sustainability of GSHP system, especially on water level change and subsurface temperature sustainability				X
Philippines	X	X	X	X	
Thailand	(e) To develop a binary system				X
Viet Nam	(e) To develop a binary system				X

HSP = ground source heat pump

resources close to the surface, several factors affect the overall cost of the geothermal generation. The levelized cost of geothermal plant is determined by the usual factors, such as installed costs, O&M costs, economic lifetime and the weighted average cost of capital. However, the analysis of geothermal is a more dynamic question than for other renewables like solar, wind and biomass. One complication is a larger uncertainty in project development, due to the risk of poorly performing production wells. Similarly over the life of project, reservoir degradation can play an important role in costs and in performance. These factors tend to introduce greater uncertainty into the development of geothermal resources and projects and may increase financing costs, compared to other technologies such as wind (GeothermEx 2010). However, this uncertainty factor is typically manageable in mature geothermal markets where financing institutions have previous experience with the industry.

2.4 Technical Barriers to Geothermal Development

Table 2.4 shows the technical barriers for geothermal power generation for three stages: exploration, installation (development) and sustainable use based on the responses from each country, as obtained through an expert study meeting organised by ERIA.

These barriers are derived from consultations with country experts and hence not listing of all barriers in a structured way. But common problems for exploration for power generation are identified as (i) drilling success: testing of new methods and applications to increase the success rate of exploration wells such as remote sensing, 3D inversion of MT, radon survey and joint geophysical imaging,

Table 2.4 Technical barriers for geothermal power generation for different stages

Country	Exploration	Installation	Sustainable use
China	Well logging instruments and circulating technique in high-temperature geothermal drilling	Domestic product limited in 5 MW, no big capacity	Sustainable reinjection has not yet done
Indonesia	Low drilling success ratio	Fluid characteristics (acidic fluid, high silica)	Decline of production well (5–10% per year in average) and reinjection well (scaling); geo-hazards (landslide, earthquake, volcanic activity)
Japan	Limit of geophysical methods (resistivity image does not always show reservoir shape)	Success rate of production well drilling Minimization of environmental impact Presence of acidic fluids	Scale, pressure decline, short circuit (reinjection fluid control)
Korea	Lack of deep well information, such as temperature, stress and fracture distribution	Lack of experience in deep drilling, measurement and reservoir engineeringdifficulty of securing proper technical services and procurements	–
Philippines	Environmental permits (tree cutting permit, access to national parks, etc.), social acceptance and access permits, insurgents, finding good permeability and high temperature for the first three exploration wells. Presence of acidic fluids	Simultaneous sustainability testing, establishing production sharing and injection interference, drilling interference. Matching of right power conversion system with reservoir characteristics to optimise resource and efficiency	Reservoir drawdown; mineral scaling in wells, surface pipeline network and reservoir; acidity of production fluids and attendant corrosion; reinjection returns; influx of shallow groundwater into reservoir; landslide risks and surface facilities' damages due to supertyphoon
Thailand	Geophysical survey and drilling technique	–	–
Viet Nam	Geophysical survey, drilling, reservoir modelling	–	–

Source: Proceedings of ERIA working meeting

(ii) lack of geophysical survey and (iii) public acceptance: national and local governments should support renewable energy projects. The common challenges associated with power generation are (i) drilling success of production well, (ii) reservoir characterisation and (iii) acidic and high-silica fluid. Common problems for sustainability of power generation are (i) sustainable reinjection: experience in different geothermal reservoirs; (ii) reinjection fluid return (short circuit);

Table 2.5 Technical barriers for GSHP for different stages

Country	Exploration	Installation	Sustainable use
China	–	–	–
Indonesia	–	–	–
Japan	Geological and hydrological database, especially, estimation of groundwater flux	Drilling cost	Control of annual heat exchange balance (extraction and/or injection)
Korea	–	Lack of information on subsurface thermal properties associated with hydrology	Lack of long-term performance analysis in conjunction with monitoring of subsurface temperature and/or water level variation
Philippines	–	–	–
Thailand	Case study	–	–
Viet Nam	Need to do the detail research	Need to have one pilot installation	–

(iii) decline of production wells (pressure drawdown); (iv) scaling in injection wells; (v) acidic fluid corrosion; (vi) shallow groundwater into reservoir; and (vii) geo-hazard (landslide, subsidence, typhoon, volcanic eruption, earthquake). For all exploration, installation and sustainable power generation, more research funding is needed. International collaborative cooperation in R&D on solving those problems above is needed.

The technical barriers related to specific technologies such as GSHP are illustrated in Table 2.5.

Challenges with GSHP are derived from (i) lack of case study (showing successful case), (ii) lack of hydrogeological database, (iii) high drilling cost and (iv) lack of information on long-term performance. More funding is needed for domestic R&D for hydrogeological studies, case studies and long-term monitoring. Also, international research collaboration is essential to share the knowledge obtained in each country. Drilling cost may be reduced by both mass production and technical improvement suitable for each local geology, which means that drilling cost can also be reduced by the accumulation of knowledge and number of installations supported by R&D on case studies, hydrogeological studies, and long-term monitoring.

Current policy frameworks to support geothermal development in East Asia countries are listed in Table 2.6.

2.5 Rethinking of Geothermal Development in East Asia Through Integrated Policies

Globally many policy instruments have developed with the goal of accelerating geothermal development. An extensive database containing examples of policies and measures is accessible and available elsewhere (Hori 1990; Korjedee 2002;

Table 2.6 Supportive measures in each country

Country	Are there FiT or RPS?		
	Power generation	Direct use	GSHP
China	No RPS for geothermal	No	Subsidy for energy saving of building. Grant for demonstration of renewable energy
Indonesia	In future (ceiling price increase at 11.8–29.6), tax incentives	No	No
Japan	Yes. >15 MW (~27JPY/kWh); <15 (~42JPY/kWh)	No	No
Korea	RPS with REC of 2.0	No	No but discussing about renewable heat obligation
Philippines	Yes, RPS but no FiT for geothermal, FiT for wind/solar	No	No
Thailand	No FiT/RPS	No	No
Viet Nam	No	No	No

RPS renewable portfolio standard, *MW* megawatt, *kWh* kilowatt hour, *JPY* Japanese yen, *REC* renewable energy certificate, *FiT* feed-in tariff

Muraoka et al. 2008; Pastor et al. 2010). In East Asia, policy instruments in support of geothermal technologies such as power generation, direct use and GSHP include financial support for university research consortia, government-sponsored R&D laboratories and public-private partnership in applied R&D. Policy instruments that help move these technologies to market may include technology transfer programmes, grants for pilot projects, loan guarantees and other financial instruments for constructing demonstration plants and industry collaborations to promulgate standards to secure technology interoperability. Each of these policies serves a specific function of a country in question. A policy tool box as shown Table 2.7 could broadly describe the set of instruments at the disposal of policy makers.

2.6 Conclusion

Geothermal energy represents one of the key options for Southeast and East Asian countries to achieve a comprehensive approach to national development, based on clean energy provision, social development and climate change. But there are many technological, managerial and financial issues that are hindering geothermal power development. This paper identified key technological barriers for geothermal development. To remove the technology barriers and enhance the uptake of geothermal energy at regional scale, a key strategy could be building a cooperation platform amongst geothermal-rich countries in East Asia, starting with China, Indonesia, the Philippines, Japan and New Zealand, and then expanding to include

Table 2.7 Policy tool box for accelerated geothermal development in East Asia

Function	Example policy tools
Creating and sharing knowledge	Subsidies and incentives for research on geothermal technologies and equipment
Building competence, awareness and human capital	Subsidies and incentives for education and training, fellowships
Knowledge diffusion/creating collaborative networks	Joining or initiating international cooperation, supporting industry association and enforcement of measures and best practices
Rapid reconnaissance surveys and assessment of potentials	Public private partnerships, incentivizing private development and investment in public energy infrastructure
Providing finance	Loan guarantees, green banks and public venture capital style funds
Establishing governance and the regulatory environment	Setting standards, setting target, taxing negative externalities, subsidising positive externalities
Creating markets	Feed-in tariffs, renewable portfolio standards, government/public procurement. Setting government requirements, taxing negative externalities, subsidising positive externalities

Korea, Viet Nama and Thailand. Countries in the Asia-Pacific region, the Philippines, Japan and New Zealand, are the largest producers of geothermal power with good record in renewable energy policy development and technology development. China and Indonesia are large emerging economies with lofty targets on low-carbon technology policies. Building a cooperation framework between these countries and relevant private sector stakeholders and academics will help shift their government's energy security agenda, drive private sector investment decisions and ensure that communities play an active role in promoting low-carbon energy supply sustainability.

References

Ardiansyah, F. & Putri, A (2013, March 14). Hot, clean and complex: Unlocking Indonesia's geothermal power. *Indonesian Journal of Leadership, Policy, and World Affairs Strategic Review* 25(1):138–147.

Bertani, R. (2015). Geothermal power generation in the world 2010–2014 update report. In *Proceedings world geothermal congress 2015*, Melbourne, 19 p.

Department of Alternative Energy Development and Efficiency (DEDE). (2012). *The renewable and alternative energy development plan for 25 percent in 10 years (AEDP 2012–2021).* Department of Alternative Energy Development and Efficiency.

Department of Energy (DOE). (2011). The national renewable energy plans and programs. *The 2012–2030 Philippine energy plan.*

Department of Energy (DOE). (2014). Indicative geothermal capacity additions. *Energy Resources.* https://www.doe.gov.ph/doe_files/pdf/04_Energy_Resources/Stat-GeoAddition.pdf

Energy Policy & Plan Office (EPPO). (2014). The Thailand energy master plan 2015–2035.

ERIA (2015). Energy saving potential and outlook of East Asia, Econoimic Reseach Institute for ASEAN and East Asia, Jakarta, 351 p.

ESMAP. (2014). Geothermal energy in Indonesia: An integrated approach to evaluating a green finance investment. ESMAP Knowledge Series report 015/13.

Fronda, A., Marasigan, M., & Lazaro, V. (2015). Geothermal development in the Philippines: Country update. In: *Proceedings of the world geothermal congress 2015*.

GeothermEx, Inc. (2010). An assessment of geothermal resource risks in Indonesia. Report prepared for The World Bank.

Hoang, H. Q. (1998). Overview of geothermal potential of Vietnam. *Geothermics, 27*(1), 109–115.

Hochstein, M. P., & Crosetti, M. (2011). Electric power potential estimates of high-temperature geothermal fields in Indonesia and the Philippines (A historical review). In *New Zealand geothermal workshop 2011 proceedings*, 6 p.

Hori, N. (1990). Cost evaluation and technology development of large scale HDR. *Chinetsu, 27*, 148–158. (in Japanese). http://www.geothermal-energy.org/pdf/IGAstandard/WGC/2010/1616.pdf.

Korjedee, T. (2002). Geothermal exploration and development in Thailand, Geothermal and Volcanological Research Report of Kyushu University, No. 11, pp. 56–66.

Lund, J. W., & Boyd, T. L. (2010). Direct utilization of geothermal energy 2015 worldwide review. In *Proceedings world geothermal congress 2015*, Melbourne, 31 p.

Lund, J. W., Freeston, D. H., & Boyd, T. L. (2010). Direct utilization of geothermal energy 2010 worldwide review. In *Proceedings of the world geothermal congress 2010*.

Meier, P., Vagliasindi, M., & Imran, M. (Eds.) (2014). Chapter 5 Case study: Indonesia. In *The design and sustainability of renewable energy incentives: An economic analysis* (pp. 125–154). Washington, DC: The World Bank.

Ministry of Energy and Mineral Resources (MEMR). (2013, December 2013). Geothermal area distribution map and its potential in Indonesia.

Muraoka, H., Sakaguchi, K., Komazawa, M., & Sasaki, S. (2008). Assessment of hydrothermal resources potential in Japan 2008. In *Abstract of GRSJ 2008 annual meeting, B01* (in Japanese).

Nguyen, T. C., Cao, D. G., & Tran, T. T. (2005). General evaluation of the geothermal potential in Vietnam and the prospect of development in the near future. In: *Proceedings of the world geothermal congress 2005*.

Pastor, M. S., Fronda, A. D., Lazaro, V. S., & Velasquez, N. B. (2010). Resource assessment of Philippine geothermal areas. In: *Proceedings of the world geothermal congress 2010*.

Sakaguchi, K., & Anbumozhi, V. (2015). *Sustainability assessment of utilizing conventional and new type of geothermal resources in East Asia, ERIA research report*. Jakarta: Economic Research Institute for ASEAN and East Asia.

Sanchez-Alfaro, P., Sielfeld, G., van Campen, B., Dobson, P., Fuentes, V., Reed, A., Palma-Behnke, R., & Morata, D. (2015). Geothermal barriers, policies and economics in Chile – Lessons for the Andes. *Renewable & Sustainable Energy Reviews, 51*, 1390–1401.

Sanyal, S.K., Morrow, J.W., Jayawardena, M.S., Berrah, N., Fei Li, S., & Suryadarma. (2014). Geothermal resource risk in Indonesia: A statistical inquiry. ESMAP report 89411.

Song Y., & Lee, T. J. (2015). Geothermal development in the Republic of Korea: Country update 2010–2014. In: *Proceedings of the world geothermal congress 2015*.

Song, Y., Baek, S.-G., Kim, H. C., & Lee, T. J. (2011). Estimation of theoretical and technical potentials of geothermal power generation using enhanced geothermal system. *Economic and Environmental Geology, 44*(2011), 513–523. (In Korean with English abstract and illustrations).

Thermal and Nuclear Power Engineering Society (TNPES). (2013). *Present status and trend of geothermal power generation 2012*. (in Japanese).

Wang, G., Li, K., Wen, D., Lin, W., Lin, L., Liu, Z., Zhang, W., Ma, F., & Wang, W. (2013). Assessment of geothermal resources in China. In *Proceedings of the 38th workshop on geothermal reservoir engineering*. Stanford University, California.

WWF. (2012). Igniting the ring of fire: A vision for developing Indonesia's geothermal power, 110 p. World Wildlife Fund.

Zheng, K., Dong, Y., Chen, Z., Tian, T., & Wang, G.. (2015). Speeding up industrialized development of geothermal resources in China – Country update report 2010–2014, In *Proceedings of the world geothermal congress 2015*.

Chapter 3
Experimental Study of Tubular Light Pipe System: Influence of Light Reflector on Its Performance

Abdus Salam Azad and Dibakar Rakshit

3.1 Introduction

In the view of energy conservation in buildings, lighting plays an important role. According to Bureau of Energy Efficiency, lighting energy end use accounts 30% of total electric energy usage in buildings (Web: BEE). Daylighting is an important passive method of building design in achieving energy conservation. Effective utilization of daylight can be achieved by the effective positioning of windows and skylights, along with the lighting controls that respond according to the daylight available. In this direction, Energy Conservation Building Code (ECBC 2007) has also recommended to employ intelligent daylight integrated lighting systems.

Tubular light guides are modern and innovative systems which provide natural illuminance to interior windowless parts of the buildings. It is one of the most promising options to provide daylight into deep spaces of not only buildings but also for lighting the road and railway tunnels (Gil-Martin et al. 2014). Light pipes integrated with lighting controls could save 20–25% of lighting energy in building (Shin et al. 2011). A study by Mohelnikova (2008) reflected that 100 light pipes replacing the fluorescent lamps could save 8 MWh of energy annually. In country like India where the solar energy is available throughout the year, use of light pipes still needs to be encouraged.

Yun et al. (2010) developed a prediction model for straight light pipe with two aspect ratios, on the basis of Zhang and Muneer and Jenkins and Muneer models for Korean climatic conditions, and found the new model as more appropriate in predicting the internal illuminance and having a good agreement with experimental values. Patil (2014) fabricated a windowless single room on the roof fitted with

A.S. Azad • D. Rakshit (✉)
Centre for Energy Studies, Indian Institute of Technology, Delhi, Hauz Khas, New Delhi, India
e-mail: dibakar@iitd.ac.in

A. Sayigh (ed.), *Transition Towards 100% Renewable Energy*,
Innovative Renewable Energy, https://doi.org/10.1007/978-3-319-69844-1_3

installed light pipe in composite climate of New Delhi, India. External and internal illuminance under overcast sky, partly overcast sky and clear sky has been measured. Effects of sky clearance index on external illuminance and on daylight penetration factor (DPF) were analysed.

Baroncini et al. (2006) conducted experimental study to determine the influence of different diffusers on the distribution of interior illuminance. Light pipes (Yanpeng 2005) were installed in different form, and their influence on indoor illuminance utilizing natural daylight was observed. The studies on light pipe systems have been carried out in different countries. But, their applications in India are very rare. Therefore, there is a need to explore more about light pipe systems and its energy conservation potential inside a building, to appreciate its utility and bring light pipes as a sustainable solar product into widespread use in India.

3.2 Methodology

3.2.1 Experimental Setup

A windowless room of size 3 × 3 m with height of 3 m (shown in the Fig. 3.1) was built on the roof top of the CES-WS block located at IIT Delhi, India. Inner wall surfaces of the test room were coated black to minimize the surface reflectance to negligible level. The reflectance of the ceiling and floor is also assumed to be negligible. A straight light pipe with 0.23 m diameter was installed at the centre of the roof of the test room.

The light pipe consists mainly of three components: receiving dome, reflecting tube and emitter (Fig. 3.2a).

Receiving part (dome): The sunlight gets collected in the top part of the hemispherical-shaped dome. It is transparent, 3 mm thick, UV resistant made up of injection-moulded acrylic material. The dome is also having a light reflector (Fig. 3.2) made of aluminium sheet of thickness 0.4 mm and positioned so that to capture low-angle sunlight.

Reflecting part (tube): The received light goes to multiple internal reflections through the reflecting tube. The inner surface of the reflecting tube is coated with highly reflective aluminium film of $\rho = 0.96$. The tubular light pipe with 0.23 m diameter is 0.8 m long, giving the aspect ratio of 3.3.

Emitting part (diffuser): The lower part of the light pipe called diffuser is responsible for light distribution inside the room.

Fig. 3.1 Test room with light pipe set-up

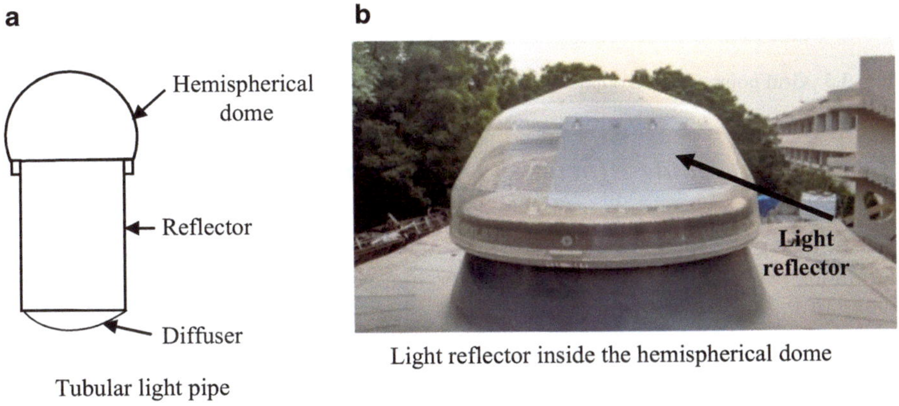

Fig. 3.2 (**a**) Tubular light pipe. (**b**) Light reflector inside the hemispherical dome

3.2.2 *Measurements*

Hourly observations of external irradiance and illuminance were taken from 8:00 AM to 5:00 PM in the composite climate of New Delhi location. The monitoring period was from 6th April to 21th April 2016. The observations have been carried out in sunny, clear sky conditions. LI 250 illuminance sensors were used to measure the internal and external illuminance levels (sensor measurement range, 0–199 klux). Measurements were taken on the horizontal working plane (floor) of the test room divided into grid of 5×5 (Fig. 3.3). The observations were taken for worst case scenario.

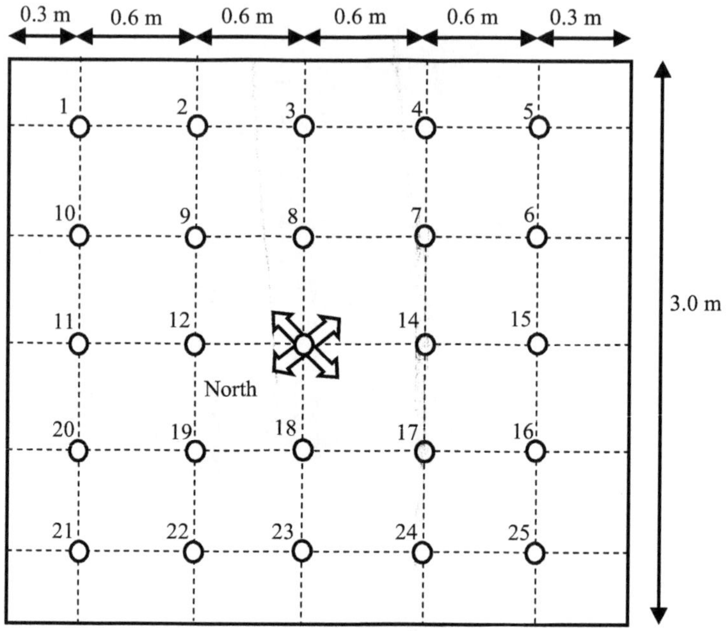

Fig. 3.3 Grid points of horizontal working plane (floor)

3.2.3 Light Guide Efficiency Determination

The efficiency of tubular light guide system can be determined as the ratio of total luminous flux leaving at the diffuser to the luminous flux entering at the receiving dome of the light pipe (Mohelnikova 2008). Mathematically,

$$\eta = \frac{\phi_{\mathrm{t}}}{\phi_{\mathrm{i}}} \tag{3.1}$$

where ϕ_{i} is the incident luminous flux and ϕ_{t} is the luminous flux transmitted through the light guide. These luminous fluxes can be easily determined from the illuminance measurements by means of the following expressions:

$$\Phi_i = L_{\mathrm{i}} S_{\mathrm{E}} \tag{3.2a}$$

$$\Phi_{\mathrm{t}} = L_{\mathrm{t}} S_{\mathrm{E}} \tag{3.2b}$$

where L_{i} is the illuminance at the inlet, L_{t} is the illuminance leaving at the light pipe and S_{E} is the outside area of the cross-section of the light pipe.

The efficiency of light guide can be expressed as the ratio of transmitted to incident illuminance, those are quantities actually measured (S_{E} is in both numerator and denominator). But, it must always express in terms of luminous fluxes because the physical meaning of luminous flux is an amount of light per unit time whereas a ratio of illuminance does not have such a physical meaning.

3.3 Results

The measurements were taken by changing the direction of light reflector inside the receiver dome. The daily average of the illuminance levels on the working plane for the different positions of light reflector is shown in the Tables 3.1 and 3.2.

It is clearly observed from the tables that the luminous flux with the reflector positioned towards south shows better illuminance as compared to the east. It is estimated that the daily average of luminous flux when reflector oriented east is 31.7 lux with $\eta = 31.11\%$. The daily average improved to 37.1 lux when reflector oriented south with $\eta = 33.40\%$. It can be easily seen that the maximum illuminance (lux) is obtained below the light diffuser (centre of the working plane). The solar irradiance variation with time for corresponding east and south orientation has been plotted for the 2 days (Fig. 3.4a, b).

3.3.1 Energy Savings Assessment

Assessment of energy saving due to light guide for substituting the standard artificial lighting systems has been carried out. Light guides were compared with existing artificial lighting systems with different luminous efficacy – see Table 3.3. Luminous flux through tubular light guide of diameter = 0.23 m and $\eta_{mean} = 33.76\%$ has been computed applying the Eq. 3.1 and shown in the Fig. 3.5. The luminous flux is determined for the observed external illuminance from 10 to 85 klux.

Comparison of light pipe of the selected diameter with artificial light sources is shown in the Table 3.4. It can be observed that for external illuminance of 20 klux, light pipe with diameter 230 mm produced 301 lux which is equivalent to luminous

Table 3.1 Illuminance (lux) on the working plane when reflector positioned towards east direction

E (lx)	**1.2**	**0.6**	**0.0**	**0.6**	**1.2**
1.2	20.28	29.88	37.5	31.25	26.31
0.6	31.57	40.31	42.58	43.08	28.67
0.0	29.29	39.68	43.31	40.3	33.89
0.6	27	32.1	37.6	32.9	27.5
1.2	21.84	25	26.71	24.8	18.59

Table 3.2 Illuminance (lux) on the working plane when reflector positioned towards south direction

E (lx)	**1.2**	**0.6**	**0.0**	**0.6**	**1.2**
1.2	24.7	36.5	48.5	37.5	25.6
0.6	38.83	49.16	52.14	53.7	32.53
0.0	34.96	45.16	50.71	47.52	40.25
0.6	31.45	37.59	41.71	37.14	29.79
1.2	24.34	28.53	29.56	28.14	21.33

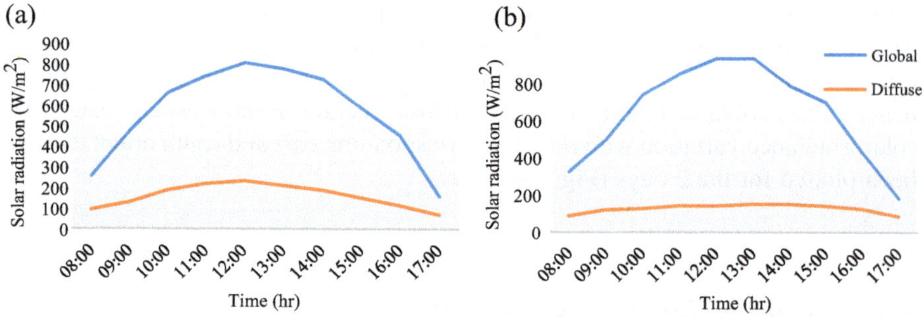

Fig. 3.4 Hourly variation of solar irradiance at New Delhi location. (**a**) On 11th April, Clear sky. (**b**) On 13th April, Clear sky

Table 3.3 Luminous efficacy of the artificial lighting systems

Artificial light	Luminous efficacy (lm/W)
Incandescent bulb	12
Fluorescent lamp	50
LED's	100

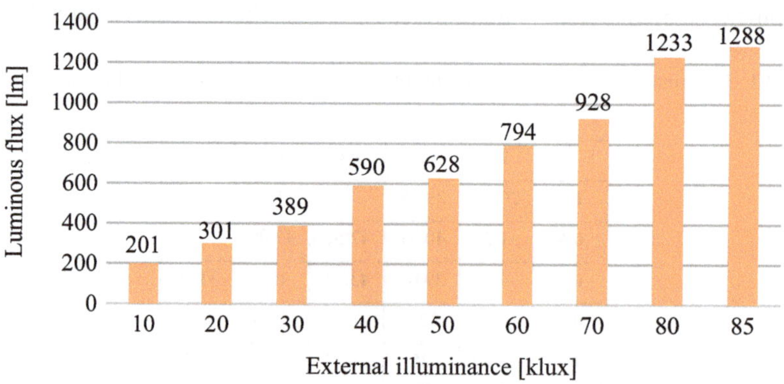

Fig. 3.5 Luminous flux output through tubular light pipe

Table 3.4 Comparison of light pipe with considered artificial lighting systems

External illuminance (klux)	Incandescent bulb (W)	Fluorescent lamp (standard) (W)	LED (W)
10	–	–	–
20	25 × 1	5 × 1	3 × 1
30	40 × 1	9 × 1	5 × 1
50	60 × 1	14 × 1	9 × 1
70	75 × 1	19 × 1	12 × 1
85	100 × 1	23 × 1	15 × 1

output of one 25 W incandescent bulb/one 5 W fluorescent lamp/one 3 W LED bulb. Therefore, even during low external illuminance, the considered light pipe is able to replace one 25 W incandescent bulb/one 5 W fluorescent lamp/one 3 W LED bulb.

Further, energy saving due to single and two light pipes for the considered room has been calculated. The room work space is assumed to be easy office work/classes type, for which recommended illuminance level is 250 lux (BEE). Monthly energy demand has been calculated maintaining required illuminance level inside the room.

The illuminance level at the floor level has been computed using the empirical relation (Web: engineering toolbox):

$$L_{\mathrm{t}} = \frac{(L_{\mathrm{l}} C_{\mathrm{u}} f_{\mathrm{lf}})}{A_{\mathrm{l}}} \tag{3.3}$$

where L_{t} is net illuminance (lux), L_{l} is luminance level per lamp (lumen), C_{u} is coefficient of utilization, f_{lf} is light loss factor and A_l is area per lamp (m^2).

In calculation, $C_{\mathrm{u}} = 0.6$ and $f_{\mathrm{lf}} = 0.8$ have been taken.

It has been determined that fluorescent lamp of 40 watt will be required to maintain the required illuminance level. During the daytime, the energy usage has been calculated for the lamp along with single and double light pipe while maintaining the illuminance level using continuous dimming control. Therefore, the monthly energy saved has been calculated with a single and two light pipes which is shown in the Table 3.5. Also, total number of hours for which illuminance demand has achieved is computed and shown in the Fig. 3.6. It has been noticed that single light pipe is not sufficient to illuminate the considered room, though two light pipes are able to meet the target. The figure shows that the maximum demand meeting hours are in the months of summer. This reduces in monsoon months and minimum during the winter. The reason for more meeting hours in summer months is availability of high external illuminance in the Indian composite climatic zones. It has also reported that the illuminance achieved is independent of number of sunshine hours.

Table 3.5 Monthly energy saving and carbon mitigation due to single and two light pipes

	Fluorescent lamp	Single light pipe			Two light pipes		
	Power required, kWh	Energy saved, kWh	Energy saved, in terms of %	Carbon mitigation, kg	Energy saved, kWh	Energy saved, in terms of %	Carbon mitigation, kg
Jan	13.52	3.2	23.8	2.6	6.4	47.7	5.2
Feb	12.32	3.7	30.1	3.0	7.4	59.7	6.0
March	14.2	5.1	35.7	4.1	9.6	67.6	7.8
April	15.6	5.5	35.4	4.5	10.2	65.3	8.2
May	16.12	6.0	37.0	4.8	11.0	68.4	8.9
June	15.88	5.3	33.2	4.3	10.0	63.3	8.1
July	16.12	4.5	27.9	3.6	8.7	53.9	7.0
Aug	16.08	4.3	26.8	3.5	8.4	52.0	6.8
Sept	14.64	4.6	31.4	3.7	8.9	60.6	7.2
Oct	14.36	4.5	31.0	3.6	8.8	61.1	7.1
Nov	12	3.5	29.2	2.8	7.0	58.3	5.7
Dec	12.4	3.2	25.4	2.6	6.3	50.9	5.1

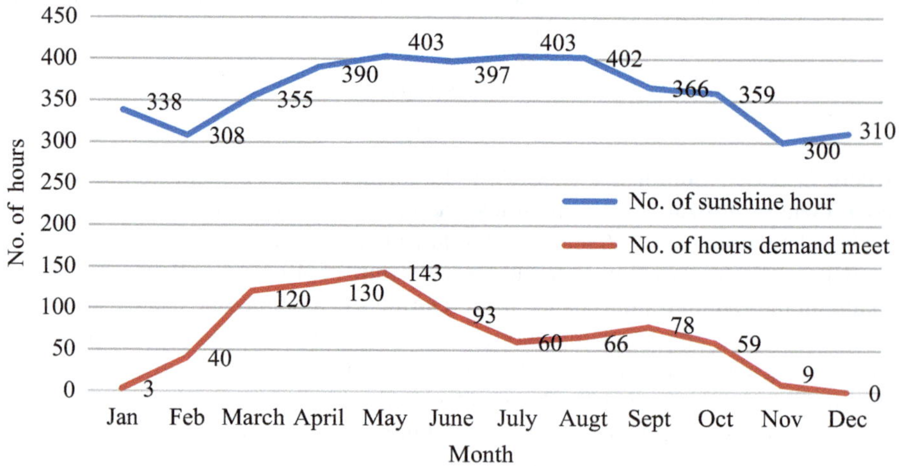

Fig. 3.6 Monthly number of sunshine hours and demand meeting hours

3.3.2 Carbon Mitigation Potential

According to the Central Electricity Authority (CEA) of India, CO_2 emission per kWh of electric power generation is higher (810 g/kWh) due to usage of low-grade coal. Assessment of carbon mitigation potential due to light pipe along with fluorescent lamp has been carried out in the study. The monthly carbon mitigation potential utilizing single and two light pipes has also been compared in the

Table 3.5. The usage of one 40-watt fluorescent lamp produces 140 kg of CO_2 annually when used only during the daytime. With the use of single light pipe, applying continuous dimming control can reduce the carbon emission by 43.2 kg/year. In a similar manner, carbon mitigation will be double with the use of two light pipes. This shows that the use of light guide system also has positive environmental impact.

3.4 Conclusions

The behaviour of light guide reflector, which reflects the low-angle sunlight, by changing its orientation, has been studied in the present communication. It has been identified that south direction is best suitable for fitting the reflector inside the receiving dome. Though it has been conducted in composite climate zone, it is valid for all Indian regions.

Our findings proved that light pipe systems can play a significant role in conserving the energy. The usage of light pipe systems will also have good environmental impact as significant amount of carbon gets mitigated. Further studies can be made to improve the performance of the light pipe in utilizing the solar energy.

After doing the energy and daylight assessment, it has demonstrated that these systems are quite useful in building purposes. Although they could not totally replace the common daylight systems of windows and skylights, they can be used along with them to provide daylight inside the interior spaces of any building.

References

Baroncini, C., Chella, F., & Zazzini, P. (2006). Experimental analysis of tubular light pipes performances: Influence of the diffuser on inside distribution of light. In *SET2006 – 5th international conference on sustainable energy technologies*. Vicenza, 30th Aug–1st Sept 2006.

Central Electricity Authority of India (CEA). Report 2009.

Energy Conservation Building Code (ECBC). (2007). *Bureau of energy efficiency*, Ministry of Power, Govt. of India.

Gil-Martin, L. M., Pena-Garcia, A., Jimenez, A., & Hernandez-Montes, E. (2014). Study of light-pipes for the use of sunlight in road tunnels: From a scale model to real tunnels. *Tunnelling and Underground Space Technology, 41*, 82–87.

Mohelnikova, J. (2008). Daylighting and energy savings with tubular light guides. *WSEAS transactions on Environment and Development, 4*(3), 200–209.

Patil, K.N. (2014). *Energy conservation studies in buildings through daylighting and natural ventilation for space conditioning*. PhD thesis, IIT-Delhi, New Delhi.

Shin, J. Y., Yun, G. Y., & Kim, J. T. (2011). Evaluation of daylighting effectiveness and energy saving potentials of light-pipe systems in buildings. *Indoor and Built Environment,* 21 (1), pp. 129-136.

Web: https://beeindia.gov.in. Last modified 18 Aug 2016.

www.engineeringtoolbox.com/light-level-rooms-d_708.html.

Yanpeng, W.U. (2005). Daylight performance of top lighting light pipes and side lighting light Pipe under sunny conditions in beijing. In *The 2005 world sustainable building conference* (pp. 1111–1116), Tokyo, 27–29 Sept 2005(SB05Tokyo).

Yun, G. Y., Shin, H. Y., & Kim, J. T. (2010). Monitoring and evaluation of a light-pipe system used in Korea. *Indoor Built Environment, 19*(1), 129–136.

Chapter 4
Techno-Economic and Environmental Implications of Electricity Generation from Solar Updraft Chimney Power Plant in Meekatharra in Western Australia

Brian Boswell and Wahidul K. Biswas

4.1 Introduction

An SCPP (solar chimney power plant) has been found to be suitable for large-scale energy production (i.e. 100 MW or more) (SBSG 2011). The construction of this plant will create employment of a local labour force and extensively use locally available materials. The solar updraft chimney, which is another name for an SCPP, can be built in desert countries and semi-arid regions like Western Australia to contribute to a significant portion of energy supply. SCPPs are particularly reliable for having few moving parts such as turbines and generators. It's a simple and robust structure which guarantees operation and requires little maintenance as there is no combustible fuel involved.

Unlike fossil fuel-fired plants and other solar-thermal power plants, an SCPP does not need cooling water, which is a key advantage in many sunny and water-scarce regions, like Western Australia, which already has major problems with water supply (SBSG 2011). In addition, an SCPP uses hot air as a medium for heat transfer instead of water. This is particularly useful in arid and semi-arid areas of Africa and Western Australia (Bansod et al. 2014). Electricity that is generated from an SCPP has been found to be the cheapest option compared to other solar power plants, but their investment cost is higher than those of conventional power plants. It is expected that the shortfall of fuel reserves in combination with a dramatically increasing demand for remote area power supply and an introduction of environmental policies could make this technology cost competitive.

B. Boswell • W.K. Biswas (✉)
School of Civil and Mechanical Engineering, Curtin University, Perth, Australia
e-mail: w.biswas@curtin.edu.au

A. Sayigh (ed.), *Transition Towards 100% Renewable Energy*,
Innovative Renewable Energy, https://doi.org/10.1007/978-3-319-69844-1_4

The high investment cost compared to the plant efficiency, and the limited height of the chimney due to technological constraints, is sometimes considered to be the main disadvantage of an SCPP (Kalash et al. 2014). The research conducted by Hamdan (2011) confirms that the cost of energy produced from an SCPP needs more investigation since it is dependent on location, labour costs and material costs, all of which vary across regions. The construction materials of an SCPP mainly glass, steel and concrete are relatively inexpensive and readily available. The main advantage of this plant lies in the low maintenance cost, the simplicity of operation and the durability of the system (Hamdan 2011). An SCPP has a longer operating life span of at least 80–100 years, and it does not require skilled staff for maintaining the plant (Ming et al. 2013; Chikere et al. 2011). Despite these advantages, the electricity produced by an SCPP in the Mediterranean region has been found to be considerably higher compared to the other power sources. Western Australia also have a Mediterranean climate with mild winters and dry summers, indicating that an investigation is required to determine whether an SCPP can generate environmentally friendly electricity in an economically feasible way.

The advantage of the use of an SCPP can be obtained if it is installed in a temperate region. As discussed earlier, other factors including costs, location of the site and the intermittent nature of solar radiation that could potentially affect the economic and technical performances of SCPP have been taken into account in the current analysis. The aim is to carry out an economic and environmental feasibility study of the substitution of diesel-generated electricity with electricity generated from a solar tower in Meekatharra, which is one of the hottest inland regions of Western Australia, with an annual average temperature of 28.8 °C (BoM 2015). Life cycle assessment (LCA) and life cycle costing (LCC) have been widely used to assess the economic and environmental performance of energy systems (Lund and Biswas 2008).

The efficiency of an SCPP is known to be low, resulting in high chimney heights and large collector areas to compensate for lack of efficiency. The consequence of this is extensive building costs, making projects subject to economy of scale. The work presented in this paper is based on research conducted using a SCPP scaled model built at Curtin University that was used to help verify the SCPP computer model. Future design changes for a location such as Meekatharra have been modelled with more confidence in determining the optimum height and collector area for a required power supply. The SimaPro computer program was used to calculate the carbon footprint of a proposed design of an SCPP, against an equivalent decentralised diesel power plant.

Firstly, this paper presents the optimum technical design for an SCPP to supply electricity to a community in Meekatharra. Based on this technical design, an LCA was carried out to determine the environmental impacts of an SCPP. Thirdly, the life cycle cost of this system using the same data that was used in the LCA analysis was then carried out to determine the cost-effectiveness of SCPP.

4.2 Methodology

This methodology section consists of the following steps:

1. Experimental setup and measurements
2. Development of a MatLab model to compare with field measurements
3. Use of computational fluid dynamics (CFD) to design a large system for Meekatharra, WA
4. Environmental and economic analyses

4.2.1 *Setup and Measurements*

The flow velocity was measured by using anemometer transducers which were located at positions 1, 2 and 3 as shown in Fig. 4.1a, b. The thermocouples were attached to the supports with one being located outside to measure the ambient temperature. All the transducers were connected to the PicoLog Data logger enabling the conditions to be monitored over a period of time; this was recorded in a notebook computer. Figure 4.1c displays the velocity data recoded for the optimum solar input of the day and demonstrates that the velocity of around 2.5 m/s is easily maintained. Table 4.1 shows data obtained from the field experiment at Curtin University's renewable energy park.

Improved efficiency is the key to reducing the height of the chimney, and to this end research has been conducted at Curtin to establish the optimum design aspects of an SCPP. Two approaches were taken; the first being the production of a reliable mathematical model of the SCPP based on previous research taken from literature review. This was then contrasted against Curtin's SCPP. The established mathematical model was used with MatLab in deriving a computer program to provide different graphical outputs of performance. Again this was checked against data obtained from Curtin's small SCPP, the aim being too allow the mathematical model to predict what would happen for any future design changes to the SCPP.

4.2.2 *Development of a MatLab Model*

The MatLab computer model has made the use of a well-established fluid mechanics formula which was then customised and applied to the SCPP, as shown below:

Calculation of power is expressed as

$$P_{\mathrm{out}} = \Delta p \times v_{\mathrm{t}} \times A_{\mathrm{c}} \times \eta_{\mathrm{tg}} \tag{4.1}$$

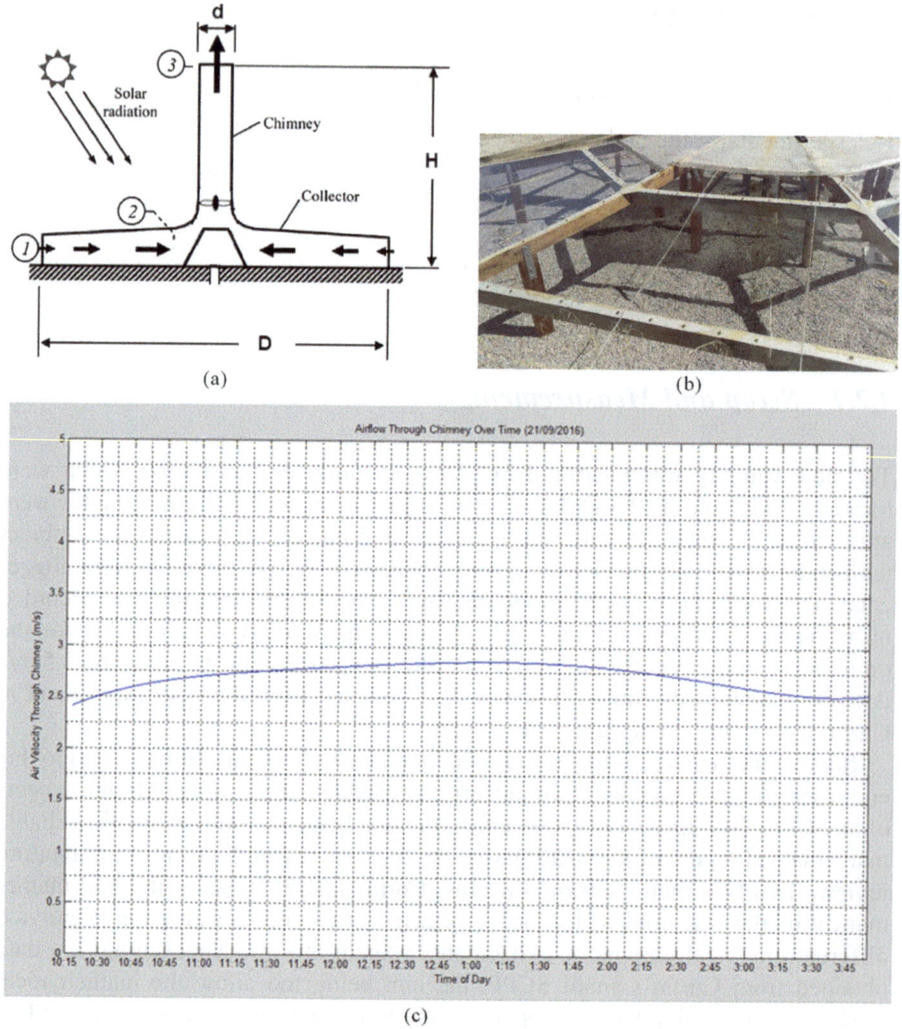

Fig. 4.1 Experimental set up and measurement. (**a**) Measurement locations on SCPP (*H* height, *D* diameter of the collector and *d* diameter of the chimney). (**b**) Picture from Curtin's experiment showing collector supports. (**c**) Velocity data recoded between 10.00 am and 4.00 pm

where

P_{out} = Electrical power output (kW)
Δp = Pressure at tower base – ambient pressure
v_t = The air velocity at tower entrance has been calculated as follows
η_{tg} = Turbine generation efficiency (%)

Table 4.1 Inputs and outputs of Curtin's SCPP at 1.00 pm on the 12/09/2016

Parameters	Value on 12/09/16	Unit
Inputs		
Collector efficiency ($\eta_{collector}$)	0.3	%
Pressure drop (Δp)	5	Pa or kg/mS^2
Temperature inside the collector (T_i)	38	°C
Ambient temperature (T_o)	20	°C
Air density at tower entrance (ρ_i)	1.127	kg/m^3
Outside air density (ρ_o)	1.2	kg/m^3
Air velocity in chimney (v_t)	2.4	m/s
Solar intensity (I_r)	760	Watt/m^2
Output		
Power (P_{out})	0.8	W

$$v_t = \sqrt{2 \times g \times H_t \times \frac{\Delta T}{T_o}} \tag{4.2}$$

where

g = Gravity (9.81 m/s^2)
ΔT = Temperature inside the collector (T_i) – ambient temperature (T_o)
T_i is calculated as follows

$$T_i = \left(\frac{I_r \times \eta_{coll} A_{coll} T_o}{\rho_i \times A_c \times C_p \times 4.43 \times H^{\frac{1}{2}}} \right)^{\frac{2}{3}} + T_o \tag{4.3}$$

$$\eta_{coll} = \frac{Q}{A_{col} \times G} \tag{4.4}$$

$$\eta_c = \frac{g \times H_t}{T_c \times C_p} \tag{4.5}$$

$$\eta_{sys} = P_{out} / \left(\pi \times I_r \times R^2 \right) \tag{4.6}$$

where

I_r = Solar radiation intensity (Watt/m^2)
η_{coll} = Collector efficiency (%)
η_c = Chimney efficiency (%)
η_{sys} = SCPP system efficiency (%)
A_{coll} = Area of the collector (m^2)

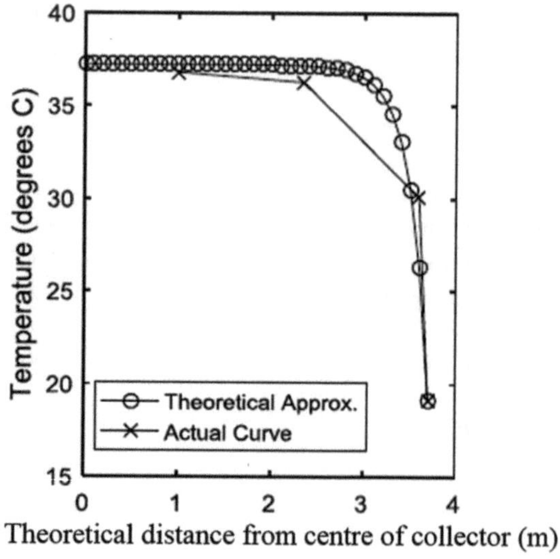

Fig. 4.2 Temperature under collector

ρ_i = Density of air inside the collector kg/m^3
A_c = Area of the chimney (m^2)
C_p= Specific heat of air (Joule/kg)
H_t = Height of the tower
R = Collector radius

Figure 4.2 shows a typical output from the MatLab model, which was then compared against Curtin's SCPP actual temperature readings, and a reasonable agreement for the MatLab model was achieved. The analytical solutions obtained from the MatLab simulation compared favourably to actual measurements taken (Fig. 4.2).

4.2.3 CFD Modelling for Designing SCPP for Meekatharra

An SCPP has to be designed to replace the diesel generating capacity of 0.3 MW in Meekatharra. CFD modelling software, which takes into account the changes of fluid flow and temperature distribution in an SCPP for future prediction purposes, has been utilised to obtain more accurate outputs for a larger system design. The power output and specifications of the experiment conducted in Curtin on an SCPP were input to a CFD to design a larger system of 0.3 MW for Meekatharra, using an area of minimum solar radiation of 2,100 kWh/m^2 (BoM 2016). The CFD analysis also compared well to actual measured data but there was a need to include some additional assumptions to predict the actual performance (Fig. 4.3). This velocity was found to be close to the velocity evaluated from the computer model.

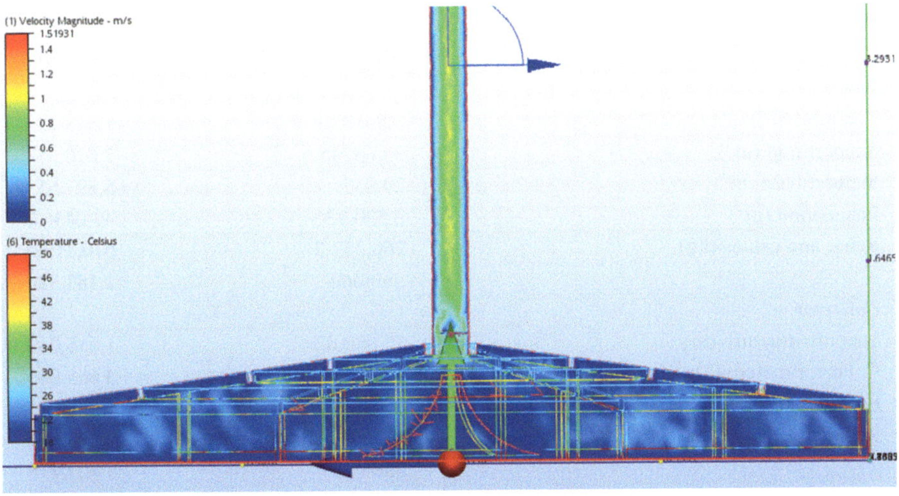

Fig. 4.3 CFD showing velocity

Incorporating new tower parameters into the computer program can produce an efficiency increase based off the original tower design.

While performing CFD, the following considerations have been made:

- Reducing the cost of construction of the plant is a necessity to ensure the SCPP is as cost efficient as possible. The cost being attributed to the height of the chimney has been kept to a minimum, this is needed to provide sufficient air buoyancy, to enable the air flow to drive the turbine. Improving the air velocity through the turbine would enable reductions to be made to the height of the chimney H. The height of the chimney is first and foremost dictated by what is practical to build and at a suitable price. Accordingly, a more practical chimney height for Meekatharra has been considered to be 200 m on the basis of recent literature (Alawin et al. 2012; Cristiana et al. 2009). The ratio of chimney height H over collector area based on the diameter D of Curtin SCPP (i.e.1.16) is being considered of this large system design. Given the height of 200 m of SCPP chimney and the lowest local solar radiation of Meekatharra (i.e. 2,100 kWh/m^2), the capacity of this large plant has been calculated as 0.298 MW using CFD modelling.
- It was also taken into consideration that the power output should obviously be able to supplement the needs of the community, improving the environmental effect of producing the required power. Therefore, the height of the chimney is shown to be a compromise between all of these requirements. Too high a chimney would be unrealistic for this project.

Table 4.2 shows the complete design parameters of 0.298 MW SCPP for Meekatharra, using inexpensive materials and proven technology and their associated costs for performing an economic analysis. The collector component consists of a transparent roof made of glass or plastic film, supported by grids, and column structures. As the hot air under the collector flows towards the chimney, it also gains

Table 4.2 Life cycle inventory and cost data of a SCPP used in Meekatharra, Western Australia

Components of an SCPP	Amount of resources	Cost (AUD)
Tower		
Top tube (m^3)	7.861836	1,548.78
Support mid (m^3)	35.63745	7,020.57
Support base (m^3)	29.937	5,897.65
Foundation (m^3)	2,530.136	498,436.7
Rebar and cables (kg)	260,357.2	670,419.9
	Subtotal	1,183,324
Collector		
Support structure (m)	15,680.02	1,615,042
Collector material (m^2)	196,349.5	3,665,846
Foundation (m^3)	280	55,160
Rebar and cables (kg)	36,400	93,730
	Subtotal	5,429,778
Power conversion unit (PCU)		
Rotor blade length (m)	2.5	647,958.4
Land		
Industrial land required (LandCorp) (m^2)	250,000	7,500,000
Transport		
Weight transported (tonne)	8,685.926	
Distance travelled (km)	764.8	
8-tonne truck (t/km)	6,642,996	1,282,704
Labour		
Overall total man-hour (hr)	219,697.4	8,897,745

radiated heat from the ground. This makes it necessary for the base to be appropriately prepared to absorb and store the energy. The collector material is a major component in determining the effectiveness and robustness of the collector. Glass is an ideal material as it needs to be luminous, durable and economical. However, it is expensive, and a compromise is a combination of two types of roof materials being used, i.e. membrane and glass to reduce cost. The collector super-structure consists of galvanized steel beams and concrete for each of the supports.

4.2.4 Environmental and Economic Assessment

Once the technical design has been performed, the next task is to assess the economic and environmental implications of supplementing a decentralised diesel power plant (DDPP) with SCPP in a semi-arid, temperate climate location in Meekatharra, Western Australia. This is a comparative assessment of determining the environmental benefits of supplementing the decentralised diesel power plant

(DDPP) with an SCPP for similar system boundaries as considered in the LCA analysis.

Environmental assessment – the LCA approach used in this paper assessed carbon footprint only – for the manufacturing, installation and generation stages of electricity generation from an SCPP and diesel power plant. According to Finkbeiner et al. (2011), this research considers carbon footprints in terms of an LCA, with the limited focus on one impact category only, i.e. climate change. All methodological requirements and principles of the LCA can be applied to estimate carbon footprints, as evidenced by local and international literature (Biswas 2014a, b; Grant and Beer 2008). This LCA follows the ISO14040-44 guidelines (ISO 2006, b) and is divided into four steps: (1) goal and scope definition, (2) inventory analysis, (3) impact assessment and (4) interpretation (as presented in Sect. 4.3 section of this paper).

The functional unit is the amount of electricity generated from SCPP over its lifetime. The project life has been considered as 25 years. Following ISO 14040-44 guidelines, this LCA approach evaluates global warming impact or carbon footprint for the following life cycle stages of an SCPP, with infrastructure impacts amortised over the infrastructure elements' useful life (Table 4.2):

Manufacturing of capital equipment includes the extraction of raw materials, transport of raw materials to the manufacturing plant, manufacturing processes of components and transport of the final product to regional storage. Table 4.2 shows the amount of components of an SCPP which were used to develop the inventory of this stage for global warming impact assessment.

Construction includes the transport of materials from regional storage to the project site, construction activities and materials and energy used to assemble the plant. The information on energy consumption for manufacturing (i.e. 7.3 MJ) and installing an SCPP (i.e. 12.54 GJ) were sourced from Kalpakjian and Schmid (2009) and Zongker (2013), respectively.

Finally the operation and maintenance that usually includes the manufacture of replacement materials and transport to the project site, and collector cleaning, has been excluded in this analysis. As stated at the beginning, the great advantage of an SCPP is that maintenance requirement for SCPP is very low, and also due to the fact that the project life has been considered as 25 years, which is much lower than the plant life (i.e. 80 years).

All inputs of the above stages were incorporated into *SimaPro* LCA software to estimate the carbon footprint associated with the production, transportation and use of these materials for constructing an SCPP. The *SimaPro* software calculated the carbon footprint or global warming impact once these material and energy inputs were linked to the relevant libraries (or emission databases). The program determined the correct chemicals and pollutants from different libraries which were selected in order to categorise each environmental impact. The Australian impact method was used to estimate the Global Warming Potential (GWP) of generating electricity from remote area power supply (RAPS) (Life Cycle Strategy Pty Ltd. 2015) as the intention is to assess whether the generation of electricity from an SCPP could potentially be used to reduce the carbon footprint.

All greenhouse gas (GHG) emissions associated with the generation of electricity generation from an SCPP have been converted to the equivalent (CO_2 equivalent) amount of GHG emissions or carbon footprint. Accordingly, the carbon footprint has been calculated as follows (Fatimah and Biswas 2016):

$$CF = \sum_{i=1}^{N} (I_i \times EF_{iCO_2} + I_i \times EF_{iCH_4} \times 28 + I_i \times EF_{iN_2O} \times 265) \quad (4.7)$$

where $i = 1, 2, 3, \ldots\ldots N$ is the input in life cycle inventory, I is the amount of input, EF_{iCO_2} is the CO_2 emission factor of input i, EF_{iCH_4} is the CH_4 emission factor of input I and EF_{iN_2O} is the N_2O emission factor of input i.

The GHG emissions or carbon footprint associated with the generation of electricity from diesel generation from a diesel generator in the remote areas was obtained directly from *SimaPro* software.

There are uncertainties associated with the use of energy and material data used for estimating a carbon footprint. In order to ascertain these uncertainties, a stochastic modelling approach was used (Clavreul et al. 2012). A Monte Carlo Simulation (MCS) was carried out to estimate the uncertainty of each of these input data and to predict the influence that a variable has on the environmental impacts (Goedkoop et al. 2013). The MCS is an iterative approach which uses an input from a probability distribution and produces a distribution of all possible values for 1,000 iterations (Goedkoop et al. 2013). MCS was conducted for a confidence interval of 95% and 1,000 iterations using the SimaPro software. It was done using a single score method.

Economic analysis – A discounted cash flow analysis (DCFA) has been carried out in order to determine the economic feasibility of an SCPP over diesel power plants. The DCFA will determine the economic indicators, such as levelised costs, and payback periods for electricity production for both renewable and nonrenewable scenarios. Accordingly, capital (plant) and operating costs-related information has been obtained through market and literature surveys. Table 4.2 shows the cost data and their sources for conducting the economic analysis. These data were obtained from BGC (2016), Scott Metal (2016) and TDC (2016). The required collector cost is \$5,429,778, and the chimney cost is estimated to be \$1,183,323 when produced from rolled galvanized steel. Additional assumption is that the land has been purchased and is part of a company's asset, i.e. for 250,000 m ^{2} land would be approximately \$7,500,000. The labour estimate factor is determined by man hours per square meter of gross building area. So the total man hours required for construction is 196,349 h, and assuming that the pay rate is \$40 per hour, the total cost of labour is estimated to be \$8.8M. The vertical axis turbine configuration utilised in this an SCPP would have the estimated cost of \$152,000. The amount of materials needed to construct the SCPP requires a transportation/handling cost of \$1.28M.

A detailed LCC analysis has been carried out following AS/NZS 4536:1999 (Standard 2014) in order to determine the cost effectiveness of an SCPP in terms of life cycle cost for an SCPP supplying electricity to the Meekatharra community. The discount factor considered for this LCC analysis is 5% per year.

The main assumptions of cost estimation of a power conversion unit (PCU) are based on Fluri et al. (2009), which are as follows:

- There is no diffuser after the turbines.
- The generator is directly driven by the main shaft (i.e. no gearbox is necessary).
- Inlet guide vanes are included, and their cost has been assumed to be equal to the total cost of the rotor blades.
- The cost for turbine ducts and the central structure are part of the PCU cost.
- The LCC of diesel generation unit was sourced from Byrnes (2014). The 2014 price of diesel-generated electricity was converted to current price by considering an escalation of diesel price (AIP 2016).

4.3 Results and Discussions

4.3.1 Carbon Footprint Analysis

The mean values of carbon footprint of the overall scheme have been estimated to be 216 kg of CO_2 equivalent and 1,179 kg of CO_2 equivalent per MWh of electricity generated from an SCPP and the diesel generation unit, respectively (Fig. 4.4). The uncertainty analysis through MCS simulation proves the validity of LCA results. This is shown as the standard deviation, only 0.32% of the mean value for an SCPP, meaning that the data is of acceptable quality and 4.05% for diesel generation unit (Goedkoop et al. 2013). The calculated value of carbon footprint of an SCPP (i.e. 228 kg CO_2 equivalent) is 5% higher than the mean value that is obtained through MCS, while it is (1,185 kg CO_2 equivalent) 0.5% higher than the mean value in the case of the diesel generation unit.

The annual load curve of Meekatharra was obtained from a local utility company Horizon Power (D. Edwards, *pers. comm.* Horizon Power, Perth). Accordingly, it was estimated that an SCPP with its current capacity (i.e. 2,614 MWh) can meet 34% of the electricity demand (7,811 MWh), and therefore it is estimated that about 37% of GHG emissions can be avoided by incorporating SCPP in the electricity mix of Meekatharra. The emission factor for SWIS electricity is 830 kg CO_2 equivalent per MWh of electricity generation (DoE 2015), which means that an SCPP produces 75% less GHG emissions than SWIS electricity.

Table 4.3 shows the breakdown of GHG emissions of the SCPP and diesel generation unit. The selection of low carbon intensive collector materials could significantly reduce the carbon footprint of SCPP. This is because the collector material alone accounts for 82% of the GHG emissions from SCPP. In the case of the diesel generation unit for a remote area power supply, the production of generating unit accounts for a very small portion (i.e. 0.2%), while diesel

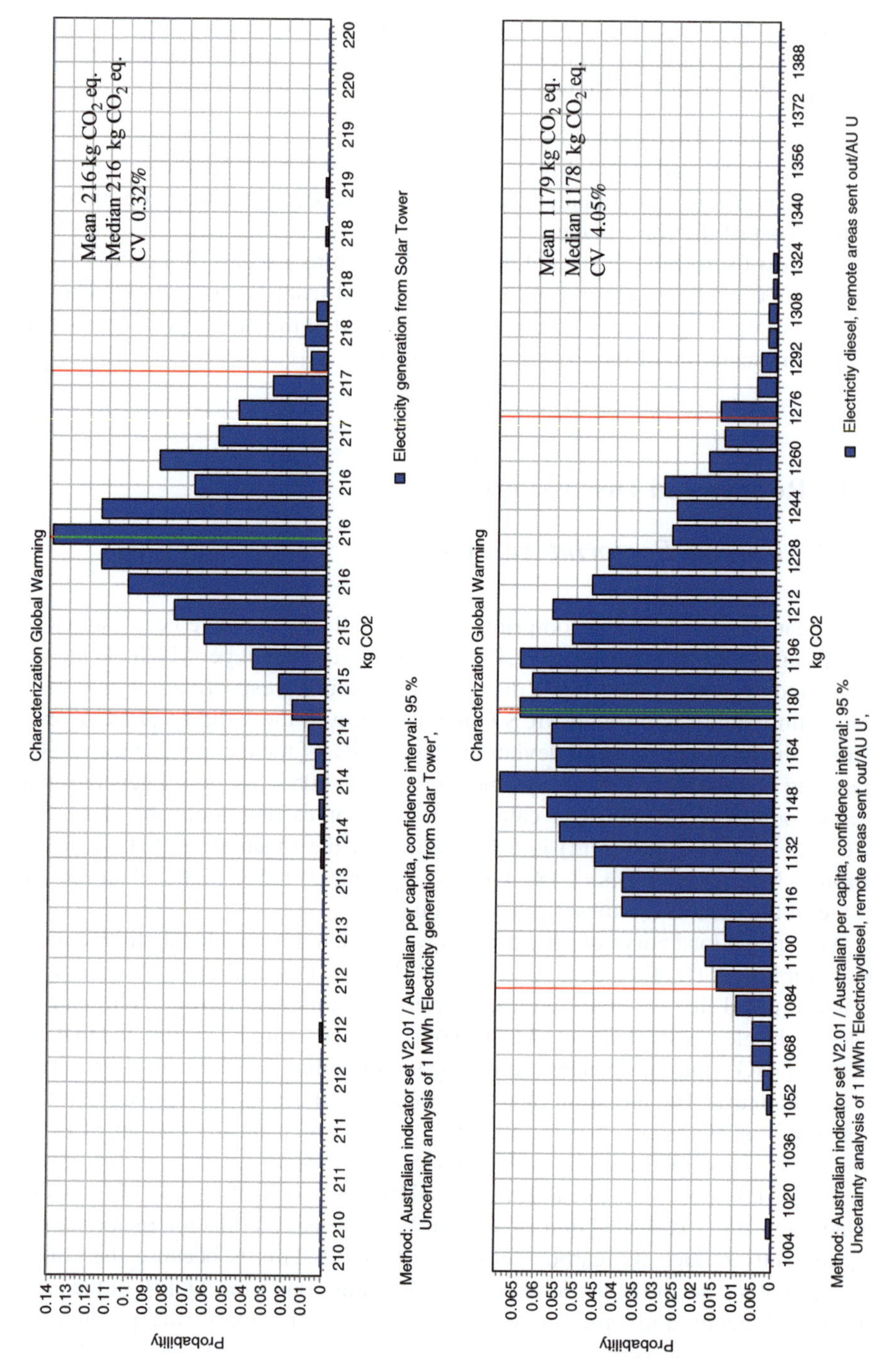

Fig. 4.4 Monte Carlo simulation results

Table 4.3 Breakdown of GHG emissions of SCPP and diesel generation unit

Solar chimney power plant	kg CO_2eq./MWh
Mining to material collector – solar collector	186.78
Mining to material production base – solar tower	19.46
Mining to material production tube – solar tower	14.19
Mining to materials support – solar collector	2.8E+00
Mining to material foundation – solar collector	2.1E+00
Mining to materials cable – solar collector	1.8E+00
Mining to material cables – solar tower	1.3E+00
Energy to constrict – digging foundation	2.0E–01
Mining to material blade – solar tower	8.9E–02
Energy to construct – lifting tower	5.1E–04
Energy for manufacturing	3.6E–04
	228.64
Diesel generation unit for remote area power supply	
Electricity diesel, remote areas sent out/AU U	965.5
Diesel, at consumer/AU U	217.9
Gas power plant/AU U	1.8
	1,185.2

Table 4.4 Summary of economic analysis of base case scenario of SCPP

Annuity	0.071
Interest rate (%)	5%
Depreciation (years)	25
Annuity on investment cost	$1,746,674.26
Annual electricity (kWh)	2,613,554
LCC ($/kWh)	0.668313926
Payback period (years)	14.1

combustion itself accounted for 81% of the total emissions, followed by diesel production (i.e. ≈ 19%).

The conclusion can be drawn that the carbon cost of infrastructure of renewable energy is much higher than a fossil-driven power generation unit due to the fact that the former is far less energy intensive than the latter.

4.3.2 Economic Feasibility

Table 4.4 shows the summary for economic analysis of the base case scenario of an SCPP for a project period of 25 years generating about 2.6 GWh of electricity per year. The LCC of an SCPP is 25% higher than that generated from the decentralised diesel power plant (DDPP) ($0.49/kWh) and 60% higher than South West Interconnected System (SWIS) (i.e. $0.26/kWh) (Byrnes 2014).

SCPP has not been found to be cost competitive with the existing diesel generation unit in Meekatharra. In order to provide power from the SCPP to a remote area, more subsidy would be required than the existing diesel generation unit. This is because the electricity price in the remote area is cross subsidised by Western Power. It means that the customers in the remote area of Meekatharra pay the same electricity tariff as those connected to the centralised electricity network. As a result, a remote area power provider like Horizon Power sells electricity at a price which is lower than their generation cost. However, Western Power covers these losses of the utility company through a Tariff Equalization Fund (TEF) (Economic Regulatory Authority, Western Australia 2010; Wood and Blowers 2015). This TEF is generated by obtaining network charges from the customer using a centralised electricity network.

Since the existing remote area power supply is already subsidised, some other economic instrument needs to be considered that could potentially make the price of electricity generated from an SCPP competitive with existing RAPS. One way to do it is by internalising the external cost which is to incorporate environmental cost into electricity price. On the basis of current LCA results, an attempt has been made to incorporate carbon cost (or tax) into the electricity price of an SCPP and existing RAPS option and SWIS. Since the SWIS electricity tariff rate is applied in the remote area, it has been considered for comparison. SCPP can only compete with DDPP if there is a carbon tax of \$100/tonne of CO_2 (Fig. 4.5a). The increase of carbon tax to \$150/tonne of CO_2 cannot even make an SCPP competitive with SWIS. Economic instrument is not enough to attain the economic viability of these emerging environmental technologies. Another study carried out by Barton et al. (2014) also found that the incorporation of a grain legume in a 2-year cropping rotation decreased GHG emissions from wheat production by 35% on a per tonne of wheat basis, but a large incentive (\$93 per tonne of carbon dioxide equivalents reduced) was required for the inclusion of grain legumes to be financially attractive. The implementation of these high carbon tax values is politically unrealistic considering the fact that introduction of carbon tax of only 23/tonne of CO_2 had been found to receive some public opposition as resource industries and exporter have been identified as the ultimate losers (Meng et al. 2011).

Since economic instrument does not work, there needs to be a massive technological development of these emerging environmental technologies like SCPP, so that the cost per unit of output can be reduced. Figure 4.5b shows that the increase of technical efficiency of SCPP from 0.63% to 1.64%, which is even far less than the efficiency of existing renewable energy technologies such as wind, solar PV, could reduce the price of electricity from SCPP from \$0.66 to more competitive price of \$0.26 per unit of electricity produced (Fig. 4.5b). Technological improvement is required to increase the efficiency by 158% in order to achieve the economic viability of an SCPP. This can be done by changing the flow of air into the chimney, while keeping tower dimensions constant as the output power is directly proportional to the square of the velocity, even a small increase in flow velocity amounts for a dramatic increase in the output power. Another key point to note is that the larger the SCPP, the more efficient it becomes. This second option would increase the need for land which may not be an issue for WA, but it will

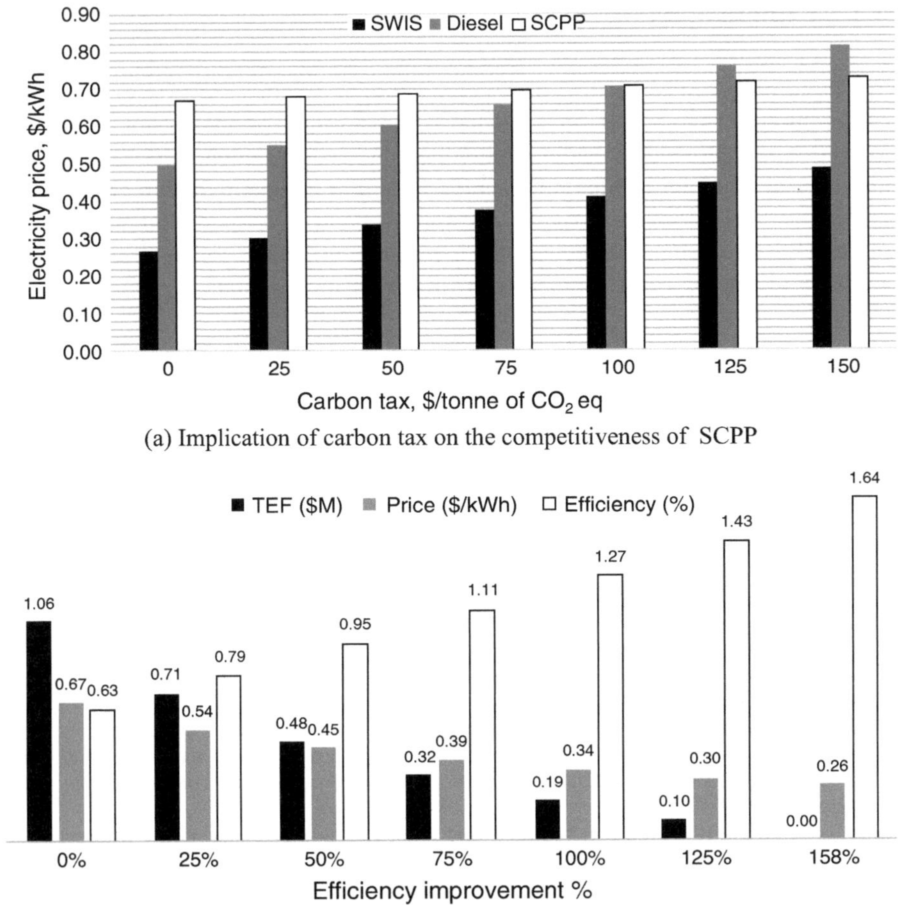

(b) Implication of efficiency on the economic viability of SCPP

Fig. 4.5 Economic viability of SCPP due to implications of the application of economic instrument and technological development. (**a**) Implication of carbon tax on the competitiveness of SCPP. (**b**) Implication of efficiency on the economic viability of SCPP

increase the cost significantly. In order to overcome these problems, many novel concepts have been proposed, such as the use of a sloped solar chimney power plant (SSCPP) to reduce the tower height for a given collector area.

Since technological development cannot take place overnight, both efficiency improvement and environmental policy instrument are considered at the same time. Our estimate based on the current data shows that the increase of carbon tax to \$75/tonne of CO_2 and an increase in an SCPP efficiency to 1.27% could make the price of electricity of SCPP (i.e. \$0.366/kWh) cheaper than SWIS (\$0.375/kWh).

4.4 Conclusions

The replacement of an existing diesel generation unit with an SCPP has significant carbon footprint saving potential (i.e. 81%). However, an SCPP has not been found to be economically viable due to its very low efficiency. Economic instruments will not work to make these emerging technologies cost competitive with existing and alternative options unless a significant improvement in technical efficiency takes place. This research has found that the increase in efficiency to 1.64% could make an SCPP technology cost competitive with the existing options.

References

AIP (Australian Institute of Petroleum). (2016). AIP annual retail price data, Australian Institute of Petroleum, Canberra City ACT 2600.

Alawin, A. A., Badran, O., Awad, A., Abdelhadi, Y., & Al-Moflesh, A. (2012). Feasibility study of a solar chimney power plant in Jordan. *Applied Solar Energy, 48*(4), 260–265.

Bansod, P. J., Thakre, S. B., & Wankhade, N. A. (2014). Solar chimney power plant – A review. *International Journal of Modern Engineering Research, 4*(11), 18–33.

Barton, L., Thamo, T., Engelbrecht, D., & Biswas, W. K. (2014). Does growing grain legumes or applying lime cost effectively lower greenhouse gas emissions from wheat production in a semi-arid climate? *The Journal of Cleaner Production, 83*, 194–203.

BGC. (2016). *BGC concrete*. http://www.bgcconcrete.com.au/. Accessed 8 Dec 2016.

Biswas, W. K. (2014a). Carbon footprint and embodied energy assessment of a civil works program in a residential estate of Western Australia. *International Journal of Life Cycle Assessment, 19*, 732–744.

Biswas, W. K. (2014b). Carbon footprint and embodied energy consumption assessment of building construction works in Western Australia. *International Journal of Sustainable Built Environment, 3*, 179–186.

BoM (Bureau of Meteorology). (2016). *Evapotranspiration calculations for Meekathara*, BoM, Australian Government. http://www.bom.gov.au/watl/eto/tables/wa/meekatharra_airport/meekatharra_airport.html

BoM (Bureau of Meteorology). (2015). Climate Statistics for Australian Llocations, Australian Government.

Byrnes, L. (2014). *The cost of failing to install renewable energy in Western Australia*. Western Australia: Department of Finance.

Chikere, A. O., Al-Kayiem, H. H., & Karim, Z. A. A. (2011). Review on the enhancement techniques and introduction of an alternate enhancement technique of solar chimney power plant. *Journal of Applied Sciences, 11*, 1877–1884.

Clavreul, J., Guyonnet, D., & Christensen, T. (2012). Quantifying uncertainty in LCA-modelling of waste management systems. *Journal of Waste Management, 32*, 2482–2495.

Cristiana, B. M., Ferreir, A. G., Valle, R. M., & Cortez, M. F. B. (2009). Theoretical evaluation of the influence of geometric parameters and materials on the behaviour of the airflow in a sloar chimney. *Computer & Fluids, 38*, 625–636.

Department of the Environment. (2015). *National greenhouse accounts factors*. Department of the Environment, Australian Government.

Economic Regulatory Authority Western Australia. (2010). Inquiry into funding arrangement to Horizon Power, Issue Paper, Economic Regulatory Authority Western Australia: Perth Business Centre WA 4869.

Fatimah, Y., & Biswas, W. K. (2016). Remanufacturing as a means for achieving low-carbon SMEs in Indonesia. *Clean Technologies and Environmental Policy, 18*, 2363–2379.

Finkbeiner, M., Tan, R., & Reginald, M. (2011). Life cycle assessment (ISO 14040/44) as basis for environmental declarations and carbon footprints products. In ISO Technical Committee 207 Workshop, Norway.

Fluri, T. P., Pretorius, J. P., Van Dyk, C., Von Backström, T. W., Kröger, D. G., & Van Zijl, G. P. A. G. (2009). Cost analysis of solar chimney power plants. *Solar Energy, 83*, 246–256.

Goedkoop, M., Oele, M., Leijting, J., Ponsioen, T., & Meijer, E. (2013). *Introduction to LCA with SimaPro*. The Netherlands: PRé.

Grant, T., & Beer, T. (2008). Life cycle assessment of greenhouse gas emissions from irrigated maize and their significance in the value chain. *Australian Journal of Experimental Agriculture, 48*, 257–270.

Hamdan, M. O. (2011). Analysis of a solar chimney power plant in the Arabian Gulf region. *Renewable Energy, 36*, 2593–2598.

ISO. (2006). *ISO 14040:2006 – Environmental management – Life cycle assessment – Principles and framework*. Geneva: International Organization for Standardization.

Kalash, S., Naimeh, W., & Ajib, S. A. (2014). Review of sloped solar updraft power technology. In *Proceedings international conference on technologies and materials for renewable energy, environment and sustainability*, TMREES14, Energy Procedia 50 (pp. 222–228).

Kalpakjian, S., & Schmid, S. R. (2009). *Manufacturing processes for engineering materials* (5th ed.). Pearson Education: London.

Life Cycle Strategy Pty Ltd. (2015). *Australian unit process LCI library and methods*. Version 2015_02_06. http://www.lifecycles.com.au

Lund C, Biswas W (2008) A review of the application of life cycle analysis to renewable energy systems. Bulletin Sci Technol Society 28(3):200–209.

Meng, S., Siriwardana, M., & McNeill, J. (2011). *Australian carbon tax – Winners and losers*. Australia: School of Business and Public Policy, The University of New England.

Ming, T., Gui, J., de Richter, R. K., Pan, Y., & Xu, G. (2013). Numerical analysis on the solar updraft power plant system with a blockage. *Solar Energy, 98*(Part A), 58–69.

Schlaich Bergermann Solar GmbH. (2011). *Solar updraft tower*. Stuttgart: Schlaich Bergermann Solar GmbH.

Scott Metals. (2016). *Steel supplies, steel fabrication and metal recycling*, Woolloongabba, Queensland. http://www.scottmetals.com.au/. Accessed 8 Dec 2016.

Standard, A. (2014). AS/NZS 4536:1999 (reconfirmed 2014) – Life cycle costing- an application guide, Standards Australia and Standards New Zealand, Australia and Wellington.

TDC (Technically Designed Concrete). (2016) Standard Concrete, TDC, Bibra Lake WA 6163, Australia.

Wood, T., & Blowers, D. (2015). Fair price for Western Australia's electricity, Grattan Institute Report no- 2015-10.

Zongker, J. (2013). Life cycle assessment of a solar chimney power plant. Unpublished MSc thesis, Mechanical Engineering, Wichita State University.

Chapter 5
Do as I Say; Don't Do as I Do, Let Alone Do as I've Done. A Study of Australian Universities' Collective Response to Climate Science

Mike Burbridge

5.1 Introduction

The global science community is united about what is forcing climate change: "among papers expressing a position on anthropogenic, or human caused, global warming (AGW), an overwhelming percentage (97.2% based on self-ratings, 97.1% based on abstract ratings) endorses the scientific consensus on AGW" (Cook 2013).

"Warming of the climate system is unequivocal, and since the 1950s, many of the observed changes are unprecedented over decades to millennia. The atmosphere and ocean have warmed, the amounts of snow and ice have diminished, and sea level has risen" (Edenhofer et al. 2014). In the Fifth Assessment Report, the IPCC states that to limit temperature increases to 2° centigrade above preindustrial levels will require a reduction in carbon (and other long-lived greenhouse gases) of between 40% and 70% reduction from 2010 levels by 2050 and near-zero emissions by 2100. To restrict global temperature increases to 1.5° centigrade would require between a 70% and 95% reduction in carbon and other long-lived greenhouse gases on 2010 levels by 2050 (Edenhofer et al. 2014).

As shown in Table 5.1, Australian universities were widely involved within this international work with over a third of Australian universities (excluding CSIRO and BOM) being involved as either authors or review editors for the three working groups that contributed to the final Fifth Assessment Report (IPCC 2014).

Universities in Australia exist to do three things: to teach, research and engage to get the research adopted (Office of Parliamentary Counsel 2017 and 2017a). Climate change penetrates all these areas – academics teach the science and social policy impacts, research the likely consequences of a changing climate and ways to

M. Burbridge (✉)
Curtin University Sustainability Policy Institute, Curtin University, Perth, Australia
e-mail: mike.burbridge@postgrad.curtin.edu.ai

A. Sayigh (ed.), *Transition Towards 100% Renewable Energy*,
Innovative Renewable Energy, https://doi.org/10.1007/978-3-319-69844-1_5

Table 5.1 Australian universities' contribution as author or review editor to IPCC Fifth Assessment Report

Working group I	Working group II	Working group III
Physical science basis	Climate change, impact and vulnerability	Mitigation of climate change
Australian National University	Griffith University	Australian National University
Antarctic Climate and Ecosystems CRC	Macquarie University	Curtin University
Monash University	University of Melbourne	Monash University
University of New England	University of New South Wales	Murdoch University
University of New South Wales	University of Queensland	University of New South Wales
University of Tasmania	University of Tasmania	
	Victoria University	
Other Australian contributors		
Bureau of Meteorology	Australia/World Bank	Department of Agriculture
CSIRO	Australian Antarctic Division	
	Bureau of Meteorology	
	Climate Risk Consultant	
	CSIRO	

Data from IPCC (2014) Fifth Assessment Report

adapt to and mitigate that change and engage in public policy debate to encourage decision makers of the need to take decisions that are cognisant of the science produced by, amongst others, Australian universities.

This paper found minimal evidence of universities adopting their own research (this paper used climate science as a test case) or using their own capital spend to embed this or any research or to deliver teaching and/or research outcomes in the operations of the university. This poses difficult questions for universities to answer – if climate science represents such an urgent threat to humanity, why are Australian universities not applying it to their operations, and why should society adopt research they do not manage to adopt themselves?

It also suggests a further difficulty for Australia to embrace the opportunities of the knowledge economy. Universities in Australia are at the bottom of the OECD rankings for their ability to partner with business for innovation. (OECD 2013). This has been recognised as a key weakness of Australia's innovation framework by both universities (ATN 2015) and in the National Innovation and Science Agenda (Commonwealth of Australia 2015). There is little evidence of universities being effective at internal partnering, and this lack of experience perhaps is one reason for their collective poor performance at external partnering as highlighted by OECD and recognised by government and universities. The government and some universities recognise this weakness by focussing on cross-cutting work in the national innovation strategy (Commonwealth of Australia 2016) and in university corporate strategies.

However, it is worth noting two important points:

- Poor performance collectively should not be interpreted as no performance collectively. There are some universities – albeit a small minority – that are taking the issues seriously.
- It is not just universities that are poor at partnering. Whilst Australian business ranks at above the OECD average for funding research at Universities (OECD 2013), it is also collectively weak at engaging with suppliers and customers for innovation (Gahan et al. 2016).

The paper concludes there are two clear benefits for universities in adopting their own research where possible – one is that they would be able to more clearly understand the difficulties organisations have in adopting or commercialising research into business models; the other is that universities would have a stronger voice if they were in a position to encourage others to do as they have done rather than encouraging others to do what the science says, but that they have not done.

5.2 Methodology

There are 43 universities in Australia. All Australian universities' websites were accessed between 18 and 20 November 2016 to download their 2015 annual report and financial statements and their forward-looking corporate strategies (dates ranging up to 2025). Also, all government/university performance agreements were accessed from the Department of Education and Training website covering 2014–2016 mission-based compacts. A limit of 30 min was allocated to each university to find and download the data. Where the data was unavailable online (either not found or not published), it was requested by email; follow-up emails were not sent.

All available annual reports and strategies were read, and the public commitments in those reports (e.g. published statements, targets and performance) to partnership and adopting research were recorded and tabulated. Research on climate change was taken as the example for adoption of research due to the agreed status of the science (Cook 2013) and the applicability of it to campus development (and therefore to capital campus projects). The documents were also searched by keywords (carbon, sustainable, greenhouse, climate change and variants) to understand the importance of sustainability and climate change to each university. They were also searched to understand the importance of collaboration within the university, to the university (words searched were cross-, inter-, multi-, transdisciplinary and variants) as well as externally with industry, society and government. Each universities' capital spend committed for 2016 was recorded to understand the potential of the sector to use their own capital spend to adopt their own research (e.g. a university with no forecast capital spend could find it more difficult to integrate teaching and research into campus development).

Due to the incomplete nature of performance data, publicly available performance data was accessed to develop an indicative understanding of the rate of change of emissions of Australian universities and thereby to understand their success in translating research into impact.

5.3 Results

Australia has 43 universities, of which three committed themselves to absolute reductions in greenhouse gas emissions up to 2020. Based on evidence from mission-based compacts covering 2014–2016, annual report and financial statements for 2016 and strategies and corporate strategies (covering a variety of start and end dates but all accessed cover years after 2016), there were essentially four groupings of universities:

1. Those with commitments to carbon constraint as well a target for absolute reductions in carbon
2. Those with commitments to carbon constraint as well a target for reductions in carbon
3. Those with commitments to carbon constraint but no evidence as to how to deliver the commitment
4. Those universities that do not mention carbon or emissions in any of their documentation

Table 5.2 may portray an unduly positive picture of universities' commitment to embracing climate science. Some commitments would not pass academic, societal or government scrutiny or commentary. Statements like "we will improve energy efficiency between 2015 and 2020", we will "adhere firmly to the principles of sustainability in all we do" or "we will reduce [the university's] operational carbon footprint" are insufficient to be convincing that such universities are genuinely tackling the issue.

There were also differences within university groupings. RMIT has "since 2008, committed to a target in partnership with Australian Technology Network of Universities to reduce greenhouse gas emissions by 25% by 2020 compared to 2007 as a base". However, this target is not shared by other ATN members (see Table 5.3).

Of the nine universities that scored the highest ranking in environmental sciences in the Australian Research Council's Excellence in Research Australia (Australian Research Council 2016), seven do not mention carbon or emissions in their corporate strategies; in the same vein, 10 of the 12 universities involved in the International Panel on Climate Change Fifth Assessment Report (IPCC 2013) did not include reference to emissions or carbon in their corporate strategies.

There are ten universities that are required to publish their emission date by the National Greenhouse and Energy Reporting Act (Office of Parliamentary Counsel

Table 5.2 Australian university commitments to carbon constraint

	No.
Universities with a science-based commitment[a]	3
Universities with a carbon commitment and published target	12
Universities who make statements but do not provide evidence as to what success looks like	15
Universities who do not mention carbon or emissions	11
Total	41[b]

Data from university annual and financial report, strategic plans and mission-based compacts
[a]This includes RMIT, UTS and Charles Sturt University. CSU is Australia's only carbon neutral university, but there was a lack of clear emission reduction targets at the same time as maintaining carbon neutrality
[b]Documentation was not available for Torrens University or UCL (Australia). Incomplete documentation for Notre Dame, University of Tasmania

Table 5.3 Commitment of ATN members to carbon reduction targets

ATN member	Target
RMIT	Reduce greenhouse gas emissions by 25% by 2020 on 2007 as a base
University of Technology Sydney	30% reduction in GHG based on 2007 by 2020/21
Queensland University of Technology	11% reduction in CO_2 per gross floor area (m^2) by 2019 on 2013
Curtin University of Technology	No target
University of South Australia	No target

Data from university annual and financial report, strategic plans and mission-based compacts

Table 5.4 Greenhouse gas emission from ten largest university emitters

	Scope 1 GHG (t CO2-e)	Scope 2 GHG (t CO2-e)	Total GHG (t CO2-e)
2010–2011	103,853	800,006	903,859
2011–2012	94,758	821,337	916,095
2012–2013	106,674	826,123	932,797
2013–2014	103,502	839,198	942,700
2014–2015	104,316	841,373	945,689
% change over 2010	+0.45%	+5.17%	+4.63%

Data from: http://www.cleanenergyregulator.gov.au/NGER/National%20greenhouse%20and%20energy%20reporting%20data

2017b). Analysis of the data shows an increase in total emissions of 4.6% between 2010/2011 and 2014/2015 as set out in Table 5.4. Currently this is the only data available.

It is interesting that the preeminent research universities working in the environmental science or climate change fields do not apply the science to their own

Table 5.5 Capital commitments of Australian universities

	Capital expenditure during 2016 (,000)
Total spend on property, plant and equipment payable within 12 months (all universities)	$1,542,878
Of which capital spend of universities committed to either using campus as a living lab or applying their knowledge to the campus (UQ, UniSA, UMelbourne, CurtinU, UCanberra, UAdelaide, RMIT, U Tasmania)	$452,737

Capital expenditure contracted for at the reporting date but not recognised as liabilities

organisations, but are committed to be part of the public debate about such grand challenges.

All universities made commitments to collaborate, debate or work in partnership with a combination of industry, government, society or the communities they serve to develop new research, translate existing research or to improve the well-being of society. This commitment to partner with organisations external to the university is a universal desire in the university sector. However, as the OECD, the university sector and the government acknowledge, Australian universities are worse than their OECD counterparts in partnering with external organisations.

The same appears to be the case when considering internal partnerships across the university or to using university spend to deliver teaching or research outcomes or commercialise research.

Australian Universities committed themselves to spend over $1.5billion on capital works during 2016 (see Table 5.5). The Australian Research Council – Australia's principle grant provider – between 2011 and 2013 granted funding worth under $100 million to universities to research the built environment and design under Field of Research code 012 (ARC 2016 p. 86). There are eight universities who have committed to use their campuses as a living lab, to apply their knowledge to the campus or to help deliver teaching and research outcomes. Those universities will be spending $452 million on property, plant and equipment in 2016. There is no evidence in the documents about how the living labs will operate or how the spend will contribute to teaching and research outcomes. The University of Queensland's corporate strategy is the only one that contains a forward commitment to use the campus as a living lab (University of Queensland 2014 p. 21). There are 17 universities that commit themselves, through their corporate strategic plans to cross-silo working. The remaining 26 universities do not make commitments to either cross-, inter-, multi- or transdisciplinary work.

5.4 Conclusions

The evidence from this study suggests that universities have some work to do to put themselves in a position where they can partner effectively with the industry or government to innovate or commercialise their research. Less than 20% of Australian universities are using their campuses to deliver teaching and research outcomes or as a living lab to innovate. Only one university is committed to doing this in the future. Yet universities will be spending over $1.5billion during 2016 on capital infrastructure. If this infrastructure spend is not used to also drive teaching and research outcomes, or to showcase how to adopt research, then it is being spent inefficiently.

Only three universities make a commitment for an absolute reduction in carbon emissions, in line with climate science. Ninety percent of universities do not make this commitment (and over 25% of universities do not mention carbon or emissions in their documentation). Yet it is the same universities that research the subject and pressure ministers and society at large to take action on climate change.

Finally, the apparent struggle universities have to partner with themselves could be dismissed as irrelevant. However, universities are being pressured into developing more effective partnerships with business; if they have little or no experience either in partnering (rather than contracting) or in implementing or commercialising research, then they will not understand the problems businesses face nor how to overcome such challenges. In an environment where government is committed to attaching an increasing share of funding to working in partnership with business, it would seem adroit for universities to get as much experience as possible in working in partnership across organisations and disciplines. This is not just a job for academics but rather for institutions.

Acknowledgement This work has been funded by a PhD scholarship from the CRC Low Carbon Living. The project is looking to develop a business model to help universities, business and government understand the benefits of high-performance, evidence-based innovation. With thanks to Abigail Burbridge for assistance in finding documentation.

Source Data

Emission data: http://www.cleanenergyregulator.gov.au/NGER/National%20greenhouse%20and%20energy%20reporting%20data

All mission-based compacts can be found at: https://docs.education.gov.au/node/34873

Uni	Annual report – 1st entry
	Strategic plan – 2nd entry
ACU	http://www.acu.edu.au/about_acu/our_university/publications/annual_reports
	https://www.acu.edu.au/about_acu/our_university/mission_and_profile/strategic_plan_2015-2020

(continued)

ANU	http://www.anu.edu.au/about/plans-reviews
	http://www.anu.edu.au/files/review/anu-2020-strategy.pdf
Bond Uni	https://bond.edu.au/about-bond/university/introducing-bond/bond-university-annual-report
	https://bond.edu.au/about-bond/university/introducing-bond/bond-mission-strategic-plan
CMU	Not found on website
CQU	https://www.cqu.edu.au/about-us/structure/governance/annual-report
	http://content.cqu.edu.au/Policy/policy_file.do?policyid=2988
CDU	http://www.cdu.edu.au/about/annual-report
	https://www.cdu.edu.au/about/strategic-plan
CSU	http://www.csu.edu.au/about/publications/annualreports
	https://www.csu.edu.au/unistats/university-strategy [2015–16]
CUT	http://about.curtin.edu.au/policy-governance/annual-reports/
	http://trove.nla.gov.au/version/196899916
Dea	http://www.deakin.edu.au/adi/strategic-plan/annual-reports
	http://www.deakin.edu.au/adi/strategic-plan
ECU	http://www.ecu.edu.au/about-ecu/reports-and-plans/annual-reports
	http://intranet.ecu.edu.au/__data/assets/pdf_file/0005/730328/ECU12439_StrategicPlan_A4_WEB.pdf
Fed	https://federation.edu.au/staff/governance/plans-publications-policies/organisational-data
	https://federation.edu.au/__data/assets/pdf_file/0010/284248/FedUni_StrategicPlan.pdf
Fli	http://www.flinders.edu.au/about/governance/annual-reports/
	http://2025.flinders.edu.au/
Gri	Https://www.griffith.edu.au/about-griffith/governance/plans-publications/annual-report
	https://www.griffith.edu.au/about-griffith/governance/plans-publications
JCU	https://www.jcu.edu.au/about-jcu/annual-report
	https://www.jcu.edu.au/about-jcu/university-plan-2013-2017
LaT	http://www.latrobe.edu.au/council/resources
	http://www.latrobe.edu.au/about/downloads/La-Trobe-Strategic-Plan-November-2015.pdf
Mac	http://www.mq.edu.au/about/about-the-university/governance/annual-reports
	http://mq.edu.au/our-university/pdf/Macquarie_University_A_Framing_of_Futures_revised.pdf
Mon	http://www.monash.edu/about/who/publications/annual-report
	http://www.monash.edu/about/who/strategic-plan
Mur	http://www.murdoch.edu.au/About-us/Annual-report/
	http://www.murdoch.edu.au/About-us/Strategic-plan/
QUT	https://www.qut.edu.au/about/governance-and-policy/annual-report
	https://www.qut.edu.au/about/strategic-ambitions/blueprint-for-the-future
RMIT	http://www.rmit.edu.au/about/governance-and-management/governance/annual-reports
	https://www.rmit.edu.au/about/our-strategy

(continued)

SCU	http://scu.edu.au/docs/annual_report/
	http://scu.edu.au/strategicplan/
Swin	http://www.swinburne.edu.au/about/strategy-initiatives/annual-report/
	http://www.swinburne.edu.au/about/strategy-initiatives/2020-plan/
Tor	Not found on website, requested 20/10/16. No response
UoA	https://www.adelaide.edu.au/publications/corporate/
	https://www.adelaide.edu.au/VCO/beacon/
UoC	https://www.canberra.edu.au/about-uc/policy-and-legislation/key-university-documents/annual-reports
	http://www.canberra.edu.au/about-uc/strategic-plan
UCL	Not found on website, requested 20/10/16. No response
UDiv	https://www.divinity.edu.au/news-events/2016/04/14/2016-annual-report/
	https://www.divinity.edu.au/university-of-divinity/about-us/our-vision-and-mission/
UMel	http://publications.unimelb.edu.au/
	https://ihouse.unimelb.edu.au/home/features/our-response-to-growing-esteem
UNE	http://www.une.edu.au/about-une/annual-reports
	http://www.une.edu.au/about-une/executive/vice-chancellor/strategic-plan/strategic-plan-2016-2020/strategic-plan-2016-2020
UNSW	http://annualreport.unsw.edu.au/2015/
	https://www.2025.unsw.edu.au/
UNew	http://www.newcastle.edu.au/about-uon/our-university/annual-report
	https://www.newcastle.edu.au/about-uon/our-university/vision-and-strategic-direction/new-futures-strategic-plan-2016-2025
UND	Not found on Notre Dame's website, requested 20/10/16. No response
UoQ	http://www.uq.edu.au/about/annual-reports
	http://www.uq.edu.au/about/docs/strategicplan/StrategicPlan2014.pdf
Uni SA	http://www.unisa.edu.au/About-UniSA/Corporate-publications/
	http://www.unisa.edu.au/About-UniSA/Strategic-action-plan-2013-2018/
USQ	https://www.usq.edu.au/about-usq/about-us/plans-reports/annual-report
	https://www.usq.edu.au/about-usq/about-us/plans-reports/strategic-plan
USC	http://www.usc.edu.au/explore/reports-and-publications/annual-report
	http://www.usc.edu.au/explore/vision/strategy-quality-and-planning/strategic-plan-2016-2020
USyd	http://sydney.edu.au/about-us/vision-and-values/annual-report.html
	http://sydney.edu.au/about-us/vision-and-values/strategy.html
UTas	http://www.utas.edu.au/university-council/university-reports
	Link to strategy not working. Copy requested 20/10/16
UTS	http://www.uts.edu.au/about/uts-governance/official-publications/uts-annual-report
	http://www.uts.edu.au/about/university/uts-strategic-direction
UWA	http://www.uts.edu.au/about/university/uts-strategic-direction
	http://www.web.uwa.edu.au/__data/assets/pdf_file/0010/2538343/114085-VICCHA-StrategicPlan-v3.pdf
UOW	http://www.uow.edu.au/about/publications/index.html
	http://www.uow.edu.au/ourpurpose/

(continued)

WSU	https://www.westernsydney.edu.au/about_uws/leadership/mission_goals_strategic_plan
	https://www.westernsydney.edu.au/about_uws/leadership/governance
VicU	https://www.vu.edu.au/about-vu/publications/annual-reports
	https://www.vu.edu.au/about-vu/vision-mission-strategy

References

ATN, 2015. Innovate and Prosper. PriceWaterhouseCoopers, Canberra

Australian Research Council. (2016). 2015–2016 State of Australian University Research Volume 1 ERA National Report. ARC, Canberra.

Department of Education and Training, Commonwealth of Australia. (2016). Universities' 2014-16 mission-based compacts. Accessed at https://docs.education.gov.au/node/34873

Commonwealth of Australia. (2015). *National innovation and science agenda*. Canberra: Commonwealth of Australia.

Edenhofer, O., Pichs-Madruga, R., Sokona, Y., Farahani, E., Kadner, S., Seyboth, K., Adler, A., Baum, I., Brunner, S., Eickemeier, P., Kriemann, B., Savolainen, J., Schlömer, S., von Stechow, C., Zwickel, T., & Minx, J. C. (2014). Summary for policymakers. In: Climate change 2014: Mitigation of climate change. Contribution of Working Group III to the Fifth Assessment Report of the Intergovernmental Panel on Climate Change Cambridge University Press, Cambridge, UK.

Gahan, P., Adamovic, M., Bevitt, A., Harley, B., Healy, J., Olsen, J.E., Theilacker, M. 2016. Leadership at Work: Do Australian leaders have what it takes? Melbourne: Centre for Workplace Leadership, University of Melbourne. Available at: workplaceleadership.com.au/sal

IPCC. (2014). *Fifth Assessment report – Authors and reviewers*. Retrieved 1 Mar 2017 from http://www.ipcc.ch/pdf/ar5/ar5_authors_review_editors_updated.pdf

Cook, J. et al. (2013). Quantifying the consensus on anthropogenic global warming in the scientific literature. *Environmental Research Letters, 8*(2), 024024.

OECD. (2013). *Science, technology and innovation scoreboard*. Paris: OECD.

Office of Parliamentary Counsel. (2017a). *Higher education support act. Compilation 62*. Canberra: Office of Parliamentary Counsel.

Office of Parliamentary Counsel. (2017b). *National greenhouse energy and reporting act, 2007. Compilation 18*. Canberra: Office of Parliamentary Counsel.

University of Queensland. (2014). *Strategic plan: 2014–2017*. Brisbane: University of Queensland. http://www.uq.edu.au/about/docs/strategicplan/StrategicPlan2014.pdf.

Chapter 6
Using Superheated Steam Dryer for Cogeneration System Improvement and Water Recovery

Somchart Chantasiriwan and Sarocha Charoenvai

6.1 Introduction

Power generation from biomass has increased its share of total power generation in many countries. Although prices of fossil fuels have dropped substantially in recent years, benefits of biomass power generation such as carbon neutrality and energy security have prevented governments of many countries from abandoning the policy of increasing dependence on biomass. Among various types of biomass conducive to power generation, sugar cane is arguably the most important due the large number of cane sugar factories around the world. Most factories use bagasse, a by-product of the raw sugar manufacturing process, as the dominant fuel in generating steam that is required for the process. A large fraction of these factories have discovered, with the improvement of the process efficiency, there is surplus bagasse that can be used to generate exportable electrical power.

From energy viewpoint, cogeneration in raw sugar manufacturing consists of four main processes: juice extraction, juice evaporation (which includes crystallization), steam generation, and turbogenerator (Chantasiriwan 2016). Sugar juice is extracted from sugar cane stalks in the juice extraction process, of which by-product is bagasse. This process requires water addition to increase juice extraction, which results in diluted sugar juice and bagasse with high moisture content. The steam generation process uses bagasse from the juice extraction process as the fuel. It consists of a boiler that can generate high-pressure steam, which may be sent to the turbogenerator to produce useful power. Low-pressure

S. Chantasiriwan (✉)
Faculty of Engineering, Thammasat University, Pathum Thani, Thailand
e-mail: somchart@engr.tu.ac.th

S. Charoenvai
Faculty of Engineering, Rajamangala University of Technology Thanyaburi, Pathum Thani, Thailand

A. Sayigh (ed.), *Transition Towards 100% Renewable Energy*,
Innovative Renewable Energy, https://doi.org/10.1007/978-3-319-69844-1_6

steam required for the juice evaporation process may be extracted from condensing-extraction steam turbine. The evaporation process produces raw sugar from diluted sugar juice. The evaporation process produces raw sugar by evaporating all water content from diluted sugar juice in multiple-effect evaporator and crystallizer. It requires low-pressure saturated steam. Since the exhausted or extracted steam produced by the steam generation process is superheated, it must be mixed with the right amount of cooling water in a desuperheater before the resulting saturated steam is sent to the evaporation process.

Bagasse leaving the juice extraction process is normally sent to the boiler in the steam generation process without being dried. Although the boiler is designed for operating with high-moisture bagasse, its efficiency is low because a substantial amount of thermal energy released from combustion is needed to evaporate water from bagasse, and the amount of excess air required for complete combustion increases with bagasse moisture, leading to higher dry flue gases loss. It has been recognized that flue gases exhausted from the steam generation process may be used to dry bagasse, which results in an increase in boiler efficiency (Maranhao 1986; Kinoshita 1991; Dixon et al. 1998; Gilberd and Sheehan 2013). However, practical problems have so far limited its use. One serious problem is a possibility of the combustion of dry bagasse in the flue gases dryer (Rein 2007). Furthermore, flue gases temperature may not be high enough to be used in the flue gases dryer, or flue gases may be more efficiently used as the heating medium for other heat exchangers in the steam generation unit.

The possibility of combustion in the dryer is eliminated if superheated steam is used as the drying medium instead of flue gases. The capability of superheated steam dryer for bagasse was demonstrated by Jensen (2003) and Morgenroth and Batstone (2005). Although superheated steam available for drying bagasse is extracted from a condensing-extraction turbine, the standard practice in a typical sugar factory does not make use of its availability. Instead the superheated steam is mixed with cooling water in a desuperheater to produce saturated steam required for the evaporation process. An obvious improvement over the standard practice is to replace the desuperheater with superheated steam dryer, which has the double benefits of producing both saturated steam and bagasse with lower moisture content. This paper presents the construction of the mathematical models of basic cogeneration system with desuperheater and improved cogeneration system with superheated steam dryer. Both models are then compared to demonstrate the advantages of the improved cogeneration system, which include not only the higher power generation efficiency but also the recovery of water that would be lost in the basic cogeneration system.

6.2 Basic Cogeneration System

Figure 6.1 shows the schematic of the basic cogeneration system in a typical sugar factory. Bagasse from the juice extraction system enters the boiler (B) along with ambient air. The dry basis moisture content of the bagasse is x_m. Thermal energy released from the combustion produces both high-pressure steam and high-temperature flue gases. The superheated steam leaving the boiler is at the design pressure p_s and the design temperature T_s. The flue gases are assumed to be maintained at a fixed temperature T_g. The type of turbine installed in this system is assumed to be the condensing-extraction steam turbine (T). The steam pressure at the extraction point is p_e, which is the same pressure required by the evaporation process (EP). The remaining steam is condensed in the condenser (C), which supplies cooling water for the desuperheater (DS). Extracted steam is superheated. Its temperature depends on the isentropic efficiency of the turbine, which is assumed to be known. The required saturated steam for EP is obtained from the mixing between superheated steam and cooling water in DS. Saturated steam will condense in EP and becomes saturated water at the exit of EP. The remaining water from C is used as feed water for the boiler.

Assume that the composition of dry bagasse is known. The higher heating value (HHV) can be determined from Dulong's formula (Keating 2007):

$$\mathrm{HHV} = x_{\mathrm{C}}\mathrm{HHV}_{\mathrm{C}} + \left(x_{\mathrm{H}} - \frac{x_{\mathrm{O}}}{8}\right)\mathrm{HHV}_{\mathrm{H}} + x_{\mathrm{S}}\mathrm{HHV}_{\mathrm{S}} \tag{6.1}$$

where x_C, x_H, x_O, and x_S are mass fraction of C, H, O, and S, respectively, in dry bagasse. Higher heating values of C, H, and S are, respectively, $\mathrm{HHV_C} = 3.396 \times 10^4$ kJ/kg, $\mathrm{HHV_H} = 1.41890 \times 10^5$ kJ/kg, and $\mathrm{HHV_S} = 9.42 \times 10^3$ kJ/kg. In order to determine the lower heating value of bagasse, the amount of water resulting from the combustion of 1 kg of dry bagasse must be known. Since the complete combustion of 1 kg of dry bagasse produces $9x_H$ kg of water, the lower heating value of dry bagasse is

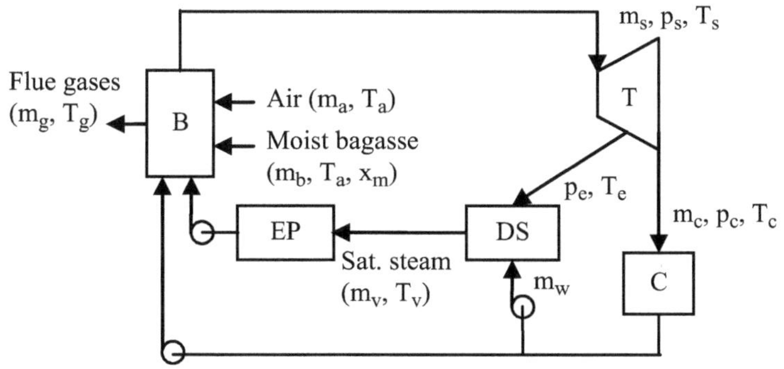

Fig. 6.1 Basic cogeneration system in sugar factory

$$\text{LHV} = \text{HHV} - 9x_{\text{H}}h_{\text{fg}} \tag{6.2}$$

where h_{fg} is the latent heat of evaporation of water at the standard state (2.44×10^3 kJ/kg).

It is assumed that combustion of bagasse is complete. The amount of excess air required for the complete combustion of bagasse depends on bagasse moisture content. According to Rein (2007), dry bagasse requires 17% of excess air ($\phi = 0.17$), and the excess air requirement of moist bagasse is given by

$$\phi = 47.57y_{\text{m}}^3 + 58.00y_{\text{m}}^2 + 23.99y_{\text{m}} - 3.18 \tag{6.3}$$

where y_{m} is the wet basis moisture content of bagasse; $y_{\text{m}} = x_{\text{m}}/(1 + x_{\text{m}})$. Equation 6.3 is to be used for $0.4 \leq y_{\text{m}} \leq 0.6$. Once the excess air is known, the mass flow rate of air (m_{a}) can be computed as follows:

$$m_{\text{a}} = (1 + \varphi)\text{AFR}m_{\text{b}} \tag{6.4}$$

where AFR is the stoichiometric air-fuel ratio:

$$\text{AFR} = 11.44x_{\text{C}} + 34.32x_{\text{H}} + 4.29(x_{\text{S}} - x_{\text{O}}) \tag{6.5}$$

The net heat input to the steam generation unit is

$$Q_{\text{in}} = (1 - \varepsilon)m_{\text{b}}\text{LHV} \tag{6.6}$$

where m_{b} is the mass flow rate of dry bagasse. The heat loss parameter ε accounts for two sources of heat losses ($\varepsilon = \varepsilon_{\text{r}} + \varepsilon_{\text{c}}$). It is assumed that the heat loss from radiation and convection between the boiler shell and the ambient air is 1.5% of the total heat released by combustion ($\varepsilon_{\text{r}} = 0.015$). In addition, there is unburned carbon that is a result of the difficulty of burning all bagasse completely. It has been found that the amount of unburned carbon depends on bagasse moisture. According to Rein (2007),

$$\varepsilon_{\text{c}} = 3.953y_{\text{m}}^3 + 5.000y_{\text{m}}^2 + 2.154y_{\text{m}} - 0.298 \tag{6.7}$$

for $0.4 \leq y_{\text{m}} \leq 0.6$. If bagasse is dry ($y_{\text{m}} = 0$), $\varepsilon_{\text{c}} = 0.024$.

Energy balance of the boiler requires that heat input is used to (6.1) increase the temperature of the moisture in bagasse to the saturation temperature, evaporate the water, and increase the temperature of the resulting vapor to T_{g}, (6.2) increase the temperature of dry flue gases to T_{g}, (6.3) increase the temperature of ash to T_{g}, and (6.4) evaporate feed water and increase its temperature to T_{s}. Therefore, the energy balance becomes

$$
\begin{aligned}
& m_b\{(1-\varepsilon)\mathrm{LHV} + [c_{pb} + (1+\phi)\mathrm{AFR}c_{pa}](T_a - T_r) - [1 + (1+\phi)\mathrm{AFR} - x_A] \\
& c_{pg}(T_g - T_r) - x_m[c_{pw}(T_r - T_a) + h_{fg} + c_{pv}(T_g - T_r)] \\
& -x_A c_{pash}(T_g - T_a)\} = m_v(h_s - c_{pw}T_v) + (m_c - m_w)(h_s - c_{pw}T_c)
\end{aligned}
\tag{6.8}
$$

where T_r is the reference temperature (25 °C). The specific heat capacities of dry bagasse (c_{pb}), water (c_{pw}), steam (c_{pv}), and ash (c_{pash}) are, respectively, 0.46, 4.18, 1.80, and 1.00 kJ/kg.K. Since the combustion of bagasse is complete, the flue gases consist of CO_2, H_2O, O_2, N_2, and SO_2. The average heat capacities (c_{pg} and c_{pa}) are determined by taking into account the variation of the heat capacity of with temperature according to Verbanck (1997).

Mass and energy balances of DS are

$$m_v = m_s + m_w - m_c \tag{6.9}$$
$$m_v h_v = (m_s - m_c)h_e + m_w c_{pw} T_c \tag{6.10}$$

where h_v is the enthalpy of the saturated steam at the exit of the desuperheater, and h_e is the enthalpy of the superheated steam at the exhaust of the turbine.

The power output of the turbine is found from

$$P = \eta_m[m_s(h_s - h_e) + m_c(h_e - h_c)] \tag{6.11}$$

where η_m is machine efficiency, which is assumed to be 0.95. If isentropic efficiencies (η_{te} and η_{tc}) are known, h_e and h_c are determined from

$$h_e = h_s - \eta_{te}(h_s - h_{es}) \tag{6.12}$$
$$h_c = h_s - \eta_{tc}(h_s - h_{cs}) \tag{6.13}$$

where h_{es} is the enthalpy of the extracted steam at pressure p_e and the same entropy as the inlet steam, and h_{cs} is the enthalpy of the exhausted steam at pressure p_c and the same entropy as the inlet steam. It can be shown that P is a function of m_v and m_c. Therefore, m_b can be found from Eq. 6.8 as a function of P and m_v.

6.3 Improved Cogeneration System

Figure 6.2 shows the schematic of the modification of the basic cogeneration system, in which desuperheater is replaced with superheated steam dryer (SSD). In this system, bagasse is divided into m_{b1}, which is fed to the boiler directly, and m_{b2}, which is passed through SSD before entering the boiler. SSD uses exhaust steam from the turbine to remove all moisture content from the bagasse so that its dry basis moisture decreases from x_m to 0. Bagasse leaving SSD is at the steam saturation temperature. Steam leaving SSD is assumed to be saturated. It should be noted that the improved system yields recovered water in the amount of $x_m m_{b2}$.

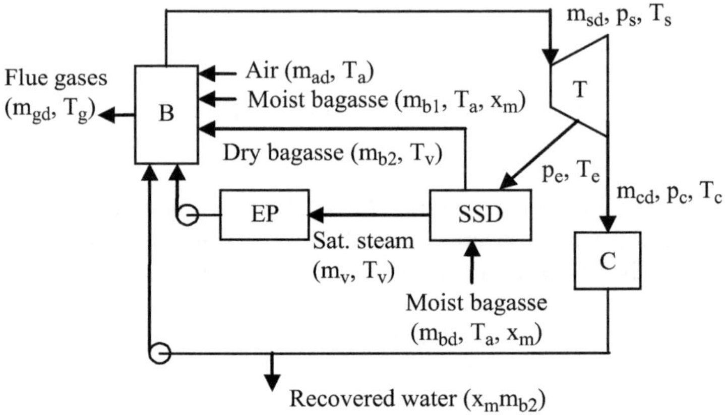

Fig. 6.2 Improved cogeneration system with superheated steam dryer

Furthermore, to allow for probable pressure loss in SSD, the extracted steam pressure must be larger than the pressure of saturated steam leaving SSD.

Mass and energy balances of SSD are

$$m_v = m_{sd} - m_{cd} + x_m m_{b2} \tag{6.14}$$

$$\begin{aligned} &m_{b2}c_{pb}(T_v - T_a) + x_m m_{b2} c_{pw}(T_v - T_a) + x_m m_{b2} h_{vl} \\ &= (m_{sd} - m_{cd}) c_{pv}(T_e - T_v) \end{aligned} \tag{6.15}$$

where h_{vl} is the latent heat of evaporation of water at p_v. It can be shown that, after eliminating $m_{sd} - m_{cd}$ from both equations, m_{b2} can be expressed as a function of m_v.

Energy balance of the boiler is

$$\begin{aligned} &m_{b1}\{(1-\varepsilon_1)\text{LHV} + c_{pb}(T_a - T_r) + (1+\phi_1)\text{AFR}c_{pa}(T_a - T_r) \\ &-[1+(1+\phi_1)\text{AFR} - x_A]c_{pg}(T_g - T_r) \\ &-x_m[c_{pw}(T_r - T_a) + h_{fg} + c_{pv}(T_g - T_r)] - x_A c_{pash}(T_g - T_a)\} = m_v(h_s - c_{pw}T_v) \\ &+m_{cd}(h_s - c_{pw}T_c) - m_{b2}\{(1-\varepsilon_2)\text{LHV} + c_{pb}(T_v - T_r) + (1+\phi_2)\text{AFR}c_{pa}(T_a - T_r) \\ &-[1+(1+\phi_2)\text{AFR} - x_A]c_{pg}(T_g - T_r) + x_m(h_s - c_{pw}T_c) - x_A c_{pash}(T_g - T_a)\} \end{aligned} \tag{6.16}$$

The power output of the turbine in this system is determined from

$$P = \eta_m[m_{sd}(h_s - h_e) + m_{cd}(h_e - h_c)] \tag{6.17}$$

It can be shown that P is a function of m_v and m_{cd}. Therefore, m_{b1} can be found from Eq. 6.16 as a function of P and m_v.

6.4 Results and Discussion

According to Yarnal and Puranik (2009), the typical composition of dry bagasse is 50% of carbon, 42% of oxygen, 6% of hydrogen, and 2% of ash. In order to carry out simulation, certain parameters of the systems must be provided. The values of these parameters are $p_s = 4000$ kPa, $T_s = 500°C$, $T_g = 150$ °C, $p_v = 200$ kPa, $p_c = 10$ kPa, $T_a = 30$ °C, and $m_v = 10$ kg/s. Furthermore, it is assumed that the turbine isentropic efficiency at the extraction point is 95% of the turbine isentropic efficiency at condensation. Therefore, if $\eta_{tc} = 0.8$, $\eta_{te} = 0.76$. It should be noted that wet basis bagasse moisture (y_m) is considered to be a free parameter, of which variation affects performances of the systems.

Consider the base case, in which $y_m = 0.50$. Simulation for the basic system yields results shown in Table 6.1. Power plant efficiency (η_p) is defined as

$$\eta_p = \frac{P}{m_b \mathrm{LHV}} \tag{6.18}$$

In the simulation for the improved cogeneration system with superheated dryer, it is assumed that $p_e = 220$ kPa to account for pressure loss in SSD. Simulation results are shown in Table 6.2. Power plant efficiency (η_{pd}) is defined as

$$\eta_{pd} = \frac{P_d}{m_{bd} \mathrm{LHV}} \tag{6.19}$$

It can be seen that the improved system with superheated dryer has a larger steam flow rate and uses less bagasse, which results in higher power plant efficiency.

Bagasse moisture content increases with the amount water added to the juice extraction process (Chantasiriwan 2016). It has been found that the optimum amount of added water results in the wet basis moisture content of about 50%. Moist bagasse is normally sent to the boiler directly. However, the amount of bagasse produced by the juice extraction process exceeds the amount of bagasse required for combustion in most sugar factories. Therefore, excess bagasse is stored for future use. Since many sugar factories store excess bagasse on an open ground, the moisture content of bagasse will increase during the rainy season. Figures 6.3 and 6.4 compare variations of the rates of dry bagasse consumption and the power generation efficiencies of, respectively, the basic system and the improved system

Table 6.1 Simulation results for the basic system

Parameters	Values
m_s (kg/s)	58.24
m_b (kg/s)	14.14
P (MW)	35.00
η_p (%)	15.68

Table 6.2 Simulation results for the improved system with superheated steam dryer

Parameters	Values
m_{sd} (kg/s)	58.85
m_{bd} (kg/s)	13.56
P (MW)	35.00
η_{pd} (%)	16.36

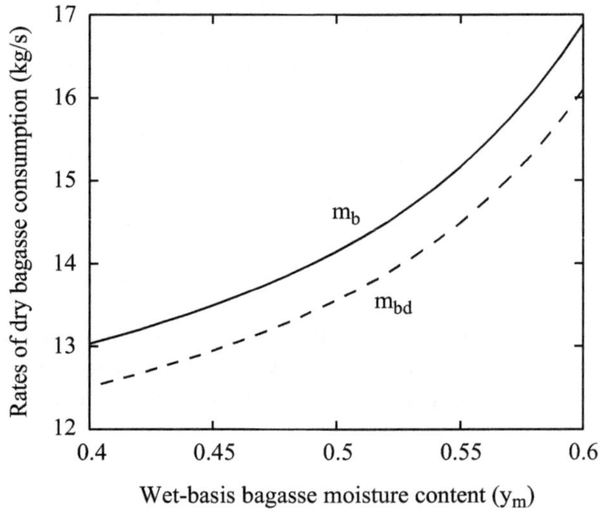

Fig. 6.3 Comparison of the rates of dry bagasse consumption of the basic cogeneration system of the improved cogeneration system

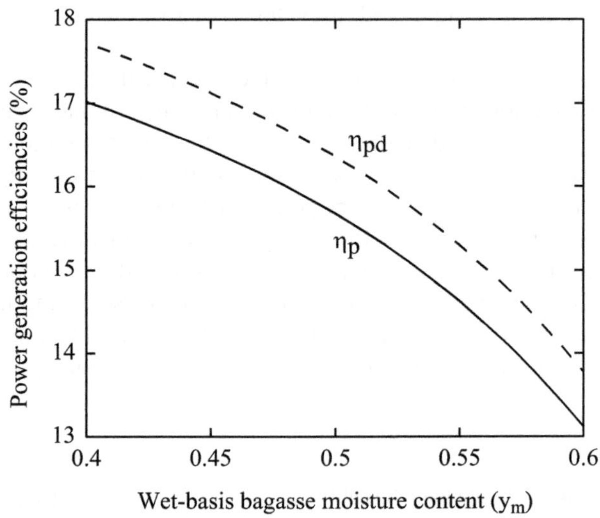

Fig. 6.4 Comparison of the power generation efficiencies of the basic cogeneration system and the improved cogeneration system

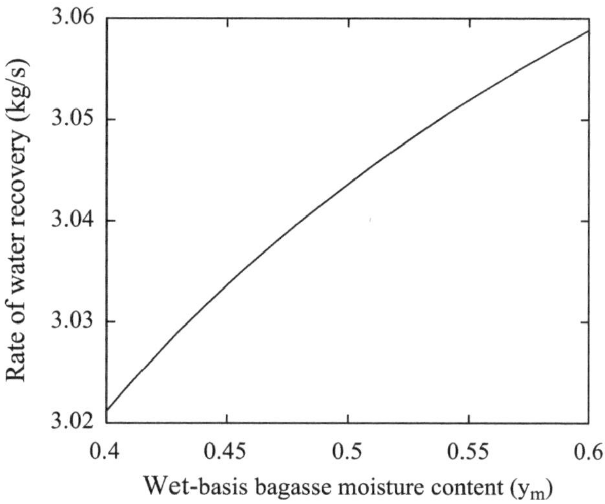

Fig. 6.5 Variation of recovered water by the improved cogeneration system with bagasse moisture content

with bagasse moisture content. It can be seen that both m_b and m_{bd} increase with y_m, which results in decreasing η_p and η_{pd}.

As bagasse is burned, its moisture becomes vapor that is lost with flue gases through the stack. Another benefit of the improved cogeneration system is the recovery of some water content in bagasse. Figure 6.2 shows that the amount of recovered water is $x_m m_{b2}$. The variation of the rate of water recovery with bagasse moisture content is plotted in Fig. 6.5. It can be seen that more water is recovered as y_m increases.

6.5 Conclusion

In the cogeneration system of sugar factory, steam extracted from condensing-extraction steam turbine must be desuperheated before being sent to the evaporation process, in which water is evaporated from sugar juice. Models are developed to simulate performances of two systems. In the first system, water is mixed with superheated steam in desuperheater. In the second system, superheated steam is used to dry bagasse in superheated steam drying. Under the condition that the amount of steam supplied to the evaporation process and generated power are fixed, the second system produces more steam flow rate, uses less dry bagasse, and has higher power generation efficiency. In addition, water that would be lost with flue gases in the first system is recovered in the second system.

References

Chantasiriwan, S. (2016). Optimum imbibition for cogeneration in sugar factories. *Applied Thermal Engineering, 103*, 1031–1038.

Dixon, T. F., Joyce, K. N., & Treloar, R. (1998). Increasing boiler capacity by dried bagasse firing. *Proceedings of the Australian Society of Sugar Cane Technologists, 20*, 445–452.

Gilberd, J., & Sheehan, M. (2013). Modelling the effects of bagasse pre-drying in sugar mill boiler systems. *Proceedings of the Australian Society of Sugar Cane Technologists, 35*, 337–343.

Jensen, A. S. (2003). Steam drying of beet pulp and bagasse. *International Sugar Journal, 105*, 83–88.

Keating, E. L. (2007). *Applied combustion*. Boca Raton: CRC Press.

Kinoshita, C. M. (1991). Flue gas drying of bagasse. *Applied Engineering in Agriculture, 7*, 729–734.

Maranhao, L. E. C. (1986). Seven years' experience with bagasse dryers. *Proceedings of the International Society of Sugar Cane Technologists, 3*, 44–61.

Morgenroth, B., & Batstone, D. (2005). Development and prospects for drying bagasse by steam. *International Sugar Journal, 107*, 410–415.

Rein, P. (2007). *Cane sugar engineering*. Berlin: Verlag.

Verbanck, H. (1997). Development of a mathematical model for watertube boiler heat transfer calculations. *Proceedings of the South African Sugar Technologists' Association, 71*, 166–171.

Yarnal, G., & Puranik, V. (2009). Energy management study in sugar industries by various bagasse drying methods. *Strategic Planning for Energy and the Environment, 29*, 56–78.

Chapter 7
Relevance and Applicability of Standards in Wind Farm Collector Circuit Design Process and Balance of Plant Selection

A.P. Clifton, Amanullah M.T. Oo, and Mohammad T. Arif

7.1 Introduction

The operation of high-voltage electricity networks has undergone significant change. Rapid increase in operational wind farms has highlighted an increased failure of medium voltage cable joints in cross-linked polyethylene (XLPE) collector circuits. Failed cable joint numbers are at the point of creating a significant financial and technical problem for existing wind farms and new wind farm developments. Current type testing for wind farm collector circuit components is not relevant to wind farm application and design practices. Electrical operating conditions (non-cyclic loading, circuit overload, harmonic unbalance or transient surges), specific to wind farms, are significantly different to traditional distribution networks (Clifton 2015).

7.2 Background

7.2.1 Wind Farm Collector Circuits

Wind farm collector circuits generally comprise several wind turbine generators (WTG's) connected in parallel to a substation. This connection acts as the point of connection to the national electricity grid. Commonly wind farm collector circuits operate at 33 kV. The electrical load in these circuits is close to component (power cables and accessories) ratings. The load is also non-cyclic when compared to

A.P. Clifton (✉) • A.M.T. Oo • M.T. Arif
School of Engineering, Faculty of Science, Engineering and Built Environment, Deakin University, Geelong, VIC, Australia
e-mail: aclifto@deakin.edu.au; aman.m@deakin.edu.au; m.arif@deakin.edu.au

A. Sayigh (ed.), *Transition Towards 100% Renewable Energy*,
Innovative Renewable Energy, https://doi.org/10.1007/978-3-319-69844-1_7

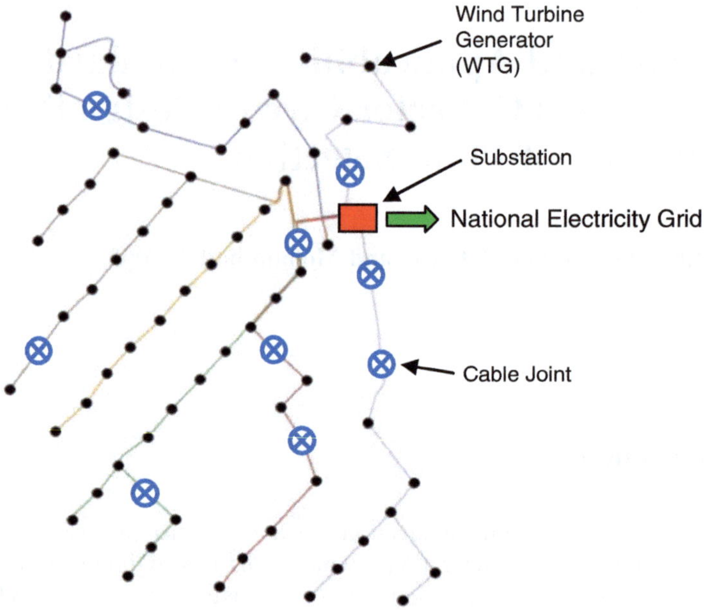

Fig. 7.1 Typical wind farm collector circuit layout showing cable joints in long cable run (Clifton 2015)

traditional distribution circuits. Wind farms share similarities with traditional distribution networks. However, there are important electrical load characteristics that may shape specific requirements for component type testing. Components of focus include extruded power cables and power cable accessories used in wind farm applications.

Figure 7.1 shows a typical wind farm collector circuit designed for Australian conditions. This wind farm comprises six collector circuits (as shown by the six colours) all connecting at a central substation. The substation is typically the connection point to the national electricity network. In Fig. 7.1 cable joints are shown by the blue circle with cross. Cable joints are required at these specific locations to allow for connection of long cable lengths to the substation.

7.2.2 Comparison of Cyclic and Non-cyclic Load Profiles

In power engineering, a load profile is a graph of the variation in the electrical load/generation versus time. A load profile will vary due to load type, generation type and temperature. Typical distribution assets utilize this information to understand and plan the power generation requirements at any given time. Comparing load profiles, it is evident that for traditional distribution assets, the load is cyclic and can be predicted with relative ease. For example, in Fig. 7.2 the load starts to build early

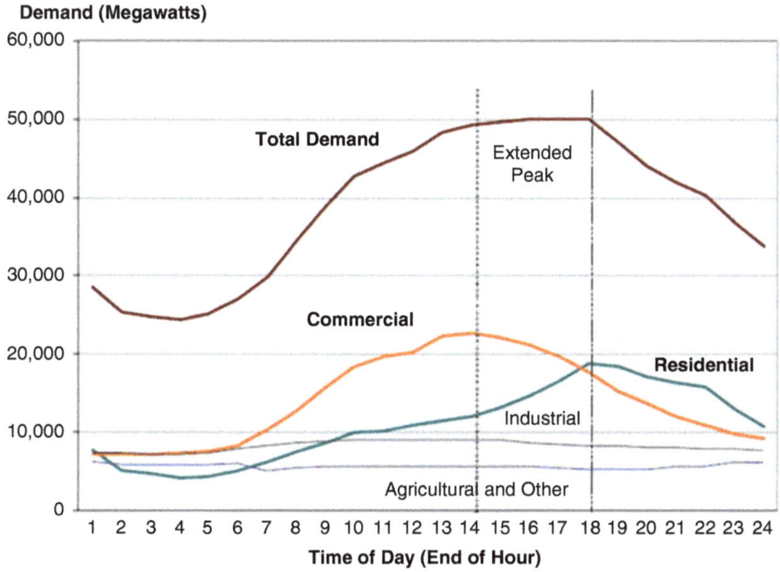

Fig. 7.2 Typical distribution load profile

morning as residential and commercial operations commence. The load continues to build throughout the day until late evening before decreasing as commercial operations cease for the day.

Wind farm load profiles can be considered non-cyclic due to the irregular nature of wind to produce energy over given time periods. A wind farm could operate at rated generation for several days before shutting down completely. Within a matter of seconds, minutes or hours operation could be at full generation again.

7.3 Literature Review

7.3.1 Standards on Which Load Rating Calculation Is Based

Selection of AC high-voltage cables and cable accessories requires guidance when identifying conductor size, insulation level and constructional requirements. To assist in this process, the current ratings and losses are calculated to ensure correctly sized equipment is utilized.

The following standards detail formulae for calculation of current ratings and losses during normal operation as well as calculation of cyclic and emergency current ratings for fault conditions (Table 7.1).

Table 7.1 Standards for calculation of current ratings

IEC 60183 guidance for the selection of high-voltage AC cable systems (IEC 2015b)
IEC 60287 electric cables – Calculation of the current rating (IEC 2015a)
IEC 60853 calculation of the cyclic and emergency current rating of cables (IEC 2002)

7.3.2 Calculation of the Permissible Current Rating, IEC

IEC 60287 series of standards provides formulae to calculate current ratings to conditions of steady state (continuous constant current or 100% load factor) operation. This applies to cables at all alternating voltages buried directly in the ground with and without partial drying out of the soil. It is necessary to calculate current ratings for conditions of partial drying out of the soil. It is also important to calculate current ratings for conditions where drying out of the soil does not occur. The lower of the two ratings is used.

The permissible current rating of an alternating current cable that is directly buried in conditions where drying out of the soil does not occur can be derived from the Eq. 7.1 (IEC 2006).

$$\Delta\theta = (I^2R + \tfrac{1}{2}W_d)T_1 + \left[I^2R(1+\lambda_1) + W_d\right]nT_2 + \left[I^2R(1+\lambda_1+\lambda_2) + W_d\right]n(T_3+T_4) \tag{7.1}$$

Rearranging for I gives,

$$I = \left[\frac{\Delta\theta - W_d[0.5T_1 + n(T_2+T_3+T_4)]}{RT_1 + nR(1+\lambda_1)T_2 + nR(1+\lambda_1+\lambda_2)(T_3+T_4)}\right]^{0.5} \tag{7.2}$$

where,

I is the current flowing in one conductor (A);
$\Delta\theta$ is the conductor temperature rise above the ambient temperature (K);
R is the alternating current resistance per unit length of the conductor at maximum operating temperature (Ω/m);
W_d is the dielectric loss per unit length for the insulation surrounding the conductor (W/m);
T_1 is the thermal resistance per unit length (1 m) between one conductor and the sheath (K·m/W);
T_2 is the thermal resistance per unit length (1 m) of the bedding between sheath and armour (K·m/W);
T_3 is the thermal resistance per unit length (1 m) of the external serving of the cable (K·m/W);
T_4 is the thermal resistance per unit length (1 m) between the cable surface and the surrounding medium, as derived from IEC 60287–2-1 Sect. 7.2.2 of part 2 (K·m/W);
n is the number of load carrying conductors in the cable (conductors of equal size and carrying the same load);

λ_1 is the ratio of losses in the metal sheath to total losses in all conductors in that cable;
λ_2 is the ratio of losses in the armouring to total losses in all conductors in that cable.

The following method for buried cables where partial drying out of the soil occurs can be applied to a single isolated cable or circuit only. This method is based on a simple two-zone approximate physical model of the soil where the zone adjacent to the cable is dried out, whilst the other zone retains the site thermal resistivity (IEC 2006):

$$I = \left[\frac{\Delta\theta - W_\mathrm{d}[0.5T_1 + n(T_2 + T_3 + vT_4)] + (v-1)\Delta\theta_\mathrm{x}}{R[T_1 + n(1+\lambda_1)T_2 + n(1+\lambda_1+\lambda_2)(T_3 + vT_4)]}\right]^{0.5} \tag{7.3}$$

where

v is the ratio of the thermal resistivity's of the dry and moist soil zones ($v = p_\mathrm{d}/p_\mathrm{w}$)
R is the AC resistance of the conductor at its maximum operating temperature (Ω/m)
p_d is the thermal resistivity of the dry soil (K·m/W)
p_w is the thermal resistivity of the moist soil (K·m/W)
θ_x is the critical temperature of the soil and temperature of the boundary between dry and moist zones (°C)
θ_a is the ambient temperature (°C)
$\Delta\theta_\mathrm{x}$ is the critical temperature rise of the soil. This is the temperature rise of the boundary between the dry and moist zones above the ambient temperature of the soil $(\theta_\mathrm{x} - \theta_\mathrm{a})$(K)
θ_x and p_d shall be determined from a knowledge of the soil conditions.

7.3.3 Standards on Which Product Testing Is Based

Electrical components are obligated to meet several requirements in areas such as functionality, technical performance, personal safety, etc. For extruded insulation power cables and cable accessories, compliance with all quality requirements is proven by type and routine testing. The relevant type and routine test standards are as follows (Table 7.2):

7.3.4 Extruded/Polymeric Insulation Power Cable

AS/NZS 1429.1 (AS/NZS 2006) specifies the requirements for cross-linked polyethylene (XLPE) and ethylene propylene rubber (EPR) insulated cables for fixed installations for electrical supply in Australia for working voltages up to and including 36 kV phase to phase. XLPE power cable undergoes routine heating

Table 7.2 Standards for power cable and power cable accessories type testing

IEC 61442: Test methods for accessories for power cables with rated voltages from 6 kV (Um = 7,2 kV) up to 30 kV (Um = 36 kV) (IEC 2005)
IEC 60502: Power cables with extruded insulation and their accessories for rate voltages form 1 kV (Um = 1,2 kV) up to 30 kV (Um = 36 kV) (IEC 2004)
AS/NZS 1429.1: Electric cables – Polymeric insulated – For working voltages 1.9/3.3 (3.6) kV up to and including 19/33 (36) kV (AS/NZS 2006)

cycle testing to determine maximum cable conductor temperature in normal operation. The test assembly is subjected to the required number of heating cycles (the test assembly is heated by passing a current through the conductor(s)) whilst being energized at rated voltage. Each cable sample is required to be heated by passing a current through the conductor (all three cores for three-core cables) until reaching and maintaining a steady temperature of between 105 and 110 °C for XLPE insulated cables. The heating cycle is required to be 8 h duration. It is necessary that the conductor temperature be maintained within stated temperature limits for at least 2 h at each heating period. This shall then be followed by at least 3 h of natural cooling in air to a conductor temperature less than 45 °C. As per the standard, the test sample is subjected to 20 cycles (AS/NZS 2006). IEC 60502-2 specifies that the sample shall be subjected to a heating current through the conductor(s) until the conductor reaches a steady temperature 5–10 K above the maximum conductor temperature in normal operation. Similar to AS/NZS 1429.1, the heating cycle will be of at least 8-h duration, whereby the conductor temperature is maintained within stated limits for at least 2 h, followed by at least 3 h of natural cooling of the conductor to within 10 K of ambient temperature. This cycle is required to be carried out 20 times (IEC 2004).

7.3.5 Extruded/Polymeric Insulation Power Cable Accessories

IEC 61442 and IEC 60502-4 specify a range of type and routine testing of accessories for power cables consisting of an array of test methods. Power cable accessories undergo routine heating cycle testing to determine maximum cable conductor temperature in normal operation. The test assembly is subjected to the required number of heating cycles (the test assembly is heated by passing a current through the conductor(s)), whilst being energized at rated voltage. Each heating cycle is required to be of at least 8 h duration with at least 2 h at a steady temperature (5–10 K above maximum cable conductor temperature in normal operation for extruded insulation cables). This is followed by at least 3 h of natural cooling to within 10 K of ambient temperature (see Fig. 7.3).

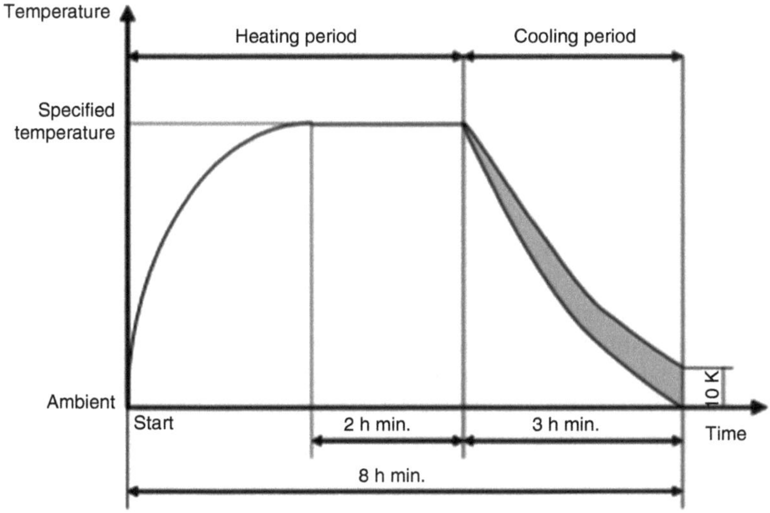

Fig. 7.3 Heating cycle (IEC 2005)

7.4 Methodology

The objective of this paper is to critically interrogate type testing parameters for extruded power cables and power cable accessories. All findings and component type test methodology and parameters were sourced from applicable IEC and AS/NZS standards as detailed in the reference list. Assessment focused on comparisons of relevance and applicability to wind farm collector circuit applications. Results allowed cataloguing and analysis across the range of identified test methodologies and parameters to draw conclusions and indicate irrelevance to wind farm applications and installation conditions. Understanding type testing methodologies provided the foundation to determine the relevance of heat cycle type testing to wind farm applications.

7.5 Findings

Design standards on which permissible current calculations are based are directly applicable to wind farm collector circuit installations. All calculations allow for selection of relevant site-specific parameters. These parameters may be divided into three groups:

- Parameters related to construction of a cable (e.g. thermal resistivity of insulating material)

- Parameters related to the surrounding condition, which may vary widely, the selection of which depends on the country in which the cables are used or are to be used
- Parameters, which result from an agreement between manufacture and end user and which involve a margin for security of service (e.g. maximum conductor temperature)

It was found that whilst widely accepted and used for load rating calculations for cables operating at 33 kV AS/NZS 3008 Electrical installations – Selection of cables. Part 1.1: Cables for alternating voltages up to and including 0.6/1 kV – typical Australian installation conditions is not suitable. As stated in the scope, the standard is applicable for cables at working alternating voltages up to 1 kV.

Design standards on which product testing is based for XLPE power cables and power cable accessories are not directly relevant to wind farm applications. The repetition of heating cycles (20 cycles for cables and 30 cycles for straight joints) utilized in the heating cycle type tests represent what's known as cyclic load profiling. This is aligned closely to the typical load profiles of traditional distribution networks.

The purpose of heating cycle type tests is to ensure both cable and accessories operate within an acceptable temperature profile for normal operation and fault conditions. Power cable and power cable accessories that undergo type testing, as detailed above, are not designed or intended for use in Australian wind farms. The duration and repetition of cycles are cyclic and not aligned with wind farm applications (Tables 7.3 and 7.4).

7.6 Conclusion

Identification of application inconsistencies forms the platform for future development of specific wind industry standards. Industry-specific standards need to include type test methodologies that utilize acceptable parameters relevant to

Table 7.3 Summary of product testing for XLPE insulated power cables

Extruded/polymeric insulation power cable					
		Duration			
Standard	Type test	Heating cycle	Steady temperature	Natural cooling	Number of cycles
AS/NZS 1429.1	Heat cycle test	8 h min	2 h minimum (between 105 and 110 °C above maximum cable conductor temperature)	3 h of natural cooling to a temperature less than 45 °C	20
IEC 60502-2	Heat cycle test	8 h min	2 h minimum (between 5 and 10 K above maximum cable conductor temperature)	3 h of natural cooling to within 10 K of ambient temperature	20

Table 7.4 Summary of product testing for XLPE insulated power cable accessories

Extruded/polymeric insulation power cable accessories					
		Duration			
Standard	Type test	Heating cycle	Steady temperature	Natural cooling	Number of cycles
IEC 60502-4	Heat cycle in air test	8 h min	2 h minimum (between 5 and 10 K above maximum cable conductor temperature)	3 h of natural cooling to within 10 K of ambient temperature	30
IEC 61442	Heat cycle voltage test	8 h min	2 h minimum (between 5 and 10 K above maximum cable conductor temperature)	3 h of natural cooling to within 10 K of ambient temperature	30

applications of wind farms. Accurate application-based type testing ensures design flaws, and inadequacies are identified prior to commercial product availability. This provides a pathway for improved product development and availability for wind farm collector circuit applications. Current design practices for wind farm applications tend to place to greater emphasis on design techniques employed for traditional distribution networks. Using current type testing methodologies, it is not possible to adequately understand and predict the performance of XLPE power cables and power cable accessories in wind farm applications. Application of type testing specific to wind farms applications may reveal the unsuitability of current power cables and accessories. This in turn may explain the large number of cable joint failures across Australian wind farms.

References

AS/NZS. (2006). AS/NZS 1429.1, electric cables—Polymeric insulated, Part 1: For working voltages 1.9/3.3, (3.6) kV up to and including 19/33 (36)kV. Australian/New Zealand Standard™ Standard.

Clifton, A. (2015). Discussion through question with industry experts in cable joint forum, Melbourne.

IEC. (2002). IEC 60853–1 Ed. 1.0, calculation of the cyclic and emergency current rating of cables – Part 1: Cyclic rating factor for cables up to and including 18/30(36) kV. IEC Standard.

IEC. (2004). IEC 60502-2: Power cables with extruded insulation and their accessories for rated voltages from 1kV (Um=1,2kV) up to 30kV (Um=36kV) – Part 2: Cables for rated voltages from 6kV (Um=7,2kV) up to 30kV (Um=36kV). Standard.

IEC. (2005). IEC 61442 test methods for accessories for power cables with rated voltages from 6 kV (Um = 7,2 kV) up to 30 kV (Um = 36 kV). IEC-Standard.

IEC. (2006). IEC 60287-1-1:2015, electric cables – Calculation of the current rating – Part 1–1: Current rating equations (100% load factor) and calculation of losses – general. IEC Standard.

IEC. (2012). IEC 60502-4 Power cables with extruded insulation and their accessories for rated voltages form 1 kV (Um = 1,2 kV) up to 30 kV (Um = 36 kV) – Part 4: Test requirements on accessories for cables with rated voltages from 6kV (Um=7,2kV) up to 30kV (Um=36kV). IEC-Standard.

IEC. (2015a). IEC 60287–2-1:2015, Electric cables – Calculation of the current rating – Part 2–1: Thermal resistance – Calculation of the thermal resistance. IEC Standard.

IEC. (2015b). IEC 60183 Guidance for the selection of high-voltage A.C cable systems. IEC-Standard.

Chapter 8
Scaling Up Miscibility Gap Alloy Thermal Storage Materials

Mark Copus, Samuel Reed, Erich Kisi, Heber Sugo, and James Bradley

8.1 Introduction

It is well established that concentrated solar power (CSP) requires reliable thermal energy storage (TES) in order to be a serious alternative to traditional electricity generation methods (Dinçer and Rosen 2011; Glatzmaier 2011; Liu et al. 2016). Currently, most TES materials suffer from poor thermal characteristics. Solid sensible heat storage materials such as concrete are inexpensive but have low energy density, low thermal conductivity and large variations in heat delivery temperature. Molten salts have moderate energy density but also have low thermal conductivity and large infrastructure costs. Inorganic phase change materials (PCMs) achieve a high energy density, in exchange for a very low thermal conductivity (Liu et al. 2016; Wang et al. 2016; Xu et al. 2015).

There are ongoing investigations of metals as PCMs for TES; however, one major problem is containment, due to the metals changing phase from solid to liquid (Risueño et al. 2016; Fukahori et al. 2016; Zhang et al. 2014). The Structure of Advanced Materials group at the University of Newcastle have developed new thermal energy storage materials known as miscibility gap alloys (MGAs) that have high energy density and moderate cost and allow very rapid storage and release of heat through thermal conductivities of up to two orders of magnitude higher than competing materials (Sugo et al. 2013; Rawson et al. 2014; Post et al. 2017).

As the name suggests, MGAs take advantage of a miscibility gap between two metals or semimetals of differing melting temperatures. The two powdered materials are pressed and sintered together so that an inverse microstructure forms, with a lower melting temperature component encapsulated as discrete particles within

M. Copus (✉) • S. Reed • E. Kisi • H. Sugo • J. Bradley
Advanced Materials Group, School of Mechanical Engineering, University of Newcastle, Callaghan, Australia
e-mail: mark.copus@uon.edu.au

A. Sayigh (ed.), *Transition Towards 100% Renewable Energy*,
Innovative Renewable Energy, https://doi.org/10.1007/978-3-319-69844-1_8

the higher melting temperature matrix (e.g. Cu in Fe). Upon heating, only the dispersed component will melt, allowing energy to be stored as the latent heat of fusion of that material.

While two metal MGA systems are able to be sintered using traditional powder metallurgy to form a completely solid and robust sample, carbon-based systems offer a more complicated challenge to attain solid storage blocks. As the melting temperature of carbon is extremely high, it is not practicable to sinter the matrix phase of carbon-based MGA, so an alternative solution is required. Additionally, the pressing of carbon in this manner is a relatively unexplored area of research. Despite this, carbon is seen as an ideal matrix material for MGA, as it is an extremely good thermal conductor, inexpensive and immiscible with a large number of metals.

As MGAs are a relatively new technology, much development is still required. The two major areas that are currently being explored are manufacture scale-up and new systems for a larger range of temperatures. This work focuses on the scaling up of MGAs, while work on creating new systems with different operation temperatures and thermal characteristics is also presented in these proceedings (Reed 2017).

8.2 Methodology

Until recently, the largest MGA sample manufactured was a flat rectangular prism with a volume of 0.26 L. In order to advance MGA technology to a form which is serviceable in an industry setting, a larger die was designed and manufactured. The design of the die and method used to make the new storage blocks are outlined below. These test storage blocks are designed to have a volume of 0.58 L, and each block can store 0.22 MJ of latent heat and a further 0.12 MJ of sensible heat when used in an operating range of 419 ± 50 °C. The blocks will be stackable, meaning any size storage block is attainable.

8.2.1 Die Design

Using FEA software Strand7, the die was modelled to determine the required wall thickness and radius in each corner in order to minimise stress concentrations (Fig. 8.1a). The die was manufactured from a solid piece of Bisplate400 steel. The opening was created via electric discharge machining and the centre preserved for use as the ram.

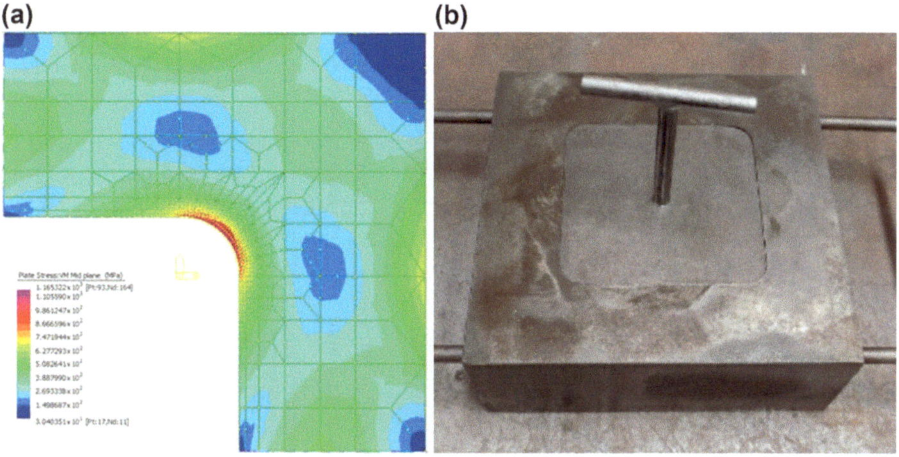

Fig. 8.1 (**a**) FEA analysis of the die with 17.5 mm radius under 200 tonnes of uniaxial pressing force; (**b**) 140 mm die used to press storage blocks

8.2.2 Storage Block Preparation

The manufacture of MGA follows the standard procedure of many powder metallurgy applications: powder mixing, loading powder into the die, compressing the sample and finally drying/sintering to achieve the final product.

All samples were prepared from zinc coarse powder (AMPS, Guildford, 0.8–2 mm) and carbon in the form of graphite (AMPS, Guildford, 45 μm) at 50 vol% Zn. Binders in the form of sodium silicate solution (3.22 weight ratio at 37.5 vol% or 2 vol% for drying in powder form) or a PVA glue solution (20 vol%) were used in the samples where binder was used. Uniaxial pressing was completed using the above die and a 300-tonne hydraulic press. All variations in method are given in Table 8.1.

The storage blocks were dried either before pressing, in powder form, or after pressing, in 80 °C and then 230 °C air ovens for 8 h each in order to remove any water still present. Sintering was completed by wrapping the samples in aluminium foil to prevent oxidation and heating them to 500 °C for 8 h. Phase identification and microscopic examination of Zn distribution were completed using a ZEISS SIGMA field emission scanning electron microscope (SEM), operated at 15 keV. Samples were also evaluated using visual inspection and mass loss measurements.

Table 8.1 Storage block preparation details

Storage block	1	2	3	4	5	6
Binder	25 wt% sodium silicate	25 wt% sodium silicate	25 wt% sodium silicate	20 wt% PVA glue	100 wt% of 2 vol% sodium silicate solution	No binder
Mixing method	Paint mixer	Paint mixer	Concrete mixer	Paint mixer	Concrete mixer	Paint mixer
Pressing	Ramp to 200 tonnes, hold for 2 min	Ramp to 200 tonnes, release. 5 repeats	Ramp to 200 tonnes, hold for 2 min	Ramp to 200 tonnes, hold for 2 min	Ramp to 200 tonnes, hold for 2 min	Ramp to 200 tonnes, release. 5 repeats

8.3 Results and Discussion

8.3.1 Weight Loss and Visual Evaluation

Weight loss for each stage and physical properties for all blocks can be found in Table 8.2. All samples exhibit at least some weight loss during the drying/sintering process, which is expected. Sample 5, manufactured without any binder, shows minimal weight loss from pressing to sintering (0.4%). All other samples experienced greater weight loss (2–7%), which correlates with their higher water content in the powder mixtures.

Densities of the storage blocks vary from 69% to 83% of the maximum theoretical density of the designed storage blocks (4700 kg/m^3). Storage blocks with lower densities (3 and 4) are generally also of a poorer quality. Their lower densities arise from a combination of poor distribution of binder and poor structural integrity, leading to large mass loss during manufacture. Figure 8.3 shows sample 4 split in two. Inevitably, some of the material was lost from the breakage.

All storage blocks which were manufactured are 140 × 140 mm on the largest face, with height varying from 23.7 to 33.5 mm. Variations in height are a result of slightly different densities in the powders after mixing influencing the die filling capacity. Different methods of pressing also had an impact on storage block height.

Figure 8.2 shows the first two samples created using the newly designed die, storage block 1 and storage block 2. Both of these samples were made using a powder mixture prepared with a paint mixer. Block 1 was pressed directly after the powders and sodium silicate solution were mixed, whereas block 2 was pressed 2 weeks after the powders were mixed. As a result, the two blocks have some differences. Storage block 1, as shown in Fig. 8.2a, illustrates a rough surface on the top edge, which is mirrored on the bottom surface. The sections of graphite missing from block 1 stuck to the die after ejection, as a result of the still wet powder mixture. Block 2 has a slightly higher quality surface on the top side, evident in Fig. 8.2b. It is hypothesised that this is a result of some of the moisture in the powder mixture evaporating during the 2 weeks standing period, resulting in a drier powder.

Table 8.2 Physical properties and weight loss of storage blocks

Storage block	1	2	3	4	5	6
Green mass (g)	2016	2162	2411	2096	1809	1737
Height (mm)	27	28.5	33.5	33	23.7	24.5
Density (kg/m^3)	3810	3872	3672	3241	3895	3617
Mass after 80 °C 24 h (g)	1964.3	2093.6	2353.1	2022.7	1808.2	N/A
Mass after 80 °C 24 h (g)	1956.9	2059.1	2349.8	2020.5	1807.9	N/A
Mass after 230 °C 12 h (g)	1950.1	2055.1	2344.2	2017.4	1807.6	N/A
Mass after 500 °C 8 h (g)	1910.3	2021.8	2294	1947.4	1803.3	1727.4
Mass after 500 °C 8 h (g)	1906.6	1985.14	N/A	N/A	1802.1	1722.3

(a) (b) (c) (d)

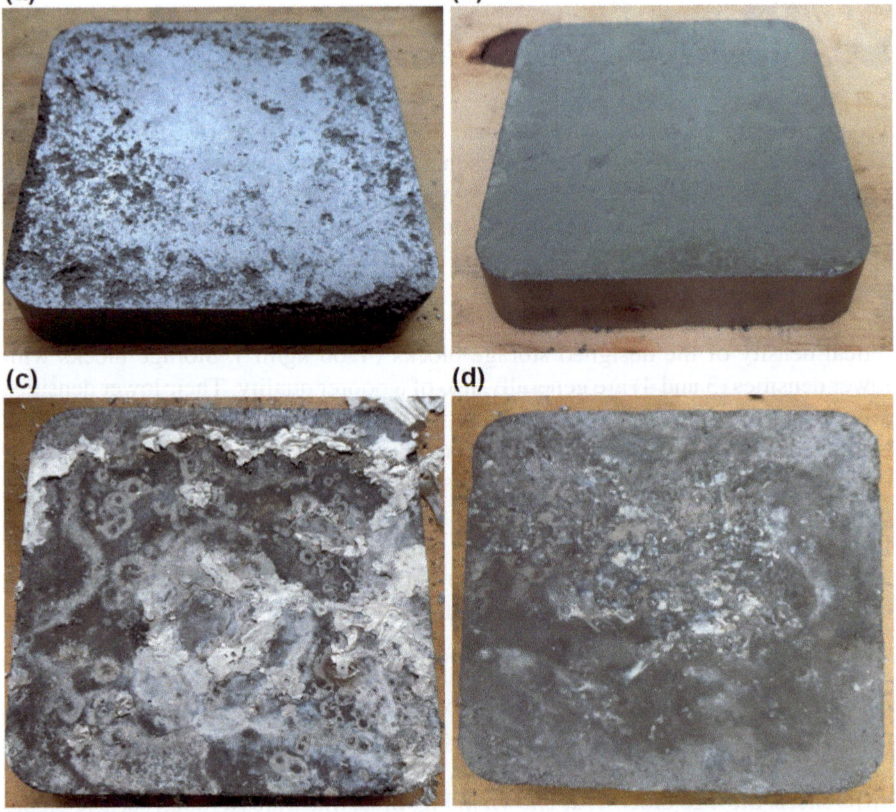

Fig. 8.2 140 × 140 × 28 mm green (non-sintered) C-Zn MGA storage blocks with 25% wt. sodium silicate solution as a binder (**a**) storage block 1; (**b**) storage block 2; (**c**) storage block 1, sintered; (**d**) storage block 2, sintered

Drying and sintering of these two blocks resulted in a substantial reduction in surface quality, with some weight loss attributed directly to this. Figure 8.2c shows foil which is trapped on the surface of the storage block. The removal of this aluminium foil resulted in small clumps of graphite/zinc coming loose.

Figure 8.3a shows the formation of spheres during mixing, trapping large amounts of sodium silicate solution, meaning that there was a nonuniform distribution of binder throughout the powder mixture and, ultimately, in the storage block. Storage block 3, shown in Fig. 8.3b, was extremely brittle and weak. Any kind of contact with the surfaces resulted in breakage and significant mass loss.

Storage block 4 split in two during ejection from the die, as shown in Fig. 8.4. Similarly to the powder mixture shown in Fig. 8.3a above, small concentrations of PVA glue formed in the powder. This prevented the PVA solution from spreading and mixing with the graphite powder, meaning that self-weight was enough to break the storage block in two.

In order to ensure an even coating of all particles, it was decided to drench graphite in a sodium silicate solution, followed by drying in an air oven. Storage

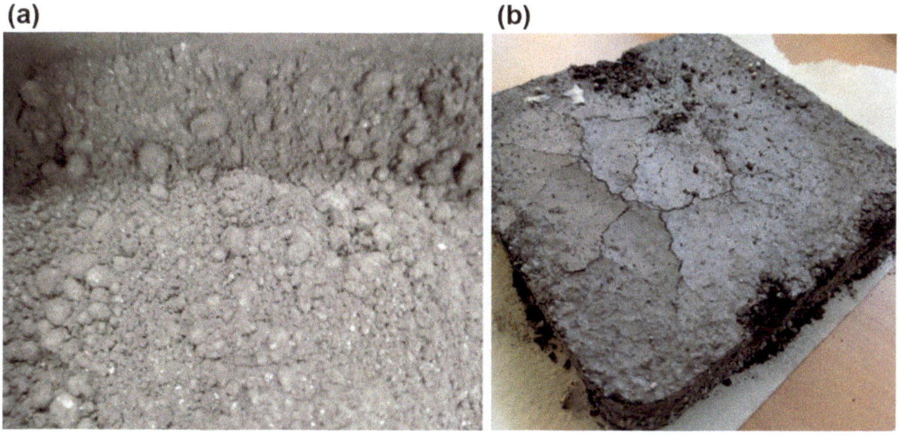

Fig. 8.3 (**a**) Powder mixed in the concrete mixer, with 25% wt. sodium silicate showing globular concentrations of graphite powder/sodium silicate solution. (**b**) Sintered storage block 3, made using the mixture in (**a**) (Note the complete failure of the edges where the sample was handled during removal from the furnace)

Fig. 8.4 PVA glue as a binder resulted in a split sample upon ejection

block 5, shown in Fig. 8.5a, displays smooth surfaces with minimal flaking as seen on other storage blocks. This method of mixing sodium silicate and graphite powder ensures that there is an even coating of all particles and the binder was able to make contact with as many graphite particles as possible. In a green state, this sample was extremely robust and uniform.

There was slight separation visible at the vertices of some of the edges, but these did not add any fragility to the block. However, after sintering, the cracks became enlarged (Fig. 8.5b) and the surface around these cracks became brittle. Despite this, the storage block is one of the best samples that come from this study.

A control block with no binder was manufactured, as shown in Fig. 8.6. The lack of moisture seemed to increase the structural integrity of storage block 6. In previous cases with smaller sample sizes (34 mm diameter), a lack of binder

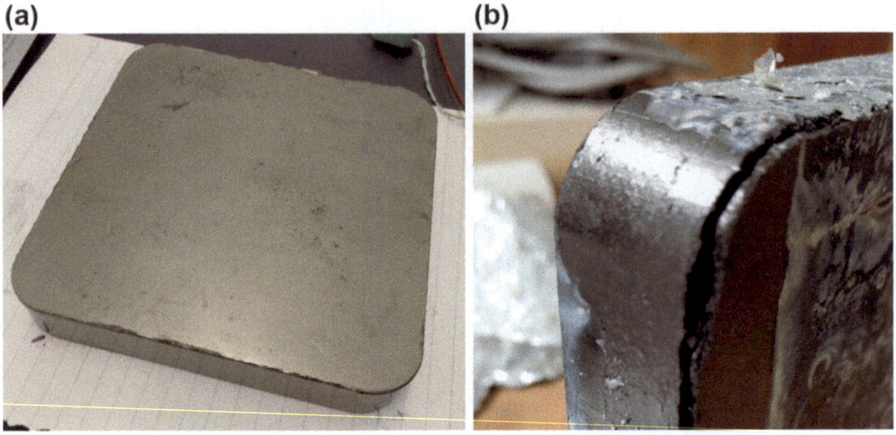

Fig. 8.5 (**a**) Green compact storage block 5; (**b**) cracks on edges exacerbated by sintering storage block 5

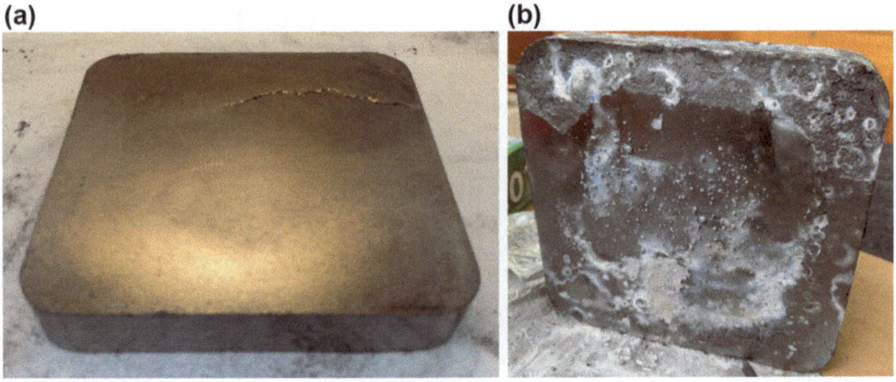

Fig. 8.6 Storage block 6 (**a**). The green sample was pressed with no binder; (**b**) sintered storage block 6

resulted in weak and brittle samples. The increase in size seems to have reduced the impact of the lack of binder on the C-Zn blocks, as this storage block is of a very high quality. Also note that storage block 6 was pressed using short, repeated loadings of 100 MPa. Further investigation will be completed on this pressing technique.

Upon sintering, storage block 6 maintained its good surface finish, with only a small amount of weight loss through surface Zn leakage. Small spheres of Zn can be seen in Fig. 8.6b, which is the result of Zn particles being in contact with the die walls, meaning they are not fully encapsulated and may exit the sample upon sintering. The white marks on the surface after sintering are the results of the leaked Zn reacting with the aluminium foil, which is minimal and occurs on all samples. After the second sintering, very little Zn is able to leave the sample.

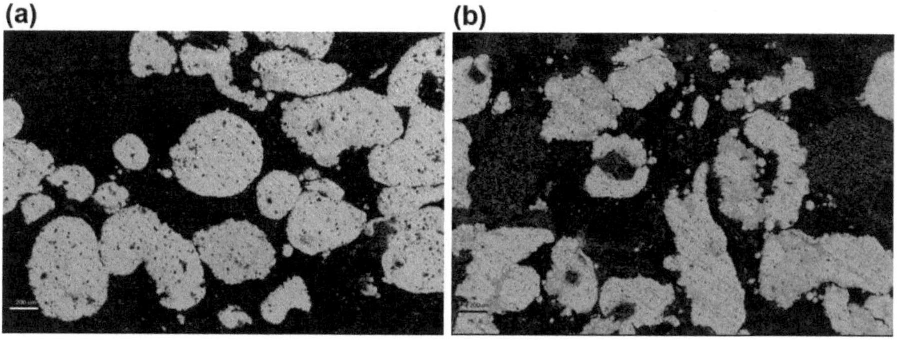

Fig. 8.7 Scanning electron microscope image of samples taken from 140 mm storage blocks. The lighter-shaded parts depict zinc and the darker areas represent the carbon matrix. (**a**) Storage block 1; (**b**) storage block 2

8.3.2 Scanning Electron Microscopy

Figure 8.7 shows typical distributions of zinc powder within the carbon matrix in an MGA storage block. The zinc powder in Fig. 8.7b exhibits an unusually large amount of inclusions of another material. At the time the SEM images were taken, the elemental analysis equipment attached to the SEM was unavailable. As a result, no elemental analysis was undertaken. Comparing these images to previous elemental analysis of C-Zn systems with a sodium silicate binder, it is reasonable to hypothesise that the inclusions are most likely a mixture of fine zinc, carbon, sodium and silicate (Copus 2015). It is possible that the use of a paint mixer caused deformation in the particles in the lower section of the mixing container, where the particles were crushed in between the mixer and container. Simple overpolishing can also result in small inclusions occurring.

Importantly, Fig. 8.7 shows that these larger-scale storage blocks have a consistent microstructure, as seen in previous, smaller-scale studies.

8.4 Conclusion

This work attempted to find the most effective method of enlarging MGA storage blocks. Initial investigations into the upscaling of the C-Zn MGA system have resulted in storage blocks more than twice as large as previous samples. It was found that mixing a smaller amount of sodium silicate solution with graphite powder is ineffective on a large scale. Conversely, it has been found that drenching graphite particles in a weaker sodium silicate solution, and subsequently drying that mixture, results in a more suitable powder mixture.

Other results also point to a dry powder being more appropriate for manufacture of MGA blocks. Storage block 6, with no binder, was surprisingly robust. Previous, smaller storage blocks made with no binder have been of a much lower quality (Copus 2015). The powder used to manufacture storage block 2 was allowed to sit for 2 weeks after the binder and graphite were combined. Again this block made from mostly dry powder resulted in a smooth surface finish and sound structural integrity. These results further encourage the use of a dry powder when manufacturing MGA storage blocks.

8.5 Future Work

A small steam turbine has been purchased for use with a large storage bank of C-Zn storage blocks as a larger-scale demonstration apparatus (Kisi et al. 2017).This will require the refinement of the manufacture process, which will be based on the work presented in this research. This future work in upscaling MGA will focus on the use of dry powders when pressing, repeated short pressings, increasing the height of the storage blocks (through extending height of die) and eventual mass production.

References

Copus, M. (2015). *Analysis and optimisation of miscibility gap alloy thermal storage demonstration apparatus*. The University of Newcastle. Newcastle.

Dinçer, İ., & Rosen, M. A. (2011). *Thermal energy storage: Systems and applications*. Hoboken: Wiley. http://encore.newcastle.edu.au/iii/encore/record/C__Rb3862149__Sthermalenergy storage__Orightresult__U__X7?lang=eng&suite=cobalt.

Fukahori, R., Nomura, T., Zhu, C., Sheng, N., Okinaka, N., & Akiyama, T. (2016). Macro-encapsulation of metallic phase change material using cylindrical-type ceramic containers for high-temperature thermal energy storage. *Applied Energy, 170*, 324–328. https://doi.org/10.1016/j.apenergy.2016.02.106.

Glatzmaier, G. (2011, May 20). Summary report for concentrating solar power thermal storage workshop: New concepts and materials for thermal energy storage and heat-transfer fluids. doi: https://doi.org/10.2172/1022291.

Kisi, E., Sugo, H, Cuskelly, D., Fiedler, T., Rawson, A., Post, A., Bradley, J., Copus, M., & Reed, S. (2017). Miscibility gap alloys – A new thermal energy storage solution. In *World renewable energy congress*, Perth W.A.

Liu, M., Steven Tay, N. H., Bell, S., Belusko, M., Jacob, R., Will, G., Saman, W., & Bruno, F. (2016). Review on concentrating solar power plants and new developments in high temperature thermal energy storage technologies. *Renewable and Sustainable Energy Reviews, 53*. Elsevier Ltd, 1411–1432. https://doi.org/10.1016/j.rser.2015.09.026.

Post, A., Rawson, A., Sugo, H., Cuskelly, D., Bradley, J., Copus, M., & Kisi, E. (2017). Price estimation for miscibility gap alloy thermal storage systems. In *World renewable energy congress*.

Rawson, A., Kisi, E., Sugo, H., & Fiedler, T. (2014). Effective conductivity of Cu–Fe and Sn–Al miscibility gap alloys. *International Journal of Heat and Mass Transfer, 77*, 395–405. https://doi.org/10.1016/j.ijheatmasstransfer.2014.05.024.

Reed, S. (2017). New highly thermally conductive thermal storage media. In *World renewable energy congress*.

Risueño, E., Doppiu, S., Rodríguez-Aseguinolaza, J., Blanco, P., Gil, A., Tello, M., Faik, A., & D'Aguanno, B. (2016). Experimental investigation of Mg-Zn-Al metal alloys for latent heat storage application. *Journal of Alloys and Compounds, 685*, 724–732. https://doi.org/10.1016/j.jallcom.2016.06.222.

Sugo, H., Kisi, E., & Cuskelly, D. (2013). Miscibility gap alloys with inverse microstructures and high thermal conductivity for high energy density thermal storage applications. *Applied Thermal Engineering, 51*(1), 1345–1350. https://doi.org/10.1016/j.applthermaleng.2012.11.029.

Wang, C., Lin, T., Li, N., & Zheng, H. (2016). Heat transfer enhancement of phase change composite material: Copper foam/paraffin. *Renewable Energy, 96*, 960–965. https://doi.org/10.1016/j.renene.2016.04.039.

Xu, B., Li, P., & Chan, C. (2015). Review: Application of phase change materials for thermal energy storage in concentrated solar thermal power plants: A review to recent developments. *Applied Energy, 160*. Elsevier Ltd, 286–307. https://doi.org/10.1016/j.apenergy.2015.09.016.

Zhang, G., Li, J., Chen, Y., Xiang, H., Ma, B., Xu, Z., & Ma, X. (2014). Encapsulation of copper-based phase change materials for high temperature thermal energy storage. *Solar Energy Materials and Solar Cells, 128*, 131–137. https://doi.org/10.1016/j.solmat.2014.05.012.

Chapter 9
Validated CFD Simulations of EWH Energy Storage Based on Tank Orientation

Roshaan de Jager, Wei Hua Ho, and Yu-Chieh J. Yen

9.1 Introduction

Before the turn of the century, hot water was seen as a luxury and not a necessity according to Weingarten (2003). Prior to the age of electrical water heating, gas, wood and coal were used to heat up water for domestic use.

In 1868, as suggested by Storify (2017), a gentleman by the name of Benjamin Maughan invented the first instant water heater which was known as a geyser whereby water was heated before entering the bath. This then inspired the invention of the electric water heater by the inventor Edwin Ruud who was a mechanical engineer by profession according to Storify (2017).

Applications for the production of heated water range from domestic household use, which includes bathing, cooking and cleaning, to industry-based needs to perform certain processes.

It is not unknown that in the twenty-first century, energy conservation is important on a global aspect. According to Foundry (2013), more than 80% of people in the world have access to electricity with an increasing demand in households. It was Foundry (2013) who also stated that a typical household in the United States consumes 11,700 kWh of electrical energy each year, with the global average of electricity consumption being 3500 kWh in 2010.

It was stated by Insights (2016) that in an average household, water heating is responsible for 12% of electrical consumption with electric water heaters being one

R. de Jager (✉) • W.H. Ho
School of Mechanical and Industrial Engineering, University of South Africa, Johannesburg, South Africa
e-mail: 46634835@mylife.unisa.za

Y.-C.J. Yen
School of Electrical & Information Engineering, University of the Witwatersrand, Johannesburg, South Africa

A. Sayigh (ed.), *Transition Towards 100% Renewable Energy*,
Innovative Renewable Energy, https://doi.org/10.1007/978-3-319-69844-1_9

of the most popular water heaters in the global market. With the current increase in the price of electricity of 9.2% according to Africa (2016), it is clear that South Africans' pockets are being stretched, thus forcing people to seek solutions to reduce their household electrical usage.

In order to reduce domestic energy consumptions, consumers who have the financial means have found alternative methods, such as solar water heaters, LGP or gas boilers. However, the capital cost on this type of equipment is quite expensive which many low-income households are unable to afford, thus analysing the effect of horizontal or vertical mounting configuration of the electric hot water geyser and concluding if either configuration allows for electrical energy saving over the other, thus maximising heat energy and electrical consumption.

According to Beeker et al. (2015), vertical cylinders used as hot water storage tanks are driven by thermal-hydraulic phenomena such as buoyancy effects and induced convection and mixing, forced convection induced by draining and associated mixing, heat diffusion as well as heat loss at the walls of the cylinder.

The stratification phenomenon which is a result of buoyancy effects according (Han et al. (2009) function), allows for high density fluid or cold water to settle at the lower part of the tank and low density or hot water at the upper part of the tank. This allows for a thermal barrier to be produced between the high and low density fluids.

It was suggested by Kandari (1990) that a horizontally mounted hot water storage tank delivers less hot water than a vertically mounted hot water storage tank. This was said to be due to the larger stratification area of the horizontal orientation than that of a vertical orientated tank.

There has been very little known research investigating the theory of analysing the effects of horizontally and vertically mounted electric domestic hot water tanks relating to energy savings and energy efficiency.

Previous studies under this umbrella include the likes of Kandari (1990), whereby thermal modelling and control of domestic water tank is discussed and verified using CFD software with regard to solar water heaters. Other studies include Beeker et al. (2015), whereby a discrete time model using three different states represents the dynamics of an electric hot water tank and analysed the behaviours of the fluid induced by stratification.

This report will therefore cover the gap of verifying using CFD software, the effects of horizontally and vertically mounted domestic electric hot water tanks and concluding if either configuration allows for a more efficient heat energy output. In order to achieve accurate results, the dimensions and specifications of an average store-bought 150 l domestic electric hot water tank will be used to create the 3D geometry which will be computed into a CFD software to be able to mimic the real-life findings of this study which was not previously known to be attempted.

9.2 Methodology

9.2.1 CFD Simulation

An average domestic electric store-bought hot water tank was used for the study. The hot water tank is a 150 l tank which is one of the most common sizes used in South African households. The 150 l Trendline electric hot water storage tank was used for the simulation. This model allows for the tank to be mounted in a horizontal and vertical configuration (dual mounted). The tank consists of an electrical spiral-shaped heating element. Power demand is 3 kW with a voltage rating of 220 V. The maximum working temperature of the tank is 70 °C. The tank has an external diameter of 460 mm and a depth of 1330 mm. The weight of the domestic electric hot water tank is approximately 44 kg.

The type of insulation used is polyutherene which is approximately 22.5 mm thick. The standing heat loss is 2.59 kWh/24 h according to the Trendline data sheet. The tank consists of a drain cock and safety valve. The safety valve is set at 600 kPa.

The simulation simulates two scenarios: a horizontally mounted geyser with a horizontal outlet and a vertically mounted geyser with a vertical outlet. By simulating these scenarios, results of each orientation allow for the analysis of the fluid flow, temperature outlet, stratification and overall electrical efficiency by comparing the two scenarios, thus concluding whether the orientation of a domestic hot water tank affects the energy usage which would result in electrical energy utilisation in a household by simply changing the mounting configuration and outlet of the hot water tank.

A domestic electric hot water tank was modelled in ANSYS Fluent using ANSYS Workbench 17.0 to simulate the 3D effects of different mounting configurations.

Each component of the electric hot water tank was modelled in Autodesk Inventor Professional 2016 according to specifications of the Heattech Trendline 150 l. Once each component of the electric geyser was modelled in Autodesk Inventor according to the specification, it was imported into ANSYS Workbench geometry where the Boolean function was used to remove the negative volume so that the model was representative of the fluid body inside the water tank.

9.2.1.1 Mesh Dependence

A mesh dependence test was conducted in order to verify that the mesh is of good quality and does not make a significant difference to the results of the simulation. In Fig. 9.1, it can be seen that mesh 4 resulted in an unstable temperature output, whereas mesh 1 resulted in a smooth transition. Mesh 1 will now be used going forward in this report. The size of the mesh used contains 223,046 elements with an aspect ratio of 1:8.

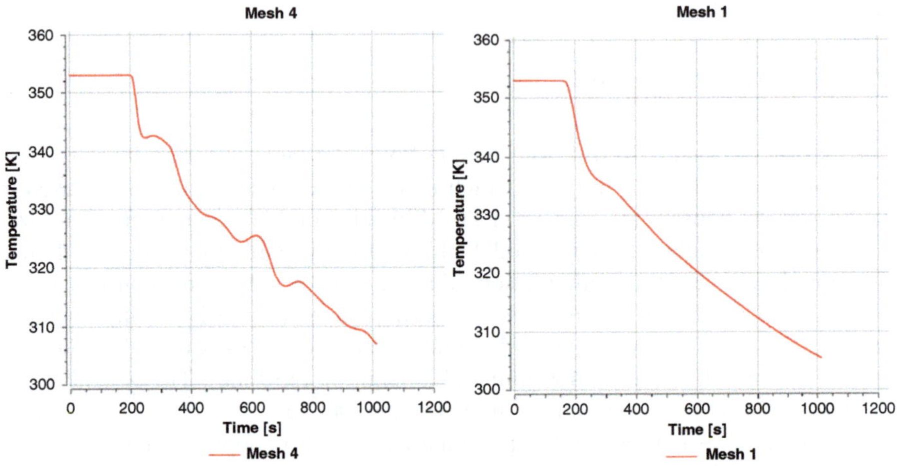

Fig. 9.1 Mesh dependence results

9.2.1.2 Fluent Setup

Once a final mesh model was selected, the transient simulations were used with a pressure-based solver. Gravity was selected according to the orientation of the geyser; $Y = -9.81$ m/s^2 for vertical orientation and $X = 9.81$ m/s^2 for horizontal orientation.

Since this simulation primarily is based upon the principle of heat transfer, the energy equation model was activated, in which Fluent solves the energy equation in the form stated in Eq. 9.1.

$$\frac{\partial}{\partial t}(\rho E) + \nabla \cdot (\vec{v}(pE + \rho)) = \nabla \cdot \left(k_{\text{eff}} \nabla T - \sum_j h_j \vec{J}_j + \left(\overline{\tau}_{\text{eff}} \cdot \vec{v} \right) \right) + S_{\text{h}} \quad (9.1)$$

where k_{eff} in Eq. 9.2 is the effective conductivity and is the sum of the turbulent conductivity kt and the conductivity of the fluid which is an empirical constant for the k-omega model.

$$k_{\text{eff}} = k + kt \quad (9.2)$$

Equation 9.1 refers to $\vec{J}_j$ which is the diffusion flux of species j and S_{h} which includes the heat of the chemical reaction and any other volumetric sources which have been defined. The first three terms of the right-hand side of Eq. 9.1 represent energy transfer due to conduction k_{eff}, species diffusion $\vec{J}_j$ and viscous dissipation $\vec{v}$, respectively, where

$$E = h - \frac{p}{\rho} + \frac{v^2}{2} \tag{9.3}$$

For this simulation, accounting for fluid flow, the k-epsilon realisable model was activated to represent the turbulent properties of the flow. This is a two equation model which allows for two extra transport equations to represent the properties of flow as well as accounts for history effects such as convection and diffusion of turbulent energy. Standard wall function was used as wall treatment and full buoyancy effects were considered.

The boundary conditions used were dependent on the tank orientation with the type of boundary condition for each outlet in use being pressure outlet. The boundary outlet condition targeted a mass flow rate: 13.8 l/min or 7.5 l/min dependent on the inlet mass flow rate condition used for each simulation. Outlets which were not being used and tapped off made use of a wall boundary type with material of aluminium, inclusive of the buffer plate, each of which classified as a stationary wall.

The heating element wall-type boundary was made from copper, with a constant temperature of 353 K as well as classified as a stationary wall. Reference values are computed from inlet conditions.

The coupled scheme was used for the pressure discretisation as it performs well under incompressible flow applications for transient simulations (ANSYS 2013). In spatial discretisation for pressure, PRESTO! scheme was used which allows for more accurate results. A second-order upwind discretisation scheme was used for all other variables. Gradients were evaluated using the least squares cell-based method. A higher order relaxation was activated which improves the start-up and general solution behaviour and has shown to prevent convergence stalling in some cases (ANSYS 2013).

Each surface monitor is plotted against flow time for every time step on an individual graph in order to analyse the solution. Each simulation was set to autosave each iteration for the purpose of analysing the solution in CFD-Post.

A standard solution initialisation was used and computed from inlet conditions. However, the temperature was used as 353 K on full tank, which would allow the tank to cool down from 353 K. Thereafter, the simulation was initialised.

For the first 10 time steps, a time step size of 0.1 s was used with 100 maximum iterations per time step. Thereafter, the maximum number of iterations per time step was changed to 3. Once this was complete, the time step size was increased to 1 s for 10 time steps and, lastly, 1 s for 1000 time steps.

9.2.1.3 CFD Postprocessing

CFD-Post in ANSYS was used to analyse the results of each simulation of each orientation. A graph of temperature vs time step was created for each simulation in order to analyse the decrease in temperature pattern.

A short animation of each simulation making use of volume rendering was created to be able to visually imitate the process of water cooling in each orientation, with each time step being saved as a frame.

A rectangular plane which dissected the outlets and the heating element was created to be able to analyse the temperature contours and velocity vectors of each simulation. A probe was created in order to read off the outlet temperatures of each orientation which was then verified against the graph of temperature vs time step.

9.3 Results and Discussion

Figures 9.2 and 9.3 indicate transient temperature contours at 13.8 l/min, vertical and horizontal orientated at the following times: $A = 60s$, $B = 120$ s, $C = 250$ s, $D = 400$ s, $E = 600$ s and $F = 1020s$.

Each tank orientation was simulated at full tank conditions at 353 K and cooled down over a period with inlet temperature taken at 273 K.

In comparing the temperature contours of Figs. 9.2 and 9.3, it is clear that at 13.8 l/min the vertical orientated tank takes longer to cool down than that of the horizontal tank. This can be verified by analysing Fig. 9.4 which is a graph of the transient outlet temperatures at a location on the outlet of each orientation.

Figures 9.2e, 9.3e and 9.4 demonstrate that the outlet temperature of the vertical tank is 335.76 K and horizontal is 324.64 K.

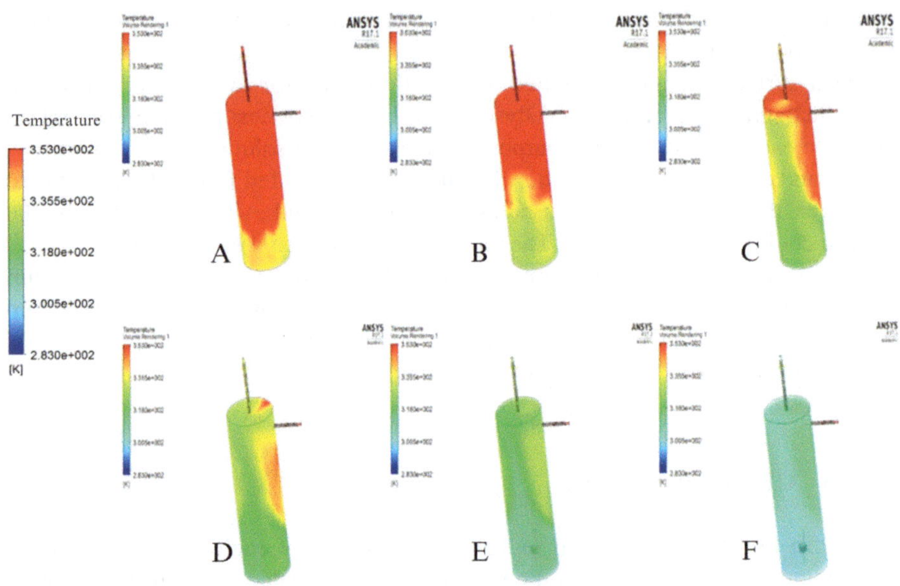

Fig. 9.2 Transient temperature contours at 13.8 l/min, vertical

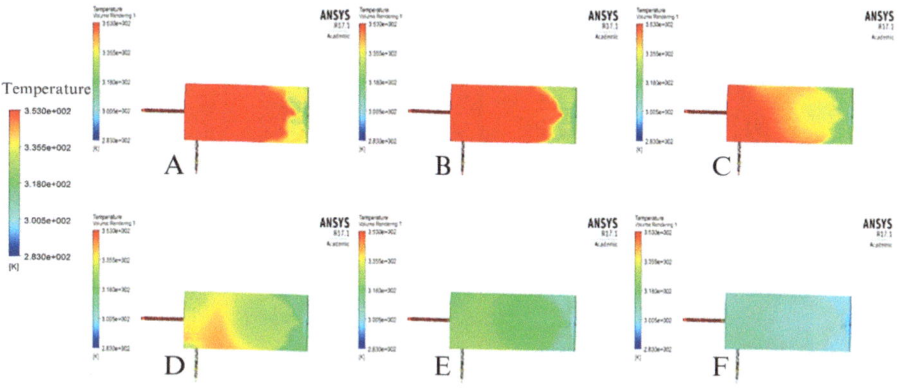

Fig. 9.3 Transient temperature contours at 13.8 l/min, horizontal

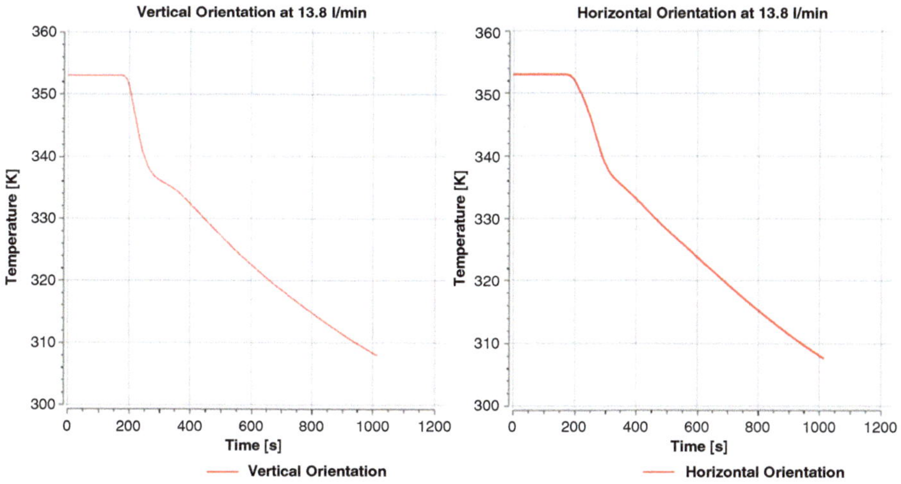

Fig. 9.4 Results of transient temperature outlet at 13.8 l/min, vertical and horizontal

Figures 9.5 and 9.6 indicate the transient temperature contours at 7.5 l/min, vertical and horizontal orientated, at the same times as Figs. 9.2 and 9.3. It can be seen that at 7.5 l/min, the horizontal orientated tank cools down slower than that of the vertical orientated tank. By analysing the temperature contours as well as the graphs in Fig. 9.7, the results indicate that the horizontal outlet transient temperature is 338.277 K and 326.26 K at the vertical orientated tank outlet.

It can be observed that the temperature stratification is a function of mass flow rate which then leads to the investigation of the flow patterns by making use of velocity vectors.

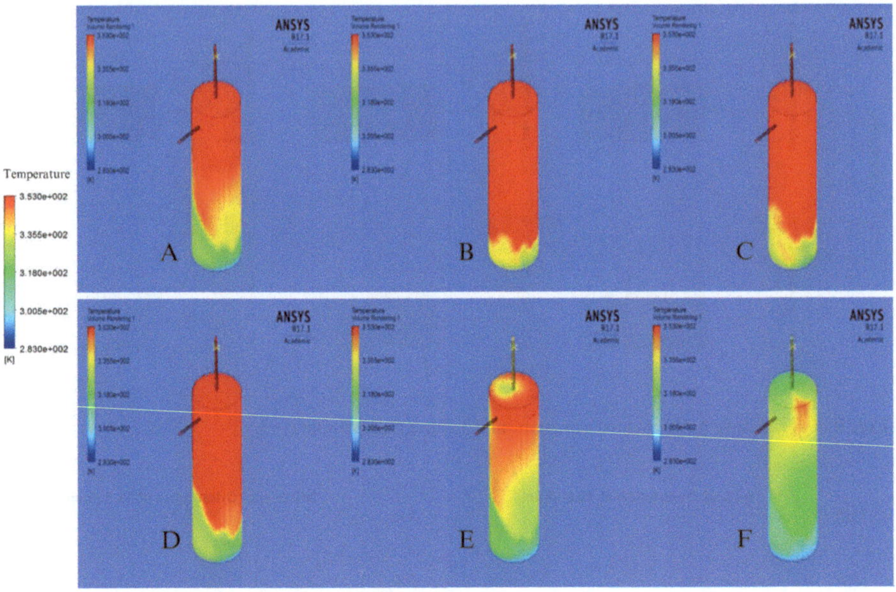

Fig. 9.5 Transient temperature contours at 7.5 l/min, vertical

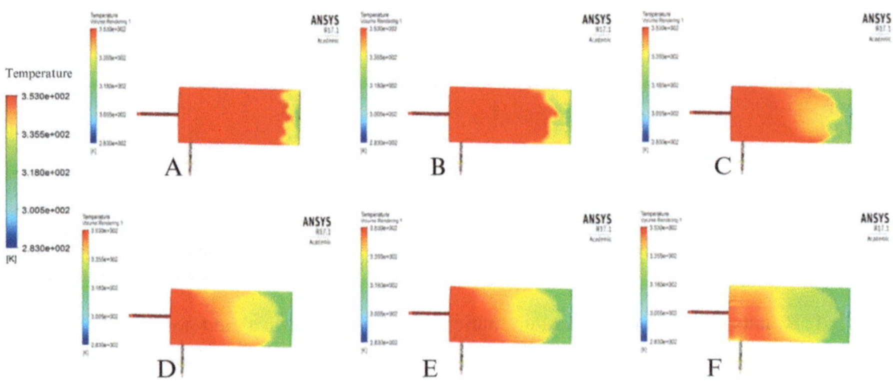

Fig. 9.6 Transient temperature contours at 7.5 l/min, horizontal

Figures 9.8 and 9.9 illustrate the flow in detail at 13.8 l/min and 7.5 l/min, respectively, for each orientation using velocity vectors. Figure 9.9 is illustrative of a higher velocity throughout the fluid body.

Figures 9.8 and 9.9 indicate the high velocities at the outlet at 600 s of cooling at the respective flow rates at 283 K. Since the plane did not cut through the inlet of the horizontal orientated tank, only the inlet velocity of the vertical orientated tank demonstrated in the figures below.

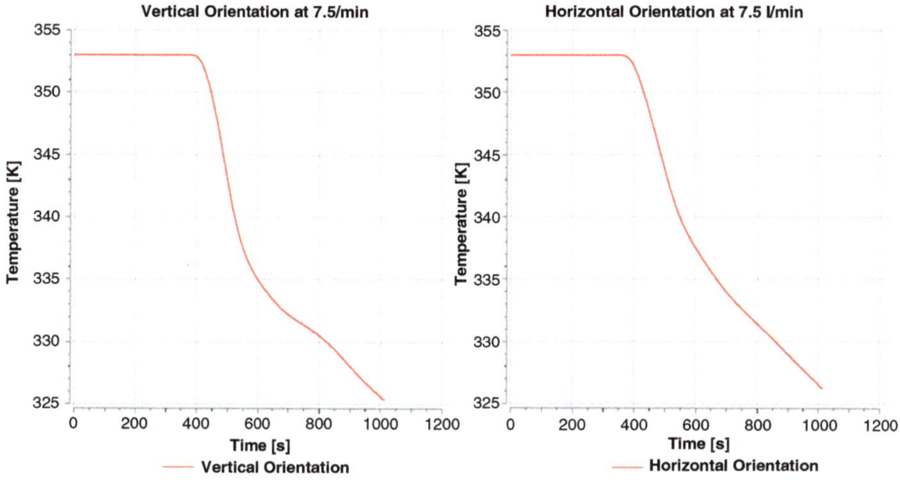

Fig. 9.7 Results of transient temperature outlet at 7.5 l/min, vertical and horizontal

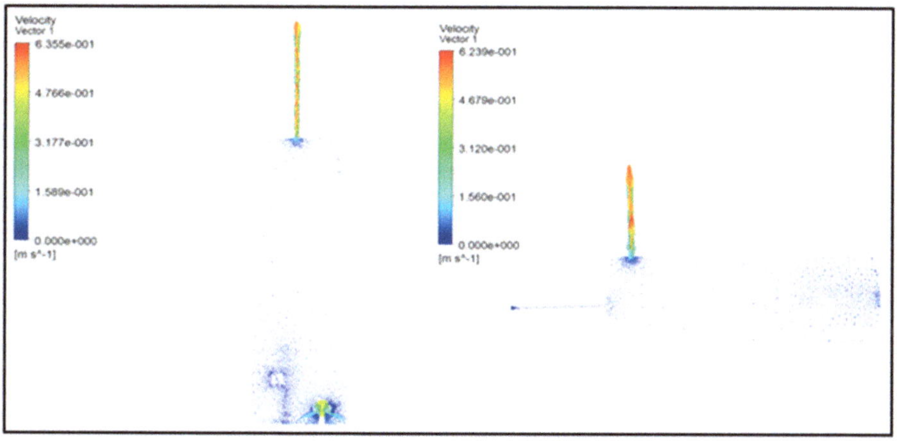

Fig. 9.8 Velocity vectors through a plane at 13.8 l/min

For each orientation the velocity vectors throughout the fluid bodies differ. According to Fig. 9.8, the velocity vectors have a low velocity of about 0.7589e-001 m/s at inlet and at outlet the velocities reach up to 5.0e-002 m/s

Figure 9.9 illustrates the velocity vectors at 7.5 l/min and averages at 4.11e-002 m/s and 3.5e-001 m/s at outlet for each horizontal and vertical. It can be seen in Figs. 9.8 and 9.9 that a decrease in temperature over time decreases the velocity of the fluid.

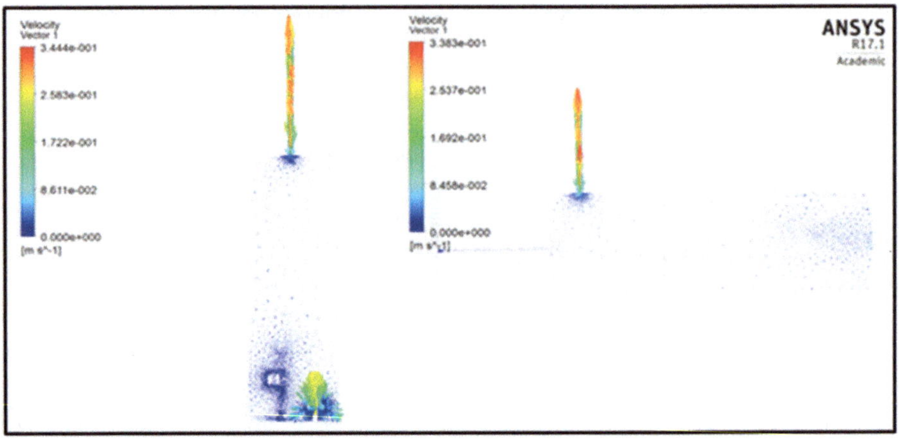

Fig. 9.9 Velocity vectors through a plane at 7.5 l/min

9.4 Conclusion

A CFD study was conducted in order to investigate if there is any difference in electrical usage on a horizontal and vertical orientated hot water storage tank. The results indicate that the orientation of the tank does not have an independent effect on the temperature output of a water storage tank. However, it was found that a vertically mounted tank cooled down less at a higher flow rate of 13.8 l/min and the reverse for the lower flow rate of 7.5 l/min. Considerations to the fluid flow rate together with the orientation of the tank should be considered in order to optimise electrical energy usage.

References

Africa, T. C. (2016). *South Africa has a new energy plan. But will it break the bank?* Available at: http://www.enca.com/opinion/south-africa-has-a-new-energy-plan-but-will-it-break-the-bank. Accessed 01 Dec 2016.

ANSYS Fluent 14.5 User's manual, Ansys Inc. (2013).

Beeker, N., Malisani, P., & Petit, N. (2015). Dynamical modeling for electric hot water tanks. *IFAC-PapersOnLine, 48*(11), 78–85. https://doi.org/10.1016/j.ifacol.2015.09.163.

Foundry, T. T. (2013). *Average household electricity use around the world.* Available at: http://shrinkthatfootprint.com/average-household-electricity-consumption. Accessed 18 Nov 2016.

Han et al. (2009) function (no date). Available at: https://www.sfu.ca/~ssurjano/hanetal09.html. Accessed 1 Dec 2016.

Insights, F. M. (2016). *Electric water heater market – Global industry analysis, size and forecast, 2015 to 2025.* Available at: http://www.futuremarketinsights.com/reports/electric-water-heater-market. Accessed 10 Oct 2016.

Kandari, A. M. (1990). Thermal stratification in hot storage-tanks. *Applied Energy, 35*(4), 299–315. https://doi.org/10.1016/0306-2619(90)90029-d.

Padovan, R., Manzan, M., Zorzi, E., Gullì, G., & Frazzica, A. (2015). Model development and validation for a tank in tank water thermal storage for domestic application. *Energy Procedia, 81*, 74–81.

Storify. (2017). *The history of who invented the electric water heater.* [Online]. Available at: https://storify.com/stingynun2966/the-history-of-who-invented-the-electric-water-hea. Accessed 15 Nov 2016.

Weingarten, L. (2003). The history of domestic water heaters. *Reeves Journal, 8*, 1.

Yearbook.enerdata.net. (2017). *World power consumption|electricity consumption|enerdata.* [Online]. Available at: https://yearbook.enerdata.net/electricity-domestic-consumption-data-by-region.html. Accessed 6 Nov 2016.

Chapter 10
Why EVs? A Comparison of Alternative Fuels to Help Australia Regain Energy Security

Alan Dunn, Martina Calais, Gareth Lee, and Trevor Pryor

10.1 Introduction

Australia is heavily dependent on imported crude oil and refined petroleum products. Since 2010–2011 it has been regularly importing over 80% of crude oil and refinery feedstock and consumes the bulk (73%) of this in transport, largely in passenger vehicles (Commonwealth of Australia 2013, 2015a).

In terms of private vehicle ownership per head of population, Australia ranks number 7 in the world, with 18.4 million vehicles, including 13.8 million light passenger cars, owned by a population of about 24 million people (ABS 2016). In average daily distances driven per person, Australian drivers average 38 km per day, plus with respect to average fuel consumption, Australian passenger cars burn on average 10.7 L per 100 km (ABS 2015), which ranks behind virtually all other industrialized nations, including the USA at 10.0 L/100 km (US Dept. of Transportation 2015) and the UK below 8.0 L/100 km (UK Dept. for Transport 2015). This means average passenger car drivers in Australia consume 4.07 L of unleaded petrol or diesel on a daily basis.

These statistics are troubling because Australia is a large industrialized nation with multiple alternative resources that has allowed itself to become foreign oil dependent at some cost to its economic security. So, how can the country transition out of its increasingly perilous situation to build capacity and sustainability in one or more of the domestically available alternative fuels?

A. Dunn (✉) • M. Calais • G. Lee • T. Pryor
School of Engineering & Information Technology, Murdoch University, Perth, Australia
e-mail: 30379386@student.murdoch.edu.au

A. Sayigh (ed.), *Transition Towards 100% Renewable Energy*,
Innovative Renewable Energy, https://doi.org/10.1007/978-3-319-69844-1_10

10.2 Declining Resources and Energy Security

The Australian domestic production of local oil and petroleum resources has been declining since the 1990s, and throughout that period energy consumption, particularly in transport, has risen steeply. In this way, the country has gone from being a marginal liquid fuel exporter to a significant importer within the space of 25 years. In 2015–2016, Australia produced 18,395 ML/year (or approximately 317,000 barrels of oil per day equivalent (BOPDe)) of crude oil and condensate, a decrease of 9% on the previous calendar year (OCE 2016). However, total petroleum product sales within Australia during the years 2015–2016 were 55,440 ML (about 955,000 BOPDe). This means that indigenous resources formed 33% of domestic consumption but less than 25% of the export and consumption total in the latest calendar year, with the difference being made up by imported crude oil and petroleum products. This decline in domestic petroleum production has been mirrored by an equivalent decline in domestic refinery output, with only four Australian refineries now operating.

In addition to the costs of fuel importation, the importation system that Australia relies on also flouts an international agreement with the OECD and potentially places the functionality of all Australian essential services (e.g. military defence, medical supplies, food distribution, public transport, etc.) in the hands of a few private fuel companies who have limited controls against supply disruption. As a member of the OECD and International Energy Agency (IEA) and a signatory to the Agreement on an International Energy Programme (the IEP Agreement), Australia has an obligation to keep a stockpile of crude oil and petroleum reserves equal to a minimum 90 days of the net import demand for the previous year (Twomey 2012). However, Australia has regularly breached this 90-day stockholding obligation since 2009 and recently has only been holding about 50 days of IEA eligible stocks – 20 days petrol, 17 days of aviation fuel and 16 days of diesel (Commonwealth of Australia 2015b).

Additionally, given this limited fuel stockholding and the 'just-in-time' logistics which private industry uses to optimize supply chains, essential services in Australia have a very restricted window of operations in the event of a major disruption to imported fuel supplies. Most petrol stations hold less than a week of fuel onsite, and once local fuel is consumed, pharmaceutical supplies would be affected after 7 days at retail outlets and 3 days in hospitals, supermarkets in just 7 days for dry foods or 9 days for frozen goods, and defence force activity would be severely limited (Blackburn 2013). For this reason, the security of liquid fuel supplies is becoming a serious concern, and the Australian government has estimated that to now meet this 90-day IEA supply obligation by increasing onshore storage capacity around the country will come at a cost of AUD $7 billion (Hale and Twomey 2013).

But these impacts on energy security need not be addressed by limiting a response in the single dimension of the oil industry. There are many alternatives which could equally and simultaneously enhance both national energy security and the national economy by putting Australians back in control of their daily fuel supplies while utilizing an abundance of domestic resources.

10.3 Alternative Solutions

A recent survey of passenger car industry stakeholders by Murdoch University (Dunn et al. 2017) found that 72% of respondents agreed with the statement that there was a growing or significant threat to energy security in Australia posed by the reliance on imported liquid petroleum products. Additionally, more than one third of respondents agreed that the situation was sufficiently alarming that immediate action was required by government and a clear plan should be developed by 2020 which enables the passenger vehicle transport industry to systematically replace conventional fuel sources with alternative domestically abundant resources.

Fortunately, Australia has many available alternatives. However, none that have advanced very far commercially due to the effective monopoly of established petroleum-fuelled vehicles. These include natural gas fuel systems, such as compressed natural gas (CNG) and liquefied petroleum gas (LPG), biofuel production and electric vehicles, alongside less mature or demand-side remedies such as hydrogen (H_2) fuel cell technology and improved public transport infrastructure. An understanding of viable solutions can be gained via a brief summary of these options below (Tables 10.1, 10.2, 10.3 and 10.4).

Option 1

Table 10.1 Convert passenger vehicles to run on compressed natural gas (CNG)

Benefits	High-energy fuel, locally abundant – Australia has significant gas resources. Low greenhouse gas emission profile compared to petroleum fuel consumption
	Decade-old technology used in commercial transport and private passenger cars, popular in Pakistan and Iran, plus parts of South America and the USA (AFDC 2014). Car models available from several vehicle manufacturers, such as Fiat
Fuel cost	Natural gas spot price is AUD $2.40 (US $1.80) per million BTUs (EIA 2016). Australia short-term trading market (STTM) gas price is AUD $4.00 per GJ = 0.95 MBTU (AEMO 2016), equivalent to 26,108 L of gas at STP or 133 L of CNG pressurized at 2,900 psi – same energy as 29 L of unleaded petrol or 26 L diesel. Therefore, average daily passenger vehicle distance of 38 km needs 18.67 L of CNG = AUD $0.56 or $205 per year
	Annual fuel saving per vehicle compared to petrol = AUD $1,725
Conversion or constraint costs	Specialized fuel tank and fuel converter in engine – AUD $13,000–$20,000. Larger tank size (3–4 times conventional tank) or reduced driving range due to fuel volume. Potential conversion subsidy cost of $10,000 per car = $10 billion per million vehicles
Infrastructure costs	Natural gas pipeline network exists across most of Australia but distributed at lower pressure in cities, about 1 MPa or 145 psi (APIA 2014). Conversion to CNG at 2,900 psi would require small compressors at homes or community refuelling stations at cost AUD $8,500 each. So, 1 million of these spread over Australia would cost $8.5 billion
Larger network cost	To provide effective security for domestic gas supply, national pipeline network required to link all states to northwest gas hubs. Northern gas pipeline project linking NT (Tennant Creek) to QLD (Mt Isa) to cost AUD $800 million (Armour Energy 2015), potential WA (Telfer) to NT (Mereenie) gas pipeline linkage of ~2,000 km could cost AUD $1.5 billion

Option 2

Table 10.2 Increase the use of LPG converted transport

Benefits	High-energy fuel, locally abundant by-product of natural gas extraction and crude oil refining. Low GHG emission profile compared to petroleum fuel consumption
	Old technology in Australia and overseas. Cars by major vehicle manufacturers, such as GM
Fuel cost	LPG commonly retails at about half the price of unleaded petrol; current retail price is approx. AUD $0.59 per L (NRMA 2016). Therefore, average daily passenger vehicle distance of 38 km would cost about AUD $2.40 or $876 per year
	Annual fuel saving per vehicle compared to petrol = AUD $1,054
Conversion or constraint costs	Specialized extra fuel tank and fuel converter in engine – AUD $2,500. Marginally reduced engine power. Past (LPGVS) subsidy cost of $1,500 per car only converted ~230,000 cars (Smit 2014) would require another $1.5 billion per million vehicles converted
Infrastructure costs	No pipeline – distributed by road transport. Therefore, continued import/transport cost
Larger network cost	To replace annual domestic road transport, consumption of conventional petroleum fuel would require additional 23 million tonnes (~31.5 GL) of LPG, via increased imports or domestic production, at cost of AUD $418 per tonne (Gas Energy Australia 2016) = AUD $9.5 billion. Unlikely due to domestic gas production and refining constraints

Option 3

Table 10.3 Increase the use of biofuel-powered transport

Benefits	High-energy fuel, common as biodiesel (B100) or ethanol (E10) – most ethanol consumed as a 10% blend in unleaded petrol
	Decade-old technology common in Australia and overseas. Low GHG emission and particulate pollution compared to standard diesel fuel consumption (Brodie 2011). Flex-fuel car models available from many major vehicle manufacturers, such as Ford
Fuel cost	Biodiesel and ethanol fuel blends have alternative fuel excise, about AUD $0.01 less than standard unleaded petrol. Pure biodiesel (B100) wholesale price is approx. 10% higher than standard diesel at AUD $1.39 per L (EcoTech 2017); retail is higher. Therefore, average daily passenger vehicle distance of 38 km would cost above AUD $5.66 or $2,065 per year
	Annual fuel cost per vehicle compared to diesel = extra AUD $135–$200. Biodiesel fuel cost estimated to drop by ~30–50% if ever manufactured on large scale (LEK 2011)
Conversion or constraint costs	No engine conversion required, can be used in flex-fuel vehicles
	Limited local production, max annual output approx. 450 ML ethanol and 430 ML biodiesel but supplied 284 ML and 114 ML, respectively (0.6% of all transport fuels) in 2012–2013 (APAC 2013) due to low demand/high feedstock costs. Consumption temporarily rose via 5%

(continued)

Table 10.3 (continued)

	biodiesel state mandate in NSW/ACT (2012) and then reduced due to competition with petrol prices
Infrastructure costs	No pipeline – distributed by road transport. Would require vastly expanded road/rail transport distribution network, therefore continued fuel transport cost
Larger network cost	Has great scope for increased production via increased high-yield cropping or algae production but isn't a current priority of industry. But requires 5–10 million hectares of land for feedstock plantations to replace petroleum fuels with 20 GL biofuel production, estimated at AUD $50 billion initial investment and major changes to land-use priorities by Australian community

Option 4 The development of hydrogen fuel cell technology.

In Australia, hydrogen (H_2) fuel cell technology is still in its infancy, with no passenger vehicles or fuel currently available for sale. It is not currently a viable option and is only mentioned briefly here. It is a high-energy fuel that requires hydrogen gas to mix with air (oxygen) in the presence of a catalyst to cause an electrochemical reaction. It is mainly created via a steam reforming process from methane and water, but for every tonne of H_2 gas formed, 9–12 tonnes of CO_2 are also formed via this reaction, depending on feedstocks (AFDC 2016). H_2 may also be produced via electrolysis of water or other chemical processes. Potential infrastructure costs for a solar electrolysis unit have been estimated at $100,000 for a five car system, and a single hydrogen refuelling station may cost AUD $3.8 million (US $2.8 million) (Melaina and Penev 2013).

Option 5 Demand side mitigation – increased use of electric public transport.

This option is based on the increased use of highly efficient public mass transport vehicles, plus other measures such as car sharing, bicycles and pedestrian traffic. By better utilization of existing and new transport infrastructure, individual consumer fuel costs and petroleum importation would be reduced. Given that it may only ever be a partial solution, this option is only briefly mentioned here. Research suggests that even with public transport options available, a large percentage (47%) of Australians would persist in using private vehicles for commuting or other small personal trips (ABS 2013). Cost estimates for electrified light rail network projects in state capital cities amount to AUD $55 billion, and additional projects such as the 1,748 km proposed high-speed rail project between Melbourne and Brisbane may cost AUD $68–$114 billion (McMah 2016). Expanded public bus systems running on electricity or alternative fuels may cost a further AUD $6 billion for 2,000 new buses plus 20 new maintenance facilities or transit interchanges across Australia.

Option 6

Table 10.4 Increased uptake of electric vehicles

Benefits	Efficiently uses existing electrical infrastructure. No combustion pollutants and low GHG emission profile if electrical power generated via renewable energy
	Old technology but new motors developed in recent decades. Car models available from many major vehicle manufacturers, such as Nissan. (Initial cost now below AUD $40,000, compared to $18,000 for equivalent petrol fuelled vehicle (Nissan 2016))
Fuel cost	Assuming grid electricity costs of AUD $0.257 per kWh (WA Govt. 2016) and 173 W/km for a small EV (DIRD 2016), e.g. Nissan Leaf has a 24 kWh battery and range of 175 km (Nissan 2016); fuel cost for 38 km average daily driving distance is $1.34 or $489 per year
	Annual fuel saving per vehicle compared to petrol = AUD $1,441. Annual maintenance costs are also negligible due to lack of moving parts in motor
Conversion or constraint costs	EV conversion of conventional vehicle, with higher motor properties but a range of only 100 km is about AUD $15,000, depending on the car (AEVA 2016). Many conversions have small batteries with driving range of 100–200 km, but larger models (e.g. Tesla) have ~500 km range
	Potential subsidy required for conversion or new vehicle purchase of AUD $10,000 would cost $10 billion per million vehicles
Small infrastructure costs	Private battery recharging via standard or three-phase wall power sockets, fast recharging systems cost AUD $1,000–$5,000 (Tesla 2016b). Many public EV recharging stations now available in capital cities and >350 exist nationally
	Australia's first regional 'EV super highway', opened in June 2015, south of Perth, W.A.(RAC 2015). This network has 10 fast recharging stations set in regional towns every 100 km, enabling EV drivers to recharge for $3 in a 30-min recharge time (Long 2015). New fast-charging highway stations are being built by Tesla in 2016–2017 between Melbourne, Sydney and Brisbane (Tesla 2016a), plus others in Queensland, increasing the range of EVs
	Each 2-vehicle fast-charging station in these networks costs about AUD $40,000 (Vorath 2014), and if this could be easily upscaled tenfold to deal with more traffic for $400,000, then installation of 3,000 EV stations on highways across regional Australia, in a coverage similar to the petrol station network, could be achieved for approximately $1.2 billion
Larger network cost	With increasingly widespread installation of grid-connected solar PV systems in homes and businesses across Australia, it is unlikely that major base-load power generation construction will be required in Australia to offset higher loads due to a rapid spike in EV consumption, so long as demand for EV recharging is properly managed (ESAA 2013). This can be done via tools such as time-of-use (TOU) or smart chargers (AEMC 2012) which encourage off-peak recharging and minimize peak demand impacts on the network. As EVs become common and vehicle-to-grid (V2G) technology improves, EV batteries will become important tools in storing and releasing power back to the grid during periods of high demand (ESAA 2013)

10.4 Electric Vehicle Synergies with Renewable Energy

As indicated in past government studies (BTE 1974), if Australian passenger vehicles became largely powered by electricity, the impact of EV battery storage could have a major effect on the operation of the Australian electricity grid, i.e. the National Electricity Market (NEM) and the geographically isolated South West Interconnected System (SWIS) in Western Australia. These systems coordinate and control the power grid in Australia, tying together the multi-faceted and geographically separate power generation from coal-fired production, gas turbines, hydropower, wind systems, solar generation, landfill and other sources. In recent years, due to inconsistent government strategy and rapid technological development in renewable options, the NEM/SWIS has had difficulty in efficiently and economically coordinating the multivariate power supplies into a stable and resilient system. As the cost of home battery power storage units (e.g. the Tesla Powerwall) have dropped sharply from US$1,000 per kWh in 2007 to US$400 per kWh in 2014, and is predicted to reach US$200 per kWh in 2020 (Stock et al. 2015), so has the uptake of both commercial and residential off-grid or grid-connected battery storage facilities. This has driven the payback period for an average solar-powered household from 12 years in 2015 to an expected 8 years in 2020, and maybe 6 years by 2030 (Stock et al. 2015). Worldwide, the market for residential solar panel and battery storage systems is expected to rise tenfold from 90 GW in 2014 to 900 GW in 2018 (IRENA and C2E2 2015), and in Australia rooftop solar with large-scale battery storage is expected to grow from 0.5 GW in 2015 to 3.4 GW in 2025 and 8 GW in 2035 (AEMO 2015).

However, these projections have not factored in the potential synergies that renewable energy has with compatible and coexistent electric vehicle development. In the same way that off-grid rooftop solar battery storage is quickly becoming affordable, the battery costs for EVs are quickly closing the gap between conventional petrol vehicle purchase prices and those of EVs, so there is real potential for the impact of EV battery storage on the NEM/SWIS to be larger and more rapid than predicted. Options for intermittent support of the grid through vocational or residential carparks that allow EV drivers to plug in to community battery banks during work or shopping hours may also have a big impact on NEM operation. It may indeed be possible in the near future for EV drivers to actually be paid by the NEM to park their car for the day and thereby plug in to recharge and stabilize the grid, while they attend work (Lovins et al. 2004)! The same could happen in private garages at night when workers return home, with power flow both into and out from the car battery into a grid-connected household residential solar battery system, during both off-peak and peak periods. These ideas suggest there is significant potential for rapid growth in both off-grid and grid-connected battery storage globally, particularly as vehicle-to-grid (V2G) options arise with increased EV uptake.

10.5 Conclusion

The brief analysis provided above shows the depths of Australia's current dependence on foreign oil and petroleum product imports. It currently relies on a daily importation of over 110 ML or 700,000 barrels per day of crude oil or petroleum products to meet domestic demand, and domestic supplies are declining. If this import dependence continues, then it will be incumbent upon the Australian government to meet their IEA agreement and stockpile a minimum 90-day supply of fuel at strategic bases around the country, at a cost of AUD $7 billion.

Fortunately, there are several alternatives that Australia could pursue to retrieve itself from the current energy insecurity. The financial benefits and costs for six of these have been compared with conventional petroleum fuel costs in the table below (Table 10.5), and each alternative has great potential to assist in solving the problem of Australia's dependence on foreign oil. However, in terms of long-term energy, financial and environmental security, it seems reasonable for government and industry to encourage those measures which are most sustainable and can be fuelled by renewable domestic resources. In this context, the rapid decrease in the price of battery power storage systems, particularly those linked with residential rooftop solar power, and the potential compatibility of these with electric vehicle battery storage and recharging systems will likely assist the adoption of both electric vehicles in Australia and a continued explosion in both grid-connected and off-grid residential solar power.

Table 10.5 Comparison of conventional and alternative fuel option costs

Fuel type	Current indigenous production level (% of domestic fuel consumption or passenger cars)	Estimated current or projected reliance on imported fuel (%) (i.e. if all cars used just one fuel type)	Relative GHG emission or pollution cost (low, med, high)	Retail cost per L equivalent (AUD$)	Average annual fuel cost to passenger vehicle drivers (AUD$)	Potential annual fuel savings compared to current petrol/diesel (AUD$ per driver/AUD $M per million drivers)	Vehicle conversion cost per driver (AUD$)	Potential cost of conversion subsidy for 1 million vehicles (AUD$ billion)	Initial cost of alternative vehicle (AUD$)	Difference in vehicle cost to conventional petrol/diesel vehicles (AUD$)	Difference in vehicle cost for 1 million vehicles (AUD$ billion)	Infrastructure cost required for change (AUD$ billion)	Overall cost of measures for change to at least 1 million vehicles (single year estimate, AUD$ billion)
Petroleum/diesel	33	75–85	High	1.30	1930	0/0	0	0	15–20,000	0	0	7.0	7.0
CNG	< 0.00001 (<1,000 cars)	0	Med	0.03	205	1,725/ 1,725 M	13–20,000	10.0	20–25,000	5,000	5.0	2.3 + 8.5 = 10.8	15.8 to 25.8–1.725/ year
LPG	1.7 (~230,000 cars)	30–50	Med	0.59	876	1,054/ 1,054 M	2,500	1.5	17–22,000	2,000	2.0	9.5	11.0 to 11.5 – 1.054/year
Biodiesel / ethanol	0.06	0	Low	1.39+	2,065	+135/ +0.135 M	0	0	17–22,000	2,000	2.0	50.0	52.0 + 0.135/ year
H_2 fuel cell	0	0	Low-med	1.33	4,025	+2,095/ +2,095 M	20,000 (PV electrolysis)	20.0	80,000+	60,000	60.0	11.2	91.2 + 2.095/ year
Improved public transport	20	40–50	Low	0	965 (assumes 50% PT use)	965/ 0.965 M	0	0	0	0	0	55.0 + 68.0 + 6.0 = 129.0	129.0–0.965/ year
EVs	0.00001 (only ~1,300 cars)	0	Low	0.33	489	1,441/ 1,441 M	15,000	10.0	40,000+	20,000	20.0	1.2	11.2 to 21.2–1.441/ year

References

ABS. (2013). *4102.0 – Australian social trends, July 2013. Car Nation*. Australian Bureau of Statistics, Canberra, Australia. Online at: http://www.abs.gov.au/AUSSTATS/abs@.nsf/Lookup/4102.0Main+Features40July+2013

ABS. (2016). *9309.0 – Motor vehicle census, Australia, 31 Jan 2016*. Australian Bureau of Statistics, Australia. Online at : http://www.abs.gov.au/ausstats/abs@.nsf/mf/9309.0

ABS. (2015). *9208.0 – Survey of motor vehicle use, Australia, 12 months ended 31 October 2014*. Australian Bureau of Statistics, Australia. Online at: http://www.abs.gov.au/ausstats/abs@.nsf/mf/9208.0/

AEMC. (2012). *Energy market arrangements for electric and natural gas vehicles. Summary of AECOM's final advice*. Australian Energy Market Commission (AEMC), Sydney, 29 August 2012. Online at: http://aemc.gov.au/getattachment/41fa2c2a-4a56-46b3-b144-452ed20dba1d/Information-Sheet-Summary-of-AECOM-s-Final-Advice.aspx

AEMO. (2015). Emerging technologies information paper. National Electricity Forecasting Report. Australian Energy Market Operator (AEMO), June 2015. Online at: https://www.aemo.com.au/-/media/Files/PDF/Emerging-Technologies-Information-Paper.pdf

AEMO. (2016). *STTM gas price and quantity*. Australian Energy Market Operator (AEMO) webpage on 29 March 2016. Online at: http://www.aemo.com.au/

AEVA. (2016). *Frequently asked questions: What performance and price can I expect?* The Australian Electric Vehicle Association webpage, 6 April 2016. Online at: http://www.aeva.asn.au/faq.php#t37n96

AFDC. (2014). *Natural gas vehicles*. US Dept. of Energy, Alternative Fuels Data Centre (AFDC), 2014, USA. Online at: http://www.afdc.energy.gov/vehicles/natural_gas.html

AFDC. (2016). *Fuel cell electric vehicles*. US Dept. of Energy, Alternative Fuels Data Centre (AFDC), 2016, USA. Online at: http://www.afdc.energy.gov/vehicles/fuel_cell.html

APAC. (2013). *Australian biofuels 2013–2014, policy and growth, study contents*. APAC Biofuel Consultants, Adelaide, South Australia, October 2013. Online at: http://www.apacbiofuel.com.au

APIA. (2014). *Factsheet 3: The gas supply chain*. The Australian Pipeline Industry Association Ltd (APIA), 2014. Online at: http://apga.org.au/wp-content/uploads/2009/10/factsheet3-Gas-Supply-Chain-110225.pdf

Armour Energy. (2015). *Jemena selected for North East gas interconnector pipeline via Northern route*. Armour Energy Limited report, 18 November 2015. Online at: http://www.armourenergy.com.au/assets/1498645.pdf

Blackburn, J. (2013). *Australia's liquid fuel security. A report for NRMA Motoring and Services*, Canberra, February 2013. Online at: https://www.mynrma.com.au/media/Fuel_Security_Report.pdf

Brodie, H. (2011). *Biofuels in Australia*. Presentation to Plastics and Chemical Industries Association (PACIA) by Biofuels Association of Australia (BAA), Melbourne, 14 April 2011. Online at: https://www.pacia.org.au/DownFile.Aspx?fileid=1320

BTE. (1974). *Electric cars*. Bureau of Transport Economics, Commonwealth Government of Australia, Australian Government Publishing Service, Canberra, June 1974. Online at: https://bitre.gov.au/publications/1974/report_013.aspx

Commonwealth of Australia. (2013). *Report on Australia's oil refinery industry*. House of Representatives Standing Committee on Economics, Parliament of the Commonwealth of Australia, January 2013. Online at: http://www.aph.gov.au/parliamentary_business/committees/house_of_representatives_committees?url=economics/oilrefineries/report.htm

Commonwealth of Australia. (2015a). *Australia's transport energy resilience and sustainability*. Senate Standing Committee on Rural and Regional Affairs and Transport, Department of the Senate, Canberra, 2015. Online at: http://www.aph.gov.au/Parliamentary_Business/Committees/Senate/Rural_and_Regional_Affairs_and_Transport/Transport_energy_resilience

Commonwealth of Australia. (2015b). *Energy white paper 2015*. Australian Government Dept. of Industry and Science, Canberra, April 2015. Online at: www.ewp.industry.gov.au

DIRD. (2016). *Green vehicle guide*. Australian Government Dept. of Infrastructure and Regional Development (DIRD), webpage, 5 April 2016. Online at: https://www.greenvehicleguide.gov.au/

Dunn, A, Calais, M., Lee, G., & Pryor, T. (2017). *Murdoch University 2015 Survey: The impact of energy security and environment concerns on the fuel mix for light passenger vehicles in Australia during the near future*. Murdoch University, Perth, 2017, unpublished.

Eco Tech. (2017). *Ecotech biodiesel terminal gate price*. Eco Tech Biodiesel Pty Ltd web page, Narangba, 4 April 2016. Online at: http://www.ecotechbiodiesel.com/terminal-gate-price

ESAA. (2013). *Sparking an electric vehicle debate in Australia. Discussion paper November 2013*. Energy Supply Association of Australia (ESAA), Canberra, November 2013. Online at: http://ewp.industry.gov.au/sites/prod.ewp/files/Sparking%20an%20Electric%20Vehicle%20Debate%20in%20Australia.pdf

EIA. (2016). *Natural gas weekly update*. U.S. Energy Information Administration, Natural Gas webpage, 29 March 2016. Online at: http://www.eia.gov/naturalgas/weekly/

Gas Energy Australia. (2016). *Saudi Aramco LPG Price*. Gas Energy Australia web page, Reports and Submissions, 3 April 2016. Online at: http://gasenergyaustralia.asn.au/reports-and-submissions/saudi-aramco-lpg-prices/

Hale, R., & Twomey, I. (2013). *Australia's emergency liquid fuel stockholding update 2013: Australia's international energy agency oil obligation. 'Main Report'*. Dept. of Industry, Canberra, October 2013. Online at: https://industry.gov.au/Energy/EnergySecurityOffice/Documents/HTMainReport2013.pdf.

IRENA and C2E2. (2015). Synergies between renewable energy and energy efficiency. Working paper, IRENA, Abu Dhabi, and C2E2, Copenhagen, 2015. Online at: http://www.irena.org/DocumentDownloads/Publications/IRENA_C2E2_Synergies_RE_EE_paper_2015.pdf

LEK. (2011). *Advanced biofuels study, strategic directions for Australia*. Summary report provided for the Australian Government, Department of Resources, Energy and Tourism by L.E.K Consulting, Sydney, 14 December 2011. Online at: http://arena.gov.au/files/2013/08/Advanced-Biofuels-Study-Appendix.pdf

Long, K. (2015). *Australia's first electric highway links Perth to South-West*. ABC News 23 June 2015. Online at: http://www.abc.net.au/news/2015-06-22/electric-highway-links-ev-drivers-from-perth-to-the-south-west/6565378

Lovins, A. B., Datta, E. K., Bustnes, O-E., Koomey, J. G., & Glasgow, N. J. (2004). *Winning the oil endgame: Innovation for profits, jobs and security*. Rocky Mountain Institute, USA, 2004. Online at: https://ia600708.us.archive.org/5/items/WinningTheOilEndgame/WinningOilEngame.pdf

McMah, L. (2016). *New push for Australia to adopt high speed rail*. News.com.au webpage, 2 March 2016. Online at: http://www.news.com.au/travel/australian-holidays/new-push-for-australia-to-adopt-highspeed-rail/news-story/d1497f88149989d0d5fb64b1e885d81f

Melaina, M., & Penev, M. (2013). *Hydrogen station cost estimates. Comparing hydrogen station cost calculator results with other recent estimates*. National Renewable Energy Laboratory (NREL), U.S. Department of Energy, Golden, September 2013. Online at: http://www.nrel.gov/docs/fy13osti/56412.pdf

Nissan. (2016). *Nissan leaf. 100% Electric. Zero petrol*. Nissan Motor Co. (Australia) Pty Ltd webpage, 5 April 2016. Online at: http://www.nissan.com.au/Cars-Vehicles/LEAF/Overview

NRMA. (2016). *Petrol prices – LPG average price*. National Roads and Motorists' Association (NRMA) webpage, Motoring Services, Petrol Prices, LPG 3 April 2016. Online at: http://www.mynrma.com.au/motoring-services/petrol-watch/fuel-prices.htm

OCE. (2016). *Australian petroleum statistics, Issue 241, August 2016*. Office of the Chief Economist, Dept. of Industry, Innovation and Science, Australian Government, Commonwealth of Australia, Canberra, 13 October 2016. Online at: https://industry.gov.au/Office-of-the-Chief-Economist/Publications/Documents/aps/2016/Australian_Petroleum_Statistics_241_August2016.pdf

RAC. (2015). *RAC electric highway an Australian first*. Royal Automobile Club (RAC) of Western Australia Inc., Media Release 22 June 2015, Perth. Online at: http://rac.com.au/cs/idcplg?IdcService=GET_FILE&dDocName=RACSTG061437&allowInterrupt=1&RevisionSelectionMethod=LatestReleased&noSaveAs=1

Smit, R. (2014). *Australian motor vehicle emission inventory for the National Pollutant Inventory (NPI)*. Prepared by Uniquest Pty Ltd for the Department of Environment, Australian Government, 2 August 2014. Online at: http://www.npi.gov.au/system/files/resources/e8311456-8a41-4473-9fa1-d2f9994ff8da/files/australian-motor-vehicle-emissions-inventory-2014.docx.

Stock, A., Stock, P., & Sahajwalla, V. (2015). Powerful potential: Battery storage for renewable energy and electric cars. Climate Council of Australia Limited, 2015. Online at: https://www.climatecouncil.org.au/uploads/ebdfcdf89a6ce85c4c19a5f6a78989d7.pdf

Tesla. (2016a). *Supercharger*. Tesla motors webpage, 6 April 2016. Online at: https://www.teslamotors.com/en_AU/supercharger?redirect=no

Tesla. (2016b). *Support – home charging installation*. Tesla Motors webpage, 6 April 2016. Online at: https://www.teslamotors.com/en_AU/support/home-charging-installation

Twomey, I. (2012). *National energy security assessment (NESA) identified issues: Australia's international energy oil obligation*. Dept. of Resources, Energy and Tourism, Canberra, July 2012. Online at: http://www.environment.gov.au/system/files/energy/files/ht2012australiasintenergyoilobligation.pdf.

UK Dept. for Transport. (2015). *ENV0103 average new car fuel consumption: Great Britain 1997–2013*. Department for Transport Statistics, UK, 2015. Online at: https://www.gov.uk/government/uploads/system/uploads/attachment_data/file/384240/env0103.xls.

US Dept. of Transportation. (2015). *Table 4–23: Average fuel efficiency of US light duty vehicles*. Bureau of Transportation Statistics, D.O.T. Online: http://www.rita.dot.gov/bts/sites/rita.dot.gov.bts/files/publications/national_transportation_statistics/html/table_04_23.html

Vorath, S. (2014). *Tritium plans Australia's largest fast-charge EV network for Qld*. Renew Economy webpage, 13 November 2014. Online at: http://reneweconomy.com.au/2014/tritium-plans-australias-largest-fast-charge-ev-network-for-qld-51842

WA Govt. (2016). *Electricity prices*. The Government of Western Australia, Department of Finance webpage, 5 April 2016. Online at: https://www.finance.wa.gov.au/cms/Public_Utilities_Office/Homes_and_Communities/Electricity/Electricity_prices.aspx

Chapter 11
The Impact of Energy Security and Environment Concerns on the Fuel Mix for Light Passenger Vehicles in Australia During the Near Future: Findings from a 2015 Murdoch University Survey

Alan Dunn, Martina Calais, Gareth Lee, and Trevor Pryor

11.1 Introduction

Australia is highly dependent on imported crude oil and refined petroleum products. It consumes 7.5 litres of oil per day per person yet produces less than 2.5 litres of oil per day per head of population – just one-third of the daily requirements (CIA 2015). Since 2010–2011, Australia has been regularly importing over 80% of crude oil and refinery feedstock (Commonwealth of Australia 2013) and consumes the bulk of the imported crude oil and products (73%) in transport, largely in passenger vehicles (Commonwealth of Australia 2015).

Australia is a large industrialized nation with multiple alternative resources that has allowed itself to become foreign oil dependent through a combination of declining local petroleum resources, complacency, and lack of planning. These deficiencies have borne a generation of passenger vehicle drivers who are prepared to pay the extra financial costs of fuel importation to keep their standard of living high, but are now challenged by the issues which fossil fuel use poses from both an environmental perspective and from the lack of critical resource control which may threaten their energy security.

Clearly, there are many possible options for the country to transition out of its increasingly perilous situation to build capacity and sustainability in one or more of the domestically available alternative fuels. In response to this issue, in 2015 a research team from Murdoch University conducted a survey to gain input from the most informed people in the Australian passenger vehicle industry, transport government regulators, and academic research specialists to gain insight into future fuel strategies and vehicle development in the light passenger vehicle industry. The results and analysis are presented here.

A. Dunn (✉) • M. Calais • G. Lee • T. Pryor
School of Engineering & Information Technology, Murdoch University, Perth, Australia
e-mail: 30379386@student.murdoch.edu.au

A. Sayigh (ed.), *Transition Towards 100% Renewable Energy*,
Innovative Renewable Energy, https://doi.org/10.1007/978-3-319-69844-1_11

11.2 Methodology

The Murdoch University survey was conducted between May and July 2015 and targeted a large group of Australian passenger vehicle specialists in order to appraise the status of development, progress, and direction from informed sector managers, engineers, and researchers with respect to the fuel choices available in current passenger vehicles and the potential impacts of these on energy security and environmental issues in Australia. Specifically, the stakeholders surveyed included leaders from all major car manufacturers, government regulators, transport academic researchers, plus associated representatives from transport industry adviser, energy suppliers, and trade groups. Overall group bias was minimized by restricting recipient numbers to 50% from industry targets and 50% from joint government (local, state, and federal bodies) and research groups, such as CSIRO and university academics. Notably, there was a distinct gender imbalance in the group, as women form far less than 50% of the senior industry leadership cohort, and although age was not specifically targeted, senior managers, engineers, and researchers with 10+ years in the industry surely represented over 50% of the targets selected.

Survey recipients were selected via online searches of publically available information for vehicle manufacturers, government and industry regulator news and conference material, and publically available industry links. Surveys were emailed to selected targets on three repeated occasions, and ultimately the responses from 47 targets were received. The survey consisted of 16 multiple-choice questions, with space permitting more considered input. Results for each question were then tabulated, with trends drawn from the most common primary and secondary answers. Popular responses from each of the three main Industry, Government and Academic groups have also been compared, along with obvious variations associated with levels of industry experience (workers with less than 10 years' experience, 11–20 years' experience, or greater than 20 years' experience, with these levels sometimes used as a proxy for age) and gender (females formed less than 20% of all respondents). Various trends and differences may be seen in the respondent answers, depending on their profession, career experience, and gender. These have been analyzed in the results provided so that some insight may be gained into the factors which may drive a popular response. The pie charts for these respondent group sizes can be seen below (Figs. 11.1, 11.2, and 11.3).

A graphic comparison of each of these features may also be seen in the graph below (Fig. 11.4).

The survey question topics were classified into three main fields. One aim of the survey was to resolve whether the continued use of fossil fuels in the passenger car industry was likely to change in the near future (e.g., before 2050) due to either a perceived threat to Australian energy security, resulting from an increasing reliance on foreign oil importation, or due to environmental impacts, such as climate change due to carbon emissions or atmospheric pollution, and if so, which was more critical, and when would this change likely occur? A second group of questions focused on investigating what changes to the fuel mix for light passenger vehicles

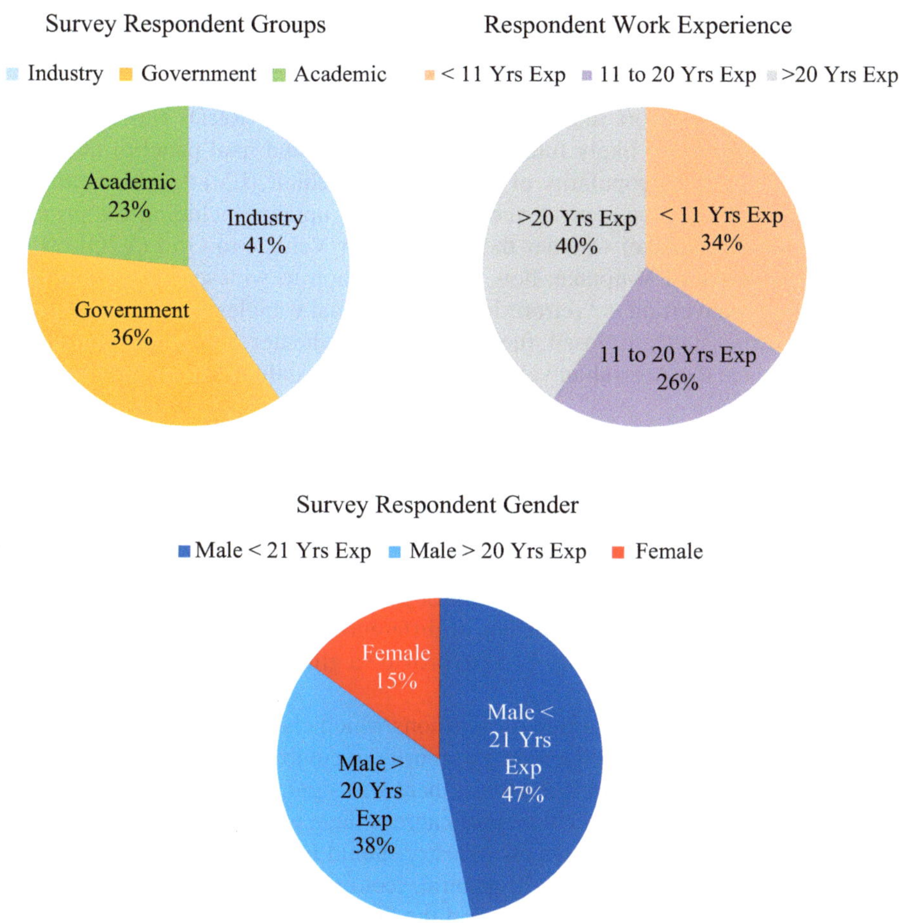

Figs. 11.1, 11.2, and 11.3 These charts compare the survey respondent group sizes in terms of their profession (Fig. 11.1), their career experience level (Fig. 11.2), and their gender (Fig. 11.3)

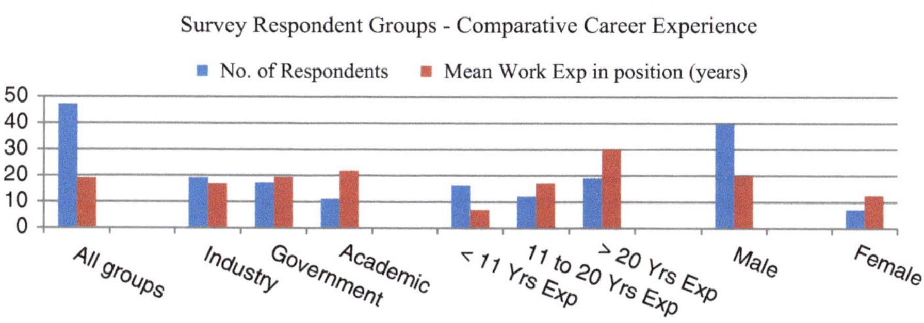

Fig. 11.4 The comparative work experience of the various survey respondent groups identified

might take place in the near future, if any, which new fuels this might include, and when this transition might take place (e.g., before 2020, between 2020 and 2030, or between 2030 and 2050)? Also, other questions investigated whether government investment into transport fuel research and development (R&D) was adequately assigned to meet these likely future changes. A third and final panel of questions related to the obvious popularity of recent electric vehicle (EV) development. The survey aimed to resolve several questions for EV uptake, including the primary obstacles for EV uptake; whether the potential for Vehicle-to-Grid (V2G) operations would impact EV uptake; how fast a transition to widespread EV use may take; whether converting or "retrofitting" conventional vehicles as EVs was a useful way to recycle standard fossil fueled cars; and whether EV production could possibly replace the Australian vehicle manufacturing industry from 2017 onward.

11.3 Results

The answers for each question were collected and grouped according to popular primary and secondary responses. Subgroups were also analyzed for trends and associations to identify any anomalies or differences. The survey results for Questions 6 are provided below (Fig. 11.5) to provide a glimpse of the analysis that was followed for each question.

With respect to the first group of questions which checked respondent attitudes to the perceived threats of either energy security or environmental damage from the continued reliance on liquid fossil fuels, there was significant agreement across all groups surveyed. Overall, a 72% majority agreed there is a growing threat to energy security, and the light passenger car industry should transition away from conventional fuels with some urgency. In general, less experienced workers (e.g., those with less than 10 years' experience) find the current imported fuel situation more alarming and the threat to energy security more pressing than more experienced

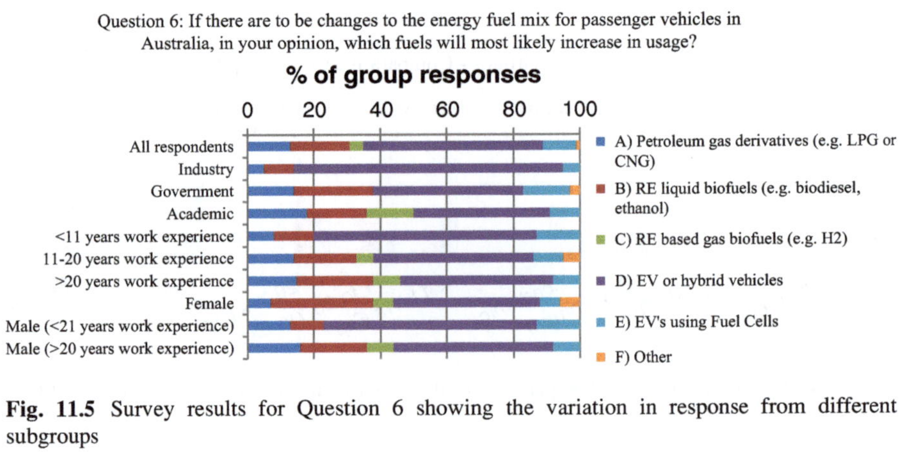

Fig. 11.5 Survey results for Question 6 showing the variation in response from different subgroups

workers (e.g., those with greater than 10 years' experience). Additionally, over 33% of respondents indicated that regardless of the urgency for energy security, the country would benefit from the Federal Government developing an alternative transport fuel strategy in the near future (e.g., before 2020) to provide options in case of a fuel security disruption.

With respect to environmental threats, the survey results indicated that more experienced people are less alarmed by the environmental effects than less experienced people. In any event, a 78% majority of respondents across all groups are greatly concerned by the impacts on the environment of continued fossil fuel use and suggest the growing threats to the environment are sufficiently urgent that we should transition away from conventional fuel usage in the passenger car industry. The group of Academic respondents, in particular, was more alarmed by the potential threat to the environment posed by continued use of hydrocarbon fuels than respondents from either the Government or Industry groups. However, almost one-third of respondents also suggested that restricting coal combustion for stationary energy generation was more immediately critical than restricting fossil fuels for transport purposes.

Comparing these threats, more experienced workers, including those from Academic and Government groups, are more concerned about the impact of the current energy fuel mix on energy security, while less experienced workers, including those from Industry, are more concerned about the impact on the environment. However, an overall 68% majority across all groups believe that Australians will see either medium or serious environmental impacts causing change to the energy fuel mix for passenger cars before the year 2050, and over one-third (including many respondents from Academic and Government groups) believe this may lead to change before 2030. Notably, almost one-quarter of respondents supplied additional comments to say that weather effects are already becoming apparent.

Furthermore, a 62% majority of all respondents believe that energy security issues will have a medium or serious impact on the energy fuel mix for light passenger cars, and more than one-quarter (including many from Academic research groups) believe this may cause changes to our fuel usage before 2030. Many respondents also provided comments which suggested that while energy security and environmental impacts were both very important, the energy fuel mix would ultimately be determined by price and financial costs to consumers, i.e., the cheaper fuels would always remain the most popular.

The second group of questions focused more on the type of passenger vehicle fuel system changes that are available to Australians if they were to try and mitigate the potential fuel security or environmental issues from continued dependence on imported fossil fuels. The overwhelming feedback from a large (83%) majority of all respondents was that if there are to be changes to the energy fuel mix in Australia, then the most likely replacement for conventionally fueled passenger vehicles will be rechargeable electric vehicles, with uptake improving as battery technology improves. Additionally, about one-quarter of respondents also stated support for renewable liquid biofuels and for compressed natural gas (CNG) and liquefied petroleum gas (LPG) fuels due to the easy conversion from existing vehicles.

With respect to the timing of these changes, a 60% majority of all respondents believe that the current conventional fuel mix will continue between now and 2020, with slowly increasing uptake of electric vehicles and hybrid passenger vehicles ahead of other alternative fuel sources. After that, a 64% majority believe that the fuel mix for light passenger vehicles will change further between 2020 and 2030, and they believe this will involve a gradual replacement of conventional Internal Combustion Engine (ICE) vehicles with electric vehicles and hybrids. Many Academic respondents also believe this period will see an increased use of biofuels. Beyond this, in the period from 2030 to 2050, a 53% majority of respondents believe that further change will involve a significant replacement of conventional ICE vehicles with electric vehicles and hybrids. Additionally, 33% of respondents, including many from Government, also suggested that although EV uptake during the period to 2050 may increase exponentially, it may then be supplanted by hydrogen fuel cell development.

With respect to current R&D investment, survey results indicate that more experienced workers (in particular many from Government) believe that current government funding for renewable fuel development should be shared with a range of new fuel technologies, while less experienced workers believe that government funding should be focused and specifically shared with electric vehicles and recharging network development. Furthermore, with respect to other government investment into the R&D of renewable fuels, more experienced respondents (particularly those from Academia) believe that government investment should be shared between renewable liquid fuels, plus both hydrogen and EV infrastructure, while less experienced people (including many from Industry) believe the investment should be focused and shared with the development of EV infrastructure ahead of other renewable fuels.

About one-third of all respondents (including many from Government groups) provided comment to state that R&D funding should be shared across all potential fuel technologies and government/industry should avoid "picking winners" before the research shows which fuel or energy source is the best solution to meet future needs. But an equal amount also suggested that investments in hydrogen fuel cell development were largely wasted as it has yielded little progress compared to other fuel alternatives.

Finally, the survey contained several questions about the potential for EV uptake, including a definition of the primary obstacles for EV uptake from industry leaders; whether the potential for Vehicle-to-Grid (V2G) operations would assist or delay EV uptake; how fast a transition to widespread EV use may take; whether converting or "retrofitting" conventional vehicles as EVs was a useful way to recycle standard ICE cars; and whether EV production could possibly replace the planned downturn in Australian vehicle manufacturing after all domestic passenger car manufacturing is closed in 2017.

In answer to this, survey results showed that less experienced workers (including many from Government) find the main obstacle to EV ownership is a high purchase price compared to conventional ICE vehicles, while more experienced people believe the main downside to EVs are their battery recharge limitations, and a large minority of all groups (particularly those from Industry and Academic groups) also find EVs are limited by their range of travel per charge. With respect to V2G impacts, less experienced people believe quite strongly (45%) that the interaction of EV technology with power grid supplies is a positive benefit for widespread use of this technology, while more experienced people (including many from Academia) believe the adoption of EVs should be more gradual until more regulations are in place to reduce EV impacts on power grids. Additionally, about 33% of respondents expressed skepticism as to whether the V2G feature of EVs would ever ultimately be utilized by either grid consumers or electricity suppliers due to variable impact this may have on grid balancing.

With respect to EV uptake, as EV prices and battery limitations become more competitive, a clear difference was shown between more experienced workers who believe that the transition from conventional ICE vehicles to EVs will be much slower than less experienced people, who believe that EV uptake in the Australian car market will be two to five times faster than their older colleagues. About one-third of respondents were quite specific, with some indicating that while 2% or better uptake by 2020 may be highly optimistic, a 10% uptake by 2030 and 50% or better by 2050 were more likely.

Concluding the survey, with respect to local replacement or retrofitting of conventional passenger vehicles by EV technology, 53% of respondents clearly believe there is no market for conventional ICE vehicle conversions to EVs, although less experienced people including some Academics and women in particular believe there may be some interest if it were possible at a low cost, such as $5000–10,000. Additionally, about 30% of respondents elaborated on various reasons as to why retrofitting was a bad idea and that it was likely to remain an uneconomic hobby of backyard enthusiasts. However, when asked whether the proposed closure of conventional car production in Australia should be resurrected via a rapid retooling for local EV production, this met an affirmative response from 57% of all age groups, particularly those from Industry and Academic groups, but only if it is strongly supported by government and if existing issues in local car manufacturing, such as uncompetitive wage costs and low productivity could be adequately addressed from the outset.

11.4 Discussion

All survey participants provided significant input about the current mix of fuel types being used by the Australian passenger car industry, particularly given the growing impacts on energy security and environmental issues, and whether there was any urgency for change in the near future, i.e. between now and periods to 2050. They also provided insightful comments on preferred alternative fuel options and potential obstacles to uptake or investment in these. This is indicated by the many questions which achieved clear overall majorities for one or two preferred options, plus others which strongly favored one or more subgroups.

Due to the large weight of support for many of overall group findings above, the impact of potential bias or error in these results seems unlikely. Similarly, although the sample size of some groups within the entire respondent class was statistically small, this is irrelevant if only the overall results of the entire group of 47 respondents are considered. A sizeable majority result from a group of that size is a statistically valid result, even if smaller internal group trends are less so. For this reason, the key findings of the survey, as listed above, should be considered as the informed perspective of passenger vehicle industry leaders, government regulators, and field researchers who have an ideal overview of potential industry problems, solutions, and development, and the views expressed are potentially critical to future industry and government discussions regarding the mitigation of transport industry-related energy security and greenhouse gas emissions in this country.

From a research perspective, the results and trends identified in this survey will be used to model likely and necessary future fuel mix scenarios for passenger vehicles in Australia. The planned modeling work will also enable the identification of potential impacts from these scenarios on economic, energy security, and environmental issues in Australia.

11.5 Conclusion

The key findings from this survey may be listed as follows:

1. A large (>72%) majority of informed specialists in the passenger vehicle industry, irrespective of age, experience, or gender, are greatly concerned that the continued reliance of Australia transport on imported fossil fuels will result in an unnecessary threat to either national energy security or environmental conditions, or both, prior to 2030. For this reason, there is clear support for the Federal Government in conjunction with industry leaders to develop a plan with some urgency (e.g., prior to 2020) which will enable Australia to transition away from this unhealthy import dependence using some of the viable and abundant domestic fuel alternatives available.
2. There is considerable agreement (83%) across the industry that the most suitable replacement option for conventional ICE vehicles will be rechargeable electric vehicles and hybrids, and uptake of these will improve alongside battery technology and recharging infrastructure, and as vehicle usage costs are reduced. The increased introduction of other alternative fuel systems, such as CNG or LPG conversions, are also expected to contribute to this transition away from crude oil or petroleum dependence, but their potential impact on the issue is considered to be less advanced.
3. A large majority (64%) of specialists expect the initial 5–10% transition of alternatively fueled vehicles will occur between 2020 and 2030, with rapidly increasing uptake of EVs continuing beyond 2030. Many specialists suggest this

transition may continue onto hydrogen fuel cell vehicles as that technology advances in development toward 2050.

4. Given this marked support for EVs in Australia, it would seem natural that increased efforts are made by government and industry to encourage the rapid development of a strategy which effectively catalyzes the transition to higher EV usage and minimizes the continued impacts of the existing problem with imported petroleum usage. This should be reflected in government policy-making, research and development investment decisions, plus transport and electric network infrastructure developments.

Given the considerable knowledge of respondents to this survey, the results listed above should be used to assist future planning that arrests the downward spiral of Australia's dependence on foreign imported oil and petroleum products as domestic supplies and fuel refining capacities become diminished. In turn, it is hoped that this may lead to a transition away from conventional fossil fuel use toward a suite of viable domestic fuel alternatives that would repair growing deficiencies in energy security, such as uncontrollable volatility in fuel control and supply, and to mitigate the potential for continued environmental damage due to increased atmospheric pollution and carbon emissions.

References

CIA. (2015). *The world factbook*. Central Intelligence Agency, USA, 2015. Online at: https://www.cia.gov/library/publications/the-world-factbook/geos/as.html

Commonwealth of Australia. (2013). *Report on Australia's oil refinery industry*. House of Representatives Standing Committee on Economics, Parliament of the Commonwealth of Australia, January 2013. Online at: http://www.aph.gov.au/parliamentary_business/committees/house_of_representatives_committees?url=economics/oilrefineries/report.htm

Commonwealth of Australia. (2015). *Australia's transport energy resilience and sustainability*. Senate Standing Committee on Rural and Regional Affairs and Transport, Department of the Senate, Canberra, Australia, 2015. Online at: http://www.aph.gov.au/Parliamentary_Business/Committees/Senate/Rural_and_Regional_Affairs_and_Transport/Transport_energy_resilience

Chapter 12
Classification of Solar Domestic Hot Water Systems

José Luis Duomarco

12.1 Introduction

For commercial reasons it is becoming more and more necessary to have an effective way of classifying solar domestic hot water (SDHW) systems. Customers need a quick advice before shopping, and impartial information printed on a label may help.

The qualification test procedures are well established and are of the pass-fail type (ISO 9806-2 1995). Instead, energy performance information adapts easily to a numeric or eventually to a color scale. At least two simulation procedures have given the raw material for such a scale. The Solar Energy Laboratory of the University of Wisconsin, UW-SEL, has developed a modular method, where each module is tested separately. The parameters obtained are then fed into especial software where annual performance is calculated. The f-chart and TRNSYS work on these trends (Duffie and Beckman 1991). The solar Collector and System Testing Group, CSTG, has followed a track quite different because no reference is made to component parameters. The system is characterized from the start as a black box with input-output parameters that are determined by an all-day test (ISO 9459-2 1995).

Until now, the main quantity used to evaluate the energetic performances of SDHW systems has been the solar fraction, defined as the percentage of the total load supplied by the sun. Solar fraction values specific to the collector area are also used (FSEC 2002; INMETRO 2012). But this information should not be enough for a customer that wants to use a constant daily volume of hot water, V_D, at a specified temperature, T_D, all through the year. The auxiliary costly energy requirement is very important and should be a part of the energy performance direct information.

J.L. Duomarco (✉)
A.I.U., Montevideo, Uruguay
e-mail: jlduomarco@gmail.com

A. Sayigh (ed.), *Transition Towards 100% Renewable Energy*,
Innovative Renewable Energy, https://doi.org/10.1007/978-3-319-69844-1_12

Two indexes, the energy factor (EF) and the figure of merit (FOM), are adequate. The EF is a metric used extensively by the US Department of Energy to compare the energy conversion efficiency of residential appliances and equipment. The EF, as applied to water heaters, indicates the amount of hot water produced per unit of fuel consumed over a typical day. For a more precise definition, when applied to SDHW systems, we suggest the EF to be the rate of the volume of hot water produced, to the auxiliary heat used in 1 year (L/kWh), calculated when the only daily evening time extraction is 2/3 of the storage volume ($2V_S/3$) and the temperature is $T_{DN} = 60$ °C. The FOM is a quantity used to characterize the performance of a device, system, or method, relative to its alternatives. In our case we define the FOM as the ratio of the EF to a theoretical reference yield (TRY); in turn ratio of the total annual volume of hot water produced, at $T_{DN} = 60$ °C, to its content in sensible heat (L/kWh), taken as the basic alternative. The sensible heat is calculated referred to the mains water temperature. The higher EF and FOM values, the better, as more hot water is produced with the unit of auxiliary energy.

The $T_{DN} = 60$ °C reference temperature has been chosen in behalf of domestic uses. The $V_{DN} = 2V_S/3$ reference volume has been arbitrarily selected with the only restriction of being less than the storage volume V_S. It is shown in Fig. 12.4 that daily extractions greater or equal than V_S tend to fade out differences between the y values of different collectors, due to a major cold-water intrusion in the tank.

12.2 Method

Two different simulation procedures are used in the evaluation of the EF and the FOM: (a) ISO simulation in 12.2.2 (Bourgues et al. 1991; Carvalho and Naron 2000; Duomarco 2015) and (b) TRNSYS simulations in 12.2.3 (TRNSYS 2000).

12.2.1 Equations Defining the EF and the FOM

- The ratio of the total annual volume of hot water produced, at $T_{DN} = 60$ °C, to the total sun-assisted auxiliary energy, and its inverse, is written as

$$v_{AUX} = \frac{365\, V_D}{\sum_{i=1}^{365} Q_{AUX,\, i}} = \frac{1}{q_{AUX}} \tag{12.1}$$

- The ratio of the total annual volume of hot water produced, at $T_{DN} = 60°$, to its content in sensible heat, is written as

$$v_{\mathrm{DN}} = \frac{365\ V_{\mathrm{D}}}{\sum_{i=1}^{365} Q_{\mathrm{DN},\ i}} = \frac{365\ V_{\mathrm{D}}}{\sum_{i=1}^{365} V_{\mathrm{D}} c_{p\omega} \rho_{\omega} (T_{\mathrm{DN}} - T_{\mathrm{MAIN},\ i})} = \frac{1}{q_{\mathrm{DN}}} = \mathrm{TRY} \tag{12.2}$$

- The ratios x and y

$$x = \frac{V_{\mathrm{D}}}{V_{\mathrm{S}}} \tag{12.3}$$

$$y = \frac{q_{\mathrm{AUX}}}{q_{\mathrm{DN}}} = \frac{v_{\mathrm{DN}}}{v_{\mathrm{AUX}}} = \frac{\sum_{i=1}^{365} Q_{\mathrm{AUX},\ i}}{\sum_{i=1}^{365} Q_{\mathrm{DN},\ i}} \tag{12.4}$$

- The energy factor

$$\mathrm{EF} = v_{\mathrm{AUX}} \qquad \text{when} \quad V_{\mathrm{D}} = V_{\mathrm{DN}} = \frac{2V_{\mathrm{S}}}{3} \tag{12.5}$$

- The figure of merit

$$\mathrm{FOM} = \frac{v_{\mathrm{AUX}}}{v_{\mathrm{DN}}} = \frac{\mathrm{EF}}{\mathrm{TRY}} = \frac{1}{y} \qquad \text{when} \quad V_{\mathrm{D}} = V_{\mathrm{DN}} = \frac{2V_{\mathrm{S}}}{3} \tag{12.6}$$

12.2.2 The ISO Simulation

A detailed description of the test method can be found in the text of the standard ISO 9459.2. The system characterization is obtained by the fulfillment of four steps:

- Input-output diagram
- Draw-off temperature profiles
- Store overnight heat loss coefficient
- Long-term performance prediction (LTPP)

The calculation details of LTPP include two load patterns:

- Load pattern 1: determined by the volume of daily hot water consumption.
- Load pattern 2: determined by a minimum useful temperature limit for the hot water consumption. When the outlet temperature is lower than this minimum value, no water is extracted.

The new ISO software has been developed with a different load pattern:

- Load pattern 3: modeled for a nominal temperature and a nominal daily hot water volume production, both constant during the year, with only one daily draw-off. According to climate conditions and hot water demand, the draw-off temperature and draw-off volume may be under or above the nominal values. If overheating is present, the excess energy over the nominal is calculated and discarded. If solar heating is under the nominal value, the auxiliary heat necessary to reach nominal settings is calculated. The calculation is extended only to 365 days assuming an annual periodicity.
- Case study systems: the performance of three different systems, System 5, System 8, and Baxiroca, were calculated and compared. The typical data of System 5 and System 8 were obtained from work at the Solar Collectors and Systems Laboratory from the National Research Centre Demokritos (Athens, Greece) (Belessiotis and Harambopoulos 1993) and similar data for Baxiroca150, from a test report emitted by the "Escuela Superior de Ingenieros" (Seville, Spain) (LCS 2009).The values are listed in Table 12.1; the draw-off profiles and the input-output plots are shown in Fig. 12.1. The daily climate data for Belo Horizonte, Salto, Montevideo, Boston, Edinburgh, and Punta Arenas were obtained from project "Surface Meteorology and Solar Energy" (NASA 2010). Simultaneous daily global solar radiation on horizontal surface, maximum and minimum temperature series for year 2010, were used in the simulation.

12.2.3 The TRNSYS Simulation

A TRNSYS model (TRNSYS 2000) has been built around component Type 45a. It requires 39 parameters, 10 inputs, and produces 12 outputs. This component models the thermosiphon solar collector system. The system consists of a flat-plate solar collector, a stratified storage tank (either vertical or horizontal cylinder) located physically above the collector plate, a check valve to prevent reverse flow, and water as the working fluid. The tank's stratification is modeled with the plug-flow concept.

Global horizontal hourly solar radiation and simultaneous hourly ambient temperature for 1 year are needed. Two such databases, from Montevideo (Ewenson 1979) and Salto (LES 2015), are used.

The daily routine consists in evaluating the storage sensible heat at 19 h (after a day's sun recollection), making the only daily constant volume draw-off at 20 h (after fixing its temperature by adding auxiliary heat or discarding overheat) and calculating the heat remains in the storage tank for the next day.

Fig. 12.2a shows hourly energy variations on May 2 in the city of Montevideo — when auxiliary heat is necessary—; Fig. 12.2b similarly on January 5 in the city of Salto—when overheat is present—; Fig. 12.3 shows the TRNSYS model layout; and Table 12.2 lists some parameters, inputs and outputs, with their definitions.

Table 12.1 Experimental results (Belessiotis and Harambopoulos 1993; LCS 2009)

System name	a1 (m^2)	a2 (MJ/K/d)	a3 (MJ/d)	Ac (m^2)	Vs_{tk} (L)	Is_{tk} (length) (m)	Ds_{tk} (diameter) (m)
System 5	1.58	0.45	−1.37	3.41	200	1.67	0.39
System 8	0.96	0.47	−1.37	2.05	160	1.34	0.39
Baxiroca	1	0.27	−0.61	1.92	150	1.25	0.39
System name	**δo = a1/Ac**	**δ1 = a2/Ac ($MJ/K/m^2/d$)**	**δ1 = a2/Ac ($W/K/m^2$)**	**a3/a2 (K)**	**As_{tk} (m^2)**	**Us_{tk} (W/K)**	**U_{SCtk} (W/K)**
System 5	0.4633	0.1320	1.5274	−3.04	2.2902	2.6	2.93
System 8	0.4683	0.2293	2.6536	−2.91	1.8799	1.81	1.87
Baxiroca	0.5208	0.1406	1.6276	−2.25	1.7773	3.23	–
System name	**Number of collectors**	**Storage type**	**Heat exchange**	**Operation type**	**V_{DN} = 2/3Vs (L/d)**	**T_{DN} (°C)**	**Yield in MVD (L/kWh)**
System 5	2	Horizontal	Tube	Thermosiphon	133	60	56.3
System 8	1	Horizontal	Double wall	Thermosiphon	107	60	33.6
Baxiroca	1	Horizontal	Tube	Thermosiphon	100	60	51.1

ISO 9459–2 – Eq. 12.2 – a_1,a_2,a_3 – determined from test results by least-squares fitting methods, Q_T net solar energy gained by the storage tank in the day, T_a ambient air temperature, T_S cold water supply temperature, A_C collector's aperture area, A_S store-tank's surface, V_S store-tank's volume, l_S store-tank's length, D_S store's tank diameter $Q_{T_i} = a_1 G_T + a_2(T_a - T_S) + a_3 = G_T A_C(\delta_0 - \delta_1 z)$

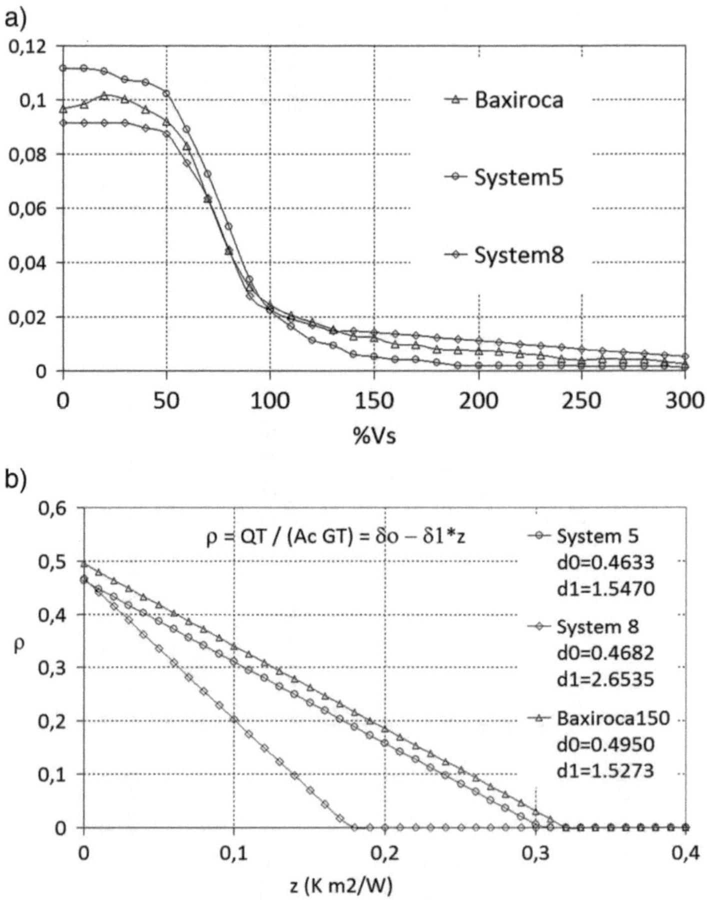

Fig. 12.1 Normalized draw-off temperature profiles (**a**) and input-output diagram (**b**) for Baxiroca, System 5, and *System 8*

12.3 Results

In Fig. 12.4, eight graphs show the index y as function of x.

The reference systems, Baxiroca in Fig. 12.4a, System 5 in Fig. 12.4b, and System 8 in Fig. 12.4c, are evaluated in the six reference cities, Belo Horizonte, Salto, Montevideo, Boston, Edinburgh, and Punta Arenas.

The reference systems, Baxiroca, System 5 and System 8, are evaluated jointly, with ISO simulation, in Boston Fig. 12.4d, Salto Fig.12.4e, Montevideo Fig.12.4f, and with TRNSYS simulation, in Montevideo Fig. 12.4g and Salto Fig. 12.4h.

For small values of x, cold-climate cities begin with high values of y, around 60 %, while mild- and warm-climate cities begin with y values between 5 % and

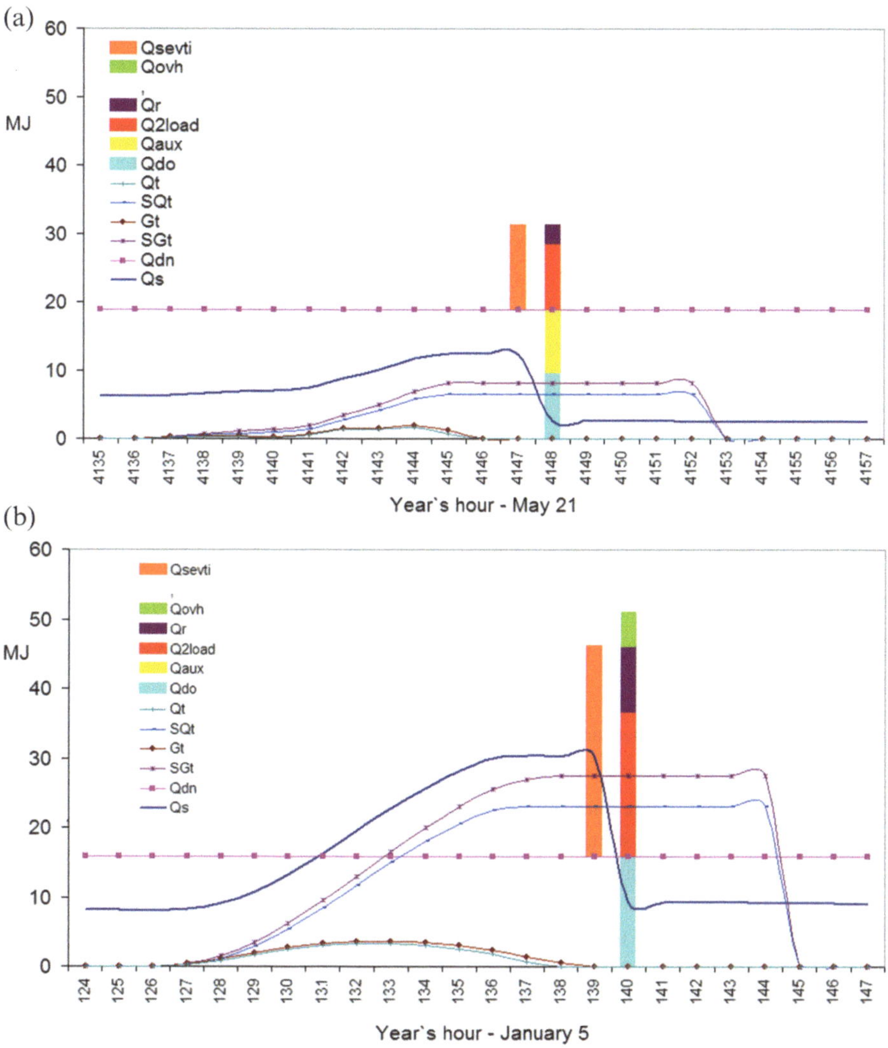

Fig. 12.2 TRNSYS simulation hourly energy variations in Salto on: (**a**) May21, with auxiliary heat and (**b**) January 5, with overheat

20 %. A null y value stands for no auxiliary heat needed. When draw-off volume V_D exceeds the storage tank volume V_S ($V_D > V_S$), all y functions asymptotically get near 100 % and q_{AUX} gets near q_{DN} from lower values.

When systems are evaluated in the same place, for V_D less than V_S ($V_D < V_S$), *System 5 leads Baxiroca and Baxiroca leads System 8*, in the sense that they need less auxiliary energy to heat up 1 L of water. In the trend of classifying solar

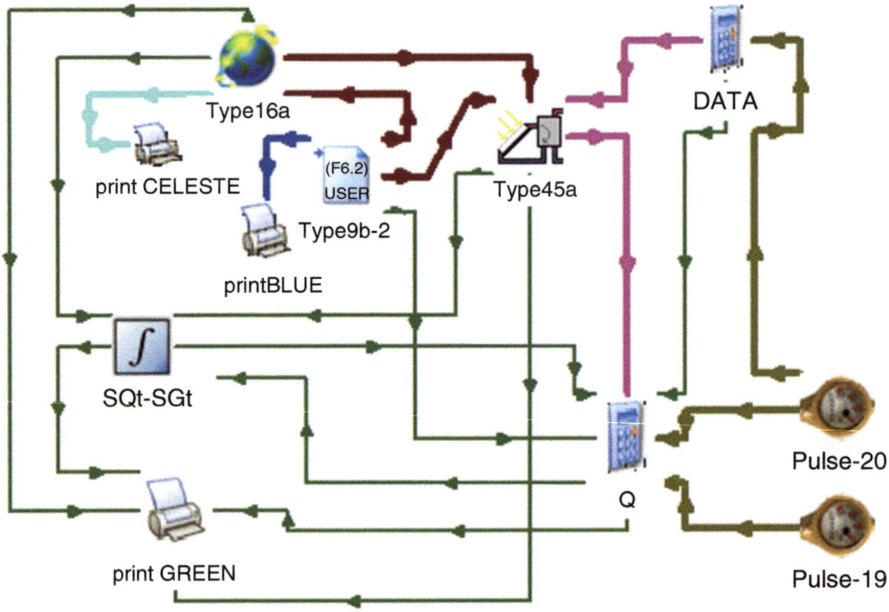

Fig. 12.3 TRNSYS model built around a thermosiphon collector-storage subsystem (Type 45a)

Table 12.2 Energy daily transactions in TRNSYS simulation

Hourly events	
Q_{DN}	Daily energy reference settings $Q_{DN.\ i} = V_D.\ 4.1868.\ (T_{DN}\text{-}T_{main,i})$ – kJ
Q_S	Sensible heat in storage tank – kJ
Q_T	The rate of energy transfer from the heat source to the storage tank – kJ/h
SQ_T	$\int Q_T\, dt$ accumulated daily energy transfer from the heat source to the storage tank – kJ
G_T	Radiation on the tilted surface (beam + sky diffuse + ground reflected diffuse) – kJ/h m^2
SG_T	$\int G_T\, dt$ accumulated daily radiation on the tilted surface – kJ/d m^2
Evening time events, at 19 h and 20 h	
Q_{SEVT}	Sensible heat in storage at evening time at 19 h – kJ
Q_{2LOAD}	Total heat to load at 20 h – kJ
Q_{OVH}	Overheat discarded at 20 h – kJ
Q_{AUX}	Auxiliary heat to load at 20 h – kJ
Q_{DO}	$(Q_{2LOAD} - Q_{OVH})$ energy removed from the tank to supply the load at 20 h – kJ
Q_R	$(SQ_{SEVT} - SQ_{2LOAD})$ remains of sensible heat in storage, after draw-off at 20 h – kJ

collector systems with only one number, we selected arbitrarily a constant daily extraction volume $V_{DN} = 2V_S/3$. In Figs. 12.5 and 12.6 and Table 12.3, EF and FOM values are calculated and plotted.

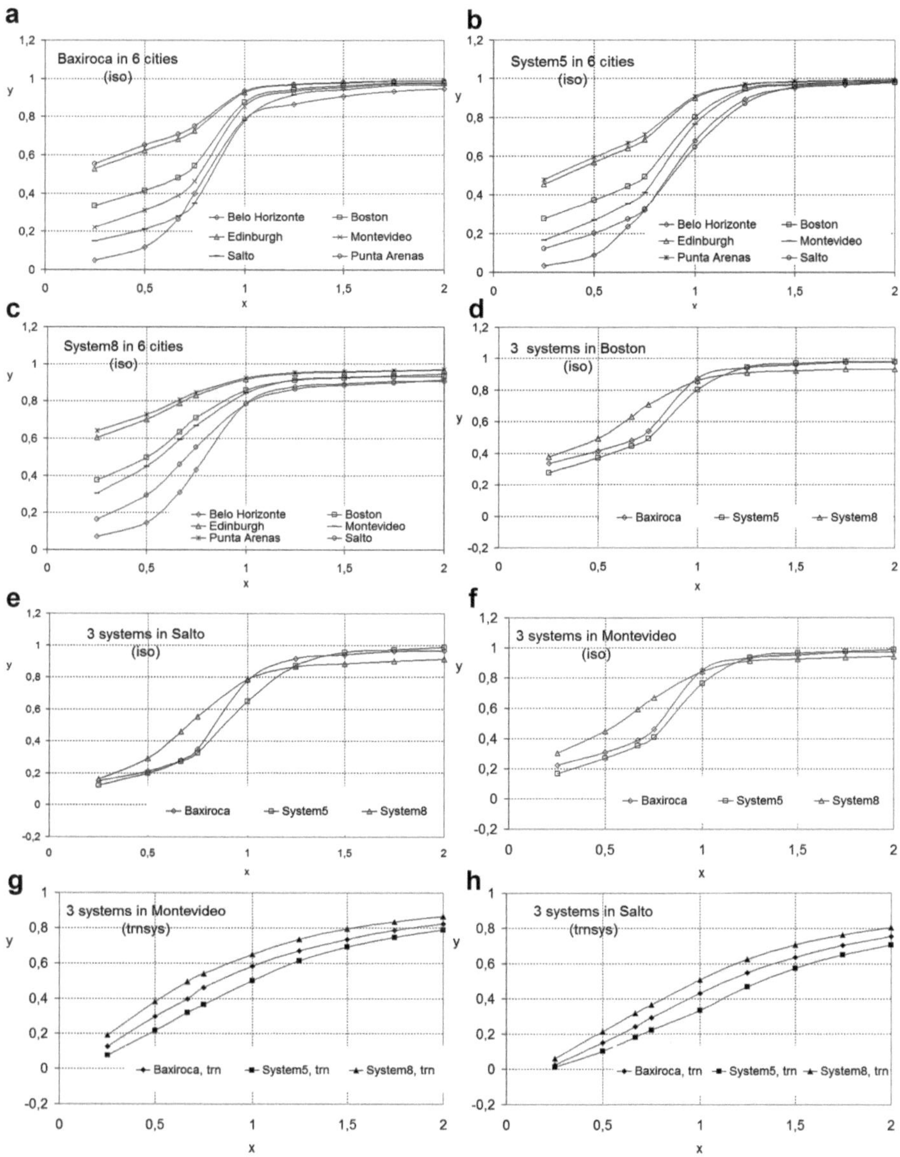

Fig. 12.4 Ratio y = qaux/qdn as function of ratio x = Vd/Vs, for System 5, System 8, and Baxiroca, using ISO and TRNSYS simulation procedures, in Belo Horizonte, Salto, Montevideo, Boston, Edinburgh, and Punta Arenas

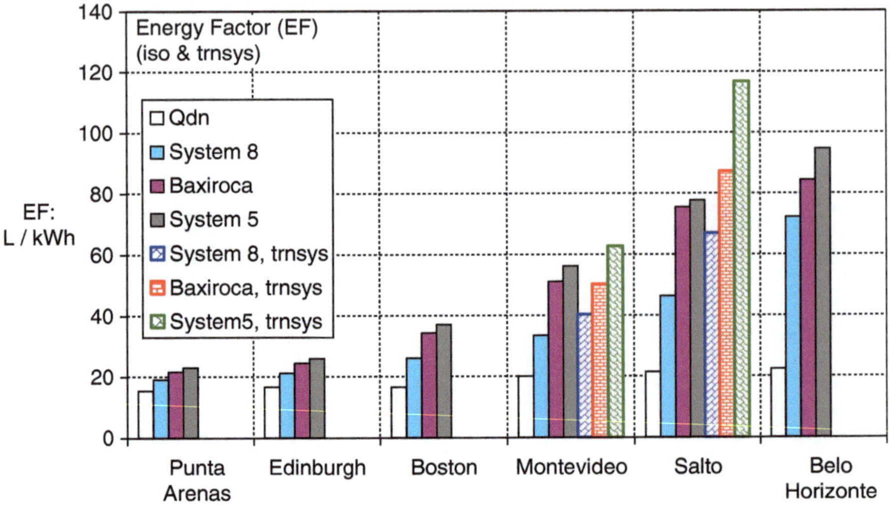

Fig. 12.5 Theoretical reference yield (TRY) and energy factor (EF) for six different cities, with ISO and TRNSYS simulations

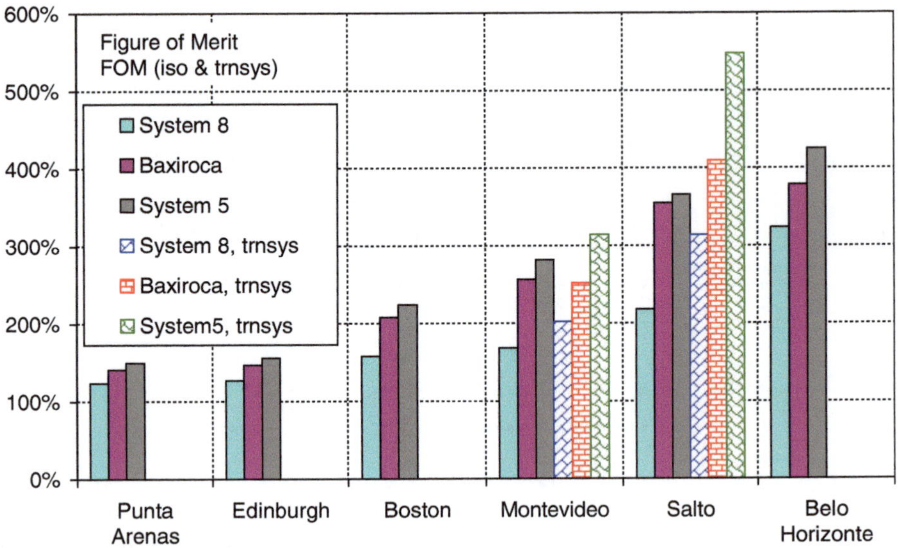

Fig. 12.6 Figure of merit FOM for six different cities with ISO and TRNSYS procedures

Table 12.3 Calculation of the EF, FOM, and q for System 5, System 8, and Baxiroca by the ISO simulation procedure in Punta Arenas, Edinburgh, Boston, Montevideo, Salto, and Belo Horizonte and by the TRNSYS simulation procedure in Montevideo and Salto

$V_{DN} = 2Vs/3$	System 5			System 8			Baxiroca		
$T_{DN} = 60$ °C	$V_{DN} = 133.3..$ L/d			$V_{DN} = 106.6...$ L/d			$V_{DN} = 100$ L/d		
	EF	q_5		EF	q_8		EF	q_{bax}	
City	L/ kWh	kWh / 100 L	FOM %	L/ kWh	kWh/ 100 L	FOM %	L/ kWh	kWh/ 100 L	FOM %
ISO									
Punta Arenas	22.9	4.354	150	19.0	5.257	124	21.6	4.616	141
Edinburgh	26.0	3.843	156	21.1	4.721	127	24.4	4.093	147
Boston	37.0	2.698	224	26.1	3.828	158	34.3	2.910	208
Montevideo	56.2	1.777	282	33.5	2.979	168	51.1	1.955	256
Salto	77.7	1.285	365	46.3	2.155	218	75.5	1.322	355
Belo Horizonte	94.6	1.056	425	72.1	1.386	324	84.4	1.184	379
TRNSYS									
Montevideo	62.7	1.595	314	40.3	2.477	202	50.3	1.989	251
Salto	116.6	0.857	547	66.9	1.495	314	87.2	1.146	409

12.4 Site-Independent Figure of Merit

From what has been said before, both EF and FOM depend on location. This hinders a universal classification of solar systems based on these indexes. It's desirable to classify SDHW systems in an intrinsic way, only dependent on their characteristics and not on their locations. Commercial confusion is likely to happen when equal indexes may address different SDHW systems measured in different places not properly recorded. An alternative classification procedure has been studied, based on a nondimensional number γ_O that gets together all three loss mechanisms:

$$\gamma_0 = 100^*\log_{10}\left[\frac{\delta_0^2}{2\delta_1\lambda_S}\left(\frac{G_0}{\Delta T_0}\right)^2\right] = 100^*\log_{10}\left[\frac{a_1^2 A_S}{2a_2 A_C U_S}\left(\frac{G_0}{\Delta T_0}\right)^2\right] \qquad (12.7)$$

where

1. $\lambda_S = U_S/A_S$ with units (W/m^{2}K) characterizes the heat loss of the storage tank.
2. $\delta_1 = a_2/A_C$ with units (W/m^{2}K) characterizes the heat loss of the collector.
3. $\delta_0 = a_1/A_C$ measures the collector's optical efficiency.
4. $\delta_0^2/(2\delta_1\lambda_S)$ gets together losses and is proportional to the surface under the input-output diagrams, Fig. 12.1b.
5. $(G_0/\Delta T_0)^2$with Go = 1367 W/m^2 and $\Delta T = 100$ K is a theory-related constant with dimensions.
6. Logarithm is taken to smooth out big variations.
7. Calculation of our reference systems is shown in Table 12.4.

Table 12.4 Site-independent nondimensional figure of merit γ_0

System 5	System 8	Baxiroca
106	90	91

12.5 Conclusions

The energy factor EF (L/kWh) depends only on end points in the line, hot water production (L), and auxiliary energy to be paid for (kWh). Free solar energy acts indirectly as a means to improve the EF. The figure of merit FOM is a nondimensional quantity used to characterize the SDHW system relative to its basic without sun's boosting, alternative.

The simulation program must be specified. Two such simulation programs have been used. TRNSYS models require numerous modules connected with plenty of parameters, acting as input and output data, between them. Sometimes it is cumbersome to set such a simulation. On the other hand, ISO has a black box layout, needs fewer experimental results, and is easier to use.

EF as FOM are site-dependent and so meaningless if measuring place is not reported. From an international commerce point of view, it would be helpful to classify SDWH systems in a site-independent way. Three independent loss mechanisms may always be identified: optical efficiency, day collector-thermal-loss, and night storage-thermal-loss. Reducing these three thermal resistances results in an increased nondimensional γ_0 number, with the additional advantage of being site-independent.

References

Belessiotis, V., & Harambopoulos, D. (1993). Testing solar water heating systems in Athens, Greece. *Solar Energy, 50*(2), 167–177.

Bourgues, B., Rabl, A., Carvalho, M. J., & Collares-Pereira, M. (1991). Accuracy of the European solar water heater test procedure. Part 1 and Part 2. *Solar Energy, 47*(1), 1–25.

Carvalho, M. J., & Naron, D. J. (2000). Comparison of test methods for evaluation of thermal performance of preheat and solar-only factory made systems. *Solar Energy, 69*(1–6), 145–156.

Duffie, & Beckman. (1991). *Solar engineering of thermal process*. New York: Wiley.

Duomarco, J. L. (2015). Figure of merit for solar domestic hot water systems. *Solar Energy, 111*, 151–156.

Ewenson, I.W. de, Facultad de Ingeniería, UdelaR. (1979). Serie horaria de radiación global de Montevideo, Uruguay.

Florida Solar Energy Center (FSEC). (2002). Operation of the Collector Certification Program. FSEC-GP-6-80-Florida.

INMETRO. (2012). Regulamento da qualidade para equipamento de aquecimento solar de agua. Brasil.

ISO 9459-2. (1995). Solar heating. Domestic water heating systems- part 2: Outdoor test methods for system performance prediction of solar-only systems.

ISO 9806-2. (1995). Test methods for solar collectors – part 2: Qualification test procedures.

Laboratorio de Captadores Solares (LCS), Escuela Superior de Ingenieros. (2009). Test Report, Baxiroca STS 150. Seville, Spain.
Laboratorio de Energía Solar (LES), UdelaR. (2015). Serie horaria de radiación global de Salto, Uruguay.
NASA. (2010). Surface meteorology and solar energy. http://eosweb.larc.nasa.gov/sse/.
Transient System Simulation Program, TRNSYS. (2000). Version 16.00.0037, Solar Energy Laboratory of the University of Wisconsin–Madison.

Chapter 13
Temperature Difference with Respect to Exposure Time for Black Paint and Galena Powder-Black Paint Composite Selective Surfaces

Iessa Sabbe Moosa and Bashar Bassam Maqableh

13.1 Introduction

In solar water heating systems for domestic use, flat-plate solar collectors can be successfully employed and normally installed facing south on a rooftop at an angle of about 40°. Tracking system in configuration with the collectors is very useful for reaching high temperature in a moderate weather, while this combination is not necessary in hot weather zones. Practical and theoretical details about history of solar energy, collectors, and applications of solar energy have been reported by Kalogirou (2004). The intensity of solar radiation that strikes the Earth at the surface of the Earth's atmosphere is around 1.36 kW/m^2 (Al-Waeli et al. 2015). This source of energy can be exploited in many ways for very long time. Very important details about using of materials for heating system by solar energy have been almost fully reported by Lenel et al. (1984). Chatterjee and Pal have announced that thin film of Galena can be used as low-cost selective absorber (Chatterjee and Pal 1993). Large-scale applications of hot water system in Europe have been reviewed by Fisch et al. (1998).

Workers in the materials science concerned with heating system by solar energy are focusing their attempts to find high absorbitivity and low emission in the solar spectrum with good stability and accessibility at using temperatures (SÜzer et al. 1998; Wazwaz et al. 2002; Bostrom et al. 2008; Shashilaka et al. 2007; Tharamani and Mayanna 2007). Many countries with abundance of sunshine started utilization of solar energy for water and air heating system for domestic applications (Kaldellis et al. 2005; Maxoulis et al. 2007; Kalogirou 2009; Miqdam et al. 2016). Addition of some metals and alloy powders to selective materials has been reported to increase

I.S. Moosa (✉) • B.B. Maqableh
College of Engineering, University of Buraimi, P.O. Box 890, P. C. 512 Al- Buraimi, Oman
e-mail: issa.m@uob.edu.om

A. Sayigh (ed.), *Transition Towards 100% Renewable Energy*,
Innovative Renewable Energy, https://doi.org/10.1007/978-3-319-69844-1_13

the thermal conductivity of these materials (AlShamaileh 2010; Chaichan and Kazem 2015; Moosa 2015, 2016).

The Sultanate of Oman is classified among countries that receive the highest solar energy in the world as that very well reported by Kazem (2011) and Miqdam et al. (2016). Hence, any work of research either theoretical or practical is very essential in the field of solar energy applications. The current work is carried out in Al Buraimi City at the northwest, Sultanate of Oman, where the solar intensity is very high. This city receives the highest sunshine hours/day compared with the other cities of the Sultanate of Oman as reported by Al-Badi et al. (2011).

The aim of this research is to study the effect of exposure time of solar radiation on water temperature and the relationship between temperature difference (ΔT) with exposure time for Galena Powder-Black Paint Composite Selective Surface.

13.2 Experimental Method

The experimental procedure was started by preparing the following stuff and equipments:

- Black paint type Berger paints, Allinone, High Gloss Enamel, product of Oman.
- Thinner Liquid type Supreme, CCIA PV 274070, product of Italy.
- Homemade experiment wooden box with double glassing front side, about 1 m of length, 15 cm of depth, and 20 cm of height. The air gap between glasses is of about 1 cm.
- Aluminum containers size of $15 \times 10 \times 4$ cm with top pipe of about 1.5 cm diameter and length of about 5 cm for temperature measurements.
- Digital thermometers, digital balance, stainless steel mortar with its hammer handle.
- Galena ingot.
- Stainless steel micro-sieves of about 63 μm and 125 μm.

Bulks of Galena with a diameter of about 2.5 cm were obtained from public market in Ajman, UAE. Figure 13.1 shows some of as received Galena ingots.

Powder of Galena was prepared by the following steps:

(i) The lumps of Galena were mechanically crushed to a diameter of about 0.3 mm and less.
(ii) The crushed small pieces were manually pulverized by using the stainless steel mortar for about 1 h.
(iii) The obtained powder was sieved by the stainless steel micro-sieves of about 65 μm, and then by 125 μm prior to mixing process with the black paint. After using the sieve of 125 μm, the collected powder particle size after this stage is within the range greater than 65–125 μm. Figure 13.1 shows some lumps of Galena ingot.

Fig. 13.1 Reveals some lumps of as received Galena ingot

A fracture surface specimen from the received Galena bulks and some produced powder were prepared for SEM tests. Mixing process of the black paint with 10 wt % percentage of prepared Galena powder was done. About 30 g of black paint was weighed by the digital balance and about 5 g of thinner liquid was added and mechanically mixed with the produced powder to be ready for painting process. Three aluminum containers were very well cleaned by using emery paper and thinner liquid; one of them was kept as it is, the second was painted by black paint, while the third was painted by the composite of the black paint with the sieved Galena powder by 125 μm sieve. The prepared containers were filled with water and then fixed in the homemade wooden box prior to water heating by solar energy. The heating process was employed by setting up the experiments proceeding to start at 9:00 am. The initial temperature T_0 was measured by using the digital thermometer, and then the wooden box was exposed to solar radiation for 7 h. The temperatures of the water in all containers as a function of exposure time were measured, while the box was directed manually to be almost perpendicular to the direction of the solar beam.

13.3 Results and Discussion

Solar energy profile of Al Buraimi City is shown in Fig. 13.2, which shows the monthly average of direct radiation, diffused radiation, and global radiation. The maximum value of solar radiation in this city is in June, which is about 9.8 kWh/m^2/day. The microstructure study of Galena ingot and the obtained powder by using the

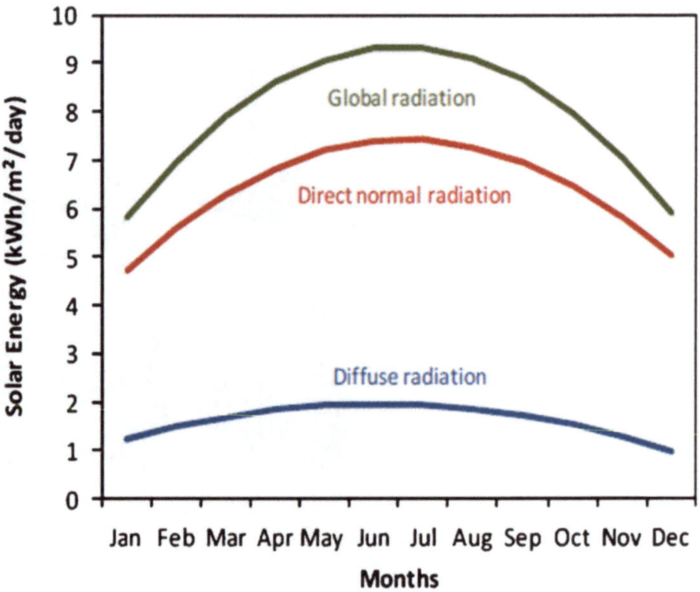

Fig. 13.2 Irradiation of Al Buraimi, Sultanate of Oman of 2012 (Moosa 2015)

SEM exhibited very brittle fracture surface as shown in Fig. 13.3. This brittleness texture is extremely important in the subsequent milling process to produce a powder from this ingot. In addition, the EDAX unit attached with the SEM showed that Galena ingot was found to be almost compound of PbS.

A run of water heating was carried out on December 2, 2015, by using three containers: pure aluminum, black painted aluminum, and the third painted with the black paint composite, which were prepared to recognize the difference between these three cases on the water temperatures. The result of this experiment is given in Fig. 13.4, from which it can be seen that the maximum temperatures of the pure aluminum, the black painted aluminum, and the black paint composite are around 75 °C, 85 °C, and 89 °C. The result is very significant, so that water with high temperature can be obtained by using pure aluminum selective surface and definitely higher temperature with black painted and/or with black paint composite. Another attempt of water heating by solar energy was conducted on June 21, 2016, with starting temperature of about 5 °C. The result of this attempt is shown in Fig. 13.5.

The maximum temperatures were about 88 °C (pure Al), 95 °C (black painted Al), and 99 °C (black paint +10 wt% Galena powder). This data reveals that the maximum water temperature in the case of the composite of Galena powder with black paint is higher than that of black paint by around 4 °C, while this value is about 6.5 °C when the particle size within the range of 60 μm and less (Moosa 2015).

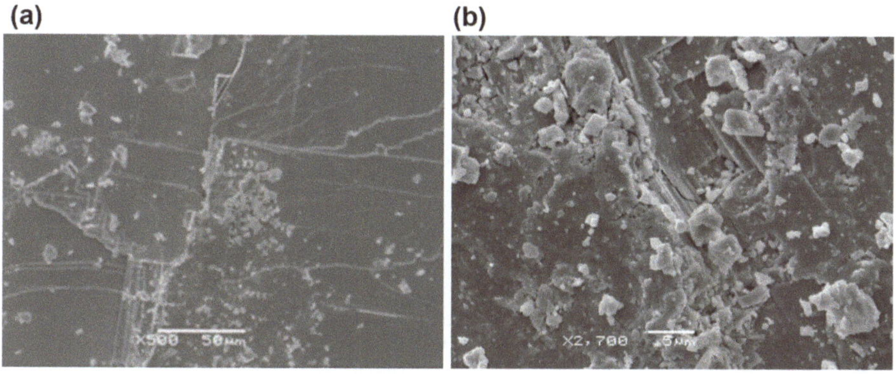

Fig. 13.3 (**a**) Secondary electron image of fracture surface of Galena ingot, 500×, (**b**) Galena particle with magnification of 700×

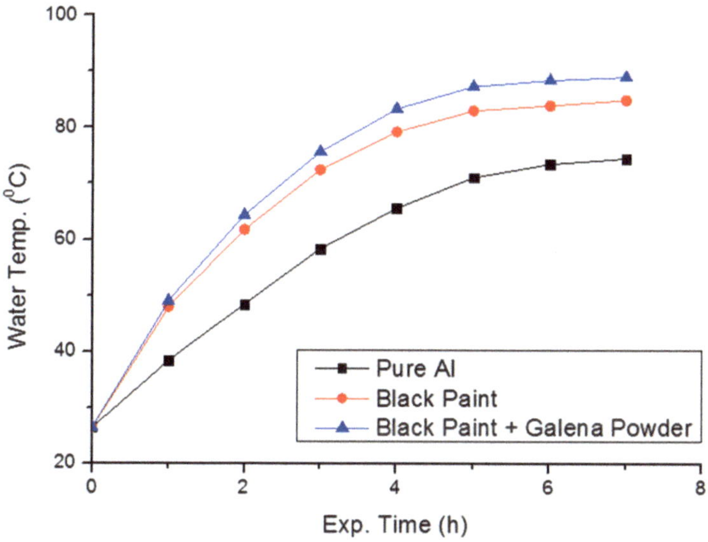

Fig. 13.4 Water temperatures against exposure time for three cases. Date, December 2, 2015; weather, sunny; weather temp., ≈28 °C, winter season in Oman

An attempt to redraw the composite case of June 21, 2016, was conducted by using the MatLab program with curve-fitting procedure for more accuracy as shown in Fig. 13.6. It seems that the relationship between the water temperature and exposure time is almost natural logarithm. Also, temperature difference (ΔT) {$\Delta T = T - T_0$, where T is the water temperature at any time, T_0 is the initial temperature} as a function of exposure time for composite selective surface was plotted as revealed in Fig. 13.7, which is similar to Fig. 13.6 except the starting point.

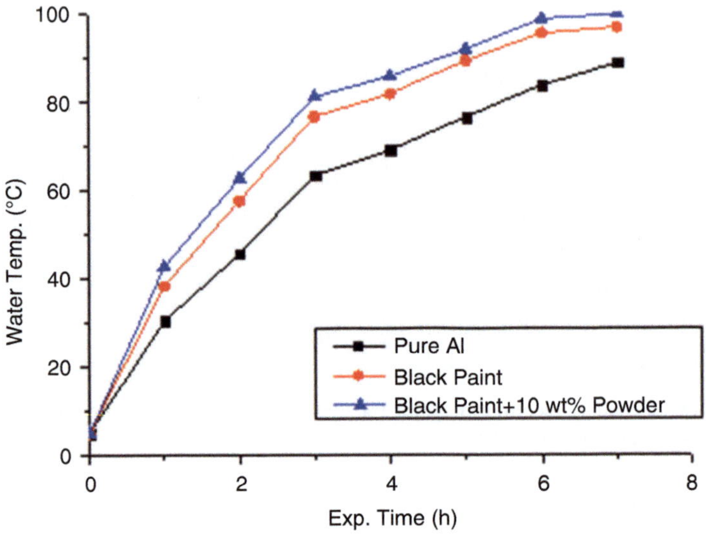

Fig. 13.5 Water temperatures against exposure time for different cases. Date: June 21, 2016; weather, very little partially cloudy; weather temp., ≈43 °C

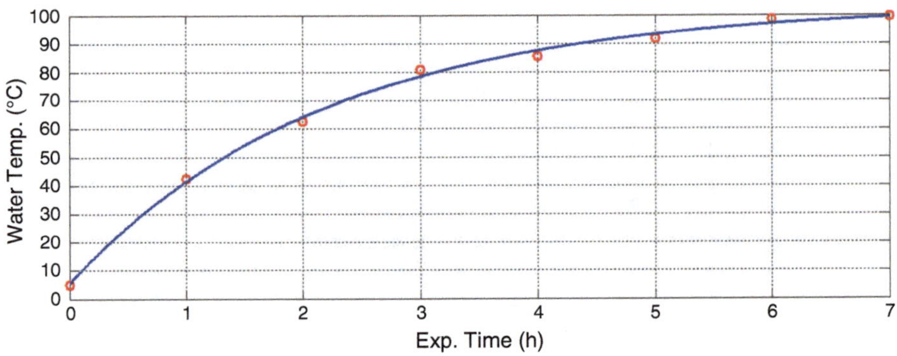

Fig. 13.6 Water temperatures against exposure time of composite case with curve fitting. Date: June 21, 2016; weather, very little partially cloudy; weather temp., ≈43 °C

Depending upon this result and by employing the MatLab program on the data of the curves above, two mathematical equations were conducted as an attempt to speculate the relationships between water temperature and ΔT with the exposure time to solar energy. The feedback of this attempt is given in the following equations:

$$T_{\text{Exp.Time}} = T_{\max}(1 - e^{-\alpha t}) + C \qquad \text{(Water Temp. versus Exp. Time)} \tag{13.1}$$

$$\Delta T_{\text{Exp.Time}} = \Delta T_{\max}(1 - e^{-\alpha t}) \qquad (\Delta T \text{ versus Exp. Time}) \tag{13.2}$$

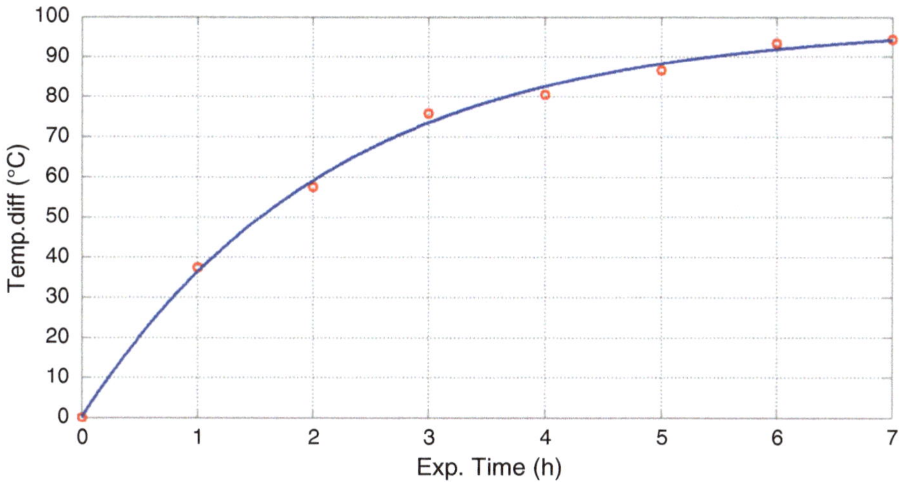

Fig. 13.7 Temp. Diff. ΔT as a function of Exp. Time, June 21, 2016, composite selective surface, weather temp.: ≈43 °C

where C is constant which represents the starting temperature and α is the natural logarithmic coefficient with a unit of h^{-1} in these relations ($T_{max} \approx 98$ °C, $\alpha \approx 0.46\ h^{-1}$, C ≈ 5.3 °C for Eq. 13.1, and $\Delta T_{max} \approx 98$ °C, $\alpha \approx 0.46\ h^{-1}$ for Eq. 13.2). These equations can be used to estimate the water temperature and ΔT as functions of exposure time. This subject needs much more work to study the effect of the weather temperatures, starting temperature, and solar intensity on the values of α.

The maximum values of temperatures obtained in this trial of research are more than enough for domestic use; consequently, black painted aluminum solar heating system is very adequate in Al Buraimi and probably in many cities in the Sultanate of Oman because of the abundance of solar energy or in any place where solar radiation is available with high intensity.

13.4 Conclusion

The most essential conclusion of this work is that even with a pure aluminum selective surface, a temperature of about 75 °C can be obtained, which is more than enough for domestic use. The results also showed that the water temperature could be increased by addition of Galena powder to black painted selective surface. A maximum temperature of about 99 °C can be achieved by the mixture of the black paint with Galena powder within the range of particle size of about greater than 63–125 μm at 10 wt%. Temperature difference as a function of exposure time is almost natural logarithm relation. Furthermore, most of the places in the Sultanate of Oman should use solar heating system for domestic use because of the great intensity of solar energy even during the winter season.

Acknowledgments The authors would like to thank Dr. Ahmed Al Rawas of the Sultan Qaboos University, Sultanate of Oman, Physics Department, for his great help to use the SEM for microstructure and chemical analysis. Also, the authors greatly appreciate Mr. Kalyan Baddipudi of the UoB for his kind support and proofreading.

References

Al-Badi, A. H., Malik, A., & Gastli, A. (2011). Sustainable energy usage in Oman—opportunities and barriers. *Renewable and Sustainable Energy Reviews, 15*, 3780–3788.

AlShamaileh, E. (2010). Testing of a new solar coating for solar water heating applications. *Solar Energy, 84*, 1637–1643.

Al-Waeli, A. H. A., Kamal El-Din, M. M., Al-Kabi, A. H. K., Al-Mamari, A., Kazem, H. A., & Chaichan, M. T. (2015). A photovoltaic application economic study in car parking lights with recycled batteries: A techno- economic study. *Australian Journal of Basic and Applied Sciences, 9*(36), 43–45.

Bostrom, T., Wackelgard, E., & Westin, G. (2008). Solution-chemical derived nickel–alumina coatings for thermal solar absorbers. *Solar Energy, 74*, 497–503.

Chaichan, M. T., & Abass, K. I. (2016). Practical investigation of effectiveness of direct solar-powered air heater. *International Journal of Advanced Engineering, Management and Science (IJAEMS), 2*(7), 1047–1053.

Chaichan, M. T., & Kazem, H. A. (2015). Using aluminium powder with PCM (paraffin wax) to enhance single slope solar water distiller productivity in Baghdad – Iraq winter weathers. *International Journal of Renewable Energy Research, 5*(1), 251–257.

Chaichan, M. T., & Kazem, H. A. (2016). Experimental analysis of solar intensity on photovoltaic in hot and humid weather conditions. *International Journal of Scientific & Engineering Research, 7*(3), 91–96.

Chatterjee, S., & Pal, U. (1993). Low cost solar selective absorbers from Indian Galena. *Optical Engineering, 32*(11), 2923–2929.

Fisch, M. N., Guigas, M., & Dalenbäck, J. O. (1998). A review of large-scale solar heating systems in Europe. *Solar Energy, 63*(6), 355–366.

Kaldellis, J. K., El-Samani, K., & Koronakis, P. (2005). Feasibility analysis of domestic solar water heating systems in Greece. *Renewable Energy, 30*, 659–682.

Kalogirou, S. A. (2004). Solar thermal collectors and applications. *Progress in Energy and Combustion Science, 30*, 231–295.

Kalogirou, S. A. (2009). Thermal performance, economic and environmental life cycle analysis of thermosiphon solar water heaters. *Solar Energy, 83*, 39–48.

Kazem, H. A. (2011). Renewable energy in Oman: Status and future prospects. *Renewable and Sustainable Energy Reviews, 15*, 3465–3469.

Lenel, U. R., & Mlid, P. R. (1984). A review of materials for solar heating systems for domestic hot water. *Solar Energy, 32*(1), 109–120.

Maxoulis, C. N., Charalampous, H. P., & Kalogirou, S. A. (2007). Cyprus solar water heating cluster: A missed opportunity. *Energy Policy, 35*, 3302–3315.

Moosa, I. S. (2015). Effect of Galena powder of 63μm micrometer particle size and less on the absorbitivity of black paint mixture. *International Journal of Advanced Research in Engineering and Technology (IJARET), 6*(4), 60–68.

Moosa, I. S. (2016). Affect of 10% wt Galena powder on the absorbitivity of black paint. *International Journal of Advanced Research in Engineering and Technology (IJARET), 7*(2), 1–8.

Shashilaka, A. R., Sharma, A. K., & Bhandari, D. R. (2007). Solar selective black nickel–cobalt coatings on aluminum alloys. *Solar Energy Materials & Solar Cells, 91*, 629–635.

SÜzer, S., Kadirgan, F., SÖhmen, H. M., Wetherilt, A. J., & TÜre, I. E. (1998). Spectroscopic characterization of Al_2O_3-Ni selective absorbers for solar collectors. *Solar Energy Materials and Solar Cells, 52*, 55–60.

Tharamani, C. N., & Mayanna, S. M. (2007). Low-cost black Cu–Ni alloy coating for solar selective applications. *Solar Energy Materials & Solar Cells, 91*, 664–669.

Wazwaz, A., Salmi, J., Hallak, H., & Bes, R. (2002). Solar thermal performance of a nickel pigmented aluminium oxide selective absorber. *Renewable Energy, 27*, 277–292.

Chapter 14
On-Track, But Off-Target: New Zealand's 90% Renewable Electricity Target and District Council Planning

Claudia Gonnelli, Hong-Key Yoon, Karen Fisher, and Julie MacArthur

14.1 Introduction

In 2007, New Zealand set a target to obtain 90% of its electricity from renewables by 2025 (hereafter referred to as NZ RET). Attaining the target appears, prima facie, readily achievable as renewables have already provided 90% of the total electricity production in the past (MBIE 2015) and the country still has enough natural resources to support extensive renewable electricity generation growth (Bertram and Clover 2009; BusinessNZ Energy Council 2015; Kelly 2011; MBIE 2015). However, the abundance of renewable electricity is also a curse for the country, as it reduces urgency to introduce significant changes in the electricity and policy system. Renewable electricity development is left in the hands of market forces, rather than guided by targeted policies. Therefore, analysts question whether government policies can effectively support this achievement given the lack of regulatory policies favoring renewables (Kelly 2011; REN21 2015:99; Chang et al. 2016).

This paper focuses on district councils as the main ultimate implementer of New Zealand's planning system. Under the premises that "decisions on environmental matters are best made by communities directly affected" (LGNZ 2013:11), New Zealand's resource management system is mostly delegated to local governments. This devolution – "the transfer of authority to make decisions on behalf of society from a higher to a lower level of government" (Kerr et al. 1998) – raises the question of whether local governments have the ability and the capacities to assume

C. Gonnelli (✉) • H.-K. Yoon • K. Fisher
School of Environment, University of Auckland, Auckland, New Zealand
e-mail: cgon517@aucklanduni.ac.nz

J. MacArthur
School of Politics and International Relations, University of Auckland, Auckland, New Zealand

A. Sayigh (ed.), *Transition Towards 100% Renewable Energy*,
Innovative Renewable Energy, https://doi.org/10.1007/978-3-319-69844-1_14

new responsibilities (Warner 1999) and of how the central government can help them perform their expanded roles as effectively as possible (Honadle 2001). Local governments' resources, capacities, and availability of qualified staff must also be considered, as well as the central government's capacity to provide clear policy direction, advisory support, and adequate funding (IRENA 2015; Raadschelders and Vigoda-Gadot 2015: 326–28).

This study seeks to contribute to the current literature by investigating local governments' implementation efforts for the National Policy Statement for Renewable Electricity Generation (NPS-REG). The aim of this investigation is to understand how the planning aspects of the NPS are being implemented by district councils. While the research focuses on the NPS-REG, it seeks to draw conclusions which may be applied to the whole New Zealand renewable electricity policy framework. The paper, hence, initially investigates the main regulatory framework in which district councils operate. This assessment then provides the backdrop for the quantitative and qualitative analysis of the impact of the NPS-REG on currently operative and proposed district plans. The combined analysis of both the regulatory framework and implementation outcomes is then used to highlight the main obstacles to the effectivity of the NPS-REG and possible solutions.

14.2 Methodology

The research was carried out in four stages. Firstly, through the analysis of the 90% renewable electricity generation target and the measures put in place to implement it. Secondly, a quantitative and qualitative review of New Zealand's 54 current and 33 proposed district plans was undertaken to critically assess their compliance with the NPS-REG. Thirdly, emails were sent to each council to confirm the results of the analysis and request additional information, particularly about future proposals. The response rate was about 80% and helped identify possible interviewees for the last stage of the research. Finally, a series of semi-structured interviews were held with a member of each councils that had expressed limited or no interest to implement the NPS-REG to establish the rationale for that choice.

14.3 Results

Contrary to common international practice, the New Zealand central government provides only minimal explicit support schemes for renewable electricity development. In this context, the achievement of the 2025 target relies mainly on market forces and few policy and legal instruments (Bertram and Clover 2009; Palmer and Grinlinton 2014). Among these policy instruments, the NPS-REG is one of the most recent and the more relevant for the planning system. The NPS was first introduced to increase planning consistency and eliminate the variable provisions for renewables in local authority policies (MfE 2011). Yet, this paper shows how the NPS-REG falls short of its ambitions due to local government bottlenecks and lack of central government's appropriate implementation guidelines.

14.3.1 The Implementation of the Target for Renewable Electricity Generation

The target for making electricity 90% renewable by 2025 (NZ RET) was first introduced in the 2007 New Zealand Energy Strategy (MfE, 2007). The central government at the time also developed two policy instruments to facilitate its achievability: a 10-year restriction on new base-load fossil-fueled thermal generation and the New Zealand Emission Trading Scheme (NZ ETS). Yet, while following governments confirmed their commitment to the target, they also significantly modified its two implementing tools and introduced a slight rewording of the target, resulting in a reduction in its scope and achievability.

In 2008, the newly elected government repealed the moratorium on fossil fuels and the NZ ETS, as they were perceived as too costly and damaging for the national security of supply (Miller 2011:132). A less rigorous version of the NZ ETS was later introduced in 2009 but has been widely criticized for lacking ambition, for failing to provide credible incentives to reduce emissions, and for creating regulatory uncertainty (Richter and Chambers 2014:58; Bailey and Inderberg 2016:3,9). Moreover, while the 2011 Energy Strategy confirmed the 2025 target, it also highlighted the importance of "proving this does not affect security of supply" (MED 2011:6). While the provision was already acknowledged in the original text, it had a more subtle tone, as the 2007 Energy Strategy recognized electricity sources as a primary provider of electricity security. Contrary to this initial understanding, the 2011 Energy Strategy emphasizes the limiting provision to the target, hence implying that increased renewable electricity generation can put security of supply at risk.

The elimination of the moratorium on fossil fuels, the weakening of the NZ RET, and the new emphasis on security of supply provision reduced the scope of the target and created a nebulous yet effective justification in case it is not achieved. These measures transformed the target into an aspirational statement that lacks specific policy tools to be achieved and allows for the continuation of the status quo. The lack of strong implementation instruments, in turn, leaves the market as the main driver of renewable electricity generation. Achieving the target becomes a matter of economic viability, in which the government does not offer direct economic incentives to renewables. Yet, reliance on market forces has already proven to be insufficient to guarantee environmental sustainability (Barton 2005:464), and its inadequacy is showing now that current slow growth in electricity demand has halted investments in the sector (MfE 2016).

In this context, the introduction of the National Policy Statement for Renewable Electricity Generation (NPS-REG) is significant, as it is one of the few regulatory tools the government uses to directly influence local governments' decision in favor of renewables. It requires, among other aspects, councils to recognize renewable electricity generation as a matter of national significance and provide for it in their plans. The following section seeks to assess the extent to which this has been the case.

14.3.2 District Councils and the National Policy Statement for Renewable Electricity Generation

Local governments are the ultimate implementers of New Zealand environment management system (Connell et al. 2009:869). They are required to produce, administer, and develop a district plan to set the legislative framework for land use management, subdivision, and development in their territory. While each plan states and reflects the respective council's objectives, policies, and rules to control the effects of activities, it also needs to be consistent with higher priority legislations, i.e., regional plans or national policy statements. This requirement seeks to guarantee local flexibility and discretion while also ensuring a cohesive approach to environment management.

The NPS-REG seeks to drive a consistent approach to planning for renewable electricity generation to contribute to the achievement of the 2025 target (MfE 2011). To achieve this objective, district plans should be timely and cohesively amended. Yet, the NPS does not provide a legislative framework consistent with these standards. Firstly, its extremely flexible timeframe negatively affects its timely implementation. For instance, while the NPS-REG was introduced in 2011, some district councils are still not required to give effect to it (MfE 2011, 2016:23). Secondly, its not-binding nature leaves the councils with a large degree of discretion in implementing it. While this flexibility allows district plans to reflect local circumstances and competing interests, it also hampers the NPS-REG overall purpose.

The lack of directive guidelines is reflected in the way in which local administration have implemented the NPS. The critical review of New Zealand's operative district plans conducted for this article shows that only around 70% of district plans (40 out of 54) include some sort of considerations on renewable electricity generation. While this percentage has increased significantly in the last decade – up from 11% in 2007 (MfE 2007) – it is still far from full implementation. In addition, there is a substantial variation in the way in which renewables are provided for in these 40 district plans. Finally, of the 14 districts that still do not give effect to the NPS-REG, 5 do not plan on addressing the issue in the foreseeable future and 3 are uncertain on whether they will (Fig. 14.1).

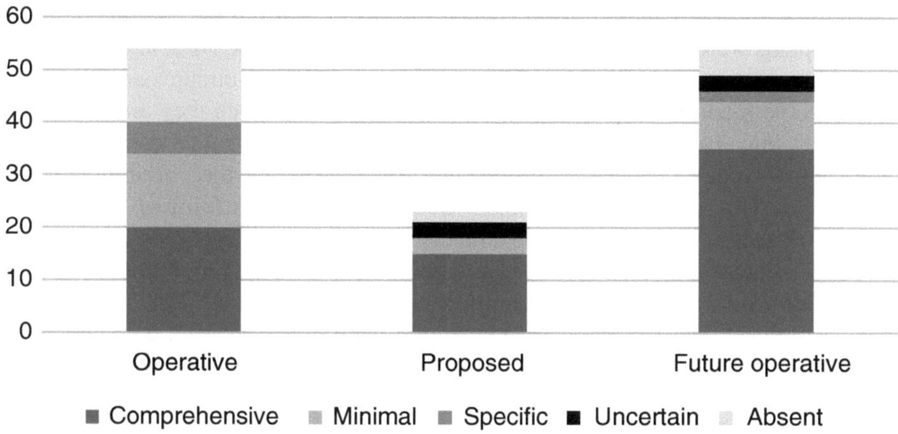

Fig. 14.1 Breakdown of the ways in which currently operative district plans, proposed district plan, and the future operative district plans (once the currently proposed one become operative) account for renewable electricity generation

The main reason for this inconsistent implementation is the decision to give effect to the NPS-REG through a comprehensive review of the district plans, which is statutorily scheduled every 10 years. The decision makes administrative sense as many district plans were already scheduled to be reviewed soon after the introduction of the NPS. Moreover, the alternative, introducing plan changes, often requires significant financial and human resources and involves a high degree of public engagement, consultation, and negotiation over a long time span. Therefore, the decision can be perceived as saving resources and avoiding duplication. However, this choice also considerably slows down the implementation of the NPS-REG.

On average, it takes over 6 years after a proposed district plan has been notified for it to become operative (LGNZ 2015:26). This means that even if all districts were to start the review process in 2011, not all of them would have become operative by now. Since some districts have yet to begin their plan review, the full NPS-REG enforcement could be postponed until 2022. While several circumstances could make this action fall under the timeline allowed under the NPS-REG, such late implementation hampers its contribution to the achievement of the 2025 target. This delay offers a significant example of how local administrative bottlenecks can slow down and limit national policy implementation. In this sense, local time and financial constrains have taken precedence over the needs to comply with higher legislative measures.

The presence of renewable electricity resources in their district also creates inconsistency in the way in which renewables are provided for in the various district plans. The absence of renewable electricity resources is used to justify some councils' decision to not introduce renewable provisions in their district plans or introduce only minimal references to them (Fig. 14.1). On the other hand, the presence of specific renewable electricity generation plants in their district has prompted some districts to introduce only specific provisions and objectives targeting exclusively these developments (Fig. 14.1). Most of the current operative district plans were implemented before the introduction of the NPS, and, therefore, their interest in renewable electricity emerges from local circumstances, which explains the inconsistency and variance in the extent to which they are accounted for. Yet, Fig. 14.1 shows how councils tend to favor a more comprehensive approach to renewable energy provisions, hence reducing the variability of the provisions. This comprehensive approach usually takes the form of a dedicated chapter or section that clearly states the district's objectives, policies, and method of implementation. However, several districts only include renewable electricity generation under the umbrella definition of Network Utilities. Yet, doubts have been raised as to whether this implementation method is consistent with and gives effect to the NPS-REG (Eccles 2013). Finally, many district plans scatter renewable electricity policies and objectives among their different sections, usually rural subdivision.

The NPS-REG is recognized as the main driver of the introduction of comprehensive renewable electricity provisions in most district plans. This inclusion, hence, is a response to the need to be consistent with higher priority documents rather than to local circumstances and needs. Yet, local bottlenecks and circumstances mean that this change is happening at a slower pace than previously envisioned, which reduces legislative flexibility, makes it harder to promote large-scale and ambitious projects, as well as responds quickly to emerging technologies.

14.4 Discussion

Contrary to other developed countries with high renewable electricity target, New Zealand has introduced only a few policy tools to achieve its renewable electricity target. Yet, 2025 is fast approaching, and there is no concrete or definitive indication that market forces alone will be enough to achieve this target. Therefore, support to existing measures should be increased while also considering the introduction of new and more incisive policy tools. To increase the efficacy of both existing and future policies, the government should address two key challenges: (1) the lack of policy implementation tools and (2) the lengthy process to implement and change district and regional plans.

The scope of the NPS-REG, one of the few instruments available for the government to directly steer local decision-making, has been halted by the same factor it tries to act upon: district plans. These obstacles undermine the symbolic value of the NPS, as they limit the extent to which councils can increase stakeholders' and public's confidence. However, far from being exclusively related to the NPS-REG, these barriers are systemic to the entire devolved resource management system. Hence, they can provide useful guidelines for future policy instruments, which should be more directive and incisive. This could be achieved by including stringent and detailed guidelines and timelines, as well as by framing them in a specific and unqualified manner.

New Zealand central government should also make the implementation of renewable electricity planning requirements mandatory, as it is already the case for the New Zealand Coastal Policy Statement. In addition, the central government should streamline plan changes for matters of national importance, such as the renewable electricity generation. This would prevent delays in the implementation of measures promoting matter of national importance, such as renewables, which is currently caused by appeals to other sections and subsections of proposed district plans. This streamlined implementation could be achieved through the introduction of an independent plan change process that is separate from other publicly notified amendments. The introduction of such measures would not hamper the protection of the local environment and interests, as they are already protected under other aspects of the resource management system. Finally, this fast-tracked approach could also be coupled with the formulation of "amendment templates" created by the central government to clearly outline the structure and the content of the expected amendments.

References

Bailey, I., & Inderberg, T. H. J. (2016). New Zealand and climate change: What are the stakes and what can New Zealand do? *Policy Quarterly, 12*(2), 3–12.

Barton, B. (2005). Electricity market liberalization and energy sustainability. In A. Bradbrook & IUCN Academy of Environmental Law (Eds.), *The law of energy for sustainable development (IUCN Academy of Environmental Law research studies)*. Cambridge: Cambridge University Press.

Bertram, G., & Clover, D. (2009). Kicking the fossil fuel habit: New Zealand's ninety percent target for renewable electricity, chapter 14. In F. P. Sioshansi (Ed.), *Generating electricity in a carbon constrained world*. Netherlands: Elsevier.

BusinessNZ Energy Council (BEC). (2015). *New Zealand energy scenarios: Navigating energy futures to 2050*. Wellington: BusinessNZ Energy Council. Available at: goo.gl/qdZVbp.

Chang, Y., Fang, Z., Li, Y. (2016). Renewable energy policies in promoting financing and investment among the East Asia Summit countries: Quantitative assessment and policy implications. *Energy Policy, 95*, 427–36.

Connell, J., Page, S. J., & Bentley, T. (2009). Towards sustainable tourism planning in New Zealand: Monitoring local government planning under the Resource Management Act. *Tourism Management, 30*, 867–877.

Eccles, G. (2013). *Network utilities: Statement of evidence of Grant Eccles for Mighty River Power*. Hamilton, NZ: Hamilton City Council. Available at: goo.gl/5af3W8.

Honadle, B. (2001). Theoretical and practical issues of local government capacity in an era of devolution. *Journal of Regional Analysis and Policy, 31*, 77–90.

IRENA. (2015). Renewable energy target setting. Available at: goo.gl/zRARUL.

Kelly, G. (2011). History and potential of renewable energy development in New Zealand. *Renewable and Sustainable Energy Reviews, 15*(5), 2501–2509.

Kerr, S., Claridge, Milicich, D. (1998). *Devolution and the New Zealand Resource Management Act*. Treasury Working Paper 98/7. Available at: goo.gl/l9AL1K.

Local Government New Zealand (LGNZ). (2013). *A journalist guide to local government*. LGNZ. Available at: goo.gl/aGPRrv.

Local Government New Zealand (LGNZ). (2015). *A 'blue skies' discussion about New Zealand's resource management system: A discussion document prepared for LGNZ by Martin Jenkins*. Auckland: LGNZ. Available at: goo.gl/BCl4iA.

Miller, C. (2011). *Implementing sustainability: The New Zealand experience (RTPI library series)*. London: Routledge.

Ministry for the Environment (MfE). (2007). *Providing national guidance on Renewable Energy Projects through the Resource Management Act 1991*. Wellington: MfE. Available at: goo.gl/O6Q9Vo.

Ministry for the Environment (MfE). (2011). *National Policy Statement for Renewable Electricity Generation 2011: Implementation guide*. Wellington: MfE. Available at: goo.gl/2WSCfZ.

Ministry for the Environment (MfE). (2016). *Report of the outcome evaluation of the national policy statement for renewable electricity generation*. Wellington: MfE.

Ministry of Business, Innovation, and Employment (MBIE). (2015). *Electricity*. Wellington: MBIE. Available at: goo.gl/tMPmXM.

Ministry of Economic Development (MED). (2007). *New Zealand energy strategy to 2050: Towards a sustainable low emissions energy system*. Wellington: MBIE. Available at: goo.gl/MQZhm4.

Ministry of Economic Development (MED). (2011). *New Zealand energy strategy 2011–2021. Developing our energy potential and the New Zealand energy efficiency and conservation strategy 2011–2016*. Available at: goo.gl/buimBE.

Palmer, K., & Grinlinton, D. (2014). Developments in renewable energy law and policy in New Zealand. *Journal of Energy & Natural Resources Law, 32*(3), 245–272.

Raadschelders, J., & Vigoda-Gadot, E. (2015). *Global dimensions of public administration and governance: A comparative voyage*. Hoboken, New Jersey : Jossey-Bass, a Wiley brand.

REN21. (2015). *The first decade: 2004–2014*. Paris: REN21. Available at: goo.gl/i8yO7z.

Richter, J., & Chambers, L. (2014). Reflections and Outlook for the New Zealand ETS must uncertain times mean uncertain measures? *Policy Quarterly, 10*(2), 57–67.

Warner, M. E. (1999). Local government financial capacity and the growing importance of state aid. *Rural Development Perspectives, 13*(3), 27–36.

Chapter 15
Exploring the Death Spiral: A System Dynamics Model of the Electricity Network in Western Australia

William Grace

15.1 Introduction

The electricity industry worldwide is talking about the so-called death spiral. Under this scenario, conventional electricity networks are undermined by customers reducing their energy demand through energy efficiency measures and/or private generation, mainly rooftop solar photovoltaic panels (PV). Both processes reduce the quantum of electricity purchased from the network, thereby reducing revenue to the network. As many of the network costs are fixed, this necessarily implies increasing unit costs and therefore increasing tariff charges for electricity. The tariff increases merely exacerbate the problem – hence the "spiral" reference.

In this study, these issues are considered in the context of Western Australia's south-west interconnected system (SWIS). The SWIS serves the south-west portion of the state, some 900,000 dwellings and 100,000 businesses. The effect of increasing rooftop solar PV and the potential in the future for private electrical storage are modelled using the system dynamics technique. System dynamics has been used previously to simulate energy systems in general and electric power systems, in particular, notably by Professor Andrew Ford of Washington State University (Ford 1997, 2005, 2008). The approach has also been used in power industry policy and strategy (Dyner and Larsen 2001; Bunn et al. 1993).

W. Grace (✉)
Australian Urban Design Research Centre, University of Western Australia,
Perth, WA, Australia
e-mail: bill.grace@uwa.edu.au

A. Sayigh (ed.), *Transition Towards 100% Renewable Energy*,
Innovative Renewable Energy, https://doi.org/10.1007/978-3-319-69844-1_15

15.1.1 The System

The system currently has a generation capacity of over 5500 MW. Despite a growing population and economy, both peak and average demand on the SWIS have plateaued in recent years (Fig. 15.1). Part of this change can be attributed to general reductions in energy intensity in the economy (i.e. energy consumed per $ gross state product).

Rooftop solar penetration has increased significantly in recent years on houses within the area served by the SWIS, reflecting that in Australia more generally (Fig. 15.2).

Households and businesses that export energy to the SWIS (at times when solar generation exceeds electrical demand) are paid in accordance with the Renewable Energy Buyback Scheme (REBS) which has been reduced from 8.85 to 7.13 c/kWh, about one-third of the household tariff. The increase in take-up of private solar PV installations coincides with a significant drop in the unit price of residential systems in recent years (Fig. 15.3).[1]

The other innovation that is likely to follow the rapid take-up of household solar energy is energy storage. In the short term, this will probably be battery storage, and lithium-ion batteries appear most likely to lead the market. Development of lithium-ion batteries is being driven by both domestic scale renewable energy and the electric vehicle industry (Fig. 15.4).[2]

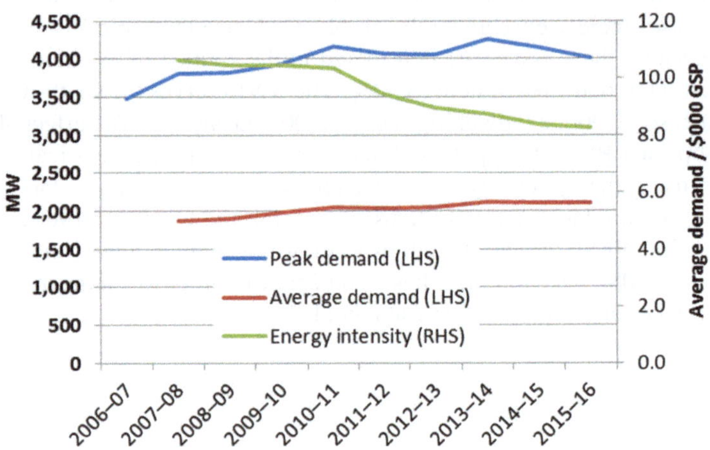

Fig. 15.1 Energy demand and intensity

[1]Source: http://www.solarchoice.net.au/blog/news/residential-solar-pv-system-prices-december-2016

[2]There several factors associated with the large range in costs for existing lithium-ion batteries, including battery size, required power demand/duration and geographical variations in manufacturing unit costs.

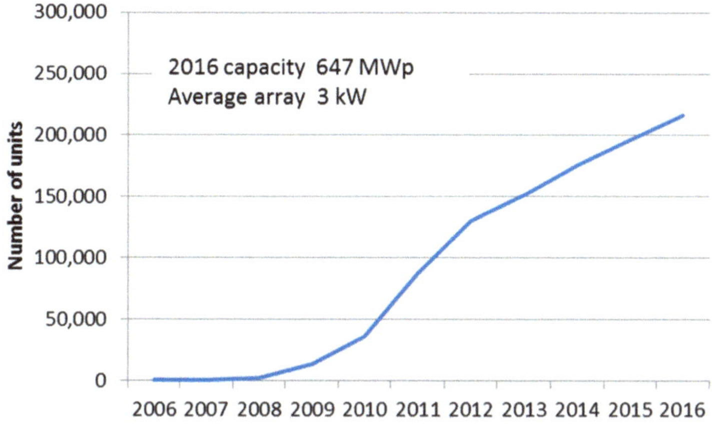

Fig. 15.2 Rooftop solar PV on the SWIS (Source: Australian Energy Market Operator)

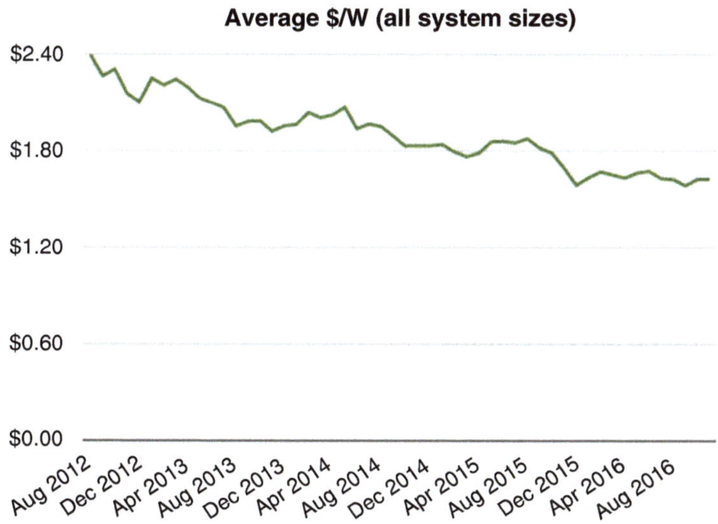

Fig. 15.3 Cost of residential solar PV in Australia ($/W)

15.2 Methodology

The purpose of the model is to explore the influence of growing solar penetration on the SWIS network. The essential structure of the model is depicted in the causal loop diagram below (Fig. 15.5).

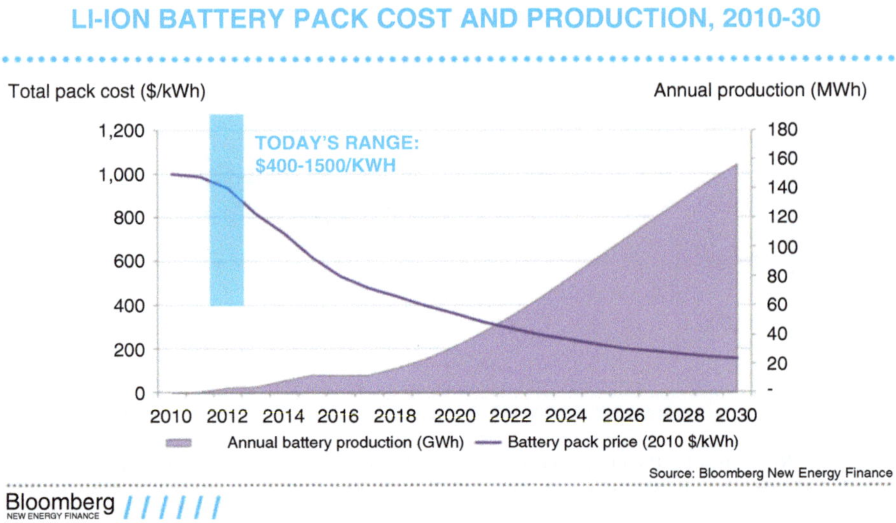

Fig. 15.4 Projected cost and production of li-ion storage (Bloomberg New Energy Finance Summit 2012, http://about.bnef.com/summit/content/uploads/sites/3/2013/11/BNEF_2012_03_19__University_Battery_Innovation.pdf)

The model uses version 6.3 of the Vensim[3] Professional software. A full explanation of the model structure, together with its documentation, is included in a detailed report on the research.[4]

15.2.1 Electricity Demand

The model determines the electricity demand arising from:

- Residential houses
- Commercial and industrial facilities

The existing residential demand and commercial demand profiles have been determined from historical half-hourly reports of total network load and presentations of the previous independent market operator (IMO WA) on residential and commercial loads.

[3]http://vensim.com/

[4]https://www.audrc.org/exploring-the-death-spiral/

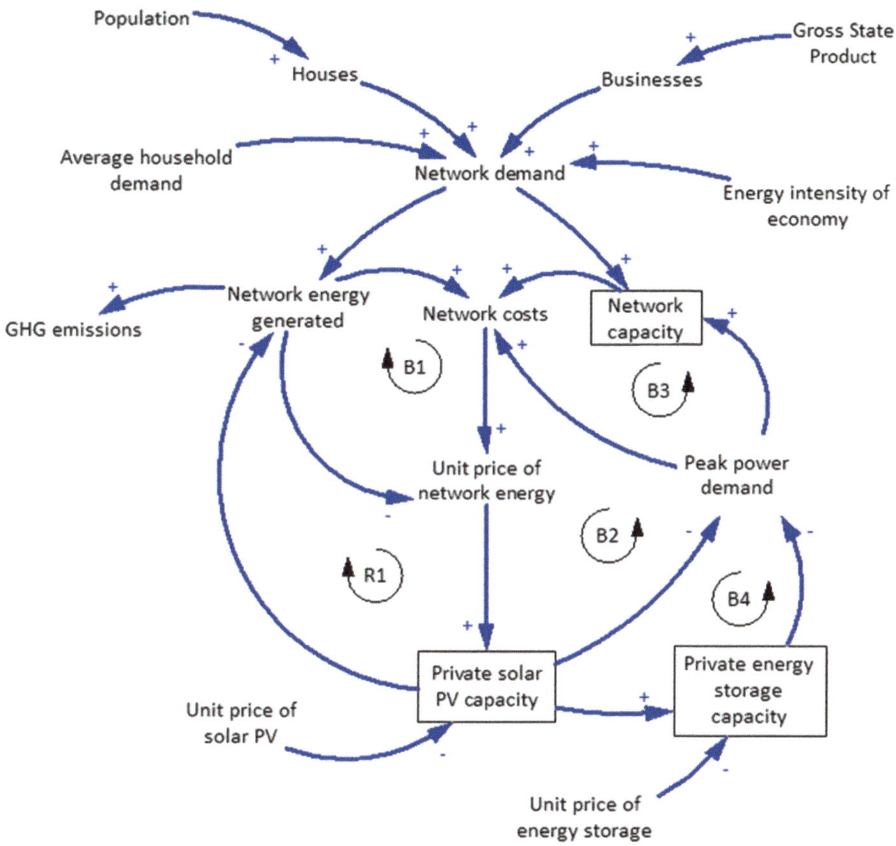

Fig. 15.5 Causal loop diagram of model

15.2.2 Solar PV and Storage

The model calculates the contribution of household scale solar energy generation, with and without energy storage. The model calculates the payback period for a household arising from:

- Avoided electricity imports from the network at the residential tariff
- Electricity exports to the network at the residential feed-in tariff (renewable buyback scheme)
- The installed cost of solar energy and battery storage

The unit cost of solar PV is modelled as a stock with an initial value reflecting present unit costs ($2,200/kW) which is the approximate installed cost of systems in Australia presently, excluding the benefit of the small-scale technology certificates (STCs). The model assumes that the unit cost transitions to a final unit cost of

\$1,000/kW.[5] Both the final unit cost and adjustment time can be easily varied in the model.

The model sets an optimum storage capacity based on the capacity of the solar array[6] which avoids multiple combinations of solar capacity and storage capacity. The model calculates the additional benefit to the householder from adding storage to their solar array. It is assumed that storage operates simply on the basis that:

- Solar generation in excess of demand is stored (up to the limit of the storage capacity)
- The storage discharges to meet demand that cannot be met by solar generation
- Remaining demand is met by the SWIS network

The incentive to add solar PV and storage is determined by a payback period calculated from the benefits noted above and the unit cost of solar storage. The latter has been determined from the technical press and assumes the present storage costs of approximately \$1,000/kWh will drop to around \$200/kWh (which may be considered conservative given the curve set out in Fig. 15.4).

The model structure for the commercial solar is identical to the residential model in all respects, except for the initial conditions. The model assumes that at the outset there is no commercial solar or storage.

15.2.3 Utility Network

The network generation capacity is used to calculate the recurrent generation costs and spot price arising from the short-term energy market (STEM). The initial capacities of each type of generation are based on the existing capacity credits allocated by the market operator and other information. Information on the capital and operating (including fuel) costs and other characteristics of each type of generation has been derived from the Australian Energy Technology Assessment (AETA) by the Australian government's Bureau of Resources and Energy Economics (BREE 2012).

The model assumes there are no additions to the existing generation capacity (including wind). As the network is presently over capacity, the model allows for retirements to each type of thermal generation. The default condition is for coal retirements at 20 MW/year throughout the simulation period, commencing in 2016.

[5]The curve is based on a review of published forecast costs including extrapolation of Fig. 15.4 and the US Sunshot Target of \$0.06/kWh – http://energy.gov/eere/sunshot/sunshot-initiative.

[6]The approach to selecting optimum storage is described in detailed report referred to above.

15.3 Results

The model has been used to investigate a number of scenarios related to the possible growth of private solar PV and storage and the impacts of this growth on the SWIS network.

15.3.1 Results: Solar Growth Without Storage

Under the model assumptions, payback periods continue to fall as unit costs reduce. This results in rapid growth in both the residential and business sectors. By 2035 around 50% of houses (3,000 MW) and 40% of businesses (6,000 MW) have solar PV (Fig. 15.6). Reducing payback periods also lead to larger installed systems with average array sizes increasing to around 4.5 kW for residential systems and 90 kW for business systems.

There are significant implications for the SWIS if this scale of growth in private solar PV occurs. By 2025, average hourly loads on the network are significantly reduced and this divergence grows thereafter. However, maximum loads are only marginally reduced by daytime solar generation. This means that the capacity of the network is required to be maintained at similar levels to the business as usual base case, with consequent capacity-based costs.

However, the most significant implication of this scale of solar generation is on minimum network loads, i.e. when solar generation is at its maximum during the middle of the day in general and in the summer, in particular. By March 2030 available solar exports exceed the total electricity demand in the middle of the day and accordingly, network loads fall to zero. The lead up to this circumstance is shown in the following figure which depicts the so-called duck curve (Fig. 15.7).

This encapsulates the problem for networks of accommodating highly varying daytime solar generation.

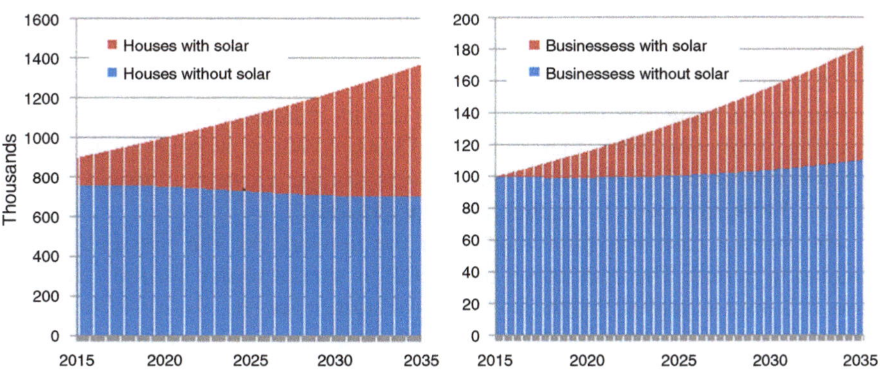

Fig. 15.6 Growth in solar PV penetration

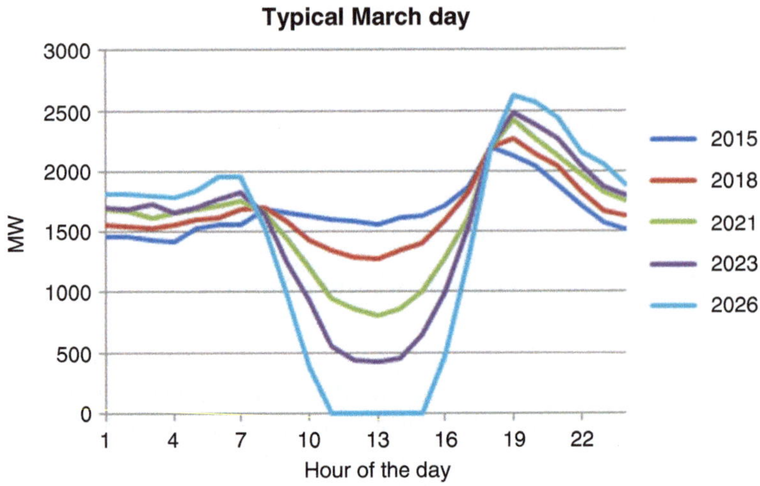

Fig. 15.7 Duck curve. Hour of the day

This suggests that there could be an intermittent over-generation problem by 2020 when minimum network loads fall below the normal operating capacity of baseload coal generation, which is intended to run consistently and cannot be readily cycled down and up in a matter of hours. Steep ramping of generation is required in the hours of declining solar generation. A similar situation has been identified by the California Independent System Operator.[7]

15.3.2 Results: Solar Growth with Storage

This case examines the implications of residential and business solar growth with storage. Storage payback periods in the model are dependent on storage costs, savings and REBS income which are all a function of solar array size. Accordingly, it takes some time for paybacks to drop to the level where take-up would be financially attractive. However, after around 2020 paybacks have dropped to the 10–15-year range.

Penetration thereafter increases steadily in both residential and business facilities (Fig. 15.8). By 2035, there is some 13,000 MWh of storage capacity, and 405,000 houses and 45,000 businesses possess storage. However, this is still only about 1.5 h of storage at the nameplate solar capacity.

The storage case has little impact on the network for the first half of the simulation period. The duck curve up to 2025 is similar to that for the solar only case. It is only thereafter that the influences of storage are felt. By 2034 storage has

[7]http://www.caiso.com/documents/flexibleresourceshelprenewables_fastfacts.pdf

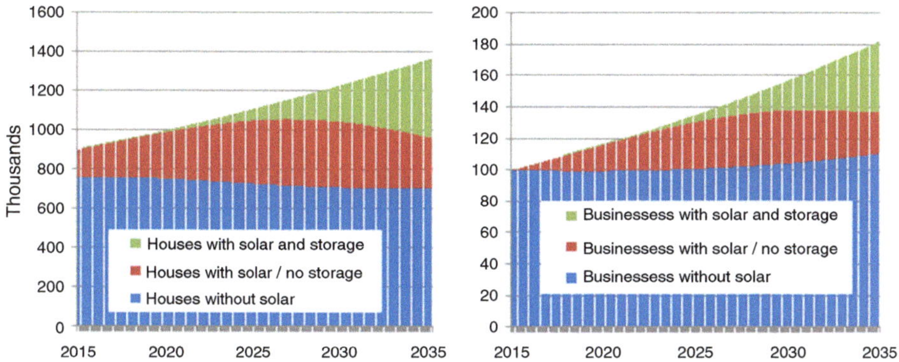

Fig. 15.8 Energy storage growth

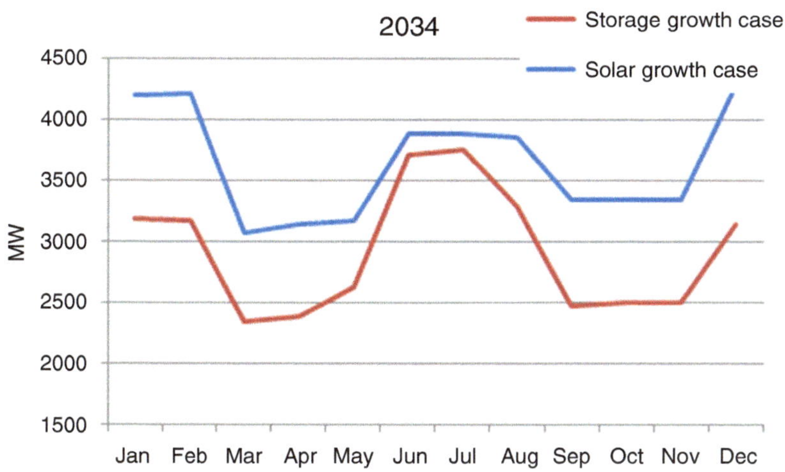

Fig. 15.9 Maximum hourly network loads

reduced maximum network loads significantly from the solar only case (Fig. 15.9). The over-generation problem is somewhat ameliorated in the storage case, but minimum network loads will still fall to 0 under certain circumstances (Fig. 15.10).

15.4 Conclusions

15.4.1 Solar Growth

In a recent report, the Australian Energy Market Operator (AEMO) forecasts the growth of private solar to grow linearly for the coming decade to reach

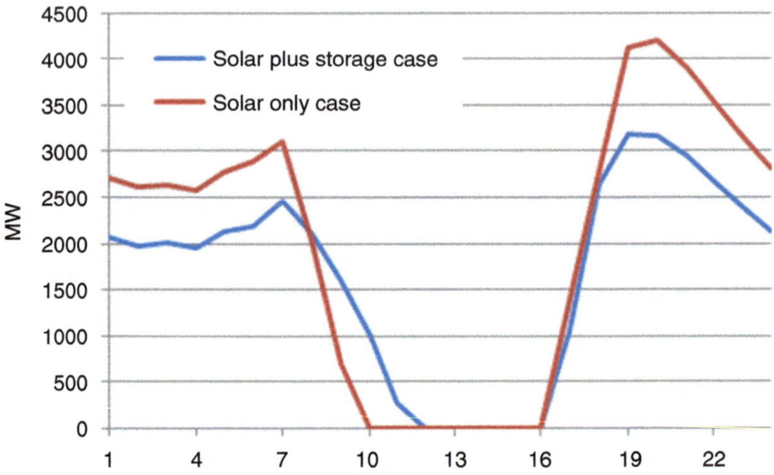

Fig. 15.10 January day network loads (2034)

approximately 1,400 MW of nameplate capacity by 2026. In contrast, the model, which is based on an increasing take-up as payback periods reduce, suggests that this figure could be as high as 3,000 MW by 2024 and three times that figure by 2035. The AEMO forecast is based on linear growth. Exponential growth is a more common phenomenon and is usually observed in the time histories of innovative technologies, including solar PV take-up in Australia.[8]

Although growth in the commercial sector has been slower than residential to date, as awareness grows and unit prices decline, this is potentially the largest contributor to growth by far. Payback periods are lower for commercial systems because there is a better match between solar generation and demand and time-of-use tariffs are higher during peak periods.

15.4.2 Network Loads

The most dramatic effect on the network of growth in solar penetration of this scale is seen in the impact on minimum loads on the network (Fig. 15.11). This illustrates the impact of growing solar exports to the network during daytime periods in general and in summer, in particular. Over the longer term, the "hollowing out" of network load would likely require a different configuration of generation type on the network.

Private energy storage ameliorates the impacts of solar generation on the network as it reduces both the peak and average energy demands, thus lowering system

[8]http://apvi.org.au/wp-content/uploads/2014/07/PV-in-Australia-Report-2013.pdf

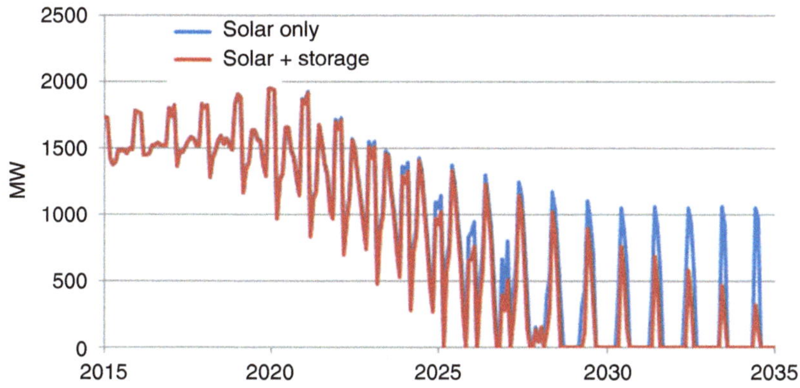

Fig. 15.11 Minimum hourly network loads

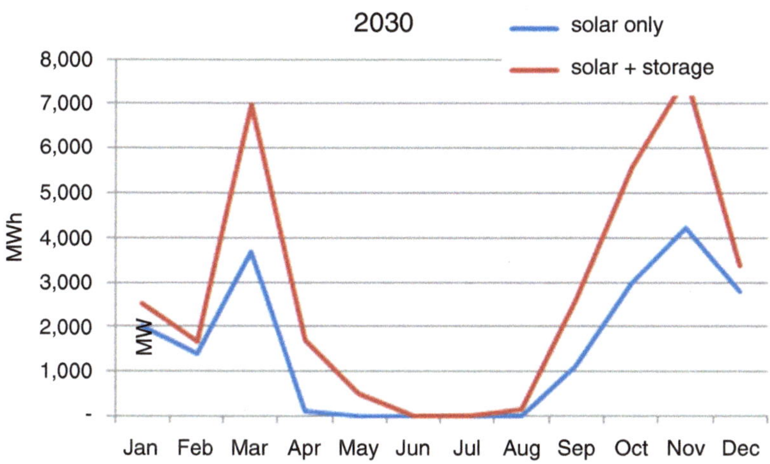

Fig. 15.12 Typical day over-generation

costs. With projected growth in storage lagging the growth in solar PV, storage only partially offsets the over-generation problem. By the end of the simulation period, there is only around 1.5 h of storage (at nameplate). This means that the amount of over-generation from solar will likely continue to increase, albeit to a lesser extent than without storage (Fig. 15.12).

15.4.3 Network Energy and System Costs

All of the cases considered result in lower annual energy required from the network than the business as usual base case. In the most likely case that business premises

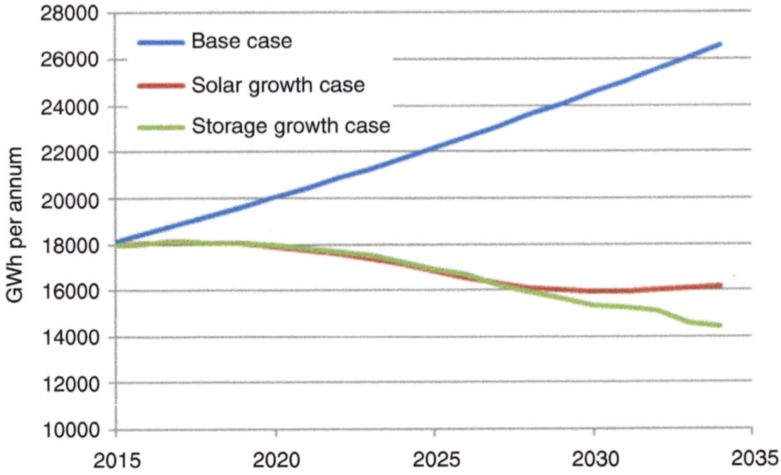

Fig. 15.13 Annual network energy

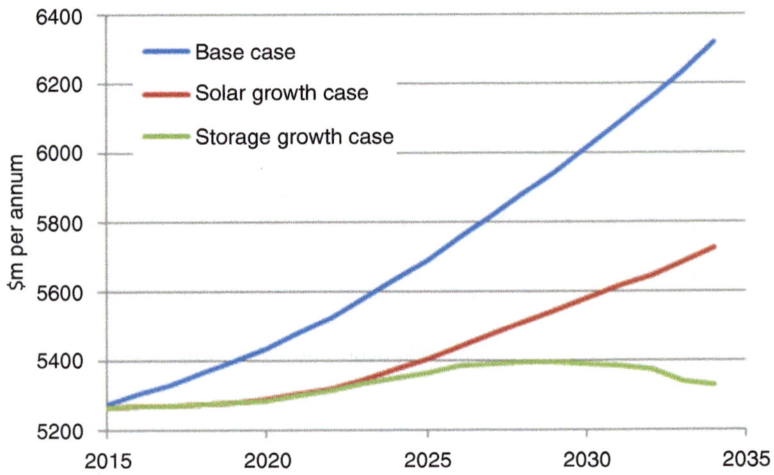

Fig. 15.14 Annual system costs

also adopt private solar and storage, there would be a dramatic reduction in required energy from the network, both in relative (45%) and absolute terms (15%) as identified in Fig. 15.13.

A reduced energy demand translates to reduced system costs. Under the storage scenario (residential and business solar and storage), system costs would rise only slowly and then potentially fall to near current costs (Fig. 15.14). This is due partly to a reduction in the capacity of the network (assumed mainly related to coal generation), which induces savings in the cost of capacity credits. However, this neglects any ongoing debt obligations associated with this generation capacity which, while not a direct cost to the SWIS, is a cost to publicly owned entities and hence the taxpayer. The costs of writing off this debt are not included in this study.

If tariffs presently delivered \$295/MWh, the SWIS would recover all costs. The model suggests that unit costs will fall from this figure under the base case conditions. However, under the other cases it is likely that, although overall system costs will decrease, unit costs and therefore tariffs will increase substantially (20% or more by 2035).

Although reduced system costs indicate that increased solar penetration will lead to an overall economic benefit, a rise in unit costs is still problematical. Individual consumers, who for one reason or other cannot reduce their network energy consumption sufficiently to offset tariff increases, will pay more for energy under this scenario. As this group will include those who are least able to absorb the additional cost, equity will become an important element of the policy response.

15.4.4 The Influence of Tariff Increases

Part of the death spiral metaphor is that increasing solar penetration will lead to higher tariffs which will only further improve the cost-benefit equation for consumers. By default, the model increases existing household and business tariffs pro-rata to system unit cost increases. However, the influence of increasing tariffs is easily tested by "switching off" tariff increases in the model. Doing so demonstrates that the tariff increases are not the main driver of reduced paybacks. The reducing unit cost of solar PV has much more influence than increasing tariffs.

15.4.5 Greenhouse Gas Emissions

The base case indicates that greenhouse gas emissions could rise by around 20% by 2035. The most likely scenario, i.e. growth in private residential and business solar and storage would reverse this and deliver reductions of around 25%. However, even this contribution would be insufficient to represent a credible emission reduction target for south-west Western Australia.

15.4.6 Policy Response

The current review of the Wholesale Electricity Market (WEM) does not address the dependence of the SWIS on fossil fuel generation, a situation that cannot continue irrespective of the current adverse political environment. Even if supply is not dampened through action on global warming, prices will inevitably and steadily rise forever after peak production in gas and coal later this century (Maggio and Cacciola 2012). Electricity industry investments are made for decades and so it is essential that the future energy generation mix and network strategy is established now to ensure that new investments minimise the risk of stranded assets.

This study identifies that the growth of renewables in the form of private solar is inevitable and will have major implications for the network irrespective of any changes likely to arise from the WEM review. The energy system will change, and therefore the implications of this study should be considered in the context of this broader transition, for which a coherent long-term energy strategy is required. The inevitable increase in the take-up of private solar PV systems in WA homes and businesses will merely hasten a transformation of the electricity network during the coming decade that is needed anyway.

References

BREE (Bureau of Resources and Energy Economics). (2012). AETA report and model version 1_0, 2012, Australia. http://www.bree.gov.au.

Bunn, D., Larsen, E., & Vlahos, K. (1993). Complementary modeling approaches for analysing several effects of privatization on electricity investment. *Journal of the Operational Research Society, 44*, 10.

Dyner, I., & Larsen, E. (2001). From planning to strategy in the electricity industry. *Energy Policy, 29*, 1145–1154.

Ford, A. (1997). System dynamics and the electric power industry. *System Dynamics Review, 13* (1), 57–85.

Ford, A. (2005). Simulating the impacts of a strategic fuel reserve in California. *Energy Policy, 33* (4), 483–498.

Ford, A. (2008). Simulation scenarios for rapid reduction in carbon dioxide emissions in the Western electricity system. *Energy Policy, 36*.

Maggio, G., & Cacciola, G. (2012). When will oil, natural gas, and coal peak? *Fuel, 98*(2012), 111–123.

Chapter 16
Efficient Seasonal Time of Use Feed-in Tariff for Residential Rooftop Solar Panels in Australian Electricity Market

Muhammad Adnan Hayat, Farhad Shahnia, and Ali Arefi

16.1 Introduction

The geographical area of Australia is extremely large and it is the sixth largest country in the world, and overall its population is very low (Countries of the world by area). Power distribution companies in Australia in particular face severe challenges of supply electricity in the regional areas as the network area to be supplied can be exceptionally high and a number of customers can be extremely low. Utilities are supplying electricity to the consumer irrespective of the location, while the cost of supply of electricity to different locations can be widely different. Therefore, the utility can be disadvantaged by the low density of consumer spread over wide areas. Another disadvantage to the utility is that the transmission and distribution network which is supplying a large area is at the constant threat of severe weather conditions and can cause high network maintenance costs. For example, in some cases, a power company may need to supply isolated communities with less than 100 people (Overview of tariff structure statement). This has resulted in high electricity bills in locations where customer density is low, or it resulted in the need of high subsidy from state government (Hayat et al. 2016).

In Australia total solar capacity has reached 6GW, and due to the rapid uptake of RSP, the total solar capacity is expected to double over the next few years (Australian solar capacity). The locations which choose to install RSP are of significant importance. In order to encourage consumer of regional areas to install RSP utility can offer high FIT to the consumers in the regional areas.

Another advantage which can be gained from high penetration of RSP is peak shaving. Households which are installing RSP can be encouraged to export electricity on the specific periods of the year and time of day to help share the extra

M.A. Hayat (✉) • F. Shahnia • A. Arefi
School of Engineering & IT, Murdoch University, Perth, Australia
e-mail: M.Hayat@murdoch.edu.au

A. Sayigh (ed.), *Transition Towards 100% Renewable Energy*,
Innovative Renewable Energy, https://doi.org/10.1007/978-3-319-69844-1_16

burden on the grid during the peak hours. Thus the immediate requirement of network upgrade and extension can be postponed, and utility can offer high FIT based on season and time of day to encourage peak shaving.

16.2 Methodology

In this paper, FIT is proposed based on the flow chart of Fig. 16.1

- Time of export of electricity generated by RSP to the grid
- Customer density of the area where RSP is installed
- Season of export of electricity generated by RSP to the grid

Western Power distribution zone is considered for the analysis of FIT. Western Power mainly covers the southwest corner of the state through 98,000 km of power lines/cables with approximately ten customers per km. Its power distribution services are spread over an area of 255,000 km^2, and it serves over one million customers (i.e., about 96% of the power consumer in the state) (About Western Power).

The Western Power Network is spread over an area from Kalbarri in the North to Albany in the South and from Kalgoorlie in the east to the metropolitan coast. The Western Power Network is segmented into 15 geographic load areas (10 Year TNDP). These 15 load areas are the basis of FIT calculation. Based on the number of customers and area covered, these load areas are divided into customer density categories of low, medium, high, and very high. Table 16.1 shows the transmission load areas categorized into different customer densities types.

This FIT is then further refined based on the season and time of day. Summer and winter season have an increasing effect because of heating and cooling load in which the major load of the residential consumer increases during these periods, whereas spring and autumn have a decreasing effect. In time of day, electricity peak periods usually occur from 3 PM to 9 PM (Smart Home Plan), and FIT has

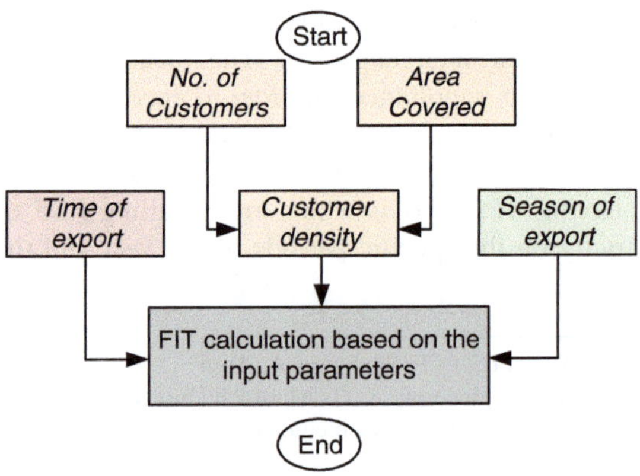

Fig. 16.1 Illustrating flow chart for calculation of FIT

Table 16.1 Feed-in tariff based on transmission load areas, where * highlights the locations on which detailed analysis is conducted.

Transmission load areas	No. of customers	Area covered	Customer density
North country*	Low	Very large	Very low
East goldfields**	Low	Large	Low
East country	Low	Very large	Very low
Muja	Medium	Very large	Low
Bunbury	Medium	Medium	Medium
Mandurah	Medium	Short	High
Kwinana	Medium	Short	High
Southern terminal	Very high	Short	High
South Fremantle	High	Short	High
Western terminal	Low	Short	High
Cannington***	High	Short	High
East Perth and CBD	Medium	Short	High
Guildford	Low	Short	High
Neerabup terminal	Very high	Medium	High
Northern terminal	Very high	Short	High

Table 16.2 Feed-in tariff based on time of day and season

Time of day	Feed-in tariff	Season	Feed-in tariff
3 PM to sunset	Increasing effect	Spring/autumn	Decreasing effect
Sunrise to 3 PM	Decreasing effect	Summer/winter	Increasing effect

increasing effect during these peak hours; 7 AM to 3 PM is usually nonpeak hours, thus FIT has a decreasing effect. Table 16.2 shows how the FIT is refined based on the parameters of season and time of day.

In this paper low range of FIT is considered to be from 5 to 8 cents per kWh, medium range is considered from 8 to 11 cents per kWh, high FIT is considered to be from range 11 to 14 cents per kWh, and very high FIT is considered to be from 14 to 17 cents per kWh.

16.3 Results

Three case studies are performed on three different suburbs in these load areas and FIT is calculated. A brief introduction of these suburbs is as follows.

Cannington load area is located in the south eastern corridor of metropolitan area and bounded by the swan and canning rivers and extends east from the CBD to Mundaring (10 Years TNDP).

Kalbarri is a regional area supplied with electricity through 150 km, radial Kalbarri feeder from Geraldton. Kalbarri feeder is exposed to severe weather

Table 16.3 Feed-in tariff calculation

Suburb	Time of day	Customer density	Season	Feed-in tariff
Cannington*** (Cannington)	3 PM to sunset	High	Spring/autumn	6
	Sunrise to 3 PM		Spring/autumn	5
	3 PM to sunset		Summer/winter	8
	Sunrise to 3 PM		Summer/winter	7
Kalbarri (North country)*	3 PM to sunset	Very low	Spring/autumn	15
	Sunrise to 3 PM		Spring/autumn	14
	3 PM to sunset		Summer/winter	17
	Sunrise to 3 PM		Summer/winter	16
Kalgoorlie (Eastern Goldfields)**	3 PM to sunset	Low	Spring/autumn	12
	Sunrise to 3 PM		Spring/autumn	11
	3 PM to sunset		Summer/winter	14
	Sunrise to 3 PM		Summer/winter	13

conditions which can lead to extended outages on the line. In Western Power network, Kalbarri is among the worst performing feeders, making outage duration greater than the average targeted by the regulator for long rural feeders (Study into the feasibility of microgrid). The North Country load area is extended to Kalbarri at the northern extremity of the Western Power Network (Annual Planning Report 2015/16).

Eastern Goldfields Load Area is comprised, in the main, of the city of Kalgoorlie. This network is very long and presents considerable operational challenges (10 Years TNDP).

Table 16.3 shows the results of three suburbs of which the FIT is calculated. These results propose that in an efficient seasonal time of use, FIT can encourage a particular location of consumers to export electricity to the grid at the time it is most valuable to the utility. Therefore, for some particular location, the FIT is generally the same; however, it varies slightly with season and time of use. The FIT mainly varies with the customer density. If customer density is low, FIT will be lower, and if customer density is higher, FIT will be higher.

16.4 Conclusions

Efficient seasonal time of use tariff will offer relatively high FIT for consumers in the regional area and encourage them to install RSP. Due to the remoteness, usually cost of RSP is more and also the transportation costs are higher in those localities; therefore, it is very reasonable to offer them FIT which ensures that their investment is returned in the reasonable time period. In addition to that, the time of day factor will encourage the export in the peak hours. This will have double benefit; the customer may shift their normal electricity consumption to other hours and export maximum to the grid during those peak hours. This will ultimately help in reducing the immediate requirement of network upgrade, increase the network reliability, decrease the transmission line losses, reduce the excessive burden on the utility, and bring economic incentives for the RSP installer. This research can be further extended by considering other parameters which are more relevant to some other utilities.

References

10 Year TNDP, Western Power Transmission Network Development Plan. (2011, September). https://www.erawa.com.au/cproot/9936/2/20111007%20%20D76352%20%20Access%20Arrangement%20Information%20-%20Appendix%20O%20%20Transmission%20Network%20Development%20Plan.PDF

About Western Power, Website. Retrieved 01-06-2016. http://www.westernpower.com.au/corporate-information-about-us.html#our_network

Annual Planning Report 2015/16. https://www.westernpower.com.au/media/1619/annual-planning-report-2015-16.pdf

Australian solar capacity. http://reneweconomy.com.au/australian-solar-capacity-now-6gw-to-double-again-by-2020-2020/

Countries of world by area. http://www.nationsonline.org/oneworld/countries_by_area.htm

Hayat, M. A., Shahnia, F., & Arefi, A. (2016). Comparison of the electricity tariffs and bills across the zones of Australian power distribution companies, 26th Australasian Universities Power Engineering Conference (AUPEC), pp. 1–6, Brisbane, Australia.

Overview of tariff structure statement. https://www.essentialenergy.com.au

Smart Home Plan. https://www.synergy.net.au/Your-home/Energy-plans/Smart-Home-Plan.

Study into the feasibility of microgrid. https://westernpower.com.au/media/2072/kalbarri-microgrid-feasibility-study.pdf.

Chapter 17
Synthesis of ZnO Hexagonal Prisms on Aluminum Substrates by the Spray Pyrolysis Technique

Shadia J. Ikhmayies and Mohamad B. Zbib

17.1 Introduction

Zinc oxide (ZnO) is a wide-bandgap compound semiconductor of bandgap energy $E_g = 3.37$ eV at room temperature and large exciton binding energy of 60 meV. It has the strongest bonding strength among the II–VI compounds. It exhibits hexagonal (wurtzite) crystal structure with $a = 3.25$ A° and $c = 5.12$ A° in its unstrained phase (Asadian 2013). It has a wide range of applications such as transparent conducting electrodes in flat panel displays, photovoltaic cells (PV cells) (Kadi Allah et al. 2010). ZnO has the ability to grow in a large variety of self-organized nanostructures, which make it one of the richest ranges of morphologies among the entire family of wide-bandgap semiconductor nanostructures (Feng et al. 2010). So, many works have been focused on the preparation of diverse morphologies of ZnO nano- and microstructures, such as nano-/micro-disks (Ngom et al. 2016; Li et al. 2004), nano-/microrods and prisms (Ekthammathat et al. 2015; Hamada et al. 2011; Ikhmayies Shadia 2016; Zhang et al. 2011), and so on. ZnO prisms are important for the design of efficient photoanode materials for dye-sensitized solar cells (DSSCs) . In addition, Prism-shaped ZnO rods are useful for optoelectronic applications because the prismatic planes of hexagonally shaped ZnO nanocrystal films serve as mirrors for laser action (Hamada et al. 2011).

There are several methods that can be used for the fabrication of ZnO micro-/nanostructures, including hydrothermal method (Lee et al. 2012; Djouadi et al. 2014), electrodeposition (Wang et al. 2016), chemical bath deposition (CBD) (Thambidurai et al. 2010), and spray pyrolysis (SP) (Ikhmayies Shadia 2016;

S.J. Ikhmayies (✉)
Department of Physics, Al Isra University, Amman, Jordan
e-mail: shadia_ikhmayiesl@yahoo.com

M.B. Zbib
Materials Engineering, Phoenicia University, Beirut, Lebanon

A. Sayigh (ed.), *Transition Towards 100% Renewable Energy*,
Innovative Renewable Energy, https://doi.org/10.1007/978-3-319-69844-1_17

Yilmaz et al. 2012; Ikhmayies Shadia et al. 2010a, b, c, 2014). Spray pyrolysis technique is widely adopted because of its low cost as there is no need of a vacuum system. In addition, its simplicity, ability to add different dopants to the precursor solution, and ability to produce large area, highly transparent, and uniform films make it attractive.

Preparations of ZnO prisms have been reported in the literature. For example, Hamada et al. (2011) prepared ZnO prisms by electrochemical deposition (ECD) on graphite substrates. Liu et al. (2005) grown ZnO hexagonal prisms from polyvinylpyrrolidone-assisted electrochemical assembly onto *p*-type Si (111) substrate. Ekthammathat et al. (2015) prepared prism ZnO microrods on Zn foils by the hydrothermal method. In this work, highly aligned ZnO hexagonal microprisms of potential use in solar cells were prepared by the low cost spray pyrolysis technique on aluminum substrates. XRD, SEM, and EDS were used in characterizing the prisms. The size analysis was performed using imageJ software.

17.2 Experimental Procedure

ZnO thin films were prepared on aluminum substrates of dimensions $10 \times 10 \times 1$ mm^3 at a deposition temperature of 350 ± 5 °C using the spray pyrolysis technique. The starting material is zinc chloride ($ZnCl_2$) of purity 99 %. The precursor solution was prepared by dissolving 2.5 g of $ZnCl_2$ in 60 ml of distilled water to get a 0.03 M solution. The solution has been sprayed vertically and intermittently with a deposition time of 10 s followed by a period of 1–2 min without spray. The time without spraying is required to avoid damage of the nano-/microstructures and to avoid strong cooling of the substrates. In addition, it is required to give enough time for the crystal growth to take place and the chemical reactions to be completed. The substrates were pre-cleaned by soap and finally rinsed in distilled water. Then, they were dried by lens paper. After finishing the deposition process, the heater was turned off, and the samples were left on the heater to slowly cool to room temperature.

The microstructure, phase identification, and orientation of crystal growth in the samples were determined using X-ray diffraction (XRD). The diffractometer used in this work is SHIMADZU XRD-7000 X-ray utilizing Cu K_α radiation ($\lambda = 1.54$ A°) with step size and scan speed of 0.02° and 2°/min, respectively. The measurements were taken in the continuous 2θ mode in the range 2–60° using a current of 30 mA and voltage of 40 KV. The morphology was examined by a FEI scanning electron microscope (SEM) (Inspect F 50) operating at 5.0 KV. The composition of the samples was examined by the same SEM system, which is supported by energy dispersion X-ray spectroscopy (EDS) . The thickness of the films was measured by the scanning electron microscope. The dimensions of the microprisms were estimated using imageJ software.

17.3 Results and Discussion

Figure 17.1 displays the XRD diffractogram for one of the films, where the wurtzite structure is apparent. It was found that all films showed preferential orientation along (002) plane, which suggests that the growth of the prisms is along the c-axis perpendicular to the substrate surface, where the c-axis has the lowest energy of the hexagonal crystalline structure (Lupanet al. 2007). The ratio of the Bragg peaks intensity between (002) and (101) determines the orientation of the ZnO growth (Van et al. 2011; Ridha Noor et al. 2013), and this ratio from Fig. 17.1 is found to be 6.037.

The diffractogram in Fig. 17.1 was used to deduce the lattice constants a and c of ZnO crystal lattice. This was done using the distance of lattice planes (d_{hkl}) from Bragg's law

$$2d_{\mathrm{hkl}} \sin \theta = \lambda \tag{17.1}$$

where θ is Bragg's angle, 2θ is the scattering angle, and λ is the wavelength of the X-ray radiation, which is 1.54 A° for the Cu K_{α} line. The relation that connects d_{hkl} with lattice constants a and c of the hexagonal structure is (Lupan et al. 2007)

$$\frac{1}{d_{\mathrm{hkl}}^2} = \frac{4\left(h^2 + \mathrm{hk} + k^2\right)}{3a^2} + \frac{l^2}{c^2} \tag{17.2}$$

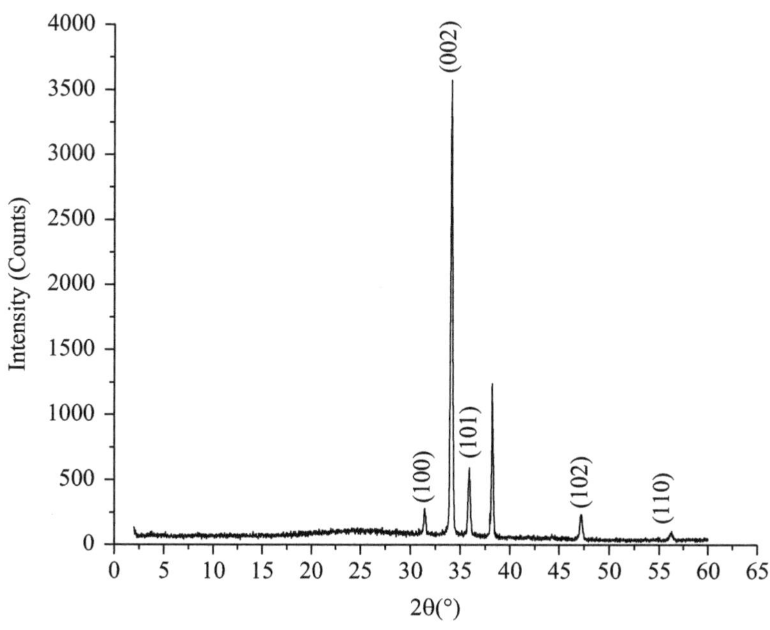

Fig. 17.1 XRD diffractogram of one of the ZnO films deposited on aluminum substrates

where h, k, and l are Miller indices. The average values are $a \pm \Delta a = 3.277 \pm 0.010\mathring{A}$, and $c \pm \Delta c = 5.234 \pm 0.020\mathring{A}$. These values are larger than those of bulk ZnO which are given in Sect. 1. The reason of this difference is that lattice parameters are usually influenced by the substrate and impurities in the samples. The use of aluminum substrates and presence of chlorine are main reasons of this discrepancy, due to the incorporation of Al and Cl in ZnO crystal lattice. Chlorine replaces oxygen, and aluminum replaces zinc. Cl has an ionic radius of 1.81 A°, while oxygen has an ionic radius of 1.35 A° (Shannon 1976), and ionic radius of Al is 0.535 Å and that of Zn is 0.74 Å (Barbalace 2016). So, incorporation of Al and Cl in ZnO crystal lattice causes expansion of the lattice parameters.

By knowing the value of c, stress can be estimated. This stress ε in the hexagonal ZnO prisms was determined using the relation (Hassan et al. 2013):

$$\varepsilon(002) = (c - c_0) \times 100\% / c_0 \tag{17.3}$$

And it is found to be 0.519 %, where c is the calculated lattice constant of the hexagonal structure of ZnO microprisms; $c_0 = 5.207$ A° is the standard (JCPDS 79-2205 card) of ZnO powder (Djouadi et al. 2014).

Surface morphology of thin films is very important tool to investigate their microstructure. Figure 17.2 depicts the SEM images of the hexagonal ZnO micro-prisms on aluminum. The films appear completely covered with material, and they are polycrystalline with the wurtzite structure and prism-like appearance. The hexagonal prisms are vertically well aligned perpendicular to the substrate, which is consistent with the XRD results, where the (002) is the preferential orientation. The prisms are uniform and the end facets and side facets are clearly identifiable, where the ends appear as clipped pyramids. These results imply that the deposited films can be used in photovoltaic devices.

The ZnO prims obtained in this work are more aligned perpendicular to the Al substrates than prisms of ZnO microrods on Zn foils synthesized by Ekthammathat et al. (2015) using the hydrothermal method. On the other hand, these micro/nano-prisms are comparable to undoped and Ni-doped ZnO microrod arrays synthesized by Yilmaz et al. (2012) using the spray pyrolysis method on glass substrates. Also they are comparable with those prepared by Kao et al. (2012) on indium titanium oxide-coated glass substrates by using the solution phase deposition method.

The composition of the samples was analyzed using X-ray energy-dispersive spectroscopy (EDS). Figure 17.3 depicts the EDS spectrum for one of the samples, and from the EDS report, it was found that the samples are rich in oxygen. Oxygen not only is chemisorped from atmosphere, since the samples are prepared in air, but also it comes from the substrate, where the aluminum substrate is oxygen rich. In addition, the films contain Cl and S, where chlorine is due to the use of $ZnCl_2$ as the starting material, and S is related to impurities in the starting material.

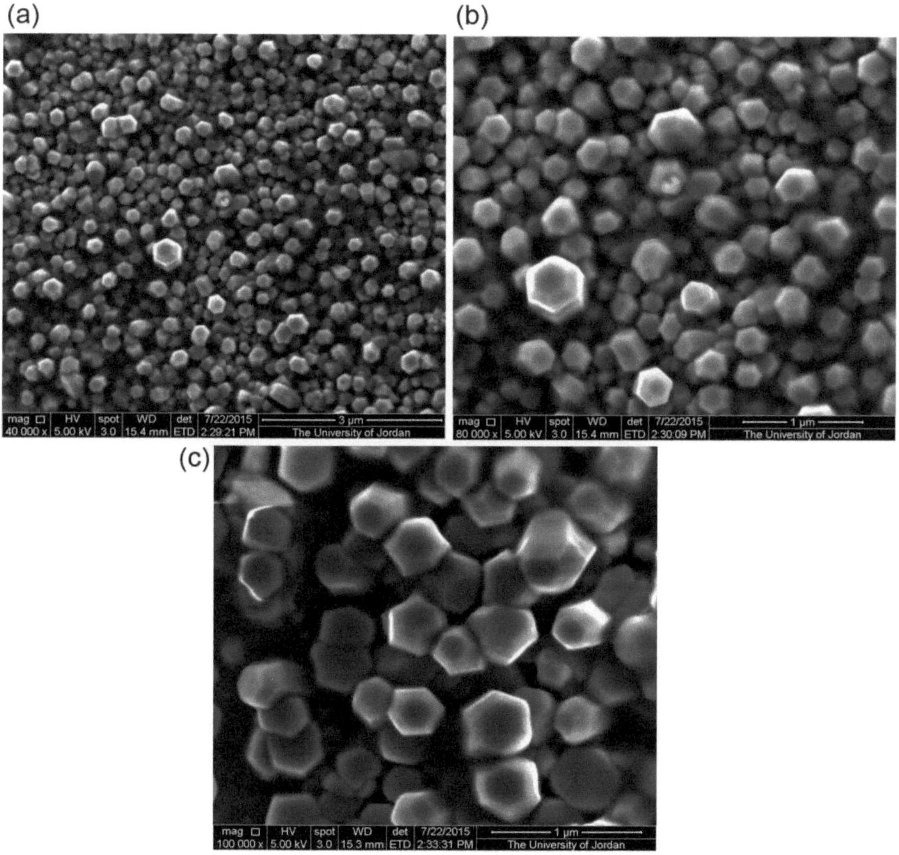

Fig. 17.2 SEM images with different magnifications of ZnO microprisms on aluminum substrates

Size analysis of the ZnO microprisms was performed using imageJ software, where Fig. 17.4a–c displays the distribution of the diameters for the images in Fig. 17.2a–c, and Table 17.1 lists the obtained parameters for the three images in Fig. 17.1 and the overall average values. For the image in Fig. 17.2a, the average diameter is 312 ± 103 nm; for the image in Fig. 17.2b, it is 321 ± 109 nm; and for that in Fig. 17.2c, it is 330 ± 119 nm. Figure 17.5 displays the overall histogram for the three images in Fig. 17.2, where the average diameter is 322 ± 109 nm. The diameters of the prisms are large such that they can provide high surface to volume ratio. The wide distribution of diameters is mainly due to the thinning of the prisms at the ends, where they take the shape of clipped pyramid, and the length of this part is different for the different prisms. Because the prisms are perpendicular to the surface, it was not possible to estimate their lengths. The solidity of the prisms was also estimated, and it was found to be the same for the three images, where the overall average value is 0.83 ± 0.01. The circularity was estimated as a measure of compactness, where it was also found to be the same for the three images, and the

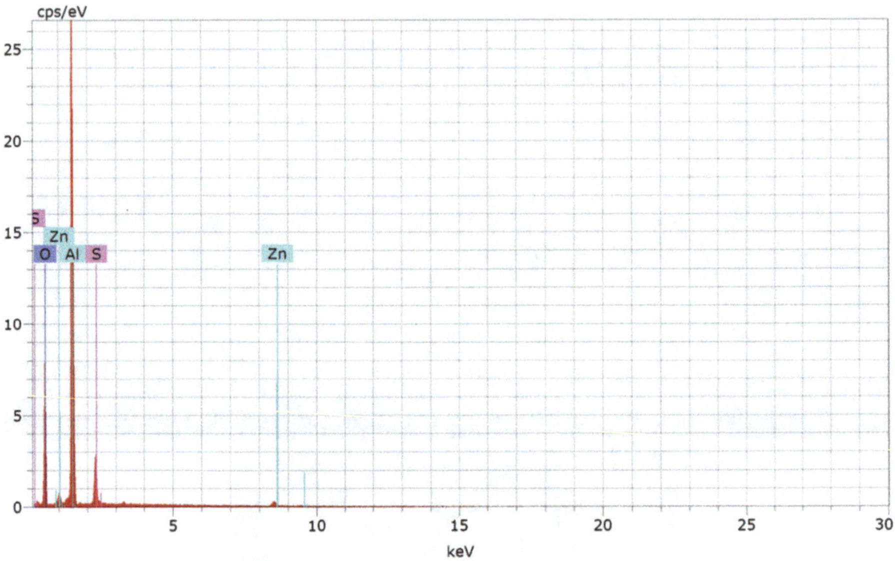

Fig. 17.3 EDS spectrum of one of the ZnO samples

overall average value is found to be 0.91 ± 0.01. Surface area was as large as 3.3×10^5 nm^2, which is adequate for use in sensitized solar cells. The wide distribution in surface area is related to the different sizes of the prisms, in addition to the end effects related to the size of the pyramidal part at the ends.

17.4 Conclusions

Directional ZnO hexagonal prisms on aluminum substrate were produced by the SP technique at a substrate temperature of about 350 °C. XRD diffraction revealed that the preferential orientation is (002), which means that the prisms are aligned on the c-axis perpendicular to the substrate. SEM images confirmed the results of XRD measurements and revealed that the prisms are uniform with apparent side and end facets. EDS analysis showed that the films are oxygen rich and they contain chlorine, and sulfur. Chlorine is due to the use of $ZnCl_2$ as the starting material, and sulfur is related to the impurities of $ZnCl_2$. Excess oxygen is mainly due to aluminum substrates, which are incorporated in Al_2O_3. The diameters of the prisms are large such that they can provide high surface to volume ratio, which make them suitable for use in solar cells, gas sensors, and other applications. These results can contribute in the development of solar cells and micro- and nano-devices.

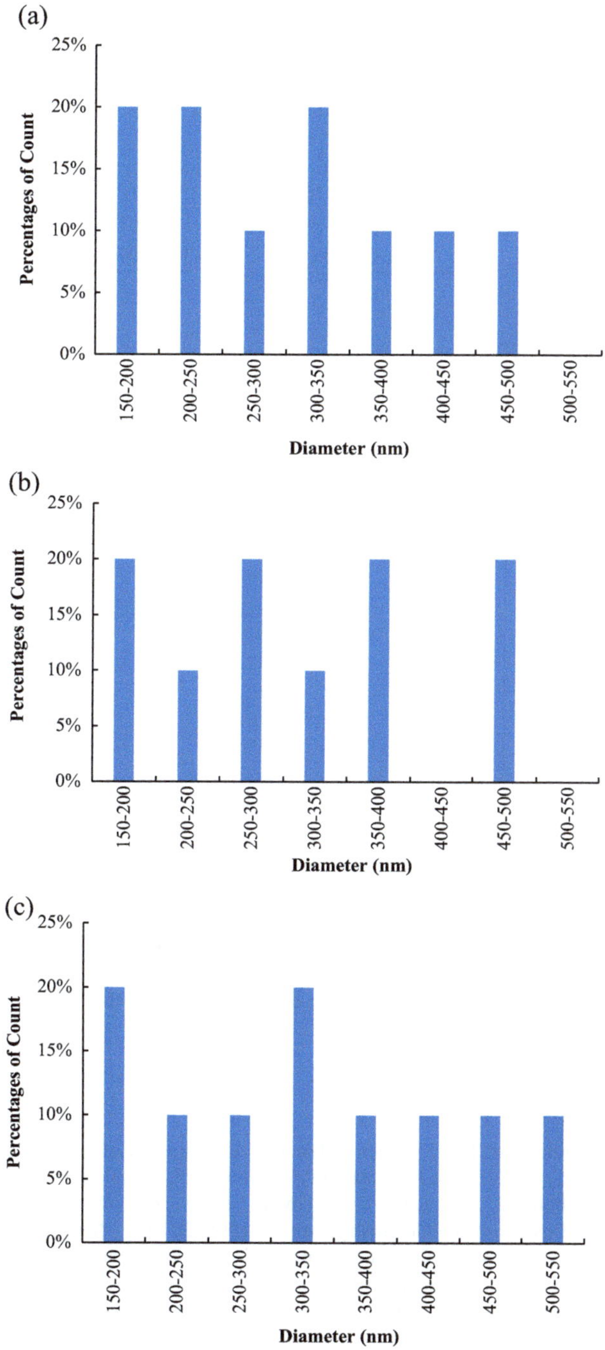

Fig. 17.4 Diameter distribution for the SEM images of ZnO prisms on Al substrates shown in Fig. 17.2. (**a**) Image in Fig. 17.2a. (**b**) Image in Fig. 17.2b. (**c**) Image in Fig. 17.2c

Table 17.1 The prism parameters obtained by imageJ software

	Image in Fig. 17.2a	Image in Fig. 17.2b	Image in Fig. 17.2c	Overall
Average diameter (nm)	312 ± 103	321 ± 109	330 ± 119	322 ± 109
Surface area (nm^2)	$3.1 \times 10^5 \pm 3.3 \times 10^4$	$3.2 \times 10^5 \pm 3.7 \times 10^4$	$3.4 \times 10^5 \pm 4.4 \times 10^4$	$3.3 \times 10^5 \pm 3.7 \times 10^4$
Circularity	0.91 ± 0.01	0.91 ± 0.01	0.91 ± 0.01	0.91 ± 0.01
Solidity	0.83 ± 0.01	0.83 ± 0.01	0.83 ± 0.01	0.83 ± 0.01

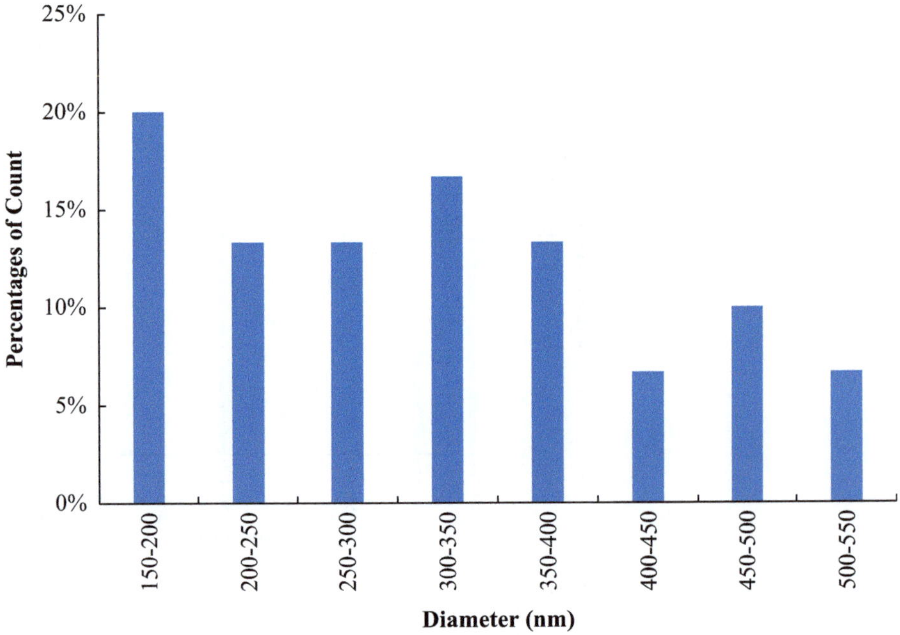

Fig. 17.5 The overall diameter distribution for the three SEM images shown in Fig. 17.2

References

Asadian, M. (2013). Thermodynamic analysis of ZnO crystal growth from the melt. *Journal of Crystallization Process and Technology, 3*, 75–80.

Barbalace, K. (2016). Periodic table of elements – sorted by ionic radius. EnvironmentalChemistry.com. 1995–2016. Accessed on-line: 7/13/2016 http://EnvironmentalChemistry.com/yogi/periodic/ionicradius.html.

Djouadi, D., Meddouri, M., Chelouche, A., Hammiche, L., & Aksas, A. (2014). Structural and morphological characterizations of ZnO nanopowder synthesized by hydrothermal route using inorganic reactants. *Journal of Semiconductors, 35*(12), 123001.

Ekthammathat, N., Phuruangrat, A., Thongtem, S., & Thongtem, T. (2015). Hydrothermal – assisted synthesis and photoluminescence of ZnO microrods. *Digest Journal of Nanomaterials and Biostructures, 10*(1), 149–153.

Feng, L., Liu, A., Ma, Y., Liu, M., & Man, B. (2010). Structural and optical properties of ZnO whiskers grown on ZnO-coated silicon substrates by non-catalytic thermal evaporation process. *Physica E, 42*, 1928–1933.
Hamada, T., Fujii, E., Chu, D., Kato, K., & Masuda, Y. (2011). Aqueous synthesis of single-crystalline ZnO prisms on graphite substrates. *Journal of Crystal Growth, 314*, 180–184.
Hassan, J. J., Mahdi, M. A., Asmiet, R., Abu Hassan, H., & Hassan, Z. (2013). Fabrication and characterization of ZnO nanorods/p-6H–SiC heterojunction LED by microwave-assisted chemical bath deposition. *Superlattices and Microstructures, 53*, 31–38.
Ikhmayies Shadia, J. (2016). Synthesis of ZnO microrods by the spray pyrolysis technique. *Journal of Electronic Materials, 45*(8), 3964–3969.
Ikhmayies Shadia, J., Abu El-Haija Naseem, M., & Ahmad-Bitar Riyad, N. (2010a). Characterization of undoped spray-deposited ZnO thin films of photovoltaic applications. *FDMP: Fluid Dynamics & Materials Processing, 6*(2), 165–178.
Ikhmayies Shadia, J., Abu El-Haija Naseem, M., & Ahmad-Bitar Riyad, N. (2010b). Electrical and optical properties of ZnO:Al thin film prepared by the spray pyrolysis technique. *Physica Scripta, 81*(1), 015703.
Ikhmayies Shadia, J., Abu El-Haija Naseem, M., & Ahmad-Bitar Riyad, N. (2010c). The Influence of annealing in nitrogen atmosphere on the electrical, optical and structural properties of spray-deposited ZnO thin films. *FDMP: Fluid Dynamics & Materials Processing, 6*(2), 219–232.
Ikhmayies Shadia, J., Abu El-Haija Naseem, M., & Ahmad-Bitar Riyad, N. (2014). A comparison between different ohmic contacts for ZnO thin films. *Journal of Semiconductors, 36*(3), 033005.
Kadi Allah, F., Cattin, L., Morsli, M., Khelil, A., Langlois, N., & Bernède, J. C. (2010). Microstructural properties of ZnO:Sn thin films deposited by intermittent spray pyrolysis process. *Journal Materials Science: Materials Electronic, 21*, 179–184.
Kao, M.-C., Chen, H.-Z., Young, S.-L., Lin, C.-C., & Kung, C.-Y. (2012). Structure and photovoltaic properties of ZnO nanowire for dye-sensitized solar cells. *Nanoscale Research Letters, 7*, 260.
Lee, S. H., Lee, J. S., Ko, W. B., Sohn, J. I., Cha, S. N., Kim, J. M., Park, Y. J., & Hong, J. P. (2012). Photoluminescence analysis of energy level on li-doped ZnO nanowires, grown by a hydrothermal method. *Applied Physics Express, 5*, 095002.
Li, F., Ding, Y., Gao, P., Xin, X., & Wang, Z. L. (2004). Single-crystal hexagonal disks and rings of ZnO: Low-temperature, large-scale synthesis and growth mechanism. *Angewandte Chemie, 116*, 350–5354.
Liu, Y. L., Liu, Y. C., Feng, W., Zhang, J. Y., Lu, Y. M., Shen, D. Z., Fan, X. W., Wang, D. J., & Zhao, Q. D. (2005). The optical properties of ZnO hexagonal prisms grown from poly (vinylpyrrolidone)-assisted electrochemical assembly onto Si (111) substrate. *The Journal of Chemical Physics, 122*, 174703.
Lupan, O., Chow, L., Chai, G., Roldan, B., Naitabdi, A., Schulte, A., & Heinrich, H. (2007). Nanofabrication and characterization of ZnO nanorod arrays and branched microrods by aqueous solution route and rapid thermal processing. *Materials Science and Engineering B, 145*, 57–66.
Ngom, B. D., Mpahane, T., Manikandan, E., & Maaza, M. (2016). ZnO nano-discs by lyophilization process: Size effects on their intrinsic luminescence. *Journal of Alloys and Compounds, 656*(25), 758–763.
Ridha Noor, J., Jumali, M. H. H., Umar, A. A., & Alosfur, F. (2013). Defects-controlled ZnO nanorods with high aspect ratio for ethanol detection. *International Journal Electrochemical Science, 8*, 4583–4594.
Shannon, R. D. (1976). Revised effective ionic radii and systematic studies of interatomic distances in halides and chalcogenides. *Acta Cryst, A32*, 751–767.
Thambidurai, M., Muthukumarasamy, N., Velauthapillai, D., & Lee, C. (2010). Chemical bath deposition of ZnO nanorods for dye sensitized solar cell applications. *Energy & Environmental Science, 3*, 789–795.

Van, Q. N., Minh, V. A., Van, L. N., Hung, V. N., & Van, H. N. (2011). Gas sensing properties at room temperature of a quartz crystal microbalance coated with ZnO nanorods. *Sensors and Actuators B: Chemical, 153*, 188–193.

Wang, M., Yi, J., Yang, S., Cao, Z., Huang, X., Li, Y., Li, H., & Zhong, J. (2016). Electrodeposition of Mg doped ZnO thin film for the window layer of CIGS solar cell. *Applied Surface Science, 382*, 217–224.

Yılmaz, S., McGlynn, E., Bacaksız, E., Cullen, J., & Chellappan, R. K. (2012). Structural, optical and magnetic properties of Ni-doped ZnO micro-rods grown by the spray pyrolysis method. *Chemical Physics Letters, 525–526*, 72–76.

Zhang, L., Zhao, J., Zhenga, J., Li, L., & Zhu, Z. (2011). Shuttle-like ZnO nano/microrods: Facile synthesis, optical characterization and high formaldehyde sensing properties. *Applied Surface Science, 258*, 711–718.

Chapter 18
Performance Evaluation of a Vertical Axis Wind Turbine Using Real-Time Measuring Wind Data

Choon-Man Jang, Chul-Kyu Kim, Sang-Moon Lee, and Sajid Ali

18.1 Introduction

Wind energy is one of the most promising renewable energy sources available that can be used instead of fossil fuels. Fossil fuels (such as oil, coal, and natural gas) are limited in amount and emit toxic gases on burning, which invokes climate change due to greenhouse gas emission. Wind energy is obtained by natural wind; therefore, it can be called clean and eco-friendly energy sources.

Recently transient performance characteristics of vertical axial wind turbine (VAWT) have been conducted by numerical simulation and experimental measurement (Danao et al. 2013, 2014). They carried out experimental study on a wind tunnel scale vertical axis wind turbine with unsteady wind condition as well as numerical analysis using Reynolds-averaged Navier-Stokes (RANS) analysis to investigate the effects of steady and unsteady wind on the performance of a wind tunnel scale VAWT. Unsteady Reynolds-averaged Navier-Stokes (URANS) analysis was reported to simulate the transient behavior of the Savonius rotor with different aspect ratios (Jaohindy et al. 2013).

In the construction of a wind turbine at the selected region, detailed analysis of the wind characteristics is important to determine the capacity and number of turbines to be installed (Qzay and Celiktas 2016). Wind characteristics like wind

C.-M. Jang (✉) • S. Ali
Environmental and Plant Engineering Research Institute, Korea Institute of Civil Engineering and Building Engineering, Goyang-Si, Republic of Korea

Construction Environment Engineering, University of Science and Technology, Goyang-Si, Republic of Korea
e-mail: jangcm@kict.re.kr

C.-K. Kim • S.-M. Lee
Environmental and Plant Engineering Research Institute, Korea Institute of Civil Engineering and Building Engineering, Goyang-Si, Republic of Korea

A. Sayigh (ed.), *Transition Towards 100% Renewable Energy*,
Innovative Renewable Energy, https://doi.org/10.1007/978-3-319-69844-1_18

speed, wind density, and wind direction depend on the geographical location. So, turbine performance is affected not only by the average wind speed but also by the terrain and surrounding structures at the installation site. Recently, various wind energy analysis methods have been studied for site installation of wind turbines (Shu et al. 2015; Tar and Szegedi 2011). They carried out wind energy analysis using various analysis techniques including Weibull distribution function.

Continuously changing angle of attack of wind on the wind turbine rotor drops its aerodynamic efficiency (Bazilevs et al. 2014). To evaluate the performance of wind turbine installed at real site, data averaging of real-time data is very important. However, it is rare to analyze the performance characteristics of the wind turbine with numerical analysis according to the different time averaging steps of wind data.

In the present study, three-dimensional unsteady flow analyses have been carried out to analyze the performance of the vertical axis wind turbine. To evaluate the effect of average time of wind data measured on the turbine performance, three different time averaging steps using real-time measuring wind data are introduced and compared to the result of URANS. Detailed flow characteristics around the turbine rotor are analyzed and compared.

18.2 Methodology

18.2.1 Vertical Axis Wind Turbine and Geographical Location

Darrieus wind turbine having three rotor blades and rated power of 1.5 kW is introduced in the present study. Figure 18.1 shows the shape of the turbine rotor and performance curve. The blade height and chord of the turbine rotor are 3 m and 0.2 m, respectively. Design blade section profile is NACA0015 while the rotational diameter of the turbine rotor is 2 m. Cut-in wind speed of the turbine is 3 m/s. Rated wind speed and rotor rotational speed are 13.5 m/s and 300 rpm, respectively.

Figure 18.2 shows the turbine location with geographical altitude and overall view of the test station. The wind turbine is installed at the "Urumsil" town of Deokjeokdo Island in South Korea. Latitude and longitude of the test station are 37°14′17.6″N and 126°07′32.0″E, respectively. The sea is less than 200 m in the east direction of wind turbine. An anemometer and anemoscope are installed at the height of 10 m above the ground.

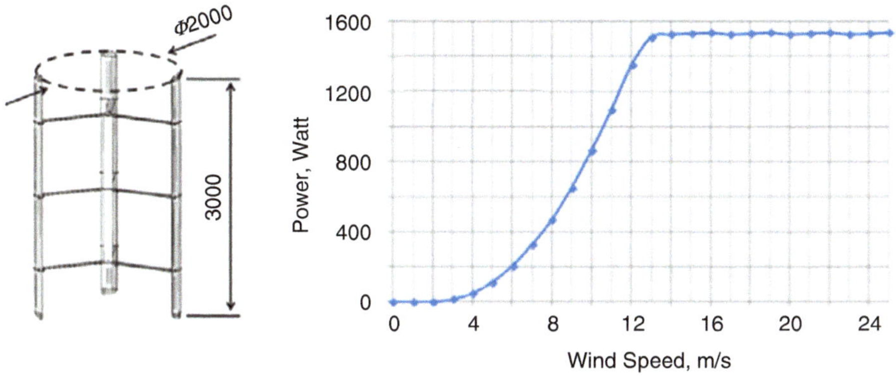

Fig. 18.1 Shape of wind turbine rotor (*left*) and turbine performance curve (*right*)

Fig. 18.2 Location of wind turbine with geographical altitude (*left*) and overall view of system (*right*)

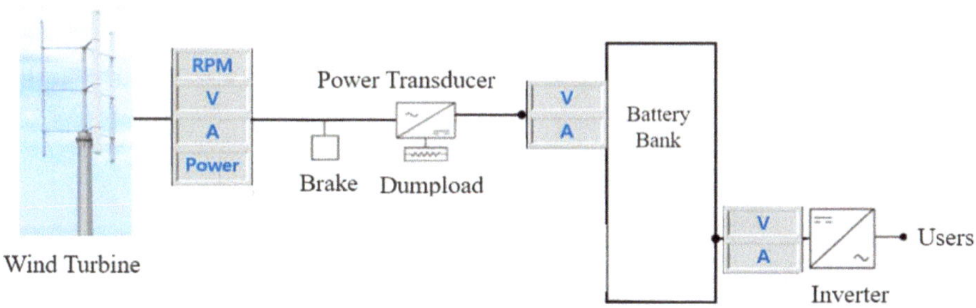

Fig. 18.3 Diagram for data acquisition system

18.2.2 Data Acquisitions

Figure 18.3 shows diagram for data acquisition system to obtain experimental data from the wind turbine and the wind master as shown in Fig. 18.2. Turbine performance data is measured between turbine and power transducer and thus contains power generator loss. Power output is stored in battery bank first and then supplied to users after converting to AC voltages.

Wind speed and direction are stored after every second in the data logger installed on the wind master. The wind master has a height of 10 m, which is same as the height of turbine rotor.

18.2.3 Numerical Analysis

The commercial code, SC/Tetra, has been employed in the present numerical simulation. It solves the governing fluid dynamics equations, which consist of continuity and unsteady Reynolds-averaged Navier-Stokes (URANS) equations. The computational domain which consists of rotational and stationary domains is shown in Fig. 18.4. Considering the wake flow formation at the downstream of the rotor, the total length of computational domain is set to be 20 times larger than rotational diameter of wind turbine.

Tetrahedral, prism, and pyramid elements have been used overall, but mostly only tetrahedral element type is employed as shown in Fig. 18.4. The unstructured meshing is expected to provide higher flexibility in automatic grid generation around complex geometries. The total number of meshing elements is around 13 million, whereas the total number of nodes is approximately 3.5 million in complete domain. Shear stress transport (SST) model (Menter 1994) with a scalable wall function is employed to estimate eddy viscosity. In terms of the boundary conditions, a velocity of 5 m/s is specified at the inlet, and natural outflow condition is imposed at the outlet.

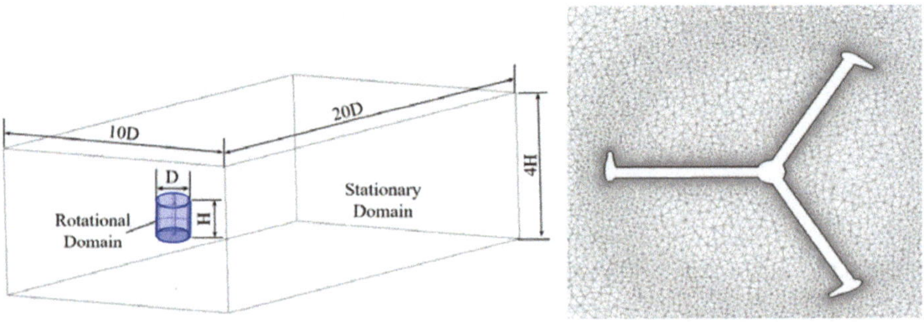

Fig. 18.4 Computational domain (*left*) and grid system around turbine rotor (*right*)

18.3 Results

18.3.1 Standard Deviation with Respect to Average Times

Figure 18.5 shows the standard deviation (SD) for three different wind speeds with respect to average time. SD is analyzed using the measured wind data at the "Urumsil" town of Deokjeokdo Island for 4 months of South Korea's winter when wind speed is relatively high. The SD is defined as follows:

$$\mathrm{SD} = \sqrt{\frac{\sum_{i=1}^{n} \left(x_i - x_{\mathrm{ave}}\right)^2}{n-1}}$$

where x_i and x_{ave} are the specific and average values of a particular variable x, respectively, and "n" is the total number of values of variable x. As shown in the figure, SD has a tendency to increase as average time increases. It is noted that at same averaged value of wind speed 10-min time step produces minimum standard deviation for all wind speed conditions. So 10-min averaged experimental data has been used to evaluate the performance of wind turbine in the present study.

18.3.2 Comparisons of Turbine Performance Between Numerical Analysis and Experiments

Figure 18.6 shows the comparisons of turbine power between numerical simulation and experimental measurement for two time averages. In the figure, turbine power obtained by numerical simulation has similar trend to the experimental result. Especially turbine power determined by 10-min average is more similar to the

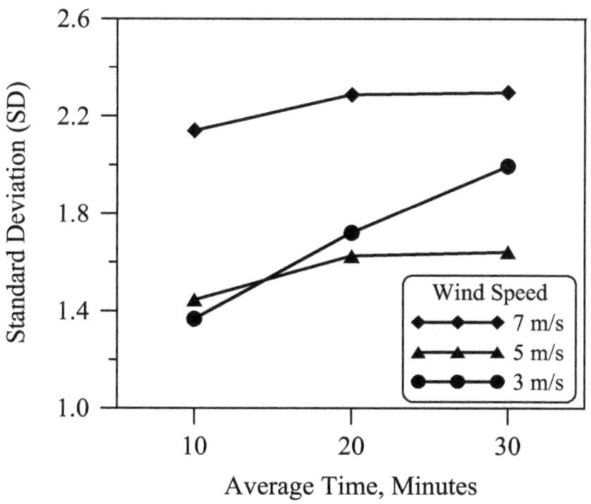

Fig. 18.5 Standard deviations of measuring wind velocities with respect to averaged time

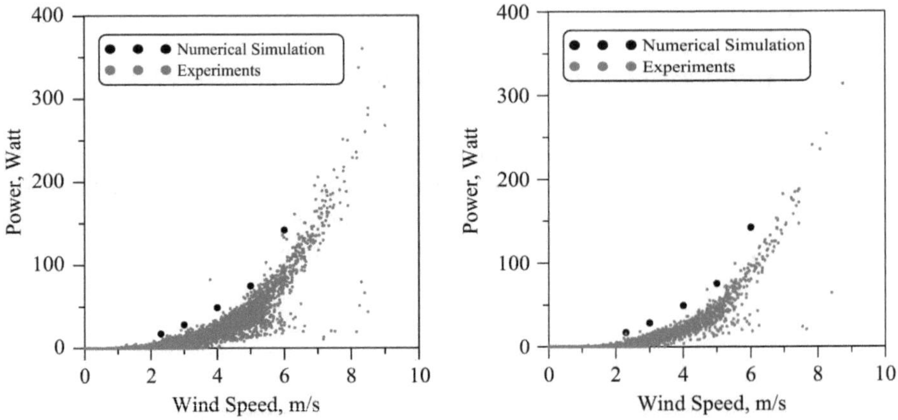

Fig. 18.6 Comparisons of turbine power between numerical simulation and experimental measurement for 10-min average (*left*) and 30-min average (*right*)

results of numerical simulation compared to 30-min average. This is considered that 10-min average step having lower SD is more effective to analyze the performance of a small vertical wind turbine. From the above comparisons, it can be said that turbine power obtained by numerical simulation is correctly analyzed.

18.3.3 Analysis of Internal Flow and Power of Wind Turbine with Respect to Rotor Rotations

Figure 18.7 shows turbine power with respect to azimuthal angle for two different tip speed ratios (TSR). Both the lines have similar trends for one rotor revolution. Local maximum power is observed whenever any of the three blades directly comes in front of wind. For two blades, one which is approaching the wind and the other one which is leaving the wind, the centroids of both are at an equal angular distance from the midpoint of a circular arc. Local minimum power is observed through centroids of both the blades. Global maximum and minimum power occur at azimuthal angle of 240° and 180°, respectively.

Figure 18.8 shows contours of wind speed and pressure around turbine rotor at two different rotation positions, where maximum and minimum values of power coefficient (Cp) occurred, during one complete revolution. Blade having maximum power surrounded by dashed line in the left side is located at the blade rotation angle of 240° where maximum wind velocity around the blade occurred without large separation flow along the blade surface. An increase in linear speed of the blade leads to increase the rotational speed of rotor, and eventually overall power output is enhanced. On the other hand, larger separated flow is observed at the blade

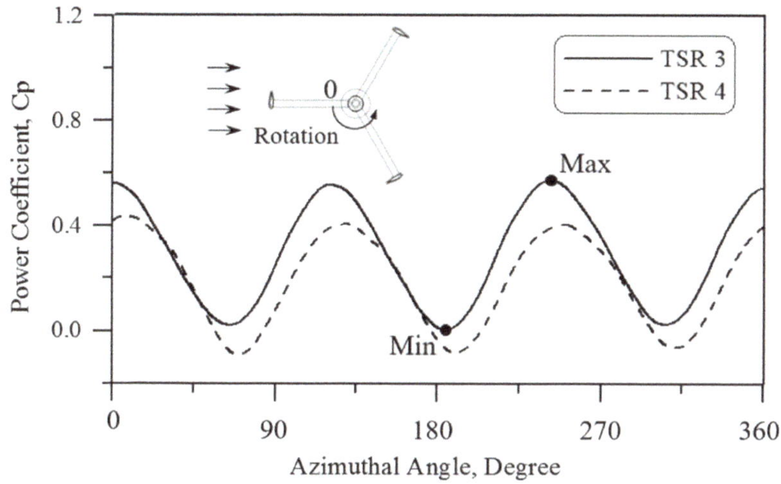

Fig. 18.7 Turbine power with respect to azimuthal angle for two TSR

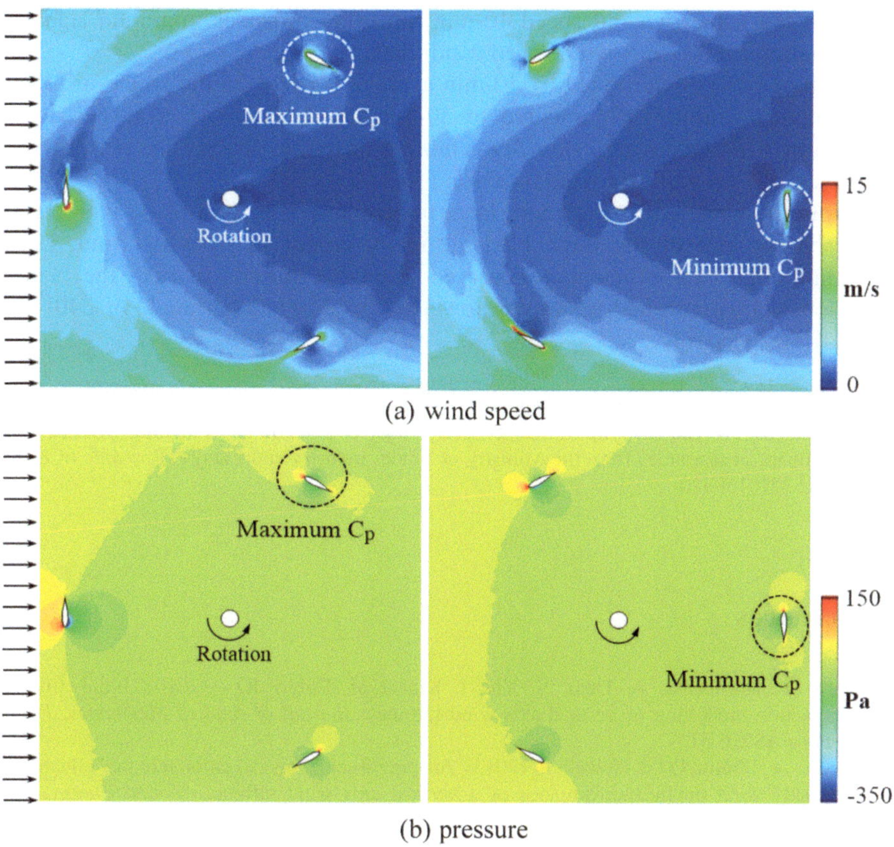

Fig. 18.8 Wind speed and pressure around turbine rotor for the blade positions of min and max power

having minimum power because inflow in front of turbine rotor directly interfaces to the blade surface. Similar pattern is repeated again and again whenever any of the three blades comes directly in front of wind.

18.4 Conclusions

Using a Darrieus wind turbine installed at the "Urumsil" town of Deokjeokdo Island in South Korea, wind turbine performance has been analyzed for three different time averaging steps of the measuring wind speed data: 10, 20, and 30 min. Three-dimensional unsteady flow analyses has also been carried out to analyze the performance of the wind turbine and compared to the experiment data. Standard deviation is used to determine optimal average time for analyzing the performance of the small wind turbine. On the basis of experimental measurement and numerical simulation, the following conclusions are drawn:

- Standard deviation has a tendency to increase as average time increases. Same averaged value of wind speed 10-min time step produces minimum standard deviation for 3–7-m/s wind speed conditions.
- Turbine power determined by 10-min average is more similar to the results of numerical simulation compared to 30-min average. This is considered that 10-min average having lower SD is more effective to analyze the performance of a small vertical axis wind turbine.
- From the numerical simulation, local maximum power of a blade is observed whenever any of the three blades directly comes in front of wind. Local minimum power efficient is observed through centroids of both the blades. Global maximum and minimum power occur at azimuthal angle of 240° and 180°, respectively.

Acknowledgments This work was supported by the new and renewable energy core technology program of the Korean Institute of Energy Technology Evaluation and Planning (KETEP) and granted financial resources from the Ministry of Trade, Industry and Energy, Republic of Korea (No. 20153010130310).

References

Bazilevs Y, Korobenko A, Deng X, Yan J, Kinzel M, Dabiri JO . (2014). Fluid-structure interaction modelling of vertical-axis wind turbines. *Journal of Applied Mechanics, Transactions* ASME 81.

Danao, L. A., Eboibi, O., & Howell, R. (2013). An experimental investigation into the influence of unsteady wind on the performance of a vertical axis wind turbine. *Applied Energy, 107*, 403–411.

Danao, L. A., Edwards, J., Eboibi, O., & Howell, R. (2014). A numerical investigation into the influence of unsteady wind on the performance and aerodynamics of a vertical axis wind turbine. *Applied Energy, 116*, 111–124.

Jaohindy, P., McTavish, S., Garde, F., & Bastide, A. (2013). An analysis of the transient force acting on Savonius rotors with different aspect ratios. *Renewable Energy, 55*, 286–295.

Menter, F. R. (1994). Two-equation Eddy-viscosity turbulence models for engineering applications. *American Institute of Aeronautics and Astronautics Journal, 32*, 1598–1605.

Qzay, C., & Celiktas, M. S. (2016). Statistical analysis of wind speed using two-parameter Weibull distribution in Alacati region. *Energy Conversion and Management, 121*, 49–54.

Shu, Z. R., Li, Q. S., & Chan, P. W. (2015). Investigation of offshore wind energy potential in Hong Kong based on Weibull distribution function. *Applied Energy, 156*, 362–373.

Tar, K., & Szegedi, S. (2011). A statistical model for estimating electricity produced by wind energy. *Renewable Energy, 36*, 823–828.

Chapter 19
SIREN: SEN's Interactive Renewable Energy Network Tool

Angus King

19.1 Introduction

The world's renewable energy resources could provide many times the amount of fossil-fuelled energy currently used. Sustainable Energy NOW Inc. (SEN) has developed an interactive computer simulation tool, SIREN, that models an electricity network and allows the user to create, cost and evaluate scenarios for supplying electricity using a mixture of renewable energy sources.

By placing virtual renewable energy plants around the area of interest, the user can determine the optimal locations to access renewable energy sources, minimise grid connection costs and meet the varying demand on the grid while achieving the best in terms of efficiency, cost-effectiveness and energy security.

SIREN uses renewable energy generation models developed as part of the System Advisor Model (SAM) from the US National Renewable Energy Laboratory (NREL) for energy calculations. "SAM represents the cost and performance of renewable energy projects using computer models developed at NREL, Sandia National Laboratories, the University of Wisconsin, and other organisations" (NREL 2010). Specifically, SIREN uses the SAM Simulation Core (SSC) software development kit (SDK). This is a collection of developer tools for creating renewable energy system models using the SSC library (NREL 2013).

A. King (✉)
Tech Team, Sustainable Energy Now Inc., Perth, WA, Australia
e-mail: siren@ozsolarwind.com

A. Sayigh (ed.), *Transition Towards 100% Renewable Energy*,
Innovative Renewable Energy, https://doi.org/10.1007/978-3-319-69844-1_19

19.2 Methodology

The objective for SIREN is to provide a tool for rapid development of renewable energy scenarios using robust models for calculating the generation from renewable energy stations. To enable this to be achieved, the tool requires a number of datasets, objects and means to create renewable energy scenarios and to invoke the SAM models to calculate renewable energy generation.

19.2.1 A Map

To help visualise the renewable energy resource, SIREN uses a map. Rather than using a sophisticated GIS system, SIREN uses a simple rectangular map image of the desired area. To be used effectively, SIREN needs to know the upper left and lower right geographic coordinates of the area to allow it to correctly overlay objects – power stations and transmission lines – on the map. SIREN uses the EPSG:3857 (WGS 84/Popular Visualisation Pseudo-Mercator) coordinate system which allows an area of the earth to be projected as a rectangle. This is a Spherical Mercator projection coordinate system popularised by Web services such as Google and later OpenStreetMap (it may be possible to use other coordinate systems but to date this has not been tested). SIREN provides a tool to download open source maps, but any suitable map image can be used.

To enhance the usefulness of SIREN, it is possible to overlay the existing electricity transmission network and power stations onto the map. SIREN uses Keyhole Markup Language (KML) files for the electricity grid. For performance reasons and to assist with adding new stations onto the existing network, the existing grid should be fairly simple and requires all transmission lines to connect accurately.

Figure 19.1 shows example maps for the South West Interconnected System (SWIS) in the south-west of Western Australia. Figure 19.1a shows the simple transmission network for the SWIS, while Fig. 19.1b shows weather resource grids for the SWIS (see below).

19.2.2 Weather Data

The SAM models use weather data to calculate renewable energy generation for a particular location. For the USA, and some other countries, this weather data is available for a range of locations, but for other parts of the world, it is not so readily available. To overcome this deficiency in available data, SEN has used Modern-Era Retrospective analysis for Research and Applications, Version 2 (MERRA-2) datasets from the National Aeronautics and Space Administration (NASA 2015).

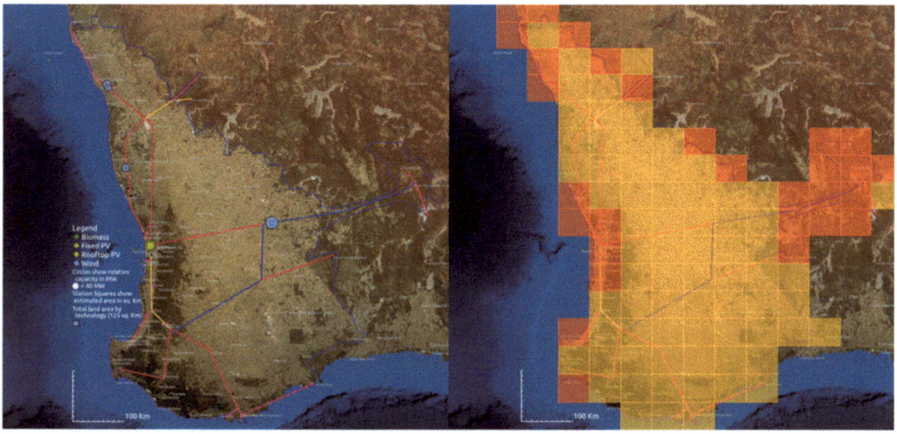

Fig. 19.1 Maps for the SWIS. (**a**) Simplified grid. (**b**) Weather resource grid (Base map data © OpenStreetMap contributors CC-BY-SA http://www.openstreetmap.org/copyright)

MERRA-2 provides data beginning in 1980 with worldwide coverage. The datasets provide spatial resolution of about 55 km in latitude (0.5° of latitude × 0.625° of longitude). SIREN makes use of two-time averaged two-dimensional MERRA-2 datasets – tavg1_2d_slv_Nx Single-Level Diagnostics and tavg1_2d_rad_Nx Radiation Diagnostics (Bosilovich et al. 2016).

The tavg1_2d_slv_Nx dataset provides hourly wind data at 50 m height which makes it suitable for input to the SAM wind model. The tavg1_2d_rad_Nx dataset provides surface net downward short wave flux which is a proxy for Global Horizontal Insolation (GHI). Using routines from NREL, it is possible to calculate Direct Normal Insolation (DNI; Maxwell 1987) and Direct Horizontal Insolation (DHI; NREL 2012) from GHI. This enables solar weather files to be created for input into the various SAM solar models.

SIREN includes tools to (a) simplify obtaining MERRA-2 data for a geographic area and timeframe and (b) to take the MERRA-2 data and generate weather files suitable for input to SAM. The weather files can also be used as input to SAM itself.

As shown in Fig. 19.1b, SIREN can overlay wind and solar resource values as a coloured grid to assist in choosing optimal locations for renewable energy stations.

Weather data that reflects actual weather conditions of the past enables any model to better map actual load demand. For example, hot weather increases air conditioning load. If load data is available for the electricity network, the weather data derived from MERRA-2 enables the model to reflect the impact of the weather on actual load.

19.2.3 Simulation Objects

SIREN has two main objects – power stations and transmission lines. Stations can be saved as scenarios in either a Microsoft Excel or comma-separated variable (CSV) file format, while transmission lines are constructed during SIREN execution rather than being saved to a file.

19.2.3.1 Stations

Stations represent either existing generation stations on the grid or new ones created during a SIREN session. The station files can be updated outside of the simulation. To simplify creation of stations, only a small number of fields are required. These include the name, technology, location and maximum capacity for stations of any technology. There are a few specific fields for some technologies, such as the type of turbine and rotor diameter for wind.

To enable a renewable electricity network to be modelled, we need a mechanism to place existing and proposed generation stations within the network. As well as having a file of stations, new stations can be easily added into the current model by clicking in the appropriate location on the map. This will present a menu to enable input of the required fields. It is possible to save these stations as part of a new scenario or save all (new) stations as a new scenario.

19.2.3.2 Transmission Lines

Transmission lines connect new stations to either existing or other new transmission lines. By default the transmission line will connect a station to the nearest point on the grid. The user can edit a transmission line to connect it elsewhere in the grid. Transmission lines are recalculated when the user adds or modifies stations but are not normally saved. However, (new) transmission lines can be listed in a table and the table saved. Again the number of fields required to define a transmission line is small. A number of other fields are created during execution. A grid is built up of a number of different capacities of transmission line to enable differing loads to be carried. Each line has a maximum carrying capacity. Greater loads require additional lines to be added. In SIREN one line may represent multiple transmission lines and carry both dispatchable and non-dispatchable transmission.

19.2.4 Using SIREN

SIREN allows the user to build scenarios for the energy mix for the area of interest. It is based around the map of the chosen area and allows the user to visualise the layout of the existing electricity network plus additional stations placed on the map to build a scenario(s). The map initially shows the main skeleton of the existing electricity grid infrastructure and current generation stations.

To add a new station, the user simply right-clicks on the map at the desired location and then updates the details for the station (name, technology, capacity and so forth) and saves these details. The user can also copy an existing station and invoke other options. Resource grids can be used to assist with optimal placement of stations.

As a scenario is built, the simulation adds additional infrastructure to connect the new generation plants in to the grid. The user navigates around the map using normal mouse movements to scroll, zoom, etc. and interact with it. Using various menu options, the user can model the whole scenario or by right-clicking deal with an individual station. There are a number of additional windows, such as menu options or resource grids, that can be opened to assist with using SIREN.

The user can save multiple scenarios and add them and remove them as they wish. When SIREN exits normally, any modified scenarios will be saved. When new stations are added to a model, they are added as a new scenario.

19.2.5 Running the Models

The strength of SIREN is in running the SAM models. Once the proposed renewable energy scenario has been defined, SIREN will run the appropriate SAM model for each station chosen or for all renewable stations in the current simulation. SAM calculates a list of power outputs for each hour of the year, ignoring leap days (8760 h).

Outputs from SIREN can include full hourly output for each station, a range of summary data and a large number of graphs to visualise the generation, load and shortfall/surplus for each hour, month, season or a broader period. Graphs can be by period or an average diurnal curve for the period. Figure 19.2 shows two sample graphs. Figure 19.2a shows an hourly plot for a 7-day period, and Fig. 19.2b shows the diurnal profile for a full year. In both instances, the solid green line shows total renewable generation, the solid black shows load for the same period and the red line shows the shortfall or excess of generation for the period. In this instance the diurnal figure shows load with a slight morning peak and larger evening peak and indicates a shortfall of generation in the early evening.

The SAM models are very comprehensive and can be quite complex. For example, there are around 200 input variables possible for the solar thermal model used by SIREN. To simplify the interface to SAM, SIREN uses a variable file for each technology to store default values for most variables and only requires a minimal number of variables to be set by the user, such as station capacity. To fully exploit the capabilities of SAM, it is possible to either create a different copy of the variable file or model a station outside of SIREN using SAM itself, save the power data produced by SAM and use it as input during SIREN execution.

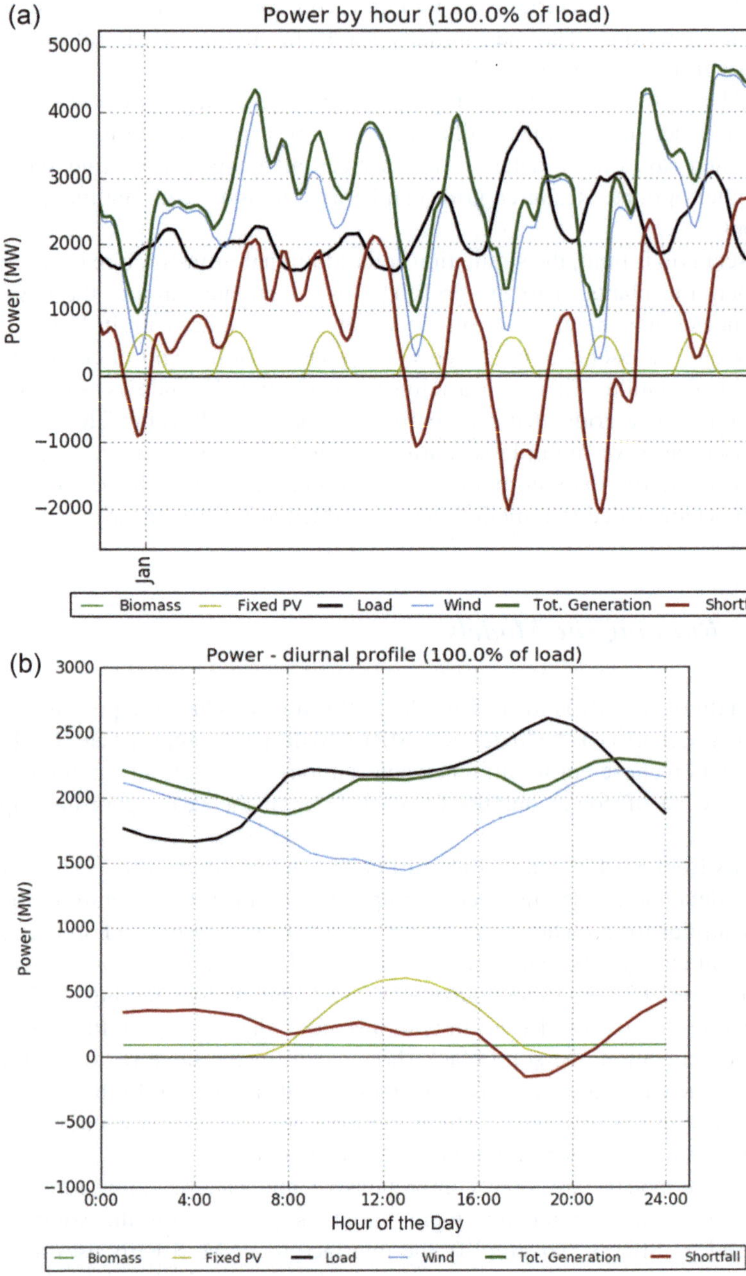

Fig. 19.2 Sample graphs. (**a**) Hourly plot for 7 days and (**b**) diurnal plot for yearly average

The models currently used by SIREN are biomass, geothermal, photovoltaic (PV), solar thermal (power tower with molten salt storage) and wind, plus two financial models. If the user chooses to run the SAM's financial models, they will be presented with an additional menu of options to produce financial outputs for SAM's Independent Power Producer (IPP) utility model.

One of the outputs from SIREN can be used in the PowerBalance part of the SIREN toolkit to "balance" supply and demand with a combination of storage, dispatchable generation and load curtailment. PowerBalance allows the determination of the total system wholesale levelised cost of electricity (LCoE) and CO2 emissions to complete a costed renewable energy scenario.

19.3 Results and Verification

SIREN has been used in the creation of a number of reports to investigate renewable generation potential within the South West Interconnected System (SWIS) in the south-west of Western Australia. As part of this process, a verification of generation calculated by the use of weather files derived from MERRA-2 data along with the SAM models was achieved by comparing it with actual generation data for the SWIS renewable energy stations for 2014. Average correlation for 2014 was 0.77, varying from 0.70 to 0.83 for wind and 0.95 for the one utility scale PV farm on the SWIS network.

Figure 19.3 illustrates this correlation by showing cumulative generation calculated by SIREN (dotted lines) with actual SCADA generation data (solid lines) overall and for each renewable generation station on the SWIS. Cumulative curves have been plotted to hide the "noise" from hourly graphs. The four graphs in the figure are (left to right, top to bottom) (i) comparison for overall renewable generation, (ii) comparison for the four largest stations (>75 GWh), (iii) the next three stations (>20 GWh; Greenough is the PV farm) and (iv) overall hourly curve for the first 7 days highlighting the "noise" (calculated in blue, SCADA in red). Average correlations are between 0.70 and 0.76 for the years 2007–2013 and 2015.

19.4 Conclusions/Discussion

The usage of SAM models and NASA-derived MERRA-2 weather files demonstrates that SIREN is a useful tool for modelling electricity networks for any geographic region (at least within latitudes 85° north and 85° south). It can be scaled from the size of an individual property to very large electricity networks and is a viable tool for initial modelling of renewable energy for all geographies.

Strong correlations between generation calculated by SAM models using NASA MERRA-2 data and actual observations from SCADA data obtained by SEN for the period 2007–2015 are enough to support the use of weather data derived from

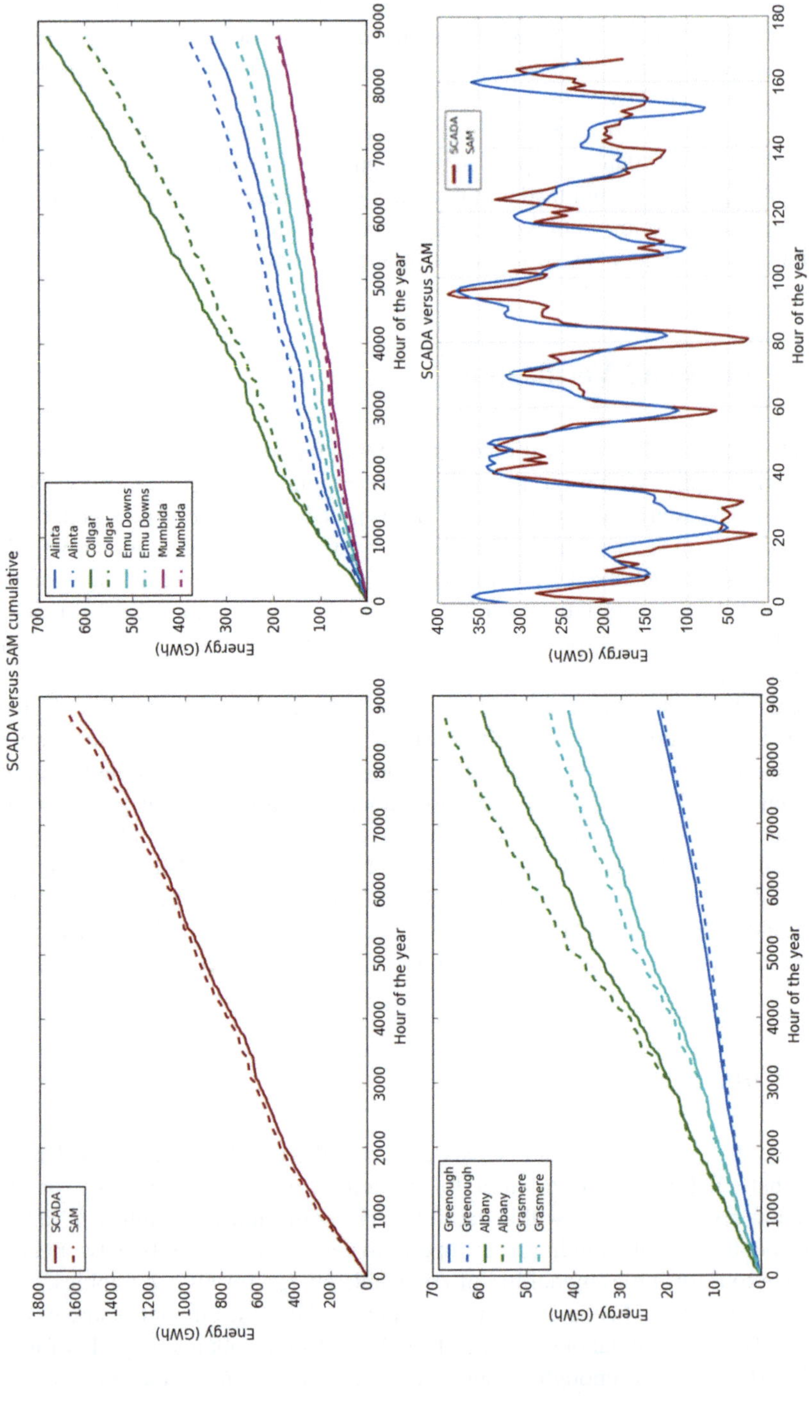

Fig. 19.3 Comparison of actual SWIS renewable generation with calculated generation for 2014

NASA MERRA-2 data with the SAM models for the purposes of modelling renewable energy systems.

SIREN has been developed in the Python programming language as an open source project and is available at https://github.com/ozsolarwind/siren (accessed November 2016). It is licensed under the GNU Affero General Public License. There is a packaged version of SIREN for Windows available as a self-extracting zip file from https://sourceforge.net/projects/sensiren/ (accessed November 2016). SIREN has also been successfully executed on Mac OSX using Wine (https://www.winehq.org/; accessed November 2016).

Feedback on the usage of SIREN or suggested improvements is welcomed.

References

Bosilovich, M. G., Lucchesi, R., Suarez, M. (2016). MERRA-2: File specification. GMAO Office Note No. 9 (Version 1.1). http://gmao.gsfc.nasa.gov/pubs/office_notes. Accessed Nov 2016.

Maxwell, E. L (1987). *A quasi-physical model for converting hourly global horizontal to direct normal insolation*. Technical Report No. SERI/TR-215-3087, Solar Energy Research Institute. Golden. Colorado. USA. http://rredc.nrel.gov/solar/pubs/pdfs/TR-215-3087.pdf. Accessed Nov 2016.

National Aeronautics and Space Administration (NASA). (2015). Modern-era retrospective analysis for research and applications, version 2. https://gmao.gsfc.nasa.gov/reanalysis/MERRA-2/ . Accessed Nov 2016.

National Renewable Energy Laboratory (NREL). (2010). Welcome to SAM. https://sam.nrel.gov/. Accessed Nov 2016.

National Renewable Energy Laboratory (NREL). (2012). DNI-GHI to DHI calculator. https://sam.nrel.gov/sites/default/files/content/documents/xls/DNI-GHI_to_DHI_Calculator.xlsx. Accessed Nov 2016.

National Renewable Energy Laboratory (NREL). (2013). SAM simulation core SDK. https://sam.nrel.gov/sdk. Accessed Nov 2016.

Chapter 20
Characterization and Alkaline Pretreatment Lignocellulose of *Cabomba caroliniana* and Its Role to Secure Sustainable Biofuel Production

Eka Razak Kurniawan, Uju, Joko Santoso, Amarulla Octavian, Yanif Dwi Kuntjoro, and Nugroho Adi Sasongko

20.1 Introduction

Bioethanol is a liquid biochemical obtained as a result of sugar fermentation from carbohydrates, mediated by microorganisms. The Renewable Fuels Association (RFA 2015) report that use of ethanol as biofuel could reduce 30% of exhaust emissions and 50% of smooth particles in exhaust emissions. Generally, bioethanol can be produced from plants that contain sugar or starch. Plants that are commonly used as a source of raw material are cassava, corn, and sugar cane. These plants, however, are also food plants that are consumed by society. Alternative raw materials for production of bioethanol can be derived from biomass. Lignocellulose has several fiber components, including cellulose, hemicellulose, and lignin (Soccol

E.R. Kurniawan (✉)
Energy Security, Faculty of Defense Management, Indonesia Defense University, Bogor, Indonesia

Aquatic Product Technology, Faculty of Fisheries and Marine Science, Bogor Agricultural University, Bogor, Indonesia
e-mail: eka.kurniawan@idu.ac.id

Uju • J. Santoso
Aquatic Product Technology, Faculty of Fisheries and Marine Science, Bogor Agricultural University, Bogor, Indonesia

A. Octavian • Y.D. Kuntjoro
Energy Security, Faculty of Defense Management, Indonesia Defense University, Bogor, Indonesia

N.A. Sasongko
Energy Security, Faculty of Defense Management, Indonesia Defense University, Bogor, Indonesia

The Agency for Assessment and Application of Technology, Central Jakarta, Indonesia
e-mail: nugroho.adi@bppt.go.id

A. Sayigh (ed.), *Transition Towards 100% Renewable Energy*,
Innovative Renewable Energy, https://doi.org/10.1007/978-3-319-69844-1_20

et al. 2011). Lignocellulose can be derived from weeds or waste from the agricultural industry.

C. caroliniana is a water plant containing fibers of lignin, cellulose, and hemicellulose. Currently, this plant causes problems in aquatic ecosystems, disturbs operational dams, influences water quality, and reduces aesthetics. This water plant has a low lignin content, is simple to cultivate, and has a high rate growth. The Cooperative Research Centres Association (CRCA 2003) reported that the rate of growth of *C. caroliniana* at Macdonald Lake, Queensland, Australia reached 55 mm per day. A high population of *C. caroliniana* can potentially ensure that the availability of raw material for bioethanol production is sustainable. Fibers from *C. caroliniana* are useful as a raw material for bioethanol production.

Cellulose and hemicellulose are covered by a lignin layer. A high lignin content can be a barrier to cellulase access, which is necessary for hydrolysis of cellulose to provide sugar monomer (Ishizaki and Keiji 2014). Plants that contain lignin require pretreatment before hydrolysis in order to break the lignin fiber and weaken hydrogen bonds. Pretreatment is carried out to decompose lignocellulose so that its polysaccharides can be broken down to sugar monomer. Biomass lignocellulose pretreatment methods can be divided into four types (Kristina et al. 2012): biological, physical, chemical, and physiochemical.

Alkaline pretreatment is one of the potential pretreatment techniques for breaking down lignin. It can change structural characteristics such as surface area and substrate crystallinity by increasing enzymatic hydrolysis and breaking down the lignin (Karunanity and Muthukumarappan 2011). During pretreatment, two reactions occur, solvation and saponification. These reactions cause the lignocellulose structure to dilate and decrease the degree of polymerization, which simplifies the lignocellulose component for easier enzymatic and fermentation processes. NaOH forms a strong alkali solution in water that can degrade lignin (Wiratmaja et al. 2011).

Using NaOH on corn stover removed 63.6% of lignin content (Zuo et al. 2012). NaOH and NH_4OH addition to bamboo removed lignin and increased cellulose concentration from 46.7% to 63.1% (Kuttiraja et al. 2013). Fockink et al. (2015) conducted pretreatments using different concentrations of NaOH (2% or 4%) at different temperatures (100 °C or 200 °C). By contrast, Cabrera et al. (2014) used NaOH at different concentrations (0.5%, 1%, or 3%) for different times (24 or 48 h). Here, we have used a low temperature and short reaction time, combining the three pretreatment factors (concentration, temperature, and time). No research was carried out regarding the characterization and potential of *C. caroliniana* as raw material for bioethanol production.

Research on structural lignocellulose characterization using alkali pretreatment to reducing total lignin is required to understand the potential of *C. caroliniana* as a raw material for bioethanol production. Such research could also help in deciding which treatment provides maximal degradation of *C. caroliniana* lignocellulose structure, and direct the role of alkaline pretreatment in energy security. The scope of this research includes raw material preparation, lignocellulose characterization, structural characterization, alkaline pretreatment, and analysis of components extracted from aqueous filtrate of *C. caroliniana* (Fig. 20.1).

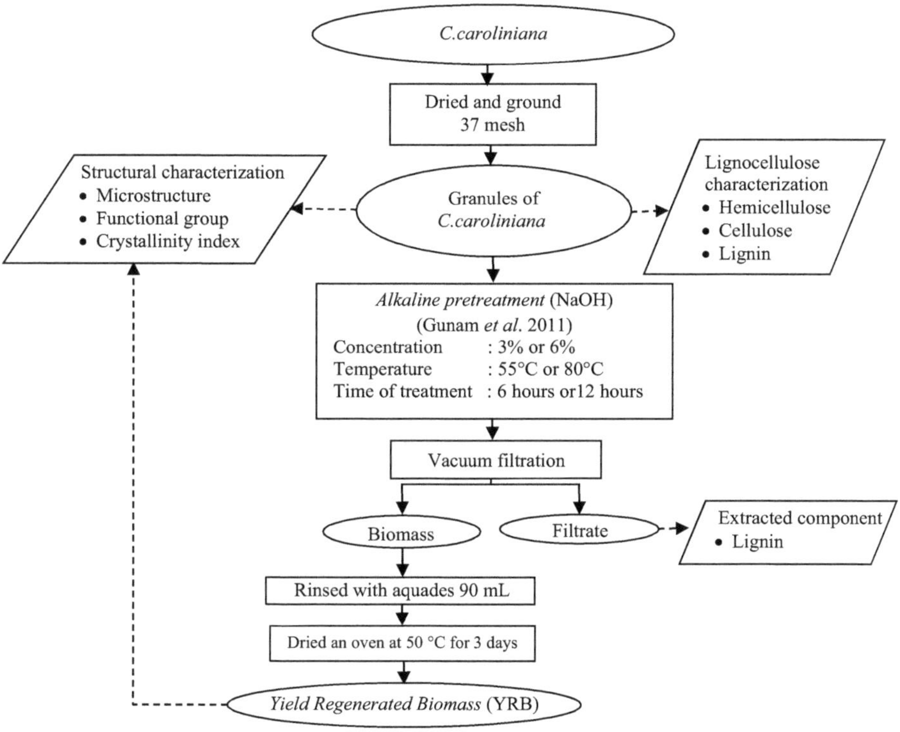

Fig. 20.1 Flowchart of research

20.2 Material and Methods

20.2.1 Materials

Fresh *C. caroliniana* was obtained from Gunung Bunder, Bogor-Indonesia, and was prepared as granules using a 37 mesh. Alkali lignin (0.5 g) was purchased from Sigma-Aldrich; 3% and 6% (w/w) NaOH and H_2SO_4 (72%) were purchased from Merck. All reagents were used as received.

20.2.2 Lignocellulose Characterization

Analysis of acid detergent fiber (ADF) was carried out according to the Van Soest (1973) protocol by applying the method of neutral and acid detergent to remove the carbohydrate.

Analysis of Acid Detergent Fiber A 1 g sample (*A*) was placed in a 600-mL glass container. To this was added 10 mL ADS solution. The sample was extracted for 60 min and then filtered using a weighed granule cup glass (*B*). The residue obtained from extraction was rinsed with hot water and acetone. The sample was dried in an oven at 105 °C for 4 h until the dry weight was stable. The cup was then cooled in a desiccator. After cooling, the cup was weighed (*C*).

Analysis of Cellulose To the weighed ADF sample (*C*) was added 72% sulfuric acid solution to cover the sample. After 3 h, the residue was rinsed with hot water and acetone. The sample was dried in an oven at 105 °C for 4 h until the dry weight was stable. The cup was then cooled in a desiccator. After cooling, the cup was taken from desiccator and weighed (*D*).

Analysis of Lignin The dried sample (*D*) was then burned in a furnace at 600 °C for 6–8 h. The sample was cooled in desiccator and weighed (*E*). ADF, cellulose, and lignin contents were calculated using (Eqs. 20.1, 20.2, and 20.3):

$$\text{ADF}\ (\%) = \frac{C - B}{A} \times 100\% \tag{20.1}$$

$$\text{Cellulose}\ (\%) = \frac{C - D}{A} \times 100\% \tag{20.2}$$

$$\text{Lignin}\ (\%) = \frac{C - E}{A} \times 100\% \tag{20.3}$$

20.2.3 C. caroliniana *Pretreatment*

Dried *C. caroliniana* was ground and screened using a 37-mesh screen. Alkaline pretreatment was as described by Gunam et al. (2011), with some modification. The sample (0.5 g) was heated on a magnetic stirrer (79-1A OSW) in 3% or 6% NaOH solution at a temperature of 55 °C or 80 °C for 6 or 12 h. The treated sample was then separated into biomass and filtrate using vacuum filtration. After separation, the biomass was rinsed with 90 mL Aquades. The resulting biomass was dried in an oven at 50 °C for 3 days. The yield of regenerated biomass (YRB) is the ratio of the mass of dried sample after and before pretreatment. Biomass extracted was calculated using (Eq. 20.4):

$$\text{YRB}\ (\%) = \frac{\text{mass of dried regenerated}}{\text{mass of dried untreated}} \times 100\% \tag{20.4}$$

20.2.4 Scanning Electron Microscopy Analysis

Characterization was carried out for the fresh sample and after pretreatment with various combinations of NaOH concentration, temperature, and time of extraction.

The sample was dried and cut into pieces of 0.5×0.5 cm^2. The sample was layered with gold using a layering tool (Quorum Q150R ES, UK). The microstructure of the gold-coated sample was viewed with an electron microscope (Zeiss EVO MA-10 Jena, Germany) at a 14 kV and magnifications of 250, 500, 1,000, 1,500, 3,000, and 10,000×.

20.2.5 Fourier-Transform Infrared Spectroscopy Analysis

C. caroliniana was analyzed by Fourier-transform infrared spectroscopy (FT-IR; Bruker Tensor 37, Kalkar, Germany) before and after pretreatment. The sample was mixed with KBr and placed in container made of tin metal sheet. The KBr-treated sample was irradiated at 3,500–450 cm^{-1} and at 4 cm^{-1} resolution to view functional groups after each different treatment.

20.2.6 X-Ray Diffraction Analysis

The crystallinity of *C. caroliniana* biomass was measured by X-ray diffraction (Bruker D8 Advance, Kalkar, Germany) using both fresh and pretreated samples. The samples, of 100 mesh, were compacted and viewed at a diffraction angle (2^{θ}) of 5–50°, with step size 0.02° at 40 kV and 35 mA.

20.2.7 Extracted Lignin Analysis

The filtrate resulting from pretreatment (as much as 2 mL) was analyzed for lignin content using a UV-visual system (Hitachi U-2900, Tokyo, Japan) at a wavelength of 283 nm. The sample was placed in a cuvette and the absorbance measured and compared with a previously prepared standard curve.

20.3 Results and Discussion

20.3.1 Characterization of C. caroliniana Lignocellulose

The measured lignin content of *C. caroliniana* was $11.14 \pm 0.68\%$; hemicellulose content was $17.30 \pm 0.13\%$ and cellulose $14.03 \pm 0.32\%$ (Fig. 20.2). Lignin content in this plant is lower than in *Pandanus tectorius*, which has a lignin content of $24 \pm 0.8\%$ (Sheltami et al. 2012). Low lignin content means that it is easier to convert cellulose and hemicellulose into sugars, which are used in fermentation. In

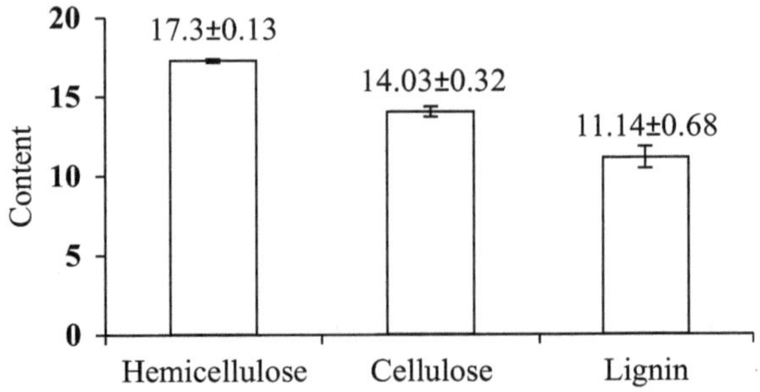

Fig. 20.2 Lignocellulose composition of *C. caroliniana*

this case, biomass conversion to sugars needs a strong pretreatment to degrade lignin and facilitate enzymatic saccharification. *C. caroliniana* has a high hemicellulose content (Fig. 20.2). Hemicellulose is a branching polymer formed of hexose and pentose. In branching polymers, the hydrogen bonds prevent the formation of a crystalline structure. This makes hemicellulose easy to hydrolyze, and many sugars are formed during fermentation.

20.3.2 Yield of Regenerated Biomass and Extracted Lignin from C. caroliniana

Lignocellulose is biomass that exists in agricultural waste, forestry industry waste, and grass. Plant biomass contains cellulose, hemicellulose, lignin, and a few small extractive components. Sindhu et al. (2015) explained that total lignocellulose measurement is influenced by alkali concentration, temperature, and time of pretreatment. The largest total lignin concentration is found in center lamella cells and is less in secondary wall cells (Sjostrom 1993). The existence of lignin gives a plant rigidity so that it can withstand large mechanical pressures.

Pretreatment with 3% NaOH at 55 °C for 6 h gave the highest YRB of $29.05 \pm 3.04\%$. Pretreatment with 6% NaOH at 80 °C for 12 h gave the lowest yield, $14.84 \pm 0.36\%$ (Fig. 20.3a). These results indicate that treatment with 6% NaOH at 80 °C for 12 h gives maximum delignification. A decrease in the final mass of sample was caused by increasing the alkali concentration and time of pretreatment. When lignin and extract components dissolve, sample mass is decreased. The decrease in mass of straw and rice husk after alkaline pretreatment was from 83.7% to 69.4% and from 75.2% to 57.6%, respectively (Cabrera et al. 2014).

The filtrate obtained after pretreatment contained lignin and extracted carbohydrates. Extracted lignin concentration increased along with increases in NaOH concentration, temperature, and time of pretreatment (Fig. 20.3b). Pretreatment with 3% NaOH at 55 °C for 6 h produced 2.4 ± 0.31 g L^{-1} of lignin concentration and pretreatment with 6% NaOH at 80 °C for 12 h produced 3.56 ± 0.03 g L^{-1}. The

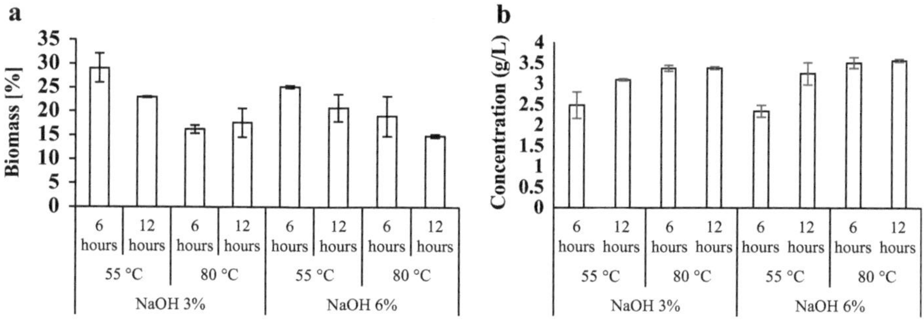

Fig. 20.3 (**a**) Yield of regenerated biomass from *C. caroliniana*. (**b**) Lignin extracted from *C. caroliniana* filtrate

amount of lignin extracted during pretreatment indicated that biomass content was low.

Lignin degradation is influenced by the content of hydroxyl ions, which increase lignin hydrophilicity. During pretreatment, degraded lignin is dissolved. NaOH forms hydroxyl ions, which react with the ether bond (lignin) and produce phenol hydroxyl groups. Degraded lignin dissolved in heated alkali turns to sodium phenoxide (Sjostorm 1993). Increasing the time and temperature of pretreatment increases the extracted lignin content (Kristina et al. 2012). As shown in Fig. 20.3, there were correlations between total extracted biomass and total extracted lignin. The lower biomass was probably a result of extraction of the lignin component and phenolic compounds in the sample. Increasing the NaOH concentration caused a higher hydroxyl ion concentration during pretreatment; thus, breakage of the lignin intramolecular bonding occurred rapidly and lignin was dissolved (Simatupang et al. 2012). Hydrogen ions were transferred from the hydroxyl group in lignin to hydroxyl ions in solution, thus decreasing biomass yield.

20.3.3 Microstructure of Pretreated Lignocellulose

Lignin is a polymer that covers hemicellulose and cellulose in plant tissue. Lignin, hemicellulose, and cellulose are connected by ether bonds (glycosidic). Lignin contains various functional groups, including aromatic rings, hydroxyl groups, carbon–carbon bonds, and ether bonds; hemicellulose includes ether bonds (glycosidic), ester bonds, and hydrogen bonds; cellulose contains ether bonds (glycosidic) and hydrogen bonds (Harmsen et al. 2010). Isolation of cellulose is conducted by degrading lignin tissue, releasing hemicellulose and cellulose from lignin bonding. Hydrolysis of pure cellulose and hemicellulose give increased total sugar production.

The structure of fresh *C. caroliniana* is compact and rigid, its density resulting from lignocellulose tissue. Pretreatment breaks the surface of plants and opens the

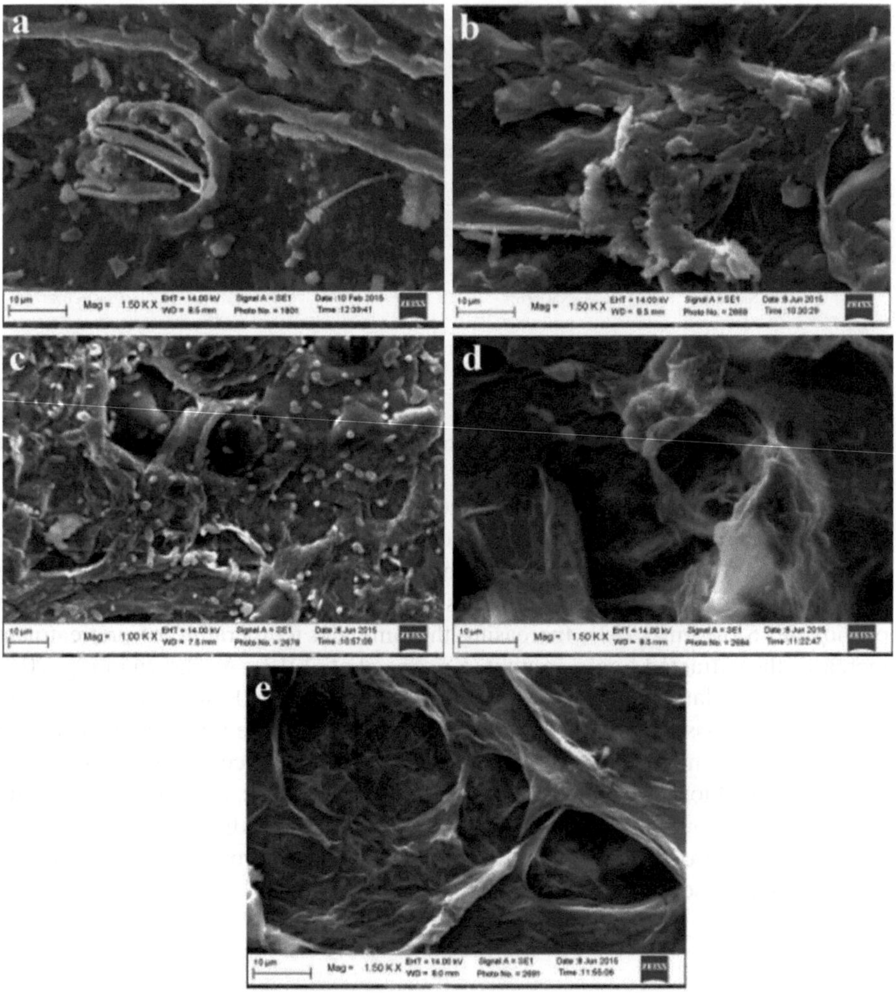

Fig. 20.4 SEM images of *Cabomba caroliniana* (**a**) before pretreament and after the following pretreatments: (**b**) 3% NaOH at 55 °C for 6 h, (**c**) 3% NaOH at 55 °C for 12 h, (**d**) 6% NaOH at 80 °C for 6 h, and (**e**) 6% NaOH at 80 °C for 12 h (Magnification 1,500×)

structure of plants, resulting in broken pores and space within tissue (Fig. 20.4a). After pretreatment, the breaking of plant tissue did not result in compaction and increased density (Fig. 20.4b–e). More broken cells were seen with increasing temperature and time of pretreatment. After alkali treatment, sugar cane appeared fractured with obvious cracks, ravines, and holes in the samples (Liu et al. 2016). The cell damage indicates that lignin was degraded in cells. Degradation increased with increasing NaOH concentration, temperature, and time of pretreatment. NaOH concentration can damage the tissue fiber structure and thus cause lignocellulose to become open and porous.

20.3.4 Functional Groups and Crystallinity Index of Lignocellulose

Cellulose, lignin, hemicellulose, pectin, and sugar are predominately carbohydrates. Each fiber has a different FT-IR spectrum. Results of FT-IR analysis (Fig. 20.5a) after sample pretreatment indicate differences between fresh and pretreated samples. The peak at wavelength 899 cm^{-1} indicates COC vibration from a β-D-cellulose group, but was not present in the fresh sample. CH vibration at 1,161 cm^{-1} is due to the alkyl group in cellulose (Fig. 20.5a) (Stuart 2004). The peak at 1,200 cm^{-1} indicates a COH bond from cellulose II groups (Oh et al. 2005), which was not formed in fresh samples. A peak at 1,543 cm^{-1} indicates C=C vibration from a lignin aromatic group (Bui et al. 2015); this was present in the fresh sample but not in the treated sample. These results indicate that damage to lignin occurs during pretreatment. The lost lignin group was caused by pretreatment. Intermolecular cellulose bonding includes hydrogen bonds, β-D-glycosidic bonds (Harmsen et al. 2010), and carboxyl groups (Krassig and Schurz 2002); hemicellulose is connected by carboxyl or ester bonds and lignin is connected by ether bonds and carbon–carbon bonds. A β-D-cellulose group and COH-cellulose II were identified in cellulose; Figure 20.5a shows the existence of peaks corresponding to these groups. An increase in cellulose concentration occurred after alkali pretreatment. During pretreatment, hydroxyl ions from the solvent damaged lignin tissue, thus freeing tissue cellulose and increasing its concentration. This result was supported by measurements for hemicellulose, which showed a peak corresponding to an alkyl cellulose group (ester bond) after each of the pretreatments, clearly indicating that lignin had been damaged by alkali solvent.

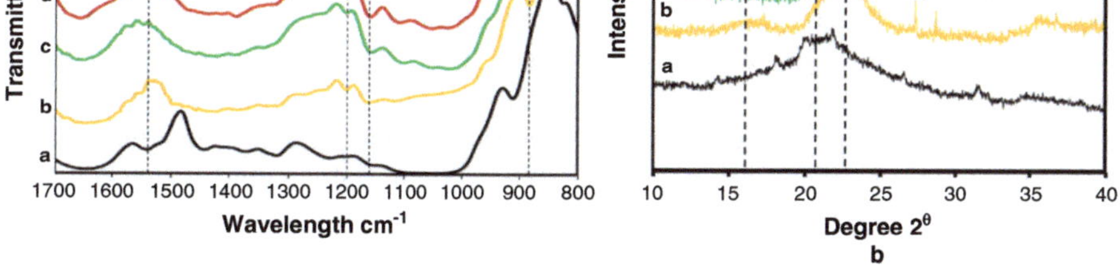

Fig. 20.5 (**a**) Infrared spectrum and (**b**) crystallinity index for *C. caroliniana*. The curves show values *a* before pretreament and after the following preatreatments: *b* 3% NaOH at 55 °C for 6 h, *c* 3% NaOH at 55 °C for 12 h, *d* 6% NaOH at 80 °C for 6 h, and *e* 6% NaOH at 80 °C for 12 h

Table 20.1 Crystallinity index for pretreated *C. caroliniana*

Pretreatment	Crystallinity index (%)
Fresh	41.1
3% NaOH at 55 °C for 6 h	43.9
3% NaOH at 55 °C for 12 h	42.9
6% NaOH at 80 °C for 6 h	54.5
6% NaOH at 80 °C for 12 h	56.0

Crystallinity index is a significant factor that influences enzymatic processing of lignocellulose. X-ray diffraction curves showed that the peak intensity of the fresh sample was at 20.6° and shifted to 22.5° after pretreatment (Fig. 20.5b). The peak corresponds to crystalline cellulose I (Uju et al. 2015). A small peak after pretreatment with 6% NaOH at 80 °C for 6 or 12 h can be seen at 16° (curves d and e in Fig. 20.5b); this corrresponds to crystalline cellulose II (Nelson and Connor 1964). Crystallinity increased along with the severity of the pretreatment process (Table 20.1). The fresh sample had the smallest crystallinity index (41.1%) and pretreatment with 6% NaOH at 80 °C for 12 h resulted in the highest crystallinity index (56%).

Increasing degree of crystallinity was observed with increasing NaOH concentration and temperature. This effect is caused by the amorphous lignin component, which dissolves during pretreatment (Zain et al. 2014). Pretreatment of amorphous hemicellulose caused increased crystallinity indices (Zhang et al. 2016). Crystallinity of mengkuang leaves (*Pandanus tectorius*) was measured before and after pretreatment. Fresh mengkuang leaves had a cristallinity index of 55.1%, which rose to 60.2% after pretreatment with 4% NaOH (Sheltami et al. 2012).

20.3.5 The Role of Alkaline Pretreatment in Sustainable Bioethanol Production

Bioethanol availability is influenced by raw materials and the production process. In Indonesia, there is great potential for bioethanol production based on lignocellulose from sources such as agricultural waste, plantations, fisheries, and grasses. Rice production in 2014 reached 70.83 million tons and corn 19.03 million tons (Statistics Indonesia 2015). In 2015, rice production reached 75.40 million tons and corn 19.61 million tons (Statistics Indonesia 2016). Rice and corn are produced each year because they are the primary food plants in Indonesia. The yields from these plants as agricultural waste or raw material for bioethanol production (based on lignocellulose content) could reached 20% of the initial weight for rice and 12% for corn (Tajalli 2015).

Indonesia is the second-largest producer of palm oil, providing 18% of global needs. Palm oil production in Indonesia was 29,278,189 tons in 2014 and 31,284,306 tons in 2015 (Directorate General of Estate Crops 2015). The yield from oil palms that produced only crude palm oil was 21–23% of fresh fruit bunches

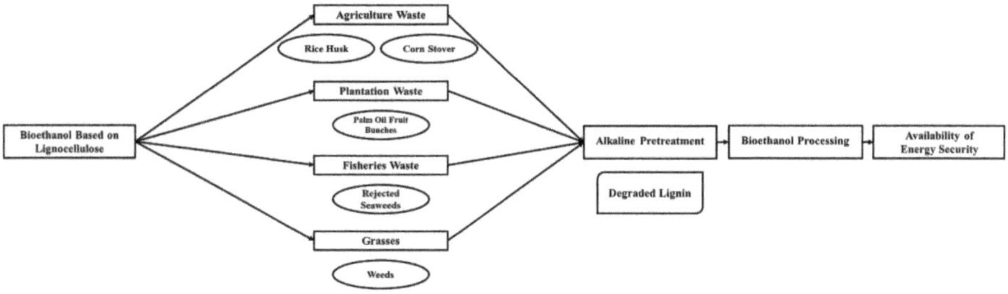

Fig. 20.6 Mind map for the role of alkaline pretreatment in energy security

and resulted in more agricultural waste. Production of seaweed in Indonesia in 2013, according to FAO (2016), reached 34.6% of global seaweed production. Grasses (weeds) such as *C. caroliniana* can reach growth rates of 55 mm day^{-1} (CRCA 2003) and crop densities of 30 plants m^{-2}. This high production produces continual waste, which is a great opportunity for Indonesia to exploit the potential of waste as a raw material or biomass energy source.

However, the problems of bioethanol production based on lignocellulose are related to the hydrogen bonds that bind the cellulose chain strongly in crystal form, which obstructs the breakdown of cellulose to glucose. Pretreatment is the initial step in separating the components of lignocellulose, which are lignin, cellulose, hemicellulose, and other extracted components.

This research examined alkaline pretreatment as the initial step in the bioethanol production process and demonstrated that alkaline pretreatment is effective at breaking down lignin. Alkaline pretreatment as part of the initial production process could resolve bioethanol production problems based on lignocellulose. Alkaline pretreatment causes breakage of the hydrogen bonds between cellulose and hemicellulose, thus enabling enzymatic processes to produce glucose effectively. There is a high potential for bioethanol fuel production in Indonesia, supported by the ready availability of raw materials based on waste, which are produced each year. Alkaline pretreatment reduces the strength of lignin binding to cellulose and hemicellulose chains; thus, conversion of cellulose and hemicellulose to glucose is greater. The availability of bioethanol fuel could increase the potential for renewable energy and offer alternatives for diversification of energy sources. From the point of view of energy security in Indonesia, it is fundamental to obtain energy from available sources (Fig. 20.6).

20.4 Conclusions

The lignocellulose composition of fresh *C. caroliniana* plant was found to be $17.30 \pm 0.13\%$ hemicellulose, $14.03 \pm 0.32\%$ cellulose, and $11.14 \pm 0.68\%$ lignin. Reduction of lignin was accomplished by addition of NaOH, with optimal adjustment of temperature and pretreatment time. Pretreatment with 6% NaOH at 80 °C for 12 h resulted in lower biomass yield and higher extracted lignin content than obtained with other pretreatments. SEM of the morphology of *C. caroliniana* after pretreatment demonstrated damaged tissue structure. Peak intensities in FT-IR spectra indicated that delignification occurs during alkaline pretreatment. The increase in crystallinity index was caused by an increase in amorphous component. Maximum delignification was obtained using a pretreatment of 6% NaOH at 80 °C for 12 h.

Alkaline pretreatment could be advantageous in bioethanol production processes and potentially increase the availability of bioethanol fuel, thus helping to realize energy security.

Acknowledgements This study was supported by the Aquatic Product Technology Department of Bogor Agricultural University, Indonesian Institute of Sciences, Institute for Research and Development of Post Harvest Agriculture and Indonesia Defense University. Thanks to Mr. Deny Verantika, Ms. Anggun Andreyani, and Ms. Asih Tri Marini for supporting the publication.

References

Bui, N. Q., Pascal, F., Franck, R., Cyril, D., Nadege, C., Christpohe, V., & Nadine, E. (2015). FTIR as a simple tool to quantify unconverted lignin from chars in biomass liquefaction process: Application to SC ethanol liquefaction of pine wood. *Fuel Processing Technology, 134*, 378–386.

Cabrera, E., Maria, J. M., Ricardo, M., Ildefonso, C., Caridad, C., & Ana, B. D. (2014). Alkaline and alkaline peroxide pretreatments at mild temperature to enhance enzymatic hydrolysis of rice hulls and straw. *Bioresource Technology, 167*, 1–7.

[CRCA] Cooperative Research Centres Association. (2003). *Weed management guide: Cabomba caroliniana* (pp. 1–6). Barton: CRCA.

Directorate General of Estate Crops. (2015). *Tree crop estate statistics of Indonesia 2014–2016 (Palm Oil)* (pp. 12–13). Jakarta: Directorate General of Estate Crops.

Fockink, D. H., Marcelo, A. C. M., & Luiz, P. R. (2015). Production of cellulosic ethanol from cotton processing residues after pretreatment with dilute sodium hydroxide and enzymatic hydrolysis. *Bioresource Technology, 187*, 91–96.

Gunam, I. B. W., Wartini, N. M., Anggreni, A. A. M. D., & Pande, M. S. (2011). Delignification sugar cane husks with a solution of sodium hydroxide before the process of saccharification in enzymatic using the enzyme cellulose rough from *Aspergillus niger* FNU 6018. *LIPI PRESS, 3*, 24–32.

Harmsen, P. F. H., Huijgen, W. J. J., Lopez, L. M. B., & Bakker, R. R. C. (2010). *Literature review of physical and chemical pretreatment processes for lignocellulosic biomass* (pp. 1–49). Wageningen: Biosynergy.

Ishizaki, H., & Keiji, H. (2014). *Ethanol production from biomass* (pp. 243–258). Oxford: Academic, Elsevier.

Karunanity, C., & Muthukumarappan, K. (2011). Optimization of alkali, big bluestem particle size and extruder parameters for maximum enzymatic sugar recovery using response surface. *BioResources, 6*(1), 762–790.

Krassig, H., & Schurz, J. (2002). *Ullmann's encyclopedia of industrial chemistry* (6th ed.). Weinheim: Wiley-VCH.

Kristina, Sari, E. R., & Novia. (2012). Alkaline pretreatment and simultaneous saccharification processes – for production of ethanol fermentation from empty oil palm bunches. *Jurnal Teknik Kimia, 18*(3), 34–42.

Kuttiraja, M., Sindhu, R., Varghese, P. E., Sandhya, S. V., Binod, P., Vani, S., Pandey, A., & Sukumaran, R. K. (2013). Bioethanol production from bamboo (*Dendrocalamus* sp.) process waste. *Biomass and Bioenergy, 59*, 142–150.

Liu, Y. Y., Jin, L. X., Yu, Z., Cui, Y. L., Min, C. H., Zhen, H. Y., & Jun, X. (2016). Reinforced alkali-pretreatment for enhancing enzymatic hydrolysis of sugarcane bagasse. *Fuel Processing Technology, 143*, 1–6.

Nelson, M. L., & O'Connor, R. T. (1964). Relation of certain infrared bands to cellulose crystallinity and crystal lattice type. Part II. A new infrared ratio for estimation of crystallinity in celluloses II and II. *Journal of Applied Polymer Science, 8*, 1325–1341.

Oh, S. Y., Dong, I. Y., Younsook, S., Hwan, C. K., Hak, Y. K., Yong, S. C., Won, H. P., & Ji, H. Y. (2005). Crystalline structure analysis of cellulose treated with sodium hydroxide and carbon dioxide by means of X-ray diffraction and FTIR spectroscopy. *Carbohydrate Research, 340*, 2376–2391.

[RFA] Renewable Fuels Association. (2015). *Ethanol facts: Environment*. http://www.ethanolrfa.org/pages/ethanol-facts-environment. 6 July 2015.

Sheltami, R. M., Ibrahim, A., Ishak, A., Alain, D., & Hanieh, K. (2012). Extraction of cellulose nanocrystals from Mengkuang leaves (*Pandanus tectorius*). *Carbohydrate Polymers, 88*, 772–779.

Simatupang, H., Nata, A., & Herlina, N. (2012). Study of isolation and yield lignin from empty oil palm bunches. *Jurnal Teknik Kimia USU, 1*(1), 20–24.

Sindhu, R., Pandey, A., & Binod, P. (2015). *Alkaline treatment* (pp. 51–60). Trivandrum: Centre for Biofuels, Biotechnology Division, CSIR. hlm.

Sjostrom, E. (1993). *Wood chemistry, fundamentals and application second edition* (pp. 1–293). London: Academic.

Soccol, C. R., Vincenza, F., Susan, K., Luciana, P. S. V., Vanete, T. S., Adenise, W., & Ashok, P. (2011). *Lignocellulosic bioethanol: Current status and future perspectives*. Oxford: Academic, Elsevier.

Statistics Indonesia. (2015). *The official statistics news XVIII edition-production of rice, corn and soybeans*. Jakarta: Statistics Indonesia.

Statistics Indonesia. (2016). *The official statistics news XIX edition-production of rice, corn and soybeans*. Jakarta: Statistics Indonesia.

Stuart, B. (2004). *Infrared spectroscopy: Fundamental and applications* (pp. 1–203). Hoboken: Wiley.

Tajalli, A. (2015). *Guide to the assessment of the potential of biomass as an alternative energy source in Indonesia* (pp. 9–11). Jakarta: Penabullu Alliance.

Uju, Wijayanta, A. T., Goto, M., & Kamiya, N. (2015). Great potency of seaweed waste biomass from carrageenan industry for bioethanol production by peracetic acid-ionic liquid pretreatment. *Biomass and Bioenergy, 81*, 63–69.

Van Soest, P. J. (1973). Collaborative study of acid-detergent fiber and lignin. *Journal of The AOAC, 56*(4), 781–784.

Wiratmaja, I. G., Gusti, B. W. K., & Nyoman, I., S. W. (2011). The manufacture of the second generation of ethanol by utilizing waste *Eucheuma cottonii* seaweed as raw material. *Jurnal Ilmiah Teknik Mesin CakraM, 5*(1), 75–84.

Zain, N. F. M., Salma, M. Y., & Ishak, A. (2014). Preparation and characterization of cellulose and nanocellulose from Pomelo (*Citrus grandis*) Albedo. *Nutrition and Food Sciences, 5*(1), 1–4.

Zhang, W., Noppadon, S., Justin, R. B., & Scott, R. (2016). Enhanced enzymatic saccharification of pretreated biomass using glycerol thermal processing (GTP). *Bioresource Technology, 199*, 148–154.

Zuo, Z., Tian, S., Chen, Z., Li, J., & Yang, X. (2012). Soaking pretreatment of corn stover for bioethanol production followed by anaerobic digestion process. *Applied Biochemistry and Biotechnology, 167*(7), 2088–2102.

Chapter 21
Competitiveness of Utility-Scale Wind Farm Development with Feed-In Tariff in Indonesia

Bimo Adi Kusumo, Akhmad Hidayatno, and Armand Omar Moeis

21.1 Introduction

The Indonesian government expressed its commitment to reduce the effects of greenhouse gas (GHG) emissions by 26% by 2020 by reducing the rate of deforestation and the conversion of fossil energy into energy that has a low carbon footprint or the new and renewable energy (Government of Indonesia 2011). In response to that commitment, the Ministry of Energy and Mineral Resources of Indonesia announced a nationwide target that in 2025 the composition of the new and renewable energy in Indonesia will be increased to 23% (Government of Indonesia 2014). On the other hand, to support national economic growth target of about 6–7% per year, Indonesian government launched an initiative to increase electricity supply by 35,000 megawatts (MW) throughout Indonesia within a period of 5 years (2014–2019). In order to meet these needs, the energy sector in Indonesia must use all its potential energy especially the clean and renewable energy such as wind energy. Wind energy utilization in Indonesia is only 1.6 MW from the total potential of 9.2 GW as claimed by Indonesian Board of Technology Assessment and Application (Martosaputro and Murti 2014).

Indonesian electricity sector is monopolized by Indonesian state electricity company (PLN), as part of government effort to maintain control of critical industry sector. PLN places itself in the full value chain of electricity sector from power generation to retail distribution. However, 25,900 MW of the 35,000 MW mega-project will be constructed and operated by independent power producer (IPP) which the electricity will be purchased by PLN per kilowatt-hour basis (Corporate

B.A. Kusumo (✉) • A. Hidayatno • A.O. Moeis
Industrial Engineering Department, Faculty of Engineering,
Universitas Indonesia, Depok, Indonesia
e-mail: bimoadikusumo@gmail.com

A. Sayigh (ed.), *Transition Towards 100% Renewable Energy*,
Innovative Renewable Energy, https://doi.org/10.1007/978-3-319-69844-1_21

Secretary PT PLN (Persero) 2016). IPP will sell the electricity to PLN to support PLN's existing generators which then will be distributed to PLN to its consumers.

The production of electricity from renewable energy especially wind energy requires a large investment in a long time with a relatively high degree of various risks (Kn et al. 2007). One of the risks is that the amount of money invested is lost because the electricity buy and sell tariff is not satisfactory which renders the project to be infeasible. Prior to the release of construction permit, IPP has to conduct wind profile measurement in the area using measurement tower for at least 1 year and conduct initial environmental examinations which require a considerable amount of money to be invested (Kementerian Energi dan Sumber Daya Mineral Republik Indonesia 2013; Sumantri et al. 2014). Furthermore, there is no assurance that after the initial development and permitting stage is completed developer can sell its electricity to PLN through power purchase agreement (PPA) because there is no regulation forcing PLN to sign the agreement. Investment assurance is needed for investors to be confident about their initial investment in wind energy development, and without a clear and fair market, the IPP's investment will be exposed to unmeasured risks. IPP will be interested to invest in the research if the tariff is determined beforehand thru a Feed-in Tariff (FIT) scheme set by the government (Fell 2009). PLN will be interested to purchase the electricity from IPP when the price of electricity offered by the IPP is equal or lower than the average generation cost of PLN's own generator. This will provide a motive to assess the competitiveness of wind energy in Indonesia compared to other generation sources.

The key factor for renewable energy penetration including wind is the competitiveness in terms of price compared to other generation type competing in the same grid (Mengal et al. 2014). Assessing competitiveness of wind energy will be done by comparing levelized cost of energy (LCOE) of wind energy compared to other generation sources in Indonesia's existing electrical grids. This paper presents the comparison of wind energy's LCOE and other electricity generation source's LCOE in Indonesia's major grids based on the wind resource and variability of capital expenditures (CAPEX) in different areas of Indonesia.

21.2 Methodology

The study was conducted at selected locations which have a high installed capacity of thermal power plant that can act as a baseload for wind farm and an area with good wind resource. In this research, the site selection in terms of wind resource follows the map published by Ministry of Energy and Mineral Resources of Indonesia (MEMR) as can be seen in Fig. 21.1, which is based on mesoscale modeling. Wind farm capacity simulated in this calculation is 20 and 50 MW with consideration of the existing demand and grid size to accommodate intermittent power from the wind farms.

LCOE is a proven method to approximate the cost to generate electricity (Ouyang and Lin 2014; Wiser et al. 2015). There are four major factors affecting

Fig. 21.1 Indonesia wind resource map. *Red* areas, wind speed average >6 m/s; *Green* areas, wind speed average 4–6 m/s (P3TKEBTKE 2014)

the LCOE which are capital expenditures (CAPEX), fixed charge rate (FCR), operation expenditures (OPEX), and annual energy production (AEP) based on formula (Eq. 21.1) used by NREL (Mone et al. 2015):

$$\mathrm{LCOE} = \frac{(\mathrm{CAPEX} \times \mathrm{FCR}) + \mathrm{OPEX}}{\mathrm{AEP}} \tag{21.1}$$

Since Indonesia is made of islands with unique geographic properties, a multiplying factor is included in the calculation of CAPEX to illustrate the unique site conditions. The factors are made based on focused group discussion conducted with wind farm developer and experts in Indonesia, and the value of the factors implemented in this research is based on actual site visit of the location. The factors can be divided into five categories as seen on Table 21.1. All the factors' multiplier for each sample location is determined based on site visit and rough approximation. Multiplier from each factor is accumulated and multiplied by initial CAPEX required to build a wind farm.

Annual energy production (AEP) is calculated from a wind resource assessment and wind farm layout optimization software "Openwind" (AWSTruepower 2014) that has been widely used to perform yield calculation and layout optimization (González-longatt et al. 2014; Wagner et al. 2013). Wind data used is published by Modern-Era Retrospective Analysis for Research and Application (MERRA) satellite managed by NASA (Rienecker et al. 2011). Openwind optimizes the turbine location to produce highest energy yield but constrained by noise and geographical feature.

FCR represents the amount of revenue required to pay the carrying charges and represented in percentage of CAPEX. FCR formula has been defined by NREL (Mone et al. 2015), which consists of discount rate (d), economic operation life (n), tax rate (T), and present value of depreciation (PVdep) as can be seen from Eq. 21.2. For research purposes, the assumption for project funding can be divided into 70% from loan with 8% interest rate and 30% from own equity with 10% return on equity. Indonesian income tax, depreciation, and amortization have been

Table 21.1 CAPEX multiplier factor due to variability in wind farm site conditions

Land price per hectare	Price	<$50,000	$50,000–100,000	$100,000–150,000	$150,000–200,000	$200,000
	Multiplier	1	1.1	1.4	1.8	2.0
Distance from port	Distance	<5 km	5–25 km	25–50 km	>50 km	Landing point
	Multiplier	1	1.15	1.3	1.5	1.7
Distance to connection point	Distance	<2 km	2–6 km	6–15 km	15–30 km	>30 km
	Multiplier	1	1.1	1.25	1.4	1.6
Road upgrade	Existing road	Concrete	Asphalt	Gravel	Dirt	No road
	Multiplier	1	1.03	1.05	1.1	1.25
Site topography	Slope (degree)	<6.5	6.5–11	>11		
	Multiplier	1	1.2	1.65		

adjusted to meet the Indonesian ministry of finance regulation on renewable energy tax easement (Ministry of Finance Republic of Indonesia 2010). Another addition to the FCR is 0.25% of the CAPEX allocated for insurance fee. Operation and maintenance fee has been previously defined for Indonesia by Asian Development Bank (Asian Development Bank 2015) for US$ 0.021 with an increment of 5% every year owing to inflation:

$$\text{FCR} = \frac{d(1+d)^n}{(1+d)^n - 1} \times \frac{1 - (T \times \text{PVdep})}{(1-T)} \tag{21.2}$$

The Feed-in Tariff policy should ideally apply for the 20 years following the program's introduction (Rigter and Vidican 2010). Feed-in Tariff is calculated by adding internal rate of return (IRR) to the LCOE which creates a fair tariff with expected return for IPP (Ashadi 2012). Based on discussion with experts in wind industry in Indonesia and sample project that is currently under construction, the IRR is defined at 14.5%.

21.3 Results and Discussion

Yield calculation in each wind farm location as calculated is shown in Fig. 21.2. Yield calculation from Openwind shows that highest energy production in Indonesia is located in Papua for both 20 and 50 MW wind farm. Contrary, Java has the lowest energy production compared to the rest of the site even though the area initially shows a good prospect for wind farm development. The difference in energy production results are primarily caused by the superior wind resource in Papua. Based on long-term wind resource assessment, wind speed at 100 m in Papua could reach up to 8.4 m/s, whereas in Java it sits around 5.9 m/s. The variation of capacity factor between 20 and 50 MW wind farm in the same location shows that energy production did not increase in direct proportion of the generation capacity. The increase of wind turbine quantity in the area will increase the wake which is generated by the turbine that reduces the wind speed in other downwind turbines and therefore reduces the capacity factor of the wind parks.

LCOE calculation result can be seen in Fig. 21.3b. Apart from Sulawesi and West Nusa Tenggara, the LCOE for wind farm in Indonesia is higher at 50 MW capacity wind farm compared to the 20 MW capacity wind farm. This indicates that the increase in energy production of a 50 MW wind farm is not proportional to the increase of cost to develop a 50 MW wind farm. This result can also indicate that if the wind farm capacity in Sumatera, Java, Kalimantan, and Papua is increased beyond 50 MW, the LCOE is predicted to be higher compared to a 50-MW-sized wind farm.

Assessing the competitiveness of utility-scale wind farm will require a comparison data of the existing average cost to generate electricity by PLN. Based on yield calculation gained and total investment cost to build wind farm in Indonesia, the

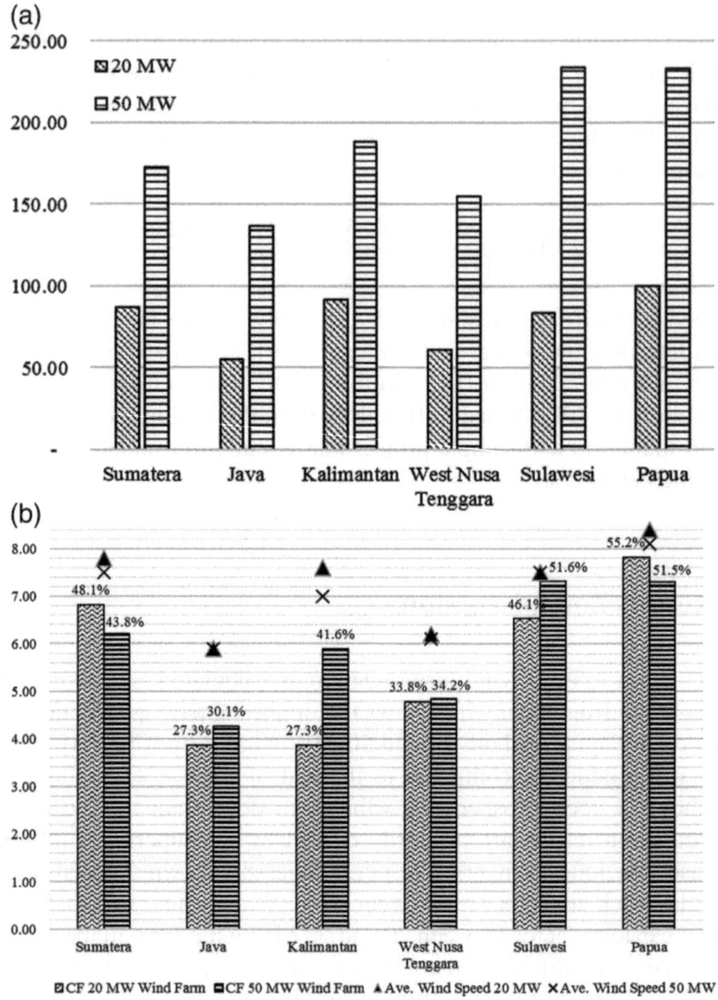

Fig. 21.2 Wind farm annual energy yield in GWh (**a**) and wind farm gross *capacity* factor and average wind speed at turbine hub height (**b**)

LCOE for wind energy sits at 0.16 US$/kWh for 20 MW wind farm and 0.17 US $/kWh for 50 MW wind farm. Wind energy ranked fifth cheapest compared to existing generation type LCOE, which points out that it is cheaper than diesel, machine gas, and gas as can be seen in Fig. 21.4a. Comparing the LCOE of wind in respective sample site by subtracting the average generation cost (AGC) with the LCOE, the gap between LCOE and AGC is produced and can be seen in Fig. 21.4b. Area in the upper part of the black line implies that the regional AGC is higher than wind LCOE which means wind is competitive enough in the region. The lower area from the black line indicates where the existing AGC in that region is already

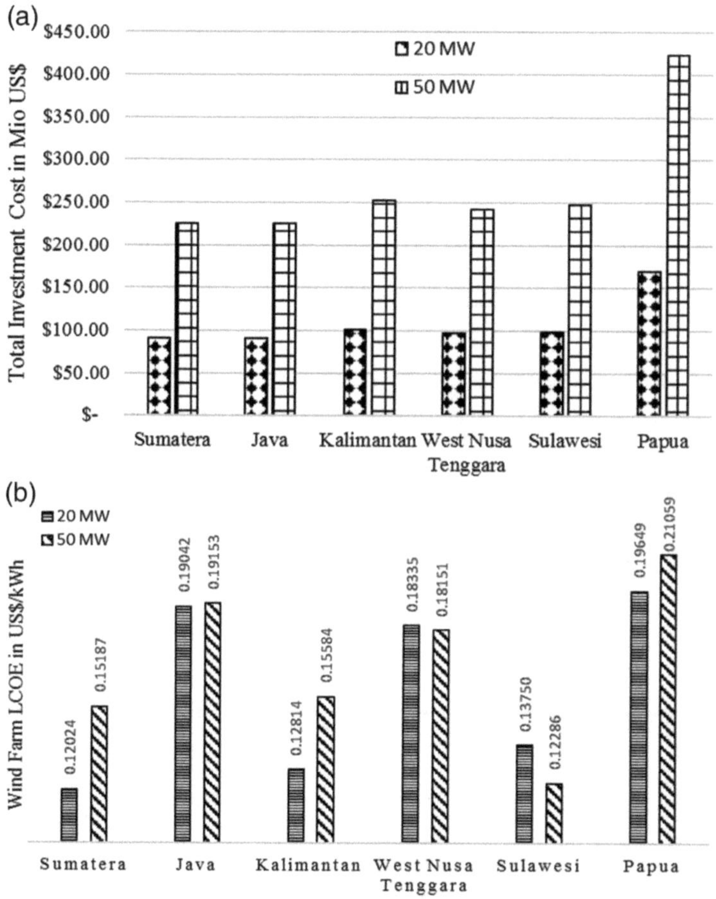

Fig. 21.3 Wind farm total investment cost in million US$ (**a**) and LCOE in US$/kWh (**b**)

cheaper than wind LCOE which makes wind energy not competitive enough in that region.

Figure 21.5 shows similar competitiveness analysis with Fig. 21.4, but instead of comparing LCOE, Fig. 21.5 compares the FiT with AGC. It shows that applying FiT will reduce the competitiveness of wind energy even further. The cost of buying electricity from wind through FiT scheme will cost almost the same with using existing diesel power plant.

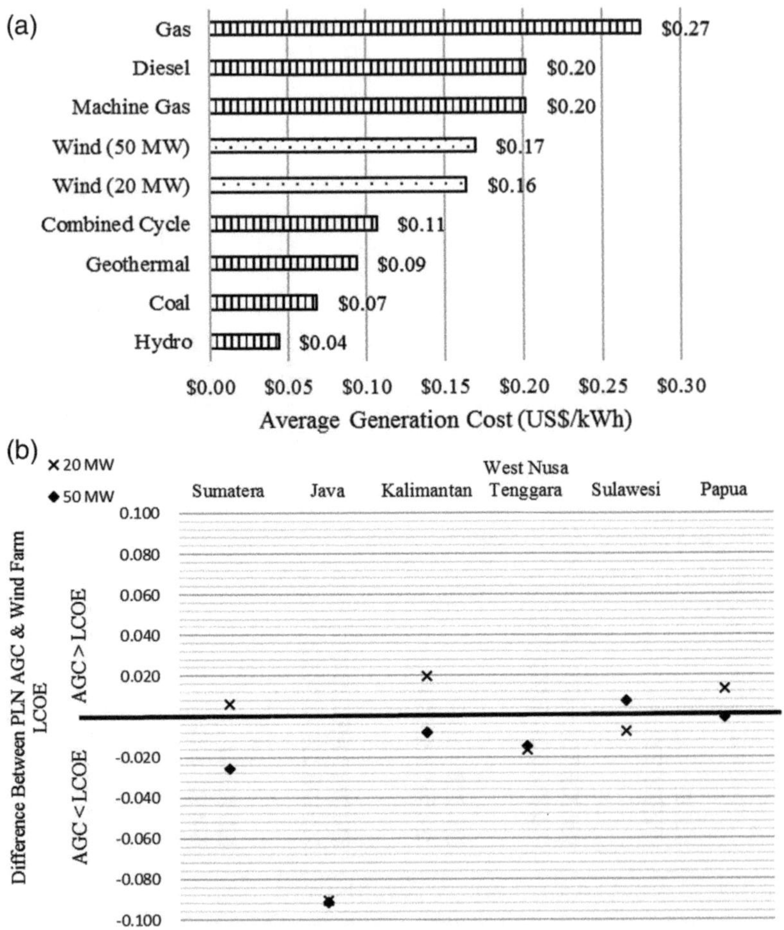

Fig. 21.4 Wind LCOE compared with other existing generation type (**a**) and difference between PLN's AGC and calculated wind farm LCOE in US$/kWh (**b**)

21.4 Conclusions

In order to gain competitiveness over other electricity generation units, wind farm must have a good wind resource and/or a low CAPEX. Competitiveness of wind energy in Indonesia is marginal at best. Only some region in Indonesia with good wind resource and low CAPEX can compete with existing generation units. The simulated Feed-in Tariff does not help this condition. By adding additional 14.5% to the LCOE, the wind energy is not competitive enough compared to existing generation. Incentive and subsidy will need to be given by the government to increase the wind energy penetration in Indonesia without sacrificing PLN as a state-owned enterprise.

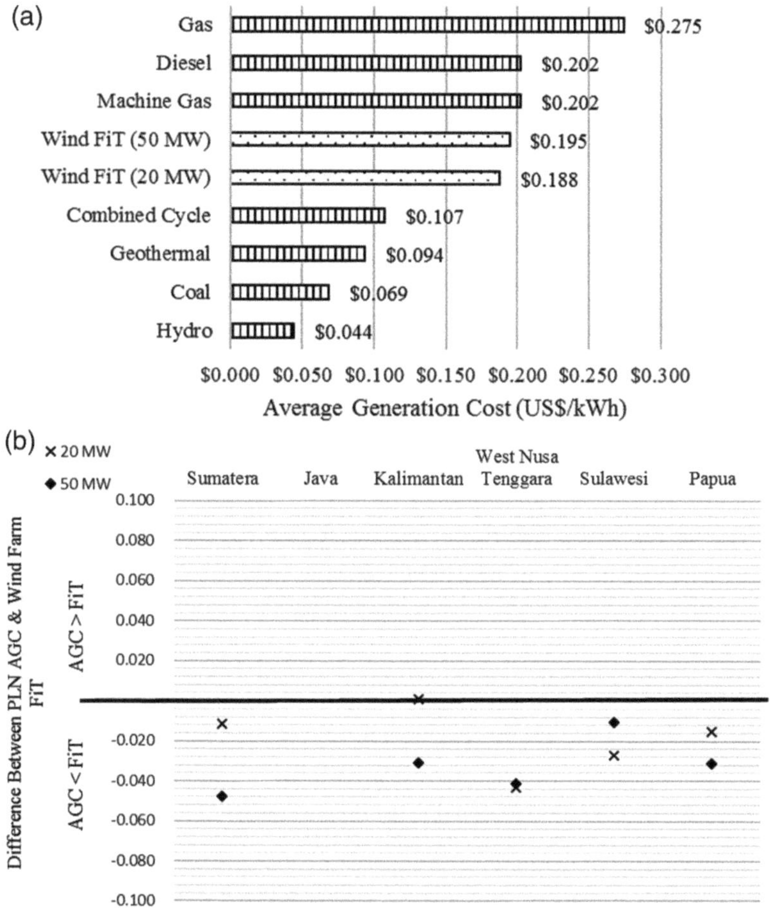

Fig. 21.5 Wind FiT compared with other existing generation type (**a**) and difference between PLN's AGC and calculated wind farm FiT in US$/kWh (**b**)

References

Ashadi. (2012). Perumusan Tarif Pembelian Listrik Pada Regulasi Feed-In Tariff untuk Teknologi Photovoltaic serta Analisa penerapannya di Indonesia.

Asian Development Bank. (2015). *Tariff support for wind power and rooftop solar PV in Indonesia*.

AWSTruepower. (2014). Openwind V 1.6.

Corporate Secretary PT PLN (Persero). (2016). *PLN statistics 2015* (1st ed.). Jakarta.

Fell, H.-J. (2009). Feed-in tariff for renewable energies: An effective stimulus package without new public borrowing. *Berlin, Germany*. Retrieved from http://extranet.ontario-sea.org/Storage/32/2364_Feed-in_Tariff_for_Renewable_Energies_-_An_Effective_Stimulus_Package_without_New_Public_Borrowing.pdf

González-longatt, F., Serrano, J., Burgos, M., Manuel, J., & Santos, R. (2014). Wind-resource atlas of Venezuela based on on-site anemometry observation. *Renewable and Sustainable Energy Reviews, 39*, 898–911. https://doi.org/10.1016/j.rser.2014.07.172.

Government of Indonesia. (2011). Presidential Regulation 61 of 2011 on National Action Plan on GHG Emissions Reduction. Retrieved from http://www.bappenas.go.id/files/6413/5228/2167/perpres-indonesia-ok__20111116110726__5.pdf

Government of Indonesia. (2014). Presidential Regulation 79 of 2014 on National Energy Policy. Retrieved from http://prokum.esdm.go.id/pp/2014/PPNomor 79 2014.pdf

Kementerian Energi dan Sumber Daya Mineral Republik Indonesia. Peraturan Menteri Energi dan Sumber Daya Mineral Republik Indonesia Nomor 35 Tahun 2013 Mengenai Tata Cara Perizinan Usaha Ketenagalistrikan (2013). jAKARTA. https://doi.org/10.1017/CBO9781107415324.004.

Kn, J., Prague, C., & Prague, C. (2007). Risk inclusion in feed-in tariffs and green bonuses calculation, 1–10.

Martosaputro, S., & Murti, N. (2014). Blowing the wind energy in Indonesia. *Energy Procedia, 47*, 273–282. https://doi.org/10.1016/j.egypro.2014.01.225.

Mengal, A., Uqaili, M. A., Harijan, K., & Ghafoor, A. (2014). Competitiveness of wind power with the conventional thermal power plants using oil and natural gas as fuel in Pakistan. *Energy Procedia, 52*, 59–67. https://doi.org/10.1016/j.egypro.2014.07.054.

Ministry of Finance Republic of Indonesia. (2010). *Ministry of finance regulation No. 21/PMK.011/2010 about tax and custom facilities for renewable energy sources utilization*. Jakarta: Ministry of Law and Human Right of Indonesia.

Mone, C., Stehly, T., Maples, B., & Settle, E. (2015, February). 2014 cost of wind energy review. *National Renewable Energy Laboratory* .

Ouyang, X., & Lin, B. (2014). Levelized cost of electricity (LCOE) of renewable energies and required subsidies in China. *Energy Policy, 70*, 64–73. https://doi.org/10.1016/j.enpol.2014.03.030.

P3TKEBTKE. (2014). Peta Potensi Angin Indonesia. Ministry of energy and mineral resources of Indonesia.

Rienecker, M., Suarez, M. J., Gelaro, R., Todling, R., Bacmeister, J., Liu, E., … Woollen, J. (2011). MERRA: NASA' s Modern-Era retrospective analysis for research and applications. *Journal of Climate, 24*, 3624–3648. https://doi.org/10.1175/JCLI-D-11-00015.1

Rigter, J., & Vidican, G. (2010). Cost and optimal feed-in tariff for small scale photovoltaic systems in China. *Energy Policy, 38*(11), 6989–7000. https://doi.org/10.1016/j.enpol.2010.07.014.

Sumantri, O., Embang, D., Rosadi, U., Trirohadi, H., Yahmadi, A., Kriswanto, … Hidayat, C. (2014). *Guide for renewable energy power plant interconnection to PLN distribution system*. Jakarta: PT. PLN (Persero).

Wagner, M., Day, J., & Neumann, F. (2013). A fast and effective local search algorithm for optimizing the placement of wind turbines. *Renewable Energy, 51*, 64–70. https://doi.org/10.1016/j.renene.2012.09.008.

Wiser, R., Bolinger, M., Barbose, G., Darghouth, N., Hoen, B., Mills, A., … Tegen, S. (2015, August). 2014 wind technologies market report. *Department of Energy*.

Chapter 22
Biogas Production from Modified Starch at the Anaerobic Digester

Rudy Laksmono, Edy Mulyadi, Soemargono, and Nugroho Adi Sasongko

22.1 Introduction

Since 1997, PT Lautan Warna Sari (LWS) has produced cassava-based modified starch raw materials to a total capacity of 200 tons/day or 40 tonnes of product. Tapioca industrial effluent is derived from a variety of existing processes in the production system. Because of this high production capacity, a large effluent treatment facility is required (six equalization basins, 3,000 m^2 each). The percentage of organic compounds in the effluent is relatively high, thus representing a high economic value when processed into biogas. However, untreated waste causes environmental pollution in the form of a bad odor in the vicinity of the plant. Earlier waste handling merely removed the smell. In fact, from field observations, formation of gas occurred, which is characterized by the emergence of gas bubbles that arise from an equalization basin. This suggested the idea for management and processing of effluent into biogas. Further study also found that in addition to the

R. Laksmono (✉)
Energy Security Program, Faculty of Defense Management, Indonesia Defense University, Bogor, Indonesia

Chemical Engineering Study Program, Engineering Faculty, UPN "Veteran" Jawa Timur, Jakarta, Indonesia
e-mail: rudy.laksmono@idu.ac.id

E. Mulyadi • Soemargono
Chemical Engineering Study Program, Engineering Faculty, UPN "Veteran" Jawa Timur, Jakarta, Indonesia

N.A. Sasongko (✉)
Energy Security Program, Faculty of Defense Management, Indonesia Defense University, Bogor, Indonesia

The Agency for Assessment and Application of Technology (BPPT), Jakarta, Indonesia
e-mail: nugroho.adi@bppt.go.id

A. Sayigh (ed.), *Transition Towards 100% Renewable Energy*,
Innovative Renewable Energy, https://doi.org/10.1007/978-3-319-69844-1_22

liquid waste, the industry generates solid waste, such as the epidermal tissue, solids, and cassava. Epidermal tissue and solids (soil and groats cassava) result from the initial washing of cassava, while cassava is produced from the starch-making process. Epidermal tissue and solid impurities have been dumped untreated to date. Cassava in the form of fiber could theoretically be processed into bioethanol through hydrolysis and fermentation. Also, to produce biogas, solid debris and sludge left behind in this process can be used as a fertilizer (Buren and Arnott 2004; Reese and Thompson 2008), and a plan was developed to integrate this with the epidermal tissue and solid debris.

Wastewater treatment conducted by PT LWS Ltd to biogas with a methane purity is relatively small (<40%), thus could not be used in the drying of modified starch products. It would require a purifying biogas digester product that has been adapted so that biogas can be used directly as a dryer for modified starch products. The sludge produced by the digester can be processed into liquid and solid organic fertilizer by using an extractor and continuous granulator. This process represents an energy saving by through the "simplicity" of the process. Such a manufacturing process does not produce waste (zero waste process). This process has been carried out and tested in the Laboratory of Appropriate Technology Teknopark UPN "Veteran" of East Java and has been applied directly to the waste processing unit PT. Color Oceans Sari in Lampung. The benefits and the expected target are to produce detailed designs and prototypes with a capacity of 17,000 m^3 digester balloon-equipped models of a moving bed of purifying biogas, which is adapted to the product's gas digester. This study improves on the gas purifier processes carried out thus far by using a compact design, adaptable to a variety of gas digester products, using local ingredients, and representing ease of handling and inexpensiveness. One possibility that has been developed is to use a model gas flow-purifying system. The process consists of only one step and produces a higher purity than the prior processes, therefore offering the possibility of obtaining a patent. With this process biogas can be directly used for drying modified starch products for PT LWS.

Tapioca industrial effluent results from various processes in the production system: the waste generated from washing cassava, the processes of extraction and separation, the reactor (for modified starch), vacuum filter, and pressing. The content of organic substances contained in waste is represented by the parameters BOD and COD. The designed system allows for designated parameter organic pollutants of up to 80–90% and the formation of methane gas from organic substance conversion. Several types of biogas reactors are in use, namely (a) a fixed dome reactor, (b) a floating reactor, (c) a balloon reactor, (d)a horizontal reactor, (e) a reactor hole, and (f) reactor ferro-cement. The most common type of biogas digesters are the fixed dome reactor floating reactor (floating drums) and the balloon reactor. The balloon-type reactor is a reactor that is widely used in the household manufacture of plastic material as the construction cost is cheaper and more efficient (Mulyadi 2010). This reactor includes a section that serves as a digester and storage of gas, each mixed in one room without barriers. Organic material is located at the bottom because it has a greater weight than the gas that fills the cavity above. Fauzi et al. (2010) developed the application of this type of reactor, but their design capacity is less than 20 tons.

One of the considerations in the development of a bio-digester is the anaerobic environment, requiring a temperature of 30–50 °C for mesophilic bacterial growth. The parameters that affect the operating conditions of the Reactor BioDigester (Dahlman 2012; Mansur 2010; Mulyadi and Billah 2011) are:

(a) The degree of acidity (pH), which should be between 6.6 and 7.0, and not below 6.2.
(b) C/N ratio, the ideal of which is between 25 and 30.
(c) Nutrition–among others, ammonia (NH_3) as a source of nitrogen and mineral nickel (Ni), copper (Cu), iron (Fe), phosphate (PO_4), and zinc (Zn).
(d) Feed levels–the biomass concentration recommended is from 0.26 to 0.3 kg/L.

According to Mulyadi (2012), stirring the digester serves to obtain a homogeneous mixture of the substrate and also increases the frequency of collisions and turbulence, so that bioconversion is increased. The processing of biomass (biodegradation) takes place in two stages (Scott and Keddy 2006), namely in anaerobic and aerobic processes.

22.1.1 Anaerobic Process

Biomass flows into and remains in an anaerobic environment (digester) for a sufficient time (±14 days). Biogas is formed in an anaerobic fermentation process through the following mechanisms:

(a) Hydrolysis of biomass containing cellulose, hemicellulose, and other materials, such as extracts of materials like proteins, carbohydrates, and lipids, which are parsed into compounds with a shorter chain.
(b) Acidification–acidogenic bacteria produce acids that function to convert to a short-chain compound hydrolysis process resulting in a fatty acid compound.

The acidogenic reaction process is represented by:

$$C_6H_{12}O_6 + 2H_2O \rightarrow 2CH_3COOH(\text{acetate}) + 2CO_2 + 4H_2 \quad (22.1)$$
$$C_6H_{12}O_6 \rightarrow CH_3CH_2CH_2COOH(\text{butyrate}) + 2CO_2 + 2H_2 \quad (22.2)$$
$$C_6H_{12}O_6 + 2H_2 \rightarrow 2CH_3CH_2COOH(\text{propionate}) + 2H_2O \quad (22.3)$$
$$CH_3CH_2CH_2COOH(\text{butyrate}) + 2H_2O \rightarrow CH_3COOH + CO_2 + 3H_2 \quad (22.4)$$
$$CH_3CH_2COOH(\text{propionate}) + 2H_2O \rightarrow 2CH_3COOH + 2H_2 \quad (22.5)$$

(c) Gas formation of methane (CH_4)–methanogenesis bacteria will form CH_4 and CO_2 from gas H_2 and CO_2, causing a methanogenic reaction, represented as:

$$CH_3COOH \rightarrow CH_4 + CO_2 \quad (22.6)$$

$$4H_2 + CO_2 \rightarrow CH4 + 2H_2O \quad (22.7)$$

We found that three groups of bacteria play a role in the formation of biogas, in addition to a group of fermentative bacteria (streptococci, bactericides, and other Enterobacteriaceae), acetogenic bacteria (desulfovibrio), and a group of methane bacteria (Methanobacterium, Methanobacellus, and Methanococcus).

22.1.2 Aerobic Process

Waste digester product still contains an organic substance that is added to an aerobic facultative with the aim of decomposing organic materials remaining in the waste water. Active microorganisms present in the pond and aerobic environment will absorb and oxidize the organic matter and then break it down into a simpler form of substance that is not harmful to the environment. The reaction that occurs is represented by:

$$\text{Organic substances} + O_2 \rightarrow CO_2 + H_2O + \text{new cells} \quad (22.8)$$

The factors that need to be considered in a pool of facultative and aerobic optimal working order are:

(a) pH: 6.5–7.5
(b) Temperature: 25–35 °C
(c) Dissolved oxygen: $\geq$ 2 mg/L

22.2 Methodology

The raw materials used in this study are derived from several parts of the waste stream from PT Tunas Jaya Lautan in Lampung. In general, the process flow diagram of making modified starch from waste is described in Fig. 22.1.

Basically, waste solids or sludge comprising the following amounts of materials are produced (raw material base cassava 50 tons/day): solid mushy/wet: 500 kg/day; solid press: 10 tons/day; powder off-grade wet: 200 kg/day; 4 groats: 1 ton/day; epidermal tissue: 4 tons/day; solids: 8 tons/day.

Research tools, such as the biodigester made in the field (industry), were prepared by PT. Tunas Jaya Lautan, Lampung, while gas purifiers were designed and made in the workshop of the appropriate technology engineering section of LPPM-UPN "Veteran" East Java. This series of anaerobic digesters is described in Fig. 22.4. The biogas reactor's upper part is made of plastic polyethylene, while the bottom part consists of the excavated soil. The digester manufacturing biogas

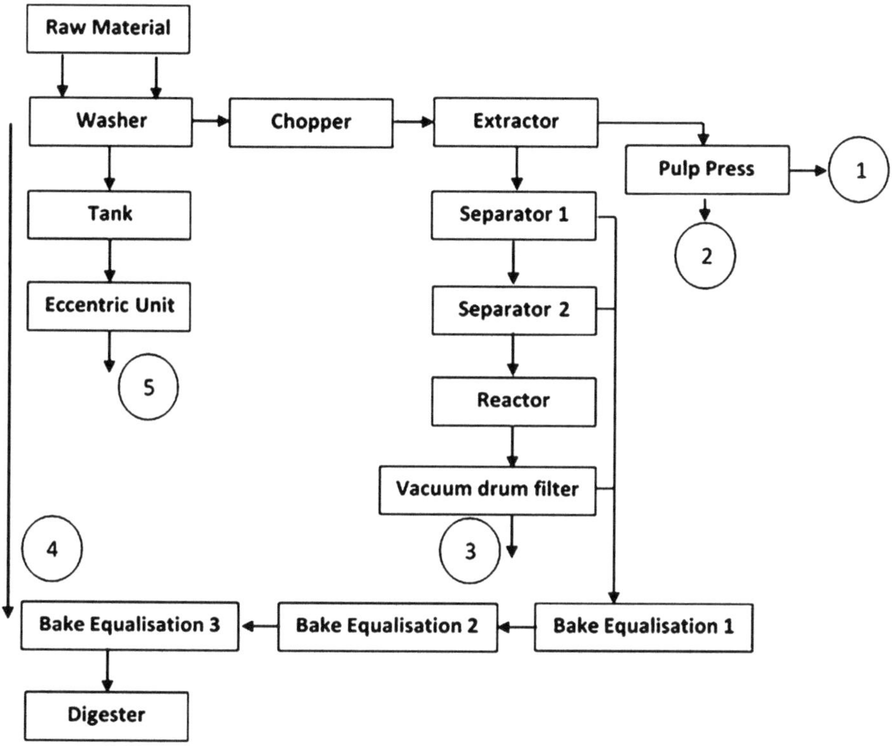

Fig. 22.1 Flow diagram of modified starch production

calculation is based on the theory of similarity according to the prototype digester unit that was created earlier. Construction or Build System Fermentation in Making Tool Biogas Capacity Volume 17,000 m^3, which consists of three parts comprising Framework Fixed anaerobic ponds that were dug a distance of 1–2 m apart, and a unit consisting of the digester balloon top, which is prepared over three times the area of the anaerobic ponds (which will be digester gas), i.e., 17,000 m^2. A fixed framework unit, consisting of the construction of buildings composed of bricks, is arranged to resemble a tank with a specific height and bottom cast with a capacity of 17,000 m^3. The skeletal unit serves as the foundation and support of the balloon digester (Fig. 22.2).

Biogas contains approximately 40% methane gas before it is used as a fuel burner; first it passes through a gas-purifying device and is then stored in a gas holder, as shown in Fig. 22.3. Other equipment involved is the gas holder (gas container), which is made from plastic, and a stabilizer tank made from SS304.

In addition, the system is also equipped with a set made up of a gas-purifying device and plumbing pipes SS304 with various sizes making up the primary network, and a secondary network of galvanized pipes that connect to the water

Fig. 22.2 An installed balloon-type anaerobic digester

Fig. 22.3 Installed biogas purifier system

heater burner. The biogas equipment production processes run semi-automatically and all equipment is designed and assembled.

Waste from industrial modified starch (PT Tunas Jaya Lautan, Teluk Dalem Ilir, District Rumbia, Central Lampung) used for biogas in the form of wastewater derived from processes such as chemical vacuum filtering, non-chemical vacuum filtering, pressed solids, and separator and washing processes is placed in three sedimentation ponds arranged in a series. The overflow liquid from the settling basin goes into the pool that serves as an anaerobic digester to produce methane gas. Based on field observations, we have identified some potential that the waste contains a large amount of biomass already encountered in equalizations I, II, and III, and an accumulator (Fig. 22.4).

The digester is in the form of a balloon reactor. Biogas is produced before being used as fuel in the burner as purified gas. This ensures that the product consists of a high purity of biogas. This product of biogas will be used by PT LWS/PT Tunas Jaya Lautan as a heat source for drying modified starch. The quality of biogas can be improved by treating some of the following parameters: The first parameter is the removal of sulfur hydrogen gas (H_2S), water content, and carbon dioxide (CO_2).

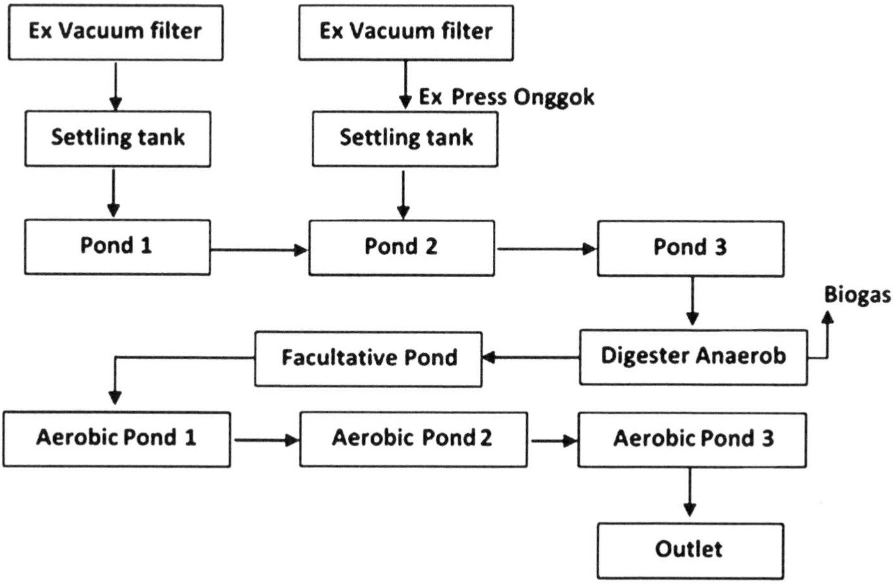

Fig. 22.4 Block diagram of tapioca wastewater treatment of PT Tunas Jaya Lautan

H_2S contains toxins and substances that cause corrosion, so the maximum allowable concentration is 5 ppm. The second parameter is the removal of carbon dioxide, the goal of which is to improve the quality so that the gas can be used to fuel vehicles. The third is that the water content in the biogas will lower the ignition point and is likely to cause corrosive biogas. The optimization stage includes:

(a) The calculation of the optimum use of root blower, as well as a test root blower.
(b) Selection of a root blower and layout purifier
(c) Optimization of the burner includes the location of the ignition control system and the lighter, as well as the type of burner used.
(d) Optimization of biogas fuel substitution and drying results.

22.3 Results and Discussion

Waste biomass is a source of raw materials that can be converted into alternative energy, biogas, ethanol, and fertilizers. Use of a bio-digester in an anaerobic lagoon can help with waste processing that produces biogas and by-products the form of organic fertilizer. In addition, the use of a bio-digester will reduce the emissions of methane (CH_4) resulting from the decomposition of biomass, as the waste from the tapioca industry is not allowed to decompose in the open but is fermented into biogas energy. Methane including greenhouse gases (greenhouse gas) along with carbon dioxide gas (CO_2) create a greenhouse effect that is causing the current global warming phenomenon. A reduction of methane gas locally could play a decisive role in solving global problems in this regard (the greenhouse effect). One type of digester design that can be used is a balloon. Bio-digester reactor design calculations were based on data that were tailored to the potential of raw materials and the amount of biomass waste to be processed. An ideal biomass composition consists of a liquid content of 80% and a solid content of 20%. The structure of the reactor (digester) is made up of two parts, namely the top of the digester, which serves as a material parse and biogas storage (gas holder), and the bottom of biogas, which is a home for bacteria, both acid-forming and methane-forming. The top pool is created using a plastic polypropylene reactor. The quality of wastewater from these processes is shown in Table 22.1.

At the start of the production biogas is formed very slowly, so that efforts are needed to accelerate growth and reaction formation. Essentially, stirring will mix the reactant materials so that a collision occurs, after which the fluid from the digester flows into the pool, which is made up water used to wash the cassava. The biodigester reactor is also equipped with a baffle, which serves as a mixture of stirred biomass from incoming feeders and the water bath, so that the mixing process can run smoothly and the gas formation process is achieved relatively quickly. The result is usually satisfactory–biogas can be produced relatively quickly. The biogas purification product before processing is still red in color and when used in a burner has a capacity of 4 m^3/min, which still burns unstably.

Table 22.1 Chemical-physical condition of laboratory analysis of tapioca from PT Tunas Jaya Lautan

No.	Source/Parameter	Unit	Result	Max limit (regulation)	Method
1.	Ex. Separator				
	pH	–	4,47	6–9	SNI 06–6889.11 – 2004
	COD	mg/l	6899,2	300	SNI 06–6889.15 – 2004
	TSS	mg/l	950	100	SNI 06–6889.3 – 2004
2.	Ex. Sedimentation basin				
	pH	–	3,41	6–9	SNI 06–6889.11 – 2004
	COD	mg/l	6522,9	300	SNI 06–6889.15 – 2004
	TSS	mg/l	110	100	SNI 06–6889.3 – 2004
3.	Ex. washing cassava				
	pH	–	6,48	6–9	SNI 06–6889.11 – 2004
	COD	mg/l	533,1	300	SNI 06–6889.15 – 2004
	TSS	mg/l	1120	100	SNI 06–6889.3 – 2004
4.	Ex. Pond 3				
	pH	–	4,83	6–9	SNI 06–6889.11 – 2004
	COD	mg/l	1285,8	300	SNI 06–6889.15 – 2004
	TSS	mg/l	830	100	SNI 06–6889.3 – 2004

Source: field sampling and laboratory analysis 2013

Table 22.2 Chemical–physical conditions resulting from laboratory analysis of anaerobic effluent

No.	Source/parameter	Unit	Result	Max. allowable limit	Method
1.	Digester effluent to aerobic ponds				
	pH	–	7,79	6–9	SNI 06–6889.11 – 2004
	COD	Mg/l	219,5	300	SNI 06–6889.15 – 2004
	TSS	Mg/l	290	100	SNI 06–6889.3 – 2004
2.	Release to river				
	pH	–	7,81	6–9	SNI 06–6889.11 – 2004
	COD	Mg/l	62,72	300	SNI 06–6889.15 – 2004
	TSS	Mg/l	130	100	SNI 06–6889.3 – 2004

Source: field sampling and laboratory analysis 2013

After leaving the anaerobic ponds, the physical chemical parameters achieved are shown in Table 22.2. The results of the laboratory analysis as shown in Table 22.2 after the anaerobic digester process, indicate that the performance of the anaerobic digester is based entirely on decomposing organic pollutants, which were measured as COD and produce gas methane (CH_4). Thus, the environmental parameters from the waste into the receiving water bodies already meet the quality standards except for total suspended solids (TSS). The process conditions were set in accordance with the ground conditions of the aerobic pond.

The results show a decent quality of water discharged into receiving water bodies in the proper way, even though recycled and thus reused, and put to use as intended. This demonstrates environmental management that leads to zero waste. The process of methane formation (CH_4), which occurs in an anaerobic digester, indicates the perfection of the process of biodegradation of organic pollutants in anaerobic-aerobic bacteria utilizing methanogenic, acidogenic, and bacterial means. In addition, the methane gas is purified and used as fuel for drying modified starch.

22.3.1 *Digester Optimization*

In this study initially characteristics of test equipment are determined and measuring instruments (flow meter and a manometer) are calibrated. Thereafter, the process of anaerobic fermentation of the biomass and the resulting products are analyzed for levels of methane gas, carbon dioxide, and other gases using gas chromatography (GC). The gases are formed as byproducts of the fermentation process, and the gas purification process is achieved by using equipment called a water trap gas absorber that absorbs the H_2S gas that is formed in the water. It is based on the solubility of H_2S gas of about 320 cc/100 cc H_2O; H_2S gas is easily

Table 22.3 Measured biogas composition

Component	Digester product composition %	Concentration after purification
CH_4	42	73
CO_2	41	6.
Water	8	0.04

diluted in water. An absorber that captivates CO_2 gas uses a solution of sodium hydroxide with a yield of 1%. CO_2 gas immediately reacts with the NaOH solution, while CH_4 gas does not react with it. The following reaction equations represent these processes:

$$CO_2(g) + NaOH(aq) \rightarrow NaHCO_3(aq) \quad (22.9)$$

$$NaOH(aq) + NaHCO_3(aq) \rightarrow Na_2CO_3(s) + H_2O(1) \quad (22.10)$$

from formula (22.9) and (22.10), which results in:

$$CO_2(g) + 2NaOH(aq) \rightarrow Na_2CO_3(s) + H_2O(1) \quad (22.11)$$

Based on this equation, there is a reduction in CO_2 concentration due to the reaction with NaOH solution, and the ratio of CH_4 to CO_2 concentration results in a greater concentration of CH_4. Absorber Column 5 comprises two pieces, each containing zeolites and silica solids, which serve to absorb residual gases that have not been absorbed in the previous stage. The energy contained in biogas depends on the concentration of methane (CH_4). The higher the concentration of methane, the greater the energy content (calorific value) for biogas, and conversely, the lower the concentration, the lower the calorific value of methane. Methane as the main component of biogas is a fuel that is used as an energy alternative because it has a calorific value that is high enough, i.e., about 4,600 kcal/m^3 before being refined, whereas pure methane gas energy contains 8,900 kcal/m^3. The average capacity of biogas production is 4.8 m^3. The quality of biogas production from three sampling times, showing the primary composition (methane, water, and carbon dioxide) averages for the biogas before and after purification, is presented in Table 22.3.

A comparison of these experimental results of the influence of biomass waste from pond III with the waste washing water to gaseous products and levels of gas formed, are shown in Table 22.4. The study was also carried out at a time of cassava production in a capacity of 80 tons/day. From the data presented in Table 22.4, it appears that the gas pressure in the reactor bio-digester, biogas production, and gas content (%) are influenced by the ratio of the addition of the wash water. The longer the fermentation process, the gases that are formed are gradually increased, so that the gas pressure in the reactor bio-digester increases sharply, and after 12 days tends to decrease. Likewise, gas products will be increased by increasing the fermentation time in comparison to certain raw materials, whereas the levels of methane (CH_4) increase significantly with the length of time of the fermentation process and on day

Table 22.4 Comparison of the effect on raw material of gas products (m^3/batch) and gas content, time series (the amount of cassava waste from 80 tons/day, with a full washing water circulation of 60 m^3)

Waste ratio with the flow of the washing water materials (part)	Time (days)	Pressure (mm air)	Gas product (m^3/min)	Gas composition (%)		
				CH_4	CO_2	Air
1:0	1	4	0,75	12	40	10
	4	22	1,1	26	32	12
	8	66	1,8	34	29	7
	12	72	2,4	40	23	12
	16	80	2,6	40	29	9
1:0.25	1	4	0,8	16	36	11
	4	34	1,2	30	22	6
	8	88	2,4	36	23	9
	12	89	3,8	45	19	7
	16	88	2,9	42	26	12
1:0,5	1	6	0,92	20	40	12
	4	75	1,5	34	30	9
	8	90	2,8	40	33	11
	12	95	4,6	48	40	10
	16	90	3,6	44	42	9
1:0.75	1	10	0,95	22	42	11
	4	77	1,5	38	36	12
	8	97	3,4	39	40	11
	12	102	4,4	44	49	13
	16	93	3,8	44	47	9
1:1	1	10	1	24	52	8
	4	80	1,4	39	50	11
	8	95	3,6	41	54	10
	12	100	4	47	59	7
	16	90	3,8	40	60	8

12 similarly decrease. The biogas thus produced is used by PT LWS as modified starch drying energy, and biogas-produced products total around 4 m^3/min. That value is equivalent to the energy needs of the plant-modified engines of Tunas Jaya PT Lautan in Lampung, i.e., 60 L/h of diesel fuel or 350 kg/h of firewood. Through this process, biogas can replace the need for diesel fuel or wood. Biogas has advantages compared to diesel or firewood that have previously been used for the drying process for flour (flash dryer), because in addition to producing a clean blue flame, it also entails a more stable process temperature so that efficiency is increased by more than 27%. Methane as the main component of biogas has a calorific value that is quite high, around 6,200 kcal/m^3, while pure methane gas contains a calorific value of 8,900 kcal/m^3. Biogas from industrial waste tapioca starch containing methane gas was produced over a 5-day period in the bio-digester,

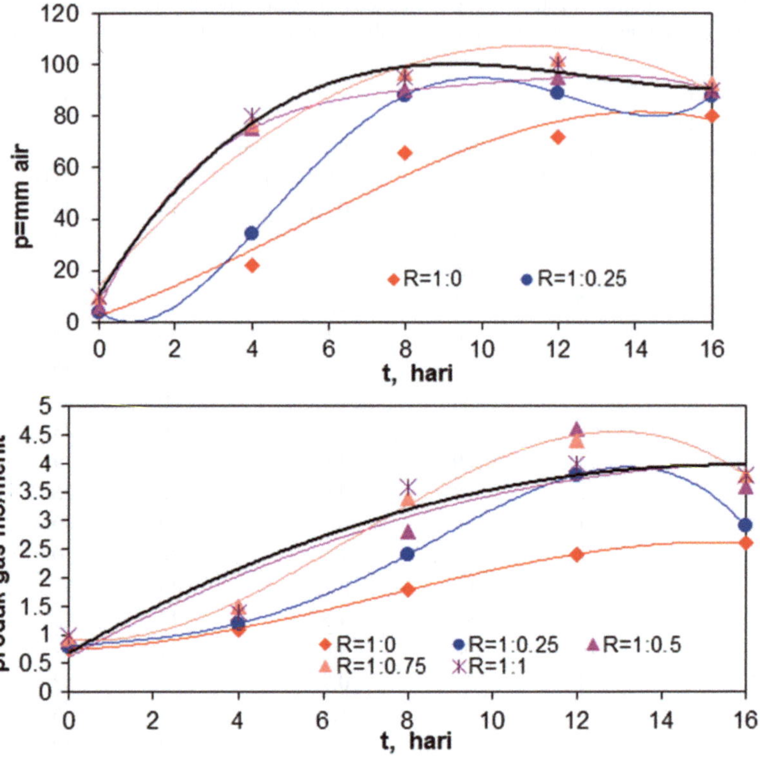

Fig. 22.5 Relationships between gas pressure (P) and gas product/process batch as a process of time series (t)

and production reached its peak by day 16. The processed biogas byproducts are used as organic fertilizer in solid or liquid form. Optimization of the digester results in energy independence for the company.

In addition, according to field measurements, the dynamic condition based on the time series of biogas production is shown in Figs. 22.5 and 22.6 below. With continuous running of the covered anaerobic pond, the biogas reaches optimum production by the tenth day.

22.4 Conclusions

1. A bio-digester designed with a capacity of 17,000 m^3 can produce methane gas (4.6 m^3/min) from tapioca effluent with constant mixing.
2. The formation of biogas in a bio-digester eliminates odor pollution around the plant and can degrade COD by more than 100-fold.
3. Purified biogas is odorless and can be used as a substitute for diesel fuel and firewood for the tapioca industry.

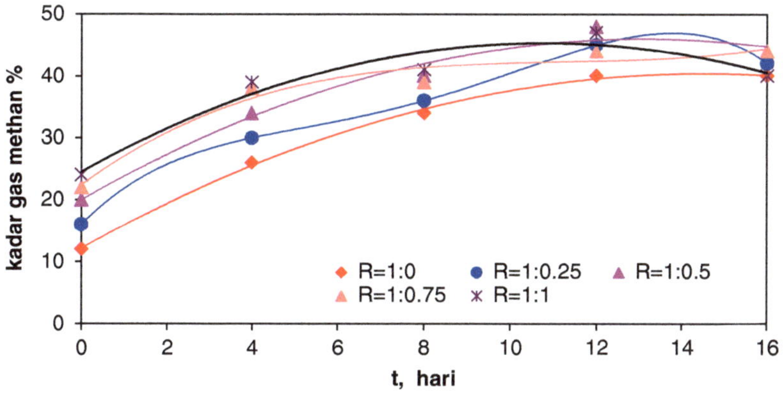

Fig. 22.6 Composition of gas methane as a digester product

References

Anonim. (2009). Classification of biogas plants; www.tutorvista.com/science/biogas-plant-design Bio digester Installation Manual, Lylian Rodriguez and T R Preston. FAO.

Buren, A., & Arnott, M. (2004). *Biogas/bio fertilizer business handbook manual; peace corps.* Chinese Biogas Manual, Intermediate Technology Publication, Ltd.

Dahlman, J. (2012). *Technologier demonstrated at echo: Floating drum biogas digester.* http://www.echonet.org

Fauzi, A., Mulyadi, E., & dan Billah, M. (2010). Teknik Pembuatan Bio Gas, Buku Seri Teknologi Kementrian Pembangunan Daerah Tertinggal (KPDT).

Mulyadi, E. (2010). Menuai Energi Alternatif dari Tebaran Limbah, Api Pembangunan, no 31, ISSN 0854–3755.

Mulyadi, E. (2012). Kajian Digester Kubah Tetap Berpiston untuk Optimasi Proses Produksi Biogas Prosiding Semnas Pemanfaat Hasil Riset untuk Menunjang Pemberdayaan Ekonomi Lokal dan Industri, ISBN 978–602–9372-49-6.

Mulyadi, E., & dan Billah, M. (2011). Fasilitasi Pemanfaatan limbah Menjadi Biogasn di Kabupaten Sigi, Sulawesi Tengah, Prosiding Semnas Pemanfaat Hasil Riset dalam Pengembangan Daya Saing Bagsa di Era Inovasi Industri, ISBN 978–602–9372-41-0.

Reese, S., & Thompson, R. S. (2008). *Hydrogen production by anaerobic fermentation using agricultural and food processing wastes utilizing a two-stage digestion system.* Utah State University.

Sadi, M. -A. (2010). *Design and building of biogas digester for organic materials gained from solid waste.* Thesis for the Degree of Master of Program in Clean Energy and Conservation Strategy Engineering, Faculty of Graduate Studies, An-Najah National University, Nablus-Palestine.

Scott, M., & Keddy, S. M. (2006). *Anaerobic digestion of the organic fraction of municipal solid waste.* University of Florida, Department of Environmental Engineering Sciences.

Zaher, U., Cheong, D. -Y., Wu, B., & Shulin Chen, S. (2007). *Producing energy and fertilizer from organic municipal solid waste.* Department of Biological Systems Engineering, Washington State University.

Chapter 23
Context and Community Renewable Energy Development in Western Australia: Towards Effective Policy and Practice

Emilia Lawonski, Nicole Hodgson, and Jonathan Whale

23.1 Introduction

Community renewable energy (CRE) projects differ from traditional energy generation models. These projects are exemplified by an approach to development that encourages open participation by local community members and creates positive shared outcomes for their communities (Walker and Devine-Wright 2008). This form of renewable energy development has made notable contributions to the clean energy transition in countries such as Scotland, Denmark and Germany (Bomberg and McEwen 2012). While currently less than 20 Australian CRE groups are generating energy (Embark n.d.), there is growing momentum spurred on by an increasingly active community of practitioners, lobby groups and activists, CRE support groups and enthusiasts. There is also increasing interest in the political sphere as CRE makes its way onto the agendas of governments and major political parties at the state and national level. CRE research and strategy development in Australia have so far focused predominantly on project development issues in the context of the National Electricity Market (NEM), the largest interconnected grid in the country that connects states in the east of Australia (see e.g. Ison et al. 2012; C4CE 2015; Mey et al. 2016). The government-funded National Community Energy Strategy discusses key policy and regulatory reform mechanisms relevant to the NEM while omitting the regulatory and market regime of Western Australia's South West Interconnected System (SWIS) (C4CE 2015). Although generalisations about CRE development have been made based on prior research, it is not a given

E. Lawonski (✉) • N. Hodgson
Sustainability, School of Business and Governance, Murdoch University, Perth, WA, Australia
e-mail: emilia_lawonski@fastmail.com.au

J. Whale
Energy Studies, School of Engineering and Information Technology, Murdoch University, Perth, WA, Australia

A. Sayigh (ed.), *Transition Towards 100% Renewable Energy*,
Innovative Renewable Energy, https://doi.org/10.1007/978-3-319-69844-1_23

that these findings can be readily extended to all Australian contexts. Given the growing interest in CRE in government and politics nationally, policy development requires inputs from a variety of Australian contexts. This paper fills an empirical gap by investigating the context of the SWIS in Western Australia (WA). The SWIS is the second largest electricity network and market in Australia and, in contrast to the NEM, is isolated from all other Australian states and has its own unique regulatory regime and market mechanisms.[1] As such it can be expected that CRE projects in WA's south west face a set of distinct conditions that warrant investigation. This paper uses case studies of CRE projects from the geographic region covered by the SWIS to investigate how context has influenced CRE development, exploring the interplay of contextual influences upon CRE development in this setting. The findings suggest that CRE development in Western Australia's south west has survived in niche pockets, which have been unusually conducive to project development in a context otherwise riddled with hurdles of a political, technical and regulatory nature.

23.2 Case Studies and Methodology

A multiple case study design (Yin 2014) is used, drawing on four projects located in a geographic area covered by the SWIS network in WA. Through an initial desktop study, three community-labelled wind generation projects in the SWIS were identified: Denmark Community Windfarm (DCW), the only case study that is both operational and has strong community credentials; Fremantle Community Wind Farm (FCWF), which is not operational and facing a major hurdle in gaining land access; and the Mt. Barker Community Windfarm (MBCW), which is an operational community-scaled wind farm but does not have as strong community credentials as DCW and FCWF. MBCW was included to provide further insights into the issues facing community-scaled projects in the SWIS and also as a contrasting case. The final case study, the Guildford Energy (GE) solar project, provides insights into a fledgling group struggling with early-stage project development issues and at a much smaller scale than the other case studies. Key case study information is presented in Table 23.1.

This research builds upon Walker and Devine-Wright's (2008) two-dimensional process and outcome framework that conceptualises CRE as a form of renewable energy development which employs open and participatory development *processes* and results in local and collective *outcomes* for their communities (Walker and Devine-Wright 2008). We developed a series of indicators to systematically

[1]An Electricity Market Review was launched by the WA Minister for Energy in early 2014. As a result a series of reforms to the market and network were proposed. A package of bills to transfer regulation of the SWIS network to the national electricity regulatory framework is currently on hold (Public Utilities Office 2017a).

Table 23.1 Overview of case study information

	DCW	FCWF	GE	MBCW
Size	2 × 800 kW wind turbines. Maximum output registered at 1.44 MW	Proposal to build 8–12 wind turbines. Total output 6.4–9.6 MW	Proposed 20–25 kW solar PV pilot project followed by 200 kW PV installation	Three 800 kW wind turbines. Maximum capacity registered at 2.43 MW
Cost	Approx. $5.8m AUD	Estimated at $16–18m AUD for eight turbines	Unknown (project abandoned late 2014)	Approx. $8.5m AUD
Location	Rezoned reserve at Wilson Head. Approx. 9 km south of Denmark townsite, Great Southern Region, WA	Rous Head at Fremantle Port. Fremantle, Perth Metropolitan Region, WA	Guildford, Perth Metropolitan Region, WA. Intention to utilise local factory roof/s and a local pub for the pilot	Private sheep farm 4 km north of Mt. Barker, Great Southern Region, WA
Status	Operational since February 2013	Seeking land access	Project abandoned late 2014	Operational since April 2011

analyse the social processes and outcomes of each project to ascertain the community value arising from each approach. A differentiation between this approach and Walker and Devine-Wright's conceptualisation (2008) is the consideration of project scale, which coupled with a project's location can imply who benefits with regard to energy generated and consumed (Hicks and Ison 2011). Our analysis draws upon a theoretical framework which conceptualises contextual influences on socio-technical innovation. Rip (2012) builds on the socio-technical transition literature (see Geels 2005), reframing the multilevel perspective on socio-technical transitions and identifying "general patterns and structures in the context" that exert an influence on project dynamics (2012). The context interacts with project development journeys across three socio-technical layers: *niches* provide breathing room for 'radical innovations' by forming 'protected spaces' for experimentation; *regimes* give the context rules and structuration; and *landscapes*, comprising social, economic, environmental and material aspects at the macro level of society, create a "backdrop of opportunities and constraints" (Rip 2012), where influences tend to be more indirect (Geels 2005). This framework is used to investigate contextual influences on both the social processes and outcomes of CRE project development and on those influences that have other material effects such as those which influence how or if a project physically eventuates in a given context. The study uses a combination of interview data corroborated with documents and a thematic network analysis approach (Attride-Stirling 2001).

23.3 Findings and Analysis

In 2003 WA led the way in CRE development in Australia with the forming of the pioneering DCW project. Since then only a few CRE projects have emerged in the state, with a majority of projects located in NSW and Victoria (C4CE 2015). Context has played a part in influencing the project development approaches of our case studies, which demonstrate the community value of CRE projects (Walker and Devine-Wright 2008). The results in Tables 23.2 and 23.3 show that against a set of process and outcome indicators distilled from the data, the development approaches of DCW, FCWF and GE are significantly community based, indicating socially transformative development approaches. MBCW is only loosely community based, having little community involvement in project development, and the financial benefits, while to some extent local, are not widely distributed throughout the community.

23.3.1 *Niche Influences: Local Contexts as CRE Incubators*

The analysis did not identify any niches in the strict sense of *actively constructed* spaces that support and protect novel developments (see Geels 2005). Instead

Table 23.2 Indicators of open and participatory project development processes across case studies

Processes	DCW	FCWF	GE	MBCW
The vision	Concept emerged from local community, led by core group of prominent locals	Concept emerged from Fremantle locals passionate about renewable energy	Concept emerged from Transition Town enthusiasts living in and around Guildford	Initiated by Perth-based renewable energy consultancy
Community engagement	Broad engagement included workshops, school engagement, site visits and bus tours to Albany Wind Farm	Extensive engagement through public meetings, workshops, presentations, stalls, events and social media	Community co-design workshops, presentations and discussions. Leaflet drop	Local stalls and newspaper advertisement
Community participation	Core group of local volunteer. Local supporters demonstrate their support publicly. Locals invest	Core project team and supportive community members volunteer, attend events and donate funds	Core group of local Transition Town enthusiasts volunteer their time, skills and resources	Little interest in participation and investment from local community. Hundreds attend official opening
Legal structure	Two entities: a not-for-profit entity and public company $500 min. investment	Cooperative with one vote per member investor	Transition Town action group with no formal structure	Private company

Table 23.3 Indicators of local and collective project outcomes across case studies

Outcomes	DCW	FCWF	GE	MBCW
Community fund	Community fund of 200,000 one dollar shares to DCW Inc.	Community fund to be established from percent of profits	Model involved fund for investing in community projects	No community fund
Investment structure	Public company. Local advertising kept investment mostly in the local community (116 shareholders)	Registered co-op limits investment to WA, (option to register in other states). Local investment sought	Project model involves local community investment (investment model not yet determined)	Private company model comprising 12 investors with ties to the local area
Other local financial benefits	Locals employed in all project phases. DCW Inc. issued $27,000 in grants as at Jan 2017 with ongoing dividends to distribute	Intention to employ local people. Tourism potential hoped to contribute to economic revitalisation	Intent to partner with local installer. Intent for local businesses to benefit from cheaper electricity	Land leased from local farmer. Some local employment during construction
Capacity building and education	Lessons shared with other groups. New local projects funded	Locals learn to build a wind farm through project participation	Core group gained knowledge about regulations and market	Development experience benefited DCW
Energy outcomes and scale	Improved local power quality. Approx 30% of local load met[a]	Project scale approximately matches load of Fremantle Port	PV to offset energy purchased from grid by local businesses	Scaled to local load. Power sold to state-owned retailer

[a]Early intention was to more closely match the town of Denmark's energy needs with a 2.4 MW wind farm; however after conducting technical studies, the network operator limited the wind farm output to 1.44 MW

certain localities possessed combinations of characteristics more likely to lead to the development of successful CRE projects. In other words these localities exhibited niche-like behaviour for the purposes of CRE development, albeit the niche conditions were established incidentally. In Denmark, features of the local historical, physical and social context interacted to create a supportive niche for the emergence of a grassroots, community-owned wind farm. A DCW director described the town of Denmark as an "incubator" for CRE development, with "a reputation for at least thirty years of being a green town", "visionary can-do people", and the right "physical environment" for a wind farm, emphasising that at the time such a project "wouldn't have happened in most other places in this state" (Interview, DCW 'D2' 2014). A community member described Denmark's interest in sustainability leadership: "there's all sorts of elements of people [who live in Denmark] wanting to show what the future more sustainable lifestyle looks like" (Interview, Denmark Community Member 2014). The impetus for DCW was about building Denmark's capacity to continue to build on its sustainability leadership: "driven by the community wanting to take charge of energy consumption in such a way that it actually fed back profit back into this community in order to grow our capacity to do more of the same" (Interview, Denmark Community Member 2014). This is reflected in the community ownership model and Community Sustainable Living Fund, which reinvests project dividends into the community via grants to local community enterprises. In contrast to DCW, while Mt. Barker had the right physical characteristics to support the development of a community-scaled wind farm, it lacked those key historical and social features of the local context that drove Denmark to establish a strongly community-driven project. Approximately 60 km from Denmark, Mt. Barker had a location with a sufficient wind resource and in close proximity to a power distribution line. Those involved in the development of MBCW characterised the town as a "conservative community", which did not show interest in being involved in the project (MBCW 'MB1', 2014). The local attitude appeared to be "great idea, just go away and do it" (MBCW 'MB2' 2014). The developer's perception of the community's attitude of "that's a good idea but don't ask us for any money" influenced the project not to pursue a strongly community-based approach: "it's not that we didn't market for it; there was no evidence that people would actually come in and do that [invest in the wind farm]" (MBCW 'MB1' 2014).

Like the town of Denmark, Fremantle also possesses key niche-like characteristics that have coincided to create the impetus and opportunity for a community-owned wind farm development in the port city. The proposed site of the wind farm at Fremantle Port is an iconic location with a proven wind resource and, being at a major port, is near transmission lines. Like Denmark, Fremantle has a reputation for its sustainably minded local and progressive community: "the City has a proud tradition of social justice and innovation in environmental management and sustainability" (Pettit 2009, 49). For example: "People [that live in Fremantle] are interested in community gardens ... [and] renewable energy" (Interview, FCWF 'F1' 2014). As a result FCWF has experienced significant community support ranging from local people volunteering their time and participating at events to public support from Fremantle's "green" mayor and the Maritime Union of Australia, which voted unanimously in favour of the project in a meeting with 700 members (Maritime Union of Australia 2012). For Guildford Energy, a protected space for conceptual project development emerged through social networks associated with the Transition Town movement. Local Transition Town events were the catalyst for project formation via the coming-together of like-minded individuals; however the niche did not extend beyond this loose social network and was not enough to sustain project development activities against the challenging backdrop of the socio-technical regime of the SWIS.

23.3.1.1 Macro-Protected Spaces

In addition to the above incubator characteristics, DCW, MBCW and FCWF experienced circumstances that provided economic protection for project development, afforded by macro-protected spaces reliant on external conditions. If niche conditions depend on a macro-protected space, such as a supportive policy or agreement, and the macro-protected space becomes unstable, the niche will be exposed (see Rip 2012). A commercial wind development was proposed at Fremantle Port as early as 1997 and received planning approval from the WA Planning Commission in 2002 (Pacific Hydro 2003) but never came to fruition due to disagreement around commercial terms for electricity purchase by the Fremantle Port Authority (FPA) (Fremantle Wind Farm 2016). Project development work completed at the time, including wind monitoring data, geotechnical drilling, grid connection studies and environmental studies, which demonstrate the feasibility of a wind farm at the site, is available to the FCWF project proponents through an agreement with the initial developers. Access to this work, worth over $500,000 is extremely valuable to the FCWF project, because as emphasised by a director: "We can never do a project like this somewhere else, because we'd have to pay that five hundred grand for that work to be done" (Interview, FCWF 'F2' 2014). The agreement to acquire this work has created a macro-protected space for a wind farm at that particular site, without which the project would not exist.

A federal grant through the Renewable Remote Power Generation Programme (RRPGP) was made available to applicants in 2006 to cover up to 50% of the capital costs of renewable energy projects in remote areas. WA was the only state to extend funding to grid-connected projects. The grant criteria limited applications to medium-sized (30 kW–2 MW) projects located within specific local government shires and preferenced fringe of grid areas within the SWIS, for which projects could demonstrate potential for improved local power quality and reduced transmission losses (Sustainable Energy Development Office 2006). DCW fit these specific criteria as an edge-of-grid town with significant power quality issues and large transmission losses, eventually securing a grant of $2.49 m. MBCW secured the grant for $4.2 m as it was within a listed shire, despite not having power quality issues or being under 2 MW. The grant was only ever made available for one round of successful applicants who applied in 2006, highlighting the intermittency of policy and the vulnerability of reliance upon macro-protected spaces. As will be seen in Sect. 23.3.2, reliance on the grant also left DCW vulnerable to changing political circumstances.

23.3.2 Socio-technical Regime Interactions

23.3.2.1 Regime Decision-Making Processes

The major hurdle for FCWF is gaining land access at the proposed site at Fremantle Port, owned by the WA Government and managed by the FPA, which refuses to lease the land. FCWF has thus embarked on a community engagement campaign to influence the FPA to allow the lease. A director stated: "It's almost become a community engagement campaign, more than a wind farm development, which fits with the community ownership model anyway, so that's great. We're building up this head of pressure of people that actually want it to go ahead" (Interview, FCWF 'F3' 2014). The strategy includes demonstrating community support to compel the state government to support the project and encourage the FPA to approve the lease. Engagement with the Barnett Government, in office in WA between September 2008 and March 2017, proved fruitless, and land access was further complicated by a bill introduced before state parliament in May 2016 to privatise the port, a move opposed by the then state opposition.

Similarly, over the course of its 10-year development, DCW contended with challenges related to changing local, state and federal governments with shifting policy and funding priorities. DCW engaged regularly with elected members: "we had to re-educate all those three levels of government every time there was an election because every time you'd get a whole bunch of people coming in who knew nothing about it. . .a pretty major logistics exercise and it took a lot of energy" (Interview, DCW 'D2' 2014). A local opposition group, the South Coast Landscape Guardians, gained the support of local councillors and the Federal Minister for the Environment, Ian Campbell, who announced that the project had divided the Denmark community and later refused to approve the RRPGP grant. In 2005 the local council voted against a town planning scheme amendment to accommodate the wind farm on the "A" Class Reserve "in order to preserve the amenity and landscape values of Wilson Head" and voted against the excision of the wind energy facility zone even after the council's 2008 Community Survey demonstrated 70% support for the wind farm (Shire of Denmark 2008). The amendment was a crucial step in securing land access from the state government, so the wind farm committee worked to convince supportive individuals to stand for the council: "we actually got a majority of councillors that were in support of the project, and that's through being proactive; we had to get people to stand for council that were pro wind farm" (Interview, DCW 'D1' 2014). The new council composition voted in favour of the excision, which was subsequently approved by state parliament. Timing was crucial due to a caveat with a deadline placed on the grant, making funding subject to the excision.

23.3.2.2 Network Access and Connection

A common thread throughout the interviews was the opinion that dealing with the state network operator with regard to gaining network access and connection is extremely difficult and expensive. A senior electrical engineer stated that the network operator has "some of the most onerous rules in the world...you'll have to do things here that you wouldn't be required to do in Europe for instance...Their engineers I'd say have got a fairly low level of understanding of inverter connected generation...so what they don't understand they throw the rule book at" (Interview, Senior Electrical Engineer 2014). A DCW director described the network operator as "incredibly obstructive. They'd get us to do studies that we would pay for and they would come back and say 'that didn't work, do you want to do another study?'...that went on three times" (Interview, DCW 'D3' 2014). A FCWF director also described the process as being "set up for project proponents with deep pockets...that fits in pretty well with...all that [project development] work being done before [by the original commercial developer] and us being able to benefit from that" (Interview, FCWF 'F3' 2014). The high cost of undertaking network studies to determine if a project can access the network and at what size is a major risk for CRE projects in the SWIS. GE's community solar project design was underpinned by advice disseminated by a CRE support organisation based in the eastern states of Australia; and the model was based on conditions in the NEM. The GE project group lost confidence in its model; "we found out it wasn't quite that simple" (Interview, GE 'G1' 2014). A significant reason being difficulties associated with network access: "Particularly once we found out that [the network operator] Western Power puts a lot more constraints on you if you're installing anything over 30 kW" (Interview, GE 'G1' 2014).

The cost of physically connecting to the network can also be significant in the SWIS as discovered by DCW, who were required to pay to replace the existing aerial distribution line which was deemed inadequate and to underground a major portion of the new power line along a snaking road reserve and up to the wind farm, a distance of 1.5 km, which was "extremely expensive" (SkyFarming 2014). Despite there being two turbines at a combined rating of 1.6 MW, the wind farm output was limited by the network operator to 1.44 MW due to the risk of voltage rise issues (SkyFarming 2014), thus reducing the potential economic output of the wind farm. For the above reasons (and more), the chosen site was described by a DCW director as "fraught": "if the government had paid generators one or two cents more per kWh we wouldn't have put it there – it was the only site that would work if we got a pittance for our energy" (Interview, DCW 'D3' 2014).

23.3.2.3 Power Sales and Lack of Retail Contestability

Customers in the SWIS consuming less than 50 MWh per year are unable to choose their electricity retailer. A FCWF director noted that "supplying energy to co-operative members ...[is] not really possible in Western Australia because of the way that our energy market is structured". (Interview, FCWF 'F1', 2014). Another director added that "if we could sell power to households it would be easier to get a community project away, because people really do want that; they really do want to be able to buy power from the wind farm" (Interview, FCWF 'F3' 2014). Both DCW and MBCW sell their power to the state-owned retailer via power purchase agreements. For GE, the ability to sell its solar power to the businesses on which it was planning to instal its solar systems became prohibitively complex. Without an electricity retail licence, GE would not be allowed to sell power directly to businesses, unless it was able to obtain an exemption from the *Electricity Industry Act 2004* (Interview, GE, 'G1' & 'G2' 2014). Furthermore, customers under the contestability threshold would not be able to choose GE as their retailer. In August 2016 a new licence exemption framework for solar power purchase agreements (solar PPAs) was introduced by the WA State Government

(Public Utilities Office 2017b). This new framework should provide greater clarity for future projects modelled on a solar PPA arrangement.

23.3.3 Landscape Constraints

The network infrastructure of the SWIS can be seen as a component of the socio-technical landscape; an "obligatory passage point" physically enabling and constraining the transport of energy through the network (see Rip 2012). As shown in Sect. 23.3.2.2, DCW had no choice but to pay to physically augment the network to enable electricity to be transported to energy consumers. In parts of the SWIS, the distribution network has reached capacity: for example, in the Wheatbelt region "capacity restraints in the distribution network are inhibiting the development of energy generation" to the extent that "Small scale renewable distributed generation is likely to be restricted to off-grid installations as the capacity of the local power distribution network limits their ability to feed power back into the grid" (Wheatbelt Development Commission 2014).

23.4 Discussion and Conclusions

This research demonstrates that CRE development in the context of the SWIS has survived under a set of very specific conditions. Particular local niche-like conditions have driven and sustained a handful of CRE projects within a socio-technical regime designed around an incumbent centralised generation model built heavily on large-scale fossil fuel generation plant owned by both the state and private corporations. The socio-technical regime faced by CRE projects in this context presents hurdles and constraints of a technical, regulatory and political nature. Macro-protected spaces have been crucial to project success, created by short-lived programmes that enabled government grants for capital works to be allocated to DCW and MBCW based on an exclusive set of conditions. Similarly, an agreement between FCWF and a commercial wind developer enabling the group to access previously completed development work at the proposed wind farm site provides a macro-protected space that makes the project economically viable. The findings indicate that while supportive local niches can sustain project activity over extended periods of time, instability in the policy sphere and resistance from within the socio-technical regime can cause significant difficulties and delays for CRE projects in the SWIS. These difficulties are highlighted by the contrasting experiences of projects proposed on state-owned versus private land. DCW and FCWF have experienced immense difficulty seeking approvals to build on state land, facing opposition from government ministers and for FCWF opposition from a state enterprise. Located on private land, MBCW gained planning approval relatively easily and did not face local opposition. This also helps explain why despite significant community support for DCW and niche-like conditions for a CRE project in the town of Denmark, a small opposition group to the wind farm gained political traction, in the otherwise progressive, "green" town, as the project was proposed on a crown reserve valued by locals. Local opposition in Denmark is in line with Bomberg and McEwen's (2012) findings that a sense of place and belonging based on a "shared geographic space" can be both a driver for CRE development and a motivator for active opposition.

Faced with significant difficulties stemming from the regime layer of the context, CRE actors in the SWIS have not been passive actors constrained by their environment. To the contrary, the project actors we interviewed have navigated and "stretched" the context (Rip 2012) with the support of their communities, by engaging with socio-technical regime stakeholders and in the case of DCW even physically augmenting the socio-technical landscape of the SWIS network infrastructure, at great cost, to accommodate their project vision. For the GE solar project, the complex regulatory regime became an impassable barrier to further project development and is a warning sign to new projects to be cautious of project models established in contexts outside the technical and regulatory jurisdictions of the SWIS, and investigate how the context may impact the project before becoming invested in a particular project design. It is clear that the SWIS regime in its current form, including its interconnectedness with the broader WA and national contexts, is not an accommodating place for CRE development. There are however signs that the SWIS is changing and may present opportunities for new CRE projects to arise. For example, the network operator is pursuing a renewable micro-grid to address power reliability issues in the edge-of-grid town of Kalbarri, stating that its engagement process "canvassed the level of support for a community-owned solution", although later resolving that the community preferred a solution led and managed by the network operator (Western Power 2016). With reliability issues and capacity constraints facing many parts of the SWIS network, win-win solutions like edge-of-grid DCW and community micro-grids have the potential to benefit communities *and* the network. If further benefits of CRE are to eventuate in WA, ongoing CRE strategy and policy development at the national level must explicitly consider the nuances of the SWIS and WA separately to the NEM, tailoring specific strategies for this context. Accordingly, CRE development in the SWIS would benefit from targeted policy development at the state government level, based on a knowledge of the intertwining contextual influences that affect CRE projects and driven by an understanding of the benefits such projects can bring to the state.

References

Attride-Stirling, J. (2001). Thematic networks: An analytic tool for qualitative research. *Qualitative Research, 1*(3), 385–405.

Bomberg, E., & McEwen, N. (2012). Mobilizing community energy. *Energy Policy, 51*, 435–444.

C4CE (Coalition for Community Energy). (2015). *National community energy strategy*, Retrieved from http://c4ce.net.au/nces/

Embark. (n.d.). *Australian community energy groups*. Retrieved from http://www.embark.com.au/display/public/Groups/Australian+community+energy+groups

Fremantle Wind Farm. (2016). *Status of the project*. Retrieved from http://www.fremantlewindfarm.com.au/?page_id=52

Geels, F. W. (2005). *Technological transitions and system innovations: A co-evolutionary and socio-technical analysis*. Cheltenham: Edward Elgar.

Hicks, J., & Ison, N. (2011). Community-owned renewable energy (CRE): Opportunities for rural Australia. *Rural Society, 20*, 244–255.

Ison, N., Hicks, J., Gilding, J., & Ross, K. (2012). *The australian community renewable energy sector: Challenges and opportunities*. Retrieved from http://cfsites1.uts.edu.au/find/isf/publications/isonetal2012ozcommyrenewablesenergy.pdf

Maritime Union of Australia. (2012). Maritime workers support Fremantle Community Wind Farm. Retrieved from http://www.mua.org.au/maritime_workers_support_fremantle_community_wind_farm

Mey, F., Diesendorf, M., & MacGill, I. (2016). Can local government play a greater role for community renewable energy? A case study from Australia. *Energy Research & Social Science, 21*, 33–43.

Pacific Hydro. (2003). *Fremantle wind project*. Retrieved from http://web.archive.org/web/20030822134217/http://www.pacifichydro.com.au/projects.asp?articleZoneID=209

Pettit, B. (2009). Embodying the past in the future? Sustainability and built heritage in Fremantle. *Historic Environment, 22*(2), 49–53.

Public Utilities Office. (2017a). *Electricity market review – phase 2*, http://www.treasury.wa.gov.au/Public-Utilities-Office/Industry-reform/Electricity-Market-Review-Phase-2/

Public Utilities Office. (2017b). *Solar power purchase agreements retail licence exemptions*. http://www.treasury.wa.gov.au/uploadedFiles/Site-content/Public_Utilities_Office/Industry_reform/Licence-Exemption-Application-Guidelines-Solar-Power-Purchase-Agreement-Providers.pdf

Rip, A. (2012). The context of innovation journeys. *Creativity and Innovation Management, 21*, 158–170.

Shire of Denmark. (2008). *2008 community needs and customer satisfaction survey result*. Denmark: Shire of Denmark.

SkyFarming. (2014). *A tale of two windfarms, Mt Barker vs Denmark*. Retrieved from http://www.skyfarming.com.au/Tale2windfarms/

Sustainable Energy Development Office. (2006). *Rural renewable energy program: Guidelines for medium projects (30kW to 2MW)*. Retrieved from http://web.archive.org/web/20070831043936/http://www1.sedo.energy.wa.gov.au/pdf/RREP_guidelines_medium_projects.pdf

Walker, G., & Devine-Wright, P. (2008). Community renewable energy: What should it mean? *Energy Policy, 36*, 497–500.

Western Power. (2016). *Study into the feasibility of a microgrid at Kalbarri*. Retrieved from https://www.westernpower.com.au/about/reports-publications/kalbarri-microgrid-study/

Wheatbelt Development Commission. (2014). *Wheatbelt snapshot series: Power and energy*. Government of Western Australia. Retrieved from http://www.wheatbelt.wa.gov.au/files/9314/0487/0817/Wheatbelt_Snapshot_Series_-_Power_and_Energy_FINAL_08072014.pdf

Yin, R. K. (2014). *Case study research: Design and methods* (5th ed.). Thousand Oaks: SAGE.

Chapter 24
Harvesting Sunshine with Smart Solar Panels and Cryptocurrencies

Colin T. Mallett

24.1 Introduction

24.1.1 New Smart Energy Ecosystems

Individuals and communities are becoming more closely involved in low-carbon energy ecosystems. At the same time, the grid and its stakeholders have to accommodate an increasingly complex mix of generation, transmission and distribution technologies. In this paper, we primarily focus on renewable energy generated close to the point of consumption as rooftop or building-integrated photovoltaics – "BIPV".

Interesting things are happening to the power distribution industry. Largely because of renewable technologies, overall demand for grid-supplied electricity (https://www.gov.uk/government/uploads/system/uploads/attachment_data/file/295225) is flat or even decreasing, but consumers still rely on it as a backup, so the peak/average ratio is increasing, resulting in significant changes in infrastructure utilisation which causes diseconomies of scale coupled with dominant fixed costs. By coupling this new reality with the need for ageing infrastructure requiring renewal, there is no wonder that utilities wish to defend their revenues to assure the sustainability of their business models by being seemingly uncooperative against the proliferation of low-cost energy suppliers.

Solar PV systems and hardware costs are plummeting. PV modules now cost around $0.57 per watt, possibly falling to $0.44 by 2020, when worldwide installed capacity may exceed 500 GW (https://www.greentechmedia.com/articles/read/Global-Solar-PV-System-Pricing-Set-to-Fall-40-by-2020). This is two billion 250 watt panels. We also expect to see the impact of price reductions of lithium-

C.T. Mallett (✉)
Trusted Renewables Ltd, Adastral Park, Martlesham Heath, Ipswich, UK
e-mail: colin@trustedrenewables.com

A. Sayigh (ed.), *Transition Towards 100% Renewable Energy*,
Innovative Renewable Energy, https://doi.org/10.1007/978-3-319-69844-1_24

ion batteries coming from the large-scale manufacturers such as Tesla Gigafactory in Reno, Nevada, and new modular storage features such as Redflow's (http://redflow.com) ZBM2 and ZCell Modules.

And, of course, everything to manage these smart devices has to be secure. It can do this by becoming "smarter". Gartner (http://www.gartner.com/newsroom/id/2636073) believes that by 2020, 26 billion devices may contain "$1 processors" and built-in M2M connectivity. Gartner also said 50 billion IoT objects could generate >£300 billion incremental revenues from products and services worldwide.

In this paper, we will not consider whether the grid really wants this power all the time. This has led to the new idea of negative feed-in tariffs (FiT) in Germany. We live in interesting times.

24.1.2 Renewable Energy Trading

Although renewable energy is essentially free at the point of use, capital and operational costs still need to be covered which means that reward or trading mechanisms are required. We are inspired by the money transfer service M-Pesa (M for mobile, pesa is Swahili for money) which allows users in Kenya to pay for goods and services with an account stored on their phones. This has given millions of people access to a new financial system and challenges the notion that value transfer mechanisms have to be via a traditional bank.

This leads to the question of how these ideas can be applied to rewarding investment in solar panels or rewarding changes in consumer or market behaviour in consumption and demand of renewable energy.

One solution is an energy-based digital currency called the SolarCoin (Gogerty et al. 2011) introduced by the SolarCoin Foundation (SCF) (https://solarcoin.org/en/front-page). The SCF has "pre-mined" about 40 years' worth of SolarCoins ready to be credited to digital wallets for every 1 MWh of verified solar electricity generated by a registered solar installation.

The SCF vision is to build a worldwide solar energy value transfer network which can be converted into money via a network of fungible value. Currently, the SolarCoin exchange rate is around 10 US cents, so this reward is still symbolic (Clapaud 2016; http://www.atelier.net/en/trends/articles/solarcoin-blockchain-service-of-renewable-energies_442008). However, at this stage, there do not appear to be any energy industry regulatory barriers to its adoption though this may not be so with financial regulations with respect to cryptocurrencies in general, especially fungible ones.

One proposed global solution that has a very compelling design is the Global 4C model of "Solar Dollars" issued as a mitigation reward, a form of complementary currency which rewards consumer or industry behaviour that verifiably offsets or mitigates climate change. This was proposed by Dr. Delton Chen et al. The Solar Dollar will have a unit of account of 100 kg of CO2, as defined under the Global

4C's "carbon monetary standard" (http://www.global4c.org/carbon-monetary-standard).

Regardless of the unit of account, there is also a growing consensus from consumers who own renewable energy installations that they should be able to sell their excess power using peer-to-peer (P2P) or local smart grid community networks by trading energy assets via smart contracts. Trials are taking place in many countries including the USA and Australia. However, it appears that vested interests are using the regulatory situation to prevent consumers from trading electricity publically. Fortunately, around the world, there appears to be general support to break down some of these policy and regulatory barriers.

24.1.3 Cryptocurrencies, Blockchains and Distributed Ledgers

Cryptocurrencies have been billed as the "biggest change to financial services since the 16th century" (http://www.adviserlounge.co.uk/2016/08/01/radical-money-is-financial-services-disruptable/). Some may disagree, but the underlying blockchain technology is definitely revolutionary, as are distributed ledgers which record every transaction made by every participant in a decentralised system.

Figure 24.1 shows a spectrum of ledger functions which can be achieved with different "degrees of centralisation". Assets recorded by the blockchain can be financial, legal, physical or electronic and can make management of distributed energy resources easier and inherently more secure. Distributed ledgers also support "smart contracts" (Distributed Ledger Technology 2017) which make them ideal for peer-to-peer of renewable energy trading and ensure aid money, perhaps for rooftop solar installations, which is spent in a traceable manner. Cryptography can keep information private, and ledgers can be "permissioned" or "unpermissioned" depending on who is allowed to make changes. On one end of the spectrum, Bitcoin ledgers are unpermissioned, so anyone is allowed to change the blockchain when the coins are spent, whereas permissioned ledgers can include rules on who is allowed to modify the ledger.

24.1.4 Ethereum and the TransActive Grid

The Ethereum distributed application platform (https://www.ethereum.org) was introduced in 2014. Developers can build applications such as smart contracts, without having to build their own blockchain and use a token called ether. Currently, transactions are stored publicly on every blockchain node, and our telecoms background suggests that as traffic grows, the open permissionless ledger architecture could suffer severe performance problems, especially latency. It is also

Fig. 24.1 Spectrum of different ledger technologies

vulnerable to cyberattacks including distributed denial of service (DDoS), and we expect the Ethereum platform will evolve in the same way that virtual private networks developed from public telecoms networks. Indeed, since its introduction, Ethereum has been forked and revised a number of times. Moves to de-bloat the ETH blockchain are already underway, and private permissioned variants of Ethereum are already being developed which would allow a more deterministic quality of service to be maintained enabled by a distribution of services to closed network participants.

24.2 Supporting Community Energy

24.2.1 Trials and Pilots

In social media discussions, community energy schemes are generally seen as a "good thing". However, translating this into reality is challenging, especially in developed countries where significant regulatory barriers and vested interests may need to be overcome. It's clear that blockchains and smart contracts will have a major part to play in this revolution, though none of us really knows how yet.

Trials and pilots help, and one of the most interesting is the Brooklyn project run by a company called L03 Energy led by Lawrence Orsini (http://lo3energy.com). Siemens is providing the community microgrid technology to enable physical exchange of electricity between local renewable energy producers and neighbouring energy consumers. This operates in parallel to the utility-provided grid so it provides a redundant local energy network which has found favour with NY Governor Cuomo. Valuable experience has come from working with regulators. L03 Energy developed the business logic for open peer-to-peer energy trading using their TransActive Grid running on an Ethereum platform. They are now transferring this knowledge to run similar trials in Australia (https://onestepoffthegrid.com.au/p2p-energy-sharing-start-brings-brooklyn-microgrid-smarts-australia).

Also in Australia is Perth company Power Ledger (https://powerledger.io) who have been running projects which demonstrate how home users could trade excess solar energy between neighbouring homes using metres linked to the EcoChain permissioned distributed ledger solution from a company called Ledger Assets (http://ledgerassets.com). Australian regulatory policies have proved difficult with the development of Power Ledger's P2P energy trading platform, and the team has turned its attention to New Zealand where it is reported that regulators have been more receptive to their ideas (https://www.vector.co.nz/innovation-news/peer-to-peer-energy-trading-to-be-trialled-in-nz).

The UK is slightly different. Through the services of Innovate UK, Trusted Renewables Ltd has benefited from significant UK government grants and support for commercial R&D in the energy and IoT sectors. The previous coalition government encouraged development of local or community public operators (https://www.gov.uk/guidance/community-energy) selling locally generated renewable energy combined with grid-sourced power purchased from wholesale energy providers. To do this, companies have to obtain a Public Energy Supply Licence (https://www.ofgem.gov.uk/licences-codes-and-standards/licences/licence-conditions), and there are many regulatory requirements to be met prior to obtaining one of these. Also, the UK FiT (feed-in tariff) scheme includes payments for all renewable energy generated, not just that which is fed into the grid.

24.2.2 New Investment Models

Eligibility to receive FiT payments can be assigned, and essentially this means that the rights to receive FiT payments for a particular scheme can be traded, and in the UK, there is nothing to stop a group of people forming a cooperative to offset local community electricity use with a shared solar panel scheme. This also led to companies installing panels on a consumer's roof and offering "free solar electricity" in return for the FiT income. This was particularly an interesting origination model for housing associations or community energy companies to reduce the energy costs for residents who might suffer fuel poverty. Community schemes often involved funding the capital costs with co-op shares or unlisted corporate mini bonds. These unregulated investments usually have to be held as a block investment until maturity (often 20 years) and cannot be traded freely. Accordingly, these schemes are generally considered subsequently illiquid and very risky.

Future solar PV investments are likely to be mostly based on the grid parity pricing boundary where the rising cost of grid-sourced electricity coupled with falling panel costs makes investment in renewables commercially viable without significant subsidies. However, this is also where SolarCoins and Solar Dollars may have an incentive-based role to play. Grants or loans are required, and the rise of online services and low-cost intermediaries has stimulated new forms of financing such as P2P lending and crowdfunding models which are trying to match lenders directly with borrowers; crowdfunding extends the idea of financing community-oriented schemes with co-op shares or mini bonds. Indeed, investors might be willing to accept minimal returns and high levels of risk if these are associated with improving the energy supply to disadvantaged communities or if investment lock-in periods can be avoided.

In 2012, the UK Crowdfunding Association was formed to support UK businesses, projects and ventures with a code of practice which seems to be well suited to smart contracts for these kinds of transactions, especially renewable schemes managed by distributed ledger technology. We think the potential impact of blockchain on crowdfunding will be significant.

In order for renewable schemes to be "investible", they must resist fraud. Grant- or loan-providing bodies expect strong asset management, and this is where DLTs based on blockchains can help to create what is essentially a secure distributed database of distributed energy resources shared across a network of multiple sites.

24.2.3 Metering and Security

A key requirement for energy trading and award of incentives such as feed-in tariffs is accurate "revenue grade" metering. In many countries, advanced metering infrastructure (AMI) or smart metering provides with two-way communications to a central server. Some smart metre programmes in Australia such as AMI Victoria (http://www.smartmeters.vic.gov.au/about-smart-meters/end-of-rollout) have struggled through the lack of transparent access and sharing of metre data with the home owner. This has arguably biased information in favour of the incumbent provider.

A typical rooftop solar installation is shown in Fig. 24.2 containing a centralised string inverter (b) and a generation metre (c) which is used to supply metering data from which PV FiT payments are normally made. Grid power is metered by a domestic utility metre (d) which may be adapted to run backwards so sometimes (c) is not required for net metering. Most inverters provide an Internet connection, so the power output can be reported back to a central management platform (i), but unfortunately most of these information feeds are not "revenue grade" or "trusted" monitoring.

This arrangement creates an opportunity for fraud through overt bypassing of the trusted metres. In places such as India (http://www.bloomberg.com/news/articles/2014-06-05/india-fights-electricity-theft-as-modi-pledges-energy-upgrade#r=nav-f-story), the grid currently encounters some of the highest rates of hijacking electric supplies and not paying for it. When "behind the metre" local power storage (g) and (h) is added and payments are made from the utilities generation metre readings (c), we think that the temptation to charge the batteries from the grid or the diesel generator and claim it as renewable energy will be very high.

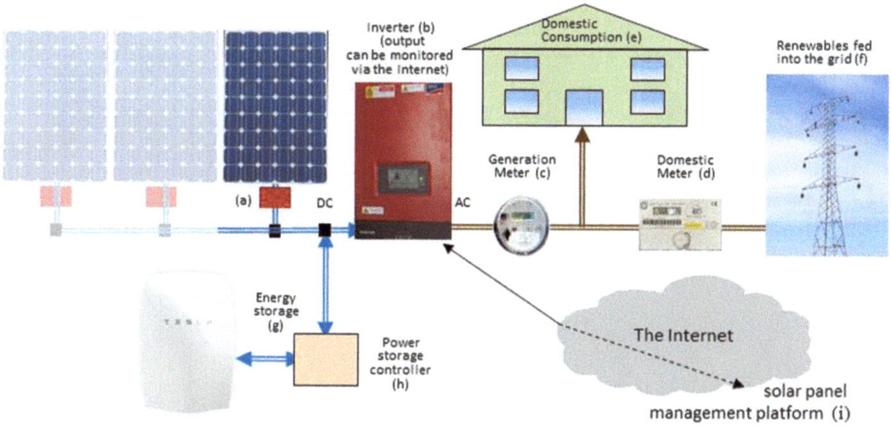

Fig. 24.2 Typical solar PV installation

24.3 Smart Solar Panels

24.3.1 Overview

Smart solar panels invented by Trusted Renewables Ltd (TRL) can mitigate elements of this fraud problem. These contain an embedded secure processor which turns each panel into a self-powered blockchain node which generates cryptographic proof that submitted data has not been tampered with or altered. These can be cross verified to a central point or by using blockchain methods, so this verification responsibility can be architected to take advantage of new types of distributed authority known as a DAO (https://techcrunch.com/2016/05/16/the-tao-of-the-dao-or-how-the-autonomous-corporation-is-already-here) – a "smart contract assisted, Distributed Autonomous Organisation".

Figure 24.3 shows an example of a number of rooftop solar systems made up of panels which all contain secure power reporting modules, the outputs of which are collated by a secure panel management unit which contains communications capability along with memory and processor chips. The security module improves resistance to cyberattacks and manages strong encryption of the various communications links. Without any additional external metering hardware, each rooftop solar system becomes a secure blockchain node which provides readings directly from each solar panel to the distributed ledgers using a cryptographic root of trust close to where the power is generated. This can be recorded in a secure distributed asset ledger fixed to a specific location reported by the embedded GPS. This can also include the status of any local energy storage installations.

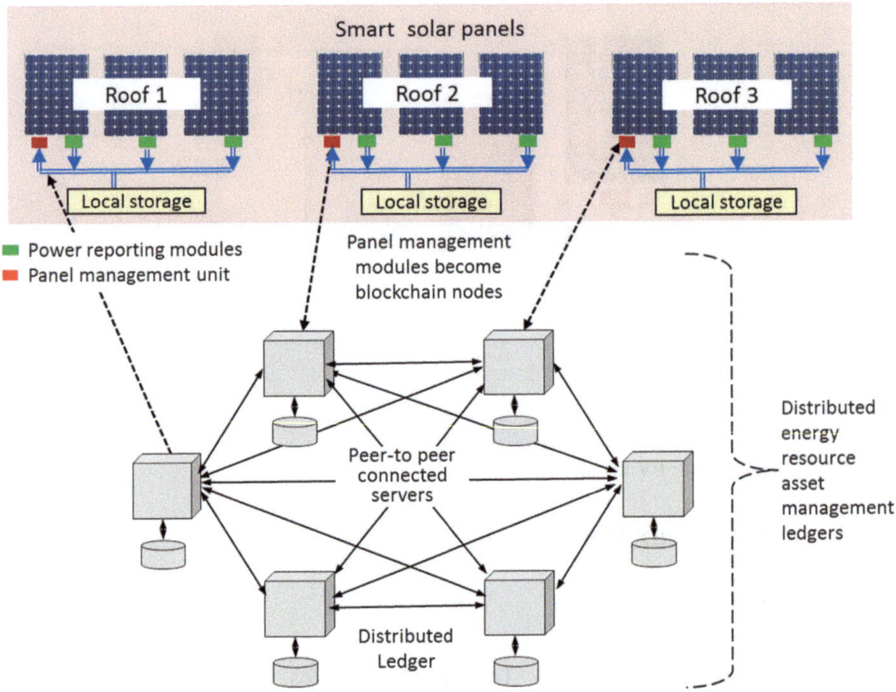

Fig. 24.3 Distributed energy resource asset management system

24.3.2 *Smart Solar Panel Modules*

These improvements are described in patent specification "Method and Apparatus for Secure Energy Management" with a priority date of 6 January 2009. This is granted or pending in >40 countries.

The UK patent GB2479324A was granted on 10 July 2012, AUD patent 0900082.9 on 22 October 2015 and the US patent 9,515,522 B2 issued on 6 December 2016. The key inventive step is to embed secure processors and communications circuits directly into solar panels. Such a module cannot be removed or altered without being destroyed, so it would be very difficult and uneconomical to falsify information coming from a smart solar panel.

One implementation of a smart solar panel is shown in Fig. 24.4. This includes two devices, the panel management "Module 400" and the power reporting unit "Module 500". Both devices include independent security modules. There are a number of options regarding the functional split and physical location of the modules. Both can be embedded in the weatherproof containment or associated with one or more wafers as shown in Fig. 24.5. This is particularly suitable for solar roof tiles which could share a panel management module to collate inputs from

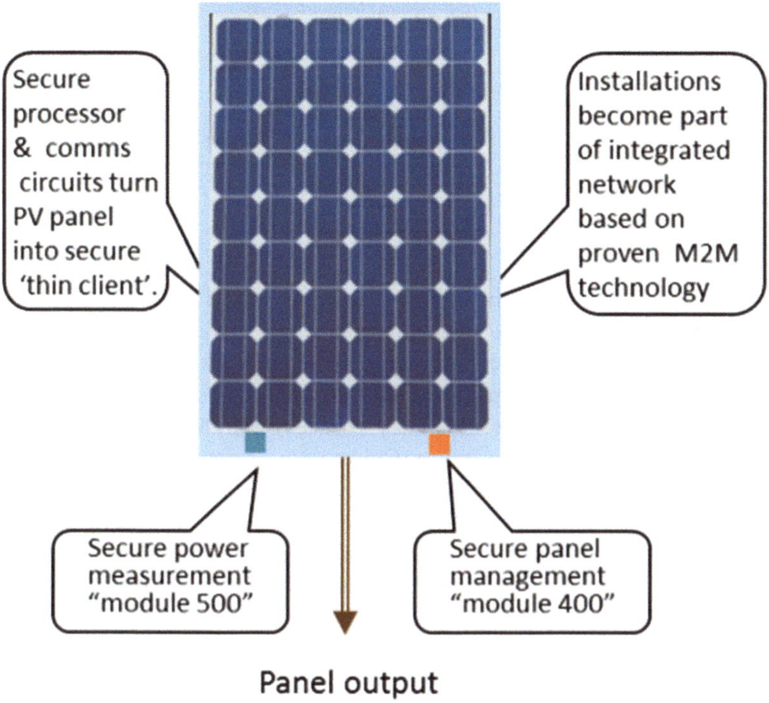

Fig. 24.4 Smart solar panels

Fig. 24.5 Embedded secure power reporting module

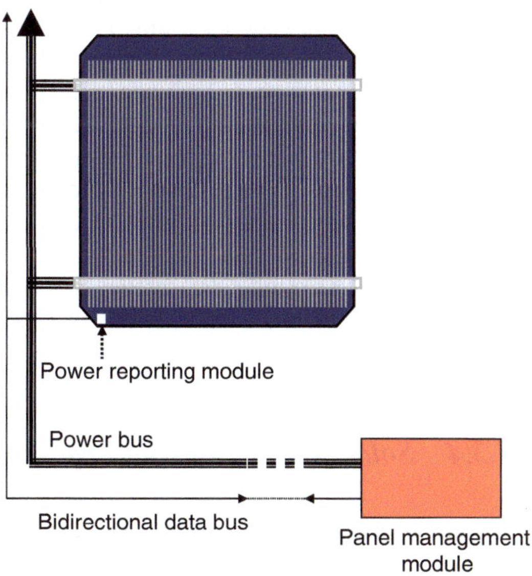

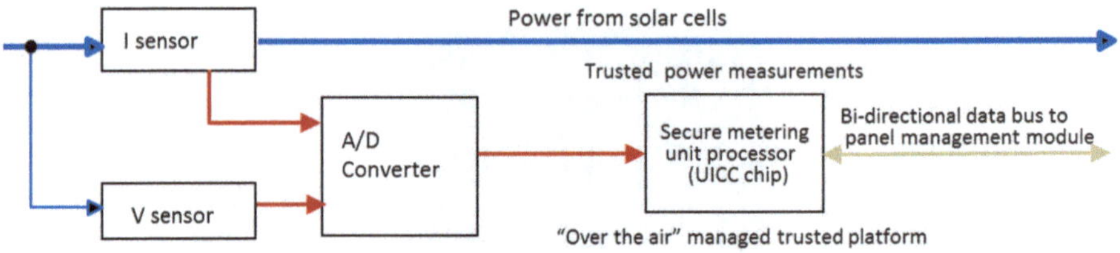

Fig. 24.6 Module 500 secure power reporting unit

power reporting modules embedded in each tile linked by an appropriate local communications network.

Figure 24.6 shows the embedded secure power reporting unit based on a secure smart card microcontroller chip. These units securely measure the power output and send metering data to Module 400. They cost less than $1.00 each (in bulk), and there are many billions of them in use. Chips were originally developed to prevent unauthorised swiping of the magnetic stripes and illegal cloning of bank and SIM cards. This greatly improves fraud resistance and makes it difficult for a user to repudiate a transaction. The basis of Module 500 is a secure microcontroller based in this technology which is configured to metre the generated power close to where it is generated and embedded into a smart solar panel.

24.3.3 Blockchain Software Stack

Smart card operating systems originally developed to support e-payment infrastructure are now quite powerful, having been extended to a multi-application framework which includes secure ID and ePassports together with on-card biometric matching. We favour any smart card OS which has the minimum of features to demonstrate smart metering applications, support public key cryptography and have low-level routines to support and host blockchain crypto-key signatures.

The panel management Module 400 also includes embedded SIM functionality which can be used to link installations together with fixed or radio wide area networks, using a cellular data network. Detailed functionality was described at the WREC XI conference (Mallett 2010).

24.3.4 Solar Allotments and Property Rights

We previously discussed regulatory challenges faced by community energy schemes, noting the apparent lack of any such barriers to prevent the adoption of rewards based on cryptocurrencies which can be converted into a fungible value. To capitalise on this, we propose the idea of a solar allotment legally secured via property rights. Rights Commerce Limited is working with Trusted Renewables on this crypto ledger and legal methodology.

A developer proposes to raise capital for a renewable scheme and then sell the right for individual panel owners to be paid for electricity from "their" smart panels. These are effectively smart contract variants "digital leases" on PV panels which do not have to be co-located. Importantly, they are linked to physical assets that can be expressed as legal contracts and secured in a property rights system. In Australia, this is called the Personal Property Securities Register (https://www.ppsr.gov.au). With this format of both hard asset and property rights legal security, investors can build portfolio of leases which can be traded, in digital, fungible and secure fashion. Our vision is for fungible digital asset trading where income rights and even future contracts can be sold, assigned or donated. In many ways, this mirrors the idea of buy-to-let where long leaseholders rent out apartments and pay "ground rent" and maintenance charges to the freeholder.

Figure 24.7 shows a simple example where adjacent panels in a single string form three separate allotment "plots". Electricity from the panels is used by local buildings, perhaps distributed by the community schemes exemplified by the TransActive Grid or Power Ledger. The advantage is that secure metering information originates independently from each smart solar panel, whilst generation metres can be used for claiming PV FiT payments for a complete site. This can be linked into the leaseholder's distributed ledger, and where appropriate, e-payments based on readings from the generation metre can be distributed in proportion to the output from each panel. Clearly, when traded, some leases would attract a premium in relative proportion to value.

It appears that solar allotment is particularly suited to building-integrated photovoltaics (BIPV) or clusters of rooftop solar installations on commercial buildings

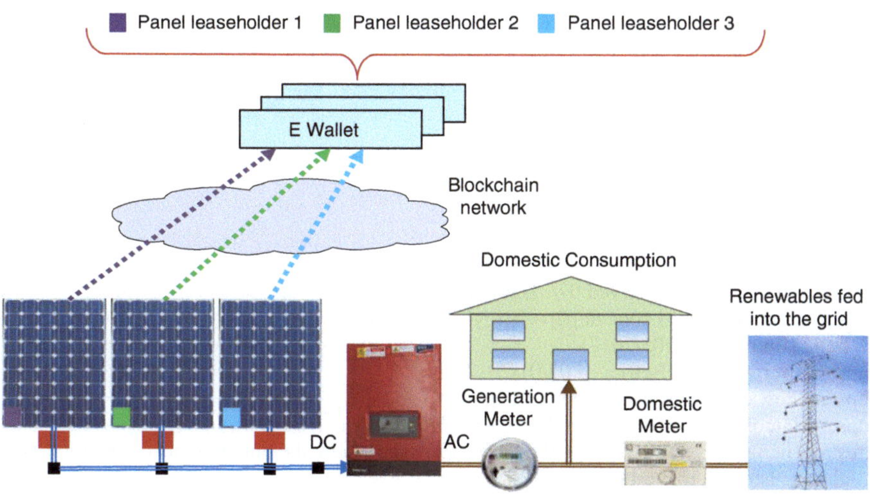

Fig. 24.7 Solar allotments

where surplus electricity is sold. It is equally compatible with P2P lending and crowdfunding managed by distributed ledger technology.

24.4 Conclusions

In this paper, we have considered new smart energy ecosystems and questioned how new ideas can be applied to rewarding investment in solar panels. We noted that energy-based digital currencies are not yet subject to regulatory barriers, which is not the case with many local energy trading schemes. Low-cost online intermediaries have stimulated P2P lending and crowdfunding models which seem to be well suited to smart contracts managed by distributed ledger technology.

A key enabler is the smart solar panel containing an embedded secure processor based on a smart card chip and operating system which supports blockchain crypto-key system services. It also allows a proprietary blockchain and cryptocurrency payment software stack to be reused. This turns PV arrays into a blockchain node which generates cryptographic proof that the submitted metering data has not been tampered with or altered. The security module improves resistance to cyberattacks and manages strong encryption of the various communications links. Inventories can be recorded in a secure distributed asset ledger and can also report the status of local energy storage installations.

Finally, we discussed the idea of solar allotments which allows individuals to access a portfolio of leases on smart solar panels where income rights can be sold, assigned or donated. Importantly, these are linked to physical assets where surplus electricity is sold to local residents. It is equally compatible with P2P lending and crowdfunding managed by distributed ledger technology.

Acknowledgment This paper has received a significant contribution from Leon-Gerard Vandenberg, Founder and CTO of Fuzo Limited, Hong Kong and Advisor to Rights Commerce Ltd, Hong Kong.

References

http://ledgerassets.com (Retrieved 07/01/2017).
http://lo3energy.com (Retrieved 07/01/2017).
https://onestepoffthegrid.com.au/p2p-energy-sharing-start-brings-brooklyn-microgrid-smarts-australia (Retrieved 07/01/2017).
https://powerledger.io (Retrieved 07/01/2017).
http://redflow.com (Retrieved 07/01/2017).
https://solarcoin.org/en/front-page (Retrieved 07/01/2017).
https://techcrunch.com/2016/05/16/the-tao-of-the-dao-or-how-the-autonomous-corporation-is-already-here (Retrieved 07/01/2017).
http://www.adviserlounge.co.uk/2016/08/01/radical-money-is-financial-services-disruptable/ (Retrieved 07/01/2017).

http://www.atelier.net/en/trends/articles/solarcoin-blockchain-service-of-renewable-energies_442008 (Retrieved 07/01/2017).

http://www.bloomberg.com/news/articles/2014-06-05/india-fights-electricity-theft-as-modi-pledges-energy-upgrade#r=nav-f-story (Retrieved 07/01/2017).

https://www.ethereum.org (Retrieved 07/01/2017).

http://www.gartner.com/newsroom/id/2636073 (Retrieved 07/01/2017).

http://www.global4c.org/carbon-monetary-standard (Retrieved 07/01/2017).

https://www.gov.uk/government/uploads/system/uploads/attachment_data/file/295225/Seasonal_variations_in_electricity_demand.pdf (Retrieved 07/01/2017).

https://www.gov.uk/guidance/community-energy (Retrieved 07/01/2017).

https://www.greentechmedia.com/articles/read/Global-Solar-PV-System-Pricing-Set-to-Fall-40-by-2020 (Retrieved 07/01/2017).

https://www.ofgem.gov.uk/licences-codes-and-standards/licences/licence-conditions (Retrieved 07/01/2017).

https://www.ppsr.gov.au (Retrieved 07/01/2017)

http://www.smartmeters.vic.gov.au/about-smart-meters/end-of-rollout (Retrieved 07/01/2017).

https://www.vector.co.nz/innovation-news/peer-to-peer-energy-trading-to-be-trialled-in-nz (Retrieved 07/01/2017).

"Distributed Ledger Technology: beyond block chain; A report by the UK Government Chief Scientific Adviser". https://www.gov.uk/government/uploads/system/uploads/attachment_data/file/492972/gs-16-1-distributed ledger-technology.pdf (Retrieved 07/01/2017).

Gogerty, Nick, Zitoli, Joseph, DeKo. An electricity-backed currency proposal. Available at SSRN: https://ssrn.com/abstract=1802166 or https://doi.org/10.2139/ssrn.18021661 (2011).

Mallett, C,T. (2010). Network-enabled intelligent photovoltaic arrays. Proceedings World Renewable Energy Congress (WREC XI) Abu Dhabi, UAE 25–30 Sept. 2010.

Chapter 25
Development of Tools for Comparative Study of Solar Cookers with Heat Storage

Maxime Mussard and Marc Clausse

25.1 Introduction

The use of firewood is an important cause of deforestation, which deteriorates the life quality of millions of people (Otte 2013). Solar cooking is an interesting option to limit deforestation, as cooking is a fundamental aspect of any culture. Solar cookers have been developed in the last few decades to overcome energy issues in developing countries. Systems such as box cookers (Saxena et al. 2011) or concentrative systems heating the food directly (Mussard et al. 2013; Farooqui 2014) have been developed.

One of the main disadvantages of direct solar cooking is the lack of thermal inertia of the system. Heat storage can help to overcome this weakness. First, by simply storing energy, the food can be kept warm for many hours. This also provides sufficient inertia to the system and help to provide a regular energy flux despite temporary shadowing such as clouds or people passing by. A more advanced system could even be based on the storage itself, making the cooking almost insensitive to solar radiation variations. Such systems may store enough energy to provide heat after an entire night.

A certain number of solar cookers with heat storage have been designed over the last 20 years (Sharma et al. 2009; Cuce and Cuce 2013). However, the number of parameters to consider when assessing a solar cooker with storage is huge.

An extensive comparison of solar cookers with heat storage has been made elsewhere (Nkhonjera et al. 2016), focusing on cookers' thermal performance. The

M. Mussard (✉)
CETHIL UMR5008, Université de Lyon, INSA-Lyon, Villeurbanne, France

School of Engineering, Nazarbayev University, Astana, Kazakhstan
e-mail: maxime.mussard@nu.edu.kz; maxime.mussard@gmail.com

M. Clausse
CETHIL UMR5008, Université de Lyon, INSA-Lyon, Villeurbanne, France

A. Sayigh (ed.), *Transition Towards 100% Renewable Energy*,
Innovative Renewable Energy, https://doi.org/10.1007/978-3-319-69844-1_25

paper presents comprehensive graphics to compare the temperature and power output of the different cookers.

The novelty of our paper lies in describing two simple methods to compare solar cookers. Several parameters are listed to assess the suitability of different solar cookers with heat storage. By analyzing the parameters for each cooker, a comparison between them is facilitated, and the advantages and weaknesses of each are enhanced. A more relevant use of the different solar cooking technologies is then expected.

25.2 Methodology

A cooker performing well in a specific environment may not be transposable into another one. On the other hand, a system that has not been successful may consider its assets and weaknesses precisely to find a better place to be tested or implemented. The potential for solar cooking is huge, and efficient technologies are available for heat storage. By comparing different parameters for a set of devices, it becomes easier to judge which technology is more relevant for a given case.

In the Methodology part, a set of various solar cookers with heat storage are selected, and the parameters for comparing them are explained. The Results part show the thermal storage potential of the different material and substances used in these cookers and the results of the comparison.

25.2.1 Interest of a Heat Storage

It is in general not possible to implement solar cookers without heat storage in standard conditions for the following (Otte 2013):

- Evening cooking habits
- Availability of a more practical energy source (fuel, coal, wood, etc.)
- Unstable climate

As a consequence, successful implementation of solar cooking is generally related to at least one of these conditions:

- Desert climate
- Existence of a backup system
- Actual improvement of the life conditions (e.g., for disabled woman)
- Nonexistence of other options (refugee camps)

Heat storage can help to overcome the main weaknesses of solar cooking.

First, by simply storing energy, it can help to keep the food warm for many hours. It can give as well enough inertia to the system and help to provide a regular energy flux despite temporary shadowing such as clouds or people passing by.

A more advanced system can even be based on the storage itself, making the cooking almost insensitive to solar radiation variations. Such systems may store enough energy to provide heat even after a full night.

Finally, systems with backup heating have the best potential so far for solar cooking. A backup system which is virtually invisible for the user is the best system so far. Indeed, it is then both easily adaptable to traditional habits and economically interesting. The solar kitchen of Mount Abu in India (Muthu and Raman 2006), able to cook up to 38,500 meals a day, is a good example of this concept. However, these technologies are not studied in this paper.

25.2.2 Different Models of Solar Cookers with Heat Storage

All the cookers studied here have functioned at least once for regular cooking, either in the laboratory or in the field. They use different methods to collect energy and transfer it to the food; the objective is to enhance the particularities of each cooker (Fig. 25.1).

Each of the devices is labelled with a letter to facilitate the comparison. The different cookers with heat storage used for the comparison are as follows:

(A) Ramadan et al. (1988); (B) Bushnell and Sohi (1992); (C) Domanski et al. (1995); (D) Buddhi and Sahoo (1997); (E) Sharma et al. (2000); (F) Buddhi et al. (2003); (G) Sharma et al. (2005); (H) Murty and Gupta (2013); (I) Nandwani et al. (1997); (J) Hussein et al. (2008); (K) Oturanç et al. (2002); (L and M) Schwarzer et al. (2008); (N) Mussard et al. (2013); (O) Tesfay et al. (2014); (P) Foong et al. (Foong 2011); and (Q) Scheffler (Kumar et al. 2017). Some of these cookers are shown in Fig. 25.1

25.2.3 Parameters for the Assessment

The relevant parameters to compare the different cookers are listed. Each parameter has its scale of evaluation, from 0 to 1, 2, or 3. The higher the grade, the most interesting the characteristic:

- The first parameter is the food load potential of each cooker (*FL*). To optimize the comparison, we compare here the maximum mass of "wet" food tested and cooked well for each cooker, independent of the cooking temperature needed. The cookers able to cook up to 2 kg of food were assigned grade 0; between 2 and 7 kg, the grade is 1; and for 7 kg and over, the grade is 2.

Fig. 25.1 (K) Oturanç et al. (2002), (Q) Scheffler (http://www.barli.org, 2017), and (M) Schwarzer and Vieira da Silva (2008) cookers

- The second parameter is the temperature of the storage (*Tp*): a high temperature may allow good cooking quality without sunshine. If Tp < 100 °C, the grade is 0, and if 100 < Tp < 170 °C, the grade is 1; if 170 < Tp < 240 °C, 2; and for over 240 °C, 3.
- Then, the cooking quality potential (*Ck*) of the cooker is assessed. The grading is as follows: 0 if the system simply helps to keep the food hot or to cook below 100 °C, 1 if it is possible to boil water (>100 °C), and 2 when the system allows frying or baking (>180 °C).
- The time scale for heat storage (*t*) is also graded. For two storages, this parameter can vary depending on the quality of the heat and insulation. If the cooker is designed to give inertia to the system only when the sun is not shining, the grade is 0, 1 if cooking during the evening is possible, and 2 if a decent amount of heat can be kept to the following morning.
- The size of the device (*Sz*) is considered as well. A small cooker is potentially more attractive due to its flexibility of installation and it being lighter. This parameter is in general directly related to the size of the collector. 0 is given for

cookers occupying 3 m^2 or more, 1 if the cooker occupies between 1.5 and 3 m^2 large, and 2 if the size is below 1.5 m^2.

- Another parameter considers safety issues (*Sf*). Two main safety concerns can be identified: the use of reactive chemicals and the use of a pressurized system. Both concepts can be a concern for non-rigorous use of the cooker. Furthermore, a pressurized system may require additional maintenance, thus decreasing the attractiveness of the cooker. The grading is as follows: 0 if both issues are involved, 1 for either one or the other, and 2 if there is no safety issue.
- Some collectors need to track the sun to be efficient enough (*Tr*). Any tracking system increases the performance, but that parameter has already been considered. However, tracking is a constraint: nowadays, they are generally automatic, but they increase the complexity of the system.

  The grading used here is 0 if a tracking system is necessary, or manual adjustment is needed every 10 min or less, and 1 for no tracking, or an adjustment over 10 min (such as box cookers).
- The concentration ratio (*C*) can be an issue as well. A high concentration ratio allows higher heat fluxes and implies a better potential for high temperatures; but parabolas and high-concentration devices can lead to eye damage and burning risks. For these reasons, the grading is then 0 if there is a high concentration ratio (>40) leading to high temperatures (typically parabolas) and 1 when the concentration is low or nonexistent (e.g., box cookers, troughs, flat panels).
- Some systems need a circulation loop with a heat transfer fluid to operate (*HTF*). This makes the system more complex to build and operate, requiring maintenance and a deep understanding of technical issues. The grading used here is 0 if there is a circulation loop and 1 if not.
- By separating the heat collection and the heat extraction part (*Sep*), the system becomes more attractive. It becomes possible to cook inside or in the shade, which is much more comfortable. The separation collection/extraction is graded as 0 if there is no separation and 1 if there is separation.
- The ease of manufacture (*E*) should be considered as well. If a system is easy to make, it can be spread or repaired much faster. 1 is given if the system can be made of locally available materials, free or available in the majority of shops, and 0 if imported or industrial-level material is necessary.
- In general, it is possible to use the cooker while it is being charged (*CkCh*); however, some cookers do not offer this opportunity: 1 is given if cooking while charging is possible and 0 if it is not possible.
- Finally, the ratio of food load/size is included (*FL/Sz*). Indeed, a cooker that remains compact while being able to cook large amounts of food is more attractive. To estimate this ratio, the values of size and food load are simply added (FL+Sz), and then 1 is given if the sum is 2 and 0 if the sum is 1.

Note that the issue of the cost is not considered here: it is indirectly included in the other parameters—size, tracking, circulation loop, and ease of manufacture.

25.3 Results

25.3.1 Sensible and Latent Heat Values for Different Materials

Different materials for heat storage are tested. Most of the systems are based on latent heat; some of them use only sensible heat. Furthermore, the melting materials have different melting temperatures.

The thermal data of each of the materials used in the different heat storage are listed together to facilitate the comparison.

For a given material, the enthalpy of fusion Δh_{fus}, in kJ/kg, provides a large amount of latent heat at melting temperature T_m. But a part of the sensible heat can be used as well; an interval of 20 °C before the melting point is then included in calculations. It facilitates as well in the comparison between sensible and latent heat-based storage materials.

It is important to notice than in practice, a storage is rarely made of a unique material; the objective here is to compare the materials chosen specifically for their heat storage due to their physical properties. The correspondence between materials and cookers is noted using the letter for each cooker (point 2.2).

For a sensible heat-based material, the sensible heat C_p (in kJ/kg.°C) is multiplied by the temperature range ΔT to calculate the heat capacity Q (Eq. 25.1). Only the data for $\Delta T = 20$ °C are listed:

$$Q = C_p \cdot \Delta T \tag{25.1}$$

When melting occurs, the latent heat Δh_{fus} is added to obtain the global heat capacity (Eq. 25.2):

$$Q = C_p \cdot \Delta T + \Delta h_{fus} \tag{25.2}$$

To compare the compactness of each material, the heat density Q_d is computed, using the density ρ (Eq. 25.3):

$$Q_d = Q \cdot \rho \tag{25.3}$$

Indeed, dense materials do not appear very competitive unless we consider the volumetric heat capacity instead of the mass one. Compact systems can be easily insulated and installed.

Finally, a comparison of the heat density Q_d on a 100 °C range is made for the relevant materials (using a value of Q calculated for a 100 °C range). Indeed, when a system can be used at high temperatures, its useful temperature range becomes potentially higher. This computation is made for the materials melting over 150 °C and for the sensible heat systems. The results are shown in Table 25.1.

Table 25.1 Heat storage potential for different substances and materials

Material	ρ (10^3 kg/m^3)	T_m (°C)	Δh_{fus} (kJ/kg)	C_p (kJ/kg.°C)	Q (kJ/kg, 20 °C)	Q_d (10^3 kJ/m^3, 20 °C)	Q_d (10^3kJ/ m^3, 100 °C)
A: Barium hydroxide octahydrate[a, b]	1.60	78	288	2.0	328	525	–
B: Pentaerythritol [c, d]	1.40	182	301	1.49	331	463	630
C, D: Stearic acid [e, f]	0.97	55	160	1.78	196	190	–
E: Acetamide[g]	0.99	92	263	1.94	302	299	–
F: Acetanilide[h]	1.21	119	222	2.0	262	317	–
G: Erythritol[i]	1.48	118	339	1.38	367	543	–
H, K: Heat transfer oil[j]	0.85	–	–	2.2	44	37	187
I: HDPE[k]	0.96	126	235	2.2	279	268	–
J: Magnesium nitrate hexahydrate[l]	1.63	89	157	2.72	211	345	–
L,M: Peanut oil[m]	0.91	–	–	2.2	44	40	200
L,M: Stones[n]	2.7	–	–	0.84	17	45	227
N, O, P: Nitrate salts[o]	1.80	22	120	1.4	148	266	468
Q: Iron[n]	7.9	–	–	0.45	9	71	355

[a]A Ramadan et al. (1988)
[b]A Li et al. (2003)
[c]B (National Institute of Standards and Technology)
[d]Bushnell and Sohi (1992)
[e]Buddhi and Sahoo (1997)
[f]Buddhi and Sharma (1999)
[g]Sharma et al. (2000)
[h]Buddhi et al. (2003)
[i]Sharma et al. (2005)
[j]Murty and Gupta (2013)
[k]Nandwani et al. (1997)
[l]www.chemistry-reference.com
[m]Fasina and Colley (2008)
[n]www.engineeringtoolbox.com
[o]Mussard et al. (2013)

25.3.2 Comparison of Solar Cookers with Heat Storage

Now that all the characteristics are selected, the different cookers are listed in a table, and the data are then converted into a corresponding grading (Table 25.2).

Starting from these data, a scatter graph is made to compare the cookers (Fig. 25.2). The "technical aspects" part is the sum of FL, Tp, Ck, and t. The "practical aspects" is the sum of Sz, Sf, Tr, C, HTF, Sep, E, CkCh, and FL/Sz.

Table 25.2 Parameter grading for each cooker

Cooker	FL	Tp	Ck	t	Sz	Sf	Tr	C	HTF	Sep	E	CkCh	FL/ Sz
A: Ramadan	0	0	0	0	2	2	1	1	1	0	1	1	1
B: Bushnell	1	2	1	2	0	1	1	0	0	1	0	1	0
C: Domanski	0	0	0	1	2	2	1	1	1	0	1	1	1
D: Buddhi-Sahoo	0	0	1	1	2	2	1	1	1	0	1	1	1
E: Sharma	0	0	0	0	2	1	1	1	1	0	0	1	1
F: Buddhi	0	1	1	1	2	2	1	1	1	0	0	1	1
G: Sharma	2	1	1	1	0	1	1	1	0	1	0	1	1
H: Murty	0	1	0	0	1	2	0	0	0	1	1	1	0
I: Nandwani	0	1	0	1	2	2	1	1	1	0	0	1	1
J: Hussein	1	0	0	1	1	2	1	1	1	1	0	1	1
K: Oturanç	0	0	0	0	2	2	1	1	1	0	1	1	1
L: Schwarzer (2 m^2)	1	2	2	1	1	2	1	1	0	1	0	1	1
M: Schwarzer (4 m^2)	2	2	2	1	0	2	1	1	0	1	0	1	1
N: Mussard	0	2	2	2	1	1	0	1	0	1	0	1	0
O: Tesfay	0	2	2	2	1	0	0	0	0	1	0	1	0
P: Foong	0	2	0	1	2	1	0	0	1	0	0	0	1
Q: Scheffler	2	3	2	2	0	1	0	0	1	1	0	0	1

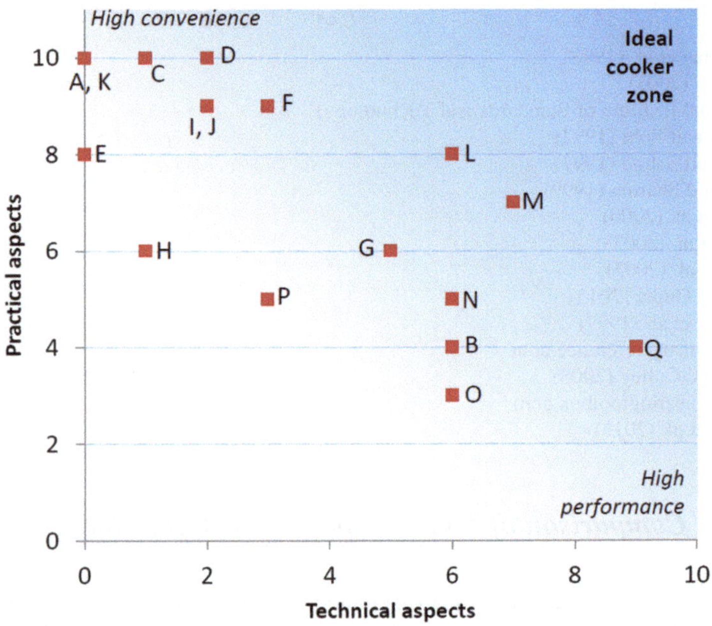

Fig. 25.2 Representation of the technical versus practical aspect parameters for each cooker

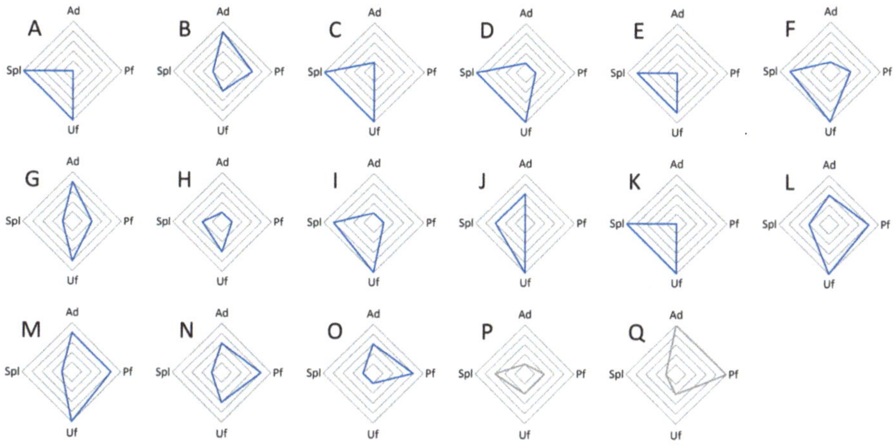

Fig. 25.3 Radar representation of the characteristics of each cooker

Another way of comparing the cookers graphically uses a radar-type graph. Four axes are chosen for these graphs:

- The "adaptation to the needs" (Ad) parameter: food load, time, and possibility of separation are summed (FL + t + Sep).
- The "performance" (Pf) parameter: the sum of the temperature of the storage and the cooking possibilities (Tp + Ck).
- The "user-friendliness" (Uf) issues: the value is obtained by adding safety, concentration ratio, cooking during charging, and ratio of food load/size (Sf + C + CkCh + FL/Sz).
- The "simplicity of manufacturing and cost" (Spl) factor includes the size, tracking, ease, and circulation loop values (Sz + Tr + E + HTF).

The results are shown in Fig. 25.3 for each of the four issues.

By comparing these two methods of comparison, it is possible to note some differences:

- It is possible to visualize the assets and weaknesses of each cooker relative to the others.
- The radar graph is slightly more complicated to read for comparing cookers but is more complete. For example, cookers I and J are similar on the scatter graph but have different radar profiles.

25.4 Conclusions

The grading scale helps to give a good overview of the relative qualities of the different models of solar cookers with heat storage. However, some parameters may be more important than others, depending on the application. For example, a nongovernmental organization looking for a system for a hospital may be more flexible on expenses but more attentive to reliability. On the other hand, a strategy of multiple decentralized cookers may focus more on the ease of fabrication and handling of the system.

For example, the Scheffler parabolas (Q) can be implemented in areas demanding high standards of cooking quality; however, a deep understanding of the system is necessary to manipulate these systems. The Schwarzer system (L and M) performs less well but is more user-friendly, which is an advantage in areas where cooking habits are less demanding. On the other hand, the Ramadan or Nandwani cookers (A and I) are attractive due to their simplicity and user-friendliness. However, they may encounter difficulties in adapting to the actual needs due to their low performance level.

A sensible definition of the important parameters is critical for each implementation procedure of solar cookers. A neglected issue may lead to incorrect judgment and implementation failure.

It could be interesting to develop other methods of comparison to refine the choice of relevant technology.

Thanks to the CETHIL laboratory for giving the possibility of making this study and to Sophie Brock from Solar Cooking International for providing information for this research.

References

Barli Development Institute for Rural Women in India. Available at: www.barli.org/solar-storage.html. Last accessed: 10/2017)

Buddhi, D., & Sahoo, L. (1997). Solar cooker with latent heat storage: Design and experimental testing. *Energy Conversion and Management, 38*, 493–498.

Buddhi, D., & Sharma, S. (1999). Measurements of transmittance of solar radiation through stearic acid: a latent heat storage material. *Energy Conversion and Management, 40*, 1979–1984.

Buddhi, D., Sharma, S., & Sharma, A. (2003). Thermal performance evaluation of a latent heat storage unit for late evening cooking in a solar cooker having three reflectors. *Energy Conversion and Management, 44*, 809–817.

Bushnell, D., & Sohi, M. (1992). A modular phase change heat exchanger for a solar oven. *Solar Energy, 49*, 235–244.

Chemistry reference. http://chemistry-reference.com/q_compounds.asp?CAS=13446-18-9

Cuce, E., & Cuce, P. (2013). A comprehensive review on solar cookers. *Applied Energy, 102*, 1399–1421.

Domanski, R., El-Sebaii, A., & Jaworski, M. (1995). Cooking during off-sunshine hours using PCMs as storage media. *Energy, 20*, 607–616.

Engineering toolbox: https://www.engineeringtoolbox.com/specific-heat-solids-d_154.html; https://www.engineeringtoolbox.com/density-solids-d_1265.html

Farooqui, S. (2014). A review of vacuum tube based solar cookers with the experimental determination of energy and exergy efficiencies of a single vacuum tube based prototype. *Renewable and Sustainable Energy Reviews, 31*, 439–445.

Fasina, O., & Colley, Z. (2008). Viscosity and specific heat of vegetable oils as a function of temperature: 35°C to 180°C. *International Journal of Food Properties, 11*, 738–746.

Foong C. (2011). Experimental and numerical investigations of a small scale double-reflector concentrating solar system with latent heat storage, *Doktoravhandlinger ved NTNU*, Trondheim, Norway, ref: 1503–8181; 313.

Hussein, H., El-Ghetany, H., & Nada, S. (2008). Experimental investigation of novel indirect solar cooker with indoor PCM thermal storage and cooking unit. *Energy Conversion and Management, 49*, 2237–2246.

Li, J., Liu, Z. and Mai, CKJ (2003). An experimental study on the stability and reliability of the thermal properties of barium hydroxide octahydrate as a phase change material. http://energy.bjut.edu.cn/

Kumar, A., Prakash, O. & Kaviti, A. (2017). A comprehensive review of Scheffler solar collector. *Renewable and Sustainable Energy Reviews, 77*, 890–898.

Murty, V., & Gupta, A. (2013). Inclined heat exchanger unit assisted SK-14 PSC for off place cooking. *International Journal of Science, Engineering and Technology Research, 2*, 1003–1006.

Mussard, M., Gueno, A., & Nydal, O. (2013). Experimental study of solar cooking using heat storage in comparison with direct heating. *Solar Energy, 98*, 375–383.

Muthu, S., Raman, R. (2006). Concentrated paraboloid solar cookers for quantity cookery. Proceedings Solar Cookers and Food Processing International Conference, Granada, Spain, 12–16 Jul 2006.

Nandwani, S., Steinhart, J., Henning, H., Rommel, M., & Wittwer, V. (1997). Experimental study of multipurpose solar hot box at Freiburg Germany. *Renewable Energy, 12*, 1–20.

National Institute of Standards and Technology. Pentaerythritol. http://webbook.nist.gov/cgi/cbook.cgi?ID=C3524683&Mask=200

Nkhonjera, L., Bello-Ochende T., John G., King'ondu C. (2016). A review of thermal energy storage designs, heat storage materials and cooking performance of solar cookers with heat storage. *Renewable and Sustainable Energy Reviews*, Article in press.

Otte, P. (2013). Solar cookers in developing countries – what is their key to success? *Energy Policy, 63*, 375–381.

Oturanç, G., Özbalta, N., & Güngör, A. (2002). Performance analysis of a solar cooker in Turkey. *International Journal of Energy Research, 26*, 105–111.

Ramadan, M., Aboul-Enein, S., & El-Sebaii, A. (1988). A model of an improved low-cost indoor solar cooker in Tanta. *Solar and Wind Technology, 5*, 387–393.

Saxena, A., Varun, P. S., & Srivastav, G. (2011). A thermodynamic review on solar box type cookers. *Renewable and Sustainable Energy Reviews, 15*, 3301–3318.

Schwarzer, K., & Vieira da Silva, M. (2008). Characterization and design methods of solar cookers. *Solar Energy, 82*, 157–163.

Sharma, S., Buddhi, D., Sawhney, R., & Sharma, A. (2000). Design, development and performance evaluation of a latent heat storage unit for evening cooking in a solar cooker. *Energy Conversion and Managmt, 41*, 1497–1508.

Sharma, S., Iwata, T., Kitano, H., & Sagara, K. (2005). Thermal performance of a solar cooker based on an evacuated tube solar collector with a PCM storage unit. *Solar Energy, 78*, 416–426.

Sharma, A., Chen, C., Murty, V., & Shukla, A. (2009). Solar cooker with latent heat storage systems: A review. *Renewable and Sustainable Energy Reviews, 13*, 1599–1605.

Tesfay, A., Kahsay, M., & Nydal, O. (2014). Solar powered heat storage for Injera baking in Ethiopia. *Energy Procedia, 57*, 1603–1612.

Chapter 26
Development of Energy Service Company (ESCO) Market to Promote Energy Efficiency Programmes in Developing Countries

Nurcahyanto and Tania Urmee

26.1 Introduction

Climate change and reducing poverty are the two important issues related to sustainable development goal 7: "ensure access to affordable, reliable, sustainable and modern energy for all" (Wu 2015; Agency 2010; Kenny 2016). To address this problem, countries need to lower their energy demand, avoid greenhouse gas emission and provide energy access in a sustainable way. Moreover, the global economy is predicted to grow 150%, and energy efficiency is crucial to limit the world energy demand to one-third by 2040 (IEA 2015). As around 80% of the world's economic growth projected from 2004 to 2030 will be contributed by non-OECD countries (Ellis 2009), energy service companies (ESCOs) can play a vital role in improving energy efficiency in these countries. ESCO can help the energy users, customers, companies, industries and commercial sectors in improving their energy efficiency by providing technical and financial services to improve their energy performance (Morgado 2014).

Most countries in the world are applying similar traditional ESCO concept from the USA (Murakoshi and Nakagami 2009). However, Hansen (2003) reported that every country has its uniqueness based on their culture, and it is not possible to replicate an ESCO model used in one country to another country. Therefore, it is essential to understand the ESCO approach in different countries in order to find the lesson learned and best practices. Furthermore, ESCO was implemented quite successfully to promote energy efficiency in majority of the European Union (EU) and developed countries such as the USA, Canada and Japan (Bertoldi et al. 2006; Vine 2005), but not many developing countries run ESCO successfully (Ellis 2009). ESCO scheme has also been used for providing energy services from

Nurcahyanto • T. Urmee (✉)
School of Engineering, Murdoch University, Perth, Australia
e-mail: t.urmee@murdoch.edu.au

A. Sayigh (ed.), *Transition Towards 100% Renewable Energy*,
Innovative Renewable Energy, https://doi.org/10.1007/978-3-319-69844-1_26

renewable energy in many countries, e.g. Fiji (Urmee et al. 2009), Kiribati (Dornan 2011), Sri Lanka (Lipp 2001), China and India (Liming 2009).

It is therefore important to review the available ESCO approaches used for energy efficiency promotions, its current situations and barriers faced during its different stages of growth to find its applicability in developing countries. This research will provide a comprehensive overview of the success and barriers to implementing ESCO programmes for energy efficiency in the developing countries. The paper discuss the available policy instruments used in different countries to promote ESCO for energy efficiency programmes and the potentials of using ESCO for energy efficiency programmes in developing countries.

26.2 ESCO Model

The concept of energy service companies (ESCOs) was found in the early of the 1980s in the North America as a result of energy crisis impacted from oil price shock in the early 1970s (Okay and Akman 2010), but its implementation started later in the 1980s (Vine 2005). An ESCO provides services to customers such as supply and management of energy, financing, consultancy and technical engineering assistance and (e.g. audits) a provision of equipment, installation and operation and maintenance including upgrade and monitoring, measurement and verification for energy savings (Morgado 2014; Bobbino et al. 2013). ESCOs are typically implemented with energy efficiency project design and development, delivering energy savings guarantee, ensuring cost-effective and optimum performance (Morgado 2014; Murakoshi and Nakagami 2009; Bertoldi et al. 2006; Okay and Akman 2010).

The ESCO models are mainly classified based on their type of financing or business. The business models that commonly used under ESCO are "guaranteed savings" and "shared savings" (Bertoldi et al. 2006; Okay and Akman 2010). Other models, e.g. outsourcing energy management and leasing model, are also popularly used in the European Union ESCOs (WorldBank 2016; Morgado 2014). In the "guaranteed savings", the ESCO will ensure the performance and saving of energy from the project, but not responsible for arranging the finance and credit/loan risk that will be taken by customers (Morgado 2014). Hence, the customers are financed directly by the banks or financial institutions as shown in Fig. 26.1. In "shared saving" mechanism, ESCO is responsible for taking both the energy saving guarantee (energy performance) and the funding (credit) risk (Commission 2016). Thus, the client will not take the financial risk and trust ESCO to run the energy efficiency programme (Morgado 2014). This model is shown in Fig. 26.2.

ESCOs conduct the project development, financing and implementation, while the payment for the projects depends on the saving share agreement between ESCOs and clients. The financial institution can give a loan to the ESCO, and ESCO can use the payment to repay some of the part of saving shares. In countries which are developing ESCO markets (e.g. China, Thailand), the "shared saving"

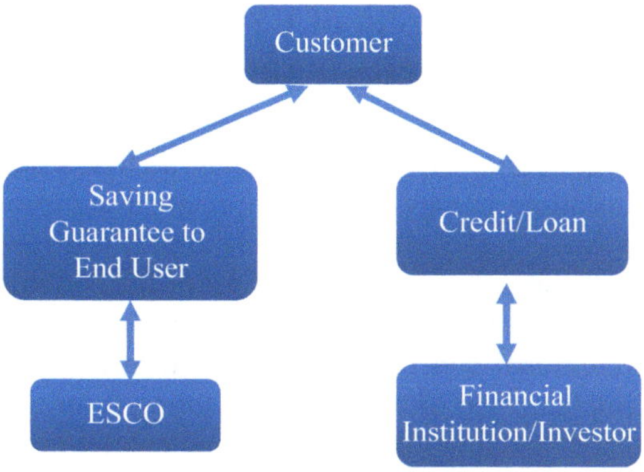

Fig. 26.1 The financing mechanism in guaranteed saving model

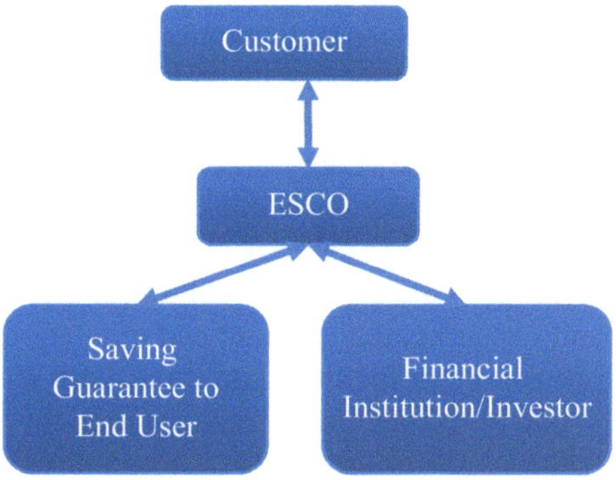

Fig. 26.2 The financing mechanism in shared savings model (Source: adapted from "Advantages and Disadvantages of the dominant world ESCO models; shared savings and guaranteed savings" (Dreessen 2003))

mechanism is more suitable since it does not require clients to assume investment-repayment risk (Okay and Akman 2010). The mechanism of leasing and outsource model is that customers are borrowing the equipment and receiving service for their energy efficiency project. In some cases, the leasing and outsource energy management can be more attractive (mostly in industrial equipment) since the lease of payment projects are sometimes lower than loan payment from banks or financial institution (Commission 2016). There are three financing mechanisms of ESCO [3]: (1) client pays for the services or the equipment based on the how much energy savings achieved, (2) financial institution gives the loan money for the customer and/or ESCO and (3) ESCO invests equipment; therefore, they can use the equipment to provide service and provide a financing/leasing to the customers.

26.3 Examples of ESCO for Energy Efficiency Developed Countries in the World

The traditional ESCO operation included the process of design, engineering and installing energy-efficient equipment and provides financing (Morgado 2014). In private sectors, the US ESCO industry is regarded as the most successful model that is commonly emulated by most of the countries in the world (Osborn et al. 2002). Thus, the ESCO operation in nearly every country is quite similar compared to the traditional concept of ESCO used in the USA. But the entire ESCO model was implemented effectively by tailoring it for that country. Hansen (2011) stated that understanding local situation and culture is essential to develop the ESCO in both developing and developed countries. Hence, although the concept of ESCO might be similar, the best implementation could be different in each country in the world.

The ESCO are implementing the following four strategies: (1) selecting funding resources and financial institution, (2) involving stakeholders or association, (3) preparing the regulation including standard and reporting and (4) setting a target and conducting the energy efficiency business based on the contract.

In Japan, the most (90%) of the ESCO followed shared savings mechanism, which have long payback period for many ESCO projects and generally applicable for large-scale customers (Murakoshi and Nakagami 2009). Japan are encouraging and strengthening standard and regulation for developing ESCO and focus more on private sector facilities particularly for industry as they are the main market driver. While in the beginning, Japan started the concept with the public facilities sector to make aware the stakeholders about ESCO concept. In terms of financing, incentives such as subsidies were introduced which facilitates ESCO projects, for example, including energy retrofit projects in Japan (Geller et al. 2006). Although Japan has an excellent implementation of large company funding mechanism, it has a lack of attractive financing for small and medium enterprises (Murakoshi and Nakagami 2009).

According to Vine (2005), Australia has implemented an energy performance contracting scheme and several regulations to promote ESCO such as creating accreditation of ESCO and standard contract, development of energy performance contract (EPC) facilitators and creating a regulatory programme for guaranteed saving loan, multiyear budget and commercial leasing. The market driver for this ESCO comes from the industry, commercial and municipal. ESCO association is also formed to promote ESCO project. However, there was a time delay in most of the project implementation because of the negotiation of contracts between companies and ESCO.

In Austria, so far around 50 ESCOs are operating to promote energy efficiency. The types of ESCO projects are space heating, air conditioning, control and automation, lighting and building envelope projects. Most of these projects are applied in the public sector buildings. These ESCOs follow shared savings mechanism. High transaction costs, small project sizes, perceived risks and lack of best practice examples are the barriers in Austria (Bertoldi et al. 2014). ESCO market in

Belgium is mainly based on the public ESCO called Fedesco. The number of ESCO companies is centred around 10–15 firms focusing on public sector buildings and private industry facility projects. The Belgian ESCO market is mainly driven by the public ESCOs that act as market facilitators. The key barriers are the lack of policy support which is a considerable lack of legislative focus on the ESCO market (Bertoldi et al. 2014). Germany has been seen as the champion amongst the world due to its maturity and the number of stakeholders in ESCO (Marino et al. 2010). Around 550 companies so far are involved with ESCO. The market volume for energy efficiency services in Germany in 2012 was estimated to be €3.5–5.0 billion/a (Seefeldt et al. 2013). Approximately 80–85% of all German ESCO projects are in the form of energy supply contracts; 8–10% of the market is covered with EPC (Bertoldi et al. 2014). National and European legislation are considered as key movers of the ESCO market in Germany (Bertoldi et al. 2014; LLC 2013). Numerous associations are available to support ESCO such as ESCO Forum (ZVEI national association for electrical and electronics industry) and VfW (association for heat supply), Forum Contracting (e.g. association of persons active in the field of energy contracting) and BDE (Bundesverband Deutscher Energiedienstleister – association of German energy service providers).

The ESCO market in the UK is amongst the most mature ones with a relatively long history of providing energy services (Marino et al. 2010). Most of the ESCO were responsible for financing, installing, operating and maintaining PV systems rather than EE. Specific regulation on ESCOs and EPC does not exist in the UK and the business model is constantly changing. Overall, the ESCO has been implemented under EPC and shared savings model in the European system. Liberalization of the electricity market and climate change campaign (Kyoto Protocol, CDM, joint implementation) are drivers for the development of ESCO market in Europe (Bertoldi et al. 2006). The implementation of ESCO in Europe, according to Bertoldi et al. (2006), through several processes and implementation such as involving Third Party Financing (TPF), started in public and building sector, creating pilot projects and standard contract for industries and building and creating online EU database.

In developing countries, the funding issue for ESCO and energy efficiency projects are still dominant. Therefore, the developing countries are trying to strengthen the ESCO cooperation by working together with the international organization to create funding mechanism and ESCO association for promoting ESCO business. The incentives and financing mechanism such as low interest, loan guarantee and revolving funds are trying to be actualized to promote the ESCO business (Murakoshi and Nakagami 2009; CCAP 2012). Furthermore, ESCO in developing countries is not applied directly in all sectors. However, it most likely uses pilot project in particular areas such as public facilities, street lighting programme, etc.

Within the developing countries, Korea is quite successful in implementing ESCO with the strategies of removing institutional barrier through allowing a multiyear procurement (Murakoshi and Nakagami 2009). Korea has implemented and created a long-term low-interest tax credit for ESCO project (Lee et al. 2003).

The barriers such as the procurement system (institutional barrier), financing barrier and low priority of energy efficiency programme are experienced in Korea (Lee et al. 2003). According to Lee et al. (2003), other challenges for developing ESCO in Korea are how to create continuity of ESCO demand and improve capacity building and financing scheme. The market driver for Korean ESCO development is the high energy prices.

In China, the implementation of ESCO association and loan-guarantee mechanism is essential to create an ESCO market (Murakoshi and Nakagami 2009). The Ministry of Finance (MoF) and National Development and Reform Commission (NDRC) introduce subsidy for the ESCO to run the energy conservation project (Lo 2014). The market driver for Chinese ESCO is potentially high for lowering emissions in industrial and commercial sector. Shared savings mechanisms are commonly used in China ESCO market. The financial barrier (lack of credit mechanism), institutional (administration and high transaction cost) and technology barrier (lack of standardized procedure) are the several obstacles that have been faced by the ESCO implementers (Da-li 2009).

Although there is no specific association, the implementation of ESCO in Thailand is successful since they have attractive financial incentives. The drivers come from small and medium enterprises, industrial sectors and building/commercial sectors ((CCAP) 2012). The project financing is one of the barriers that should be taken into account for ESCO project. To remove this barrier, Thailand introduces Energy Conservation Promotion Funds (ENCON Funds) which provide incentives for capital investment and technical assistance to create energy efficiency market (Murakoshi and Nakagami 2009). The ENCON Fund is gained from the initial capital outlay from oil and fossil fuel funds. ESCO in India were started by conducting small project ESCO for municipal retrofit lighting, improving the energy efficiency of hotels, and demand side management for industrial operations (Murakoshi and Nakagami 2009). The ESCO barriers in India are lack of development skill, public awareness, financing and policy regulation (Murakoshi and Nakagami 2009).

In the Philippines, the implementation of ESCO for street lighting retrofitting project was effective to introduce the ESCO scheme. This project was supported by the Development Bank of the Philippines to showcase a benchmark for the industry (Vine 2005). Lack of business concept of ESCO is still the biggest problem, so that the successful pilot project can be helpful to run the other ESCO projects.

In Africa (Egypt, Kenya and South Africa), conducting cooperation with the international institution for creating funding, increasing public awareness and creating ESCO association are the strategies used for promoting ESCO in their countries (Vine 2005).

In Brazil, the development of ESCO association and cooperation with the Canadian government had enhanced the ESCO development. Brazil also has an active ESCO Association which is growing and still potentially to be larger (Ellis 2009). The national electricity conservation programme (PROCEL) provided fund or co-funds to ESCO to promote energy efficiency project for research and development, education, standard and label and also pilot projects (Taylor et al. 2008).

Table 26.1 The implementation ESCO in developed countries

Country and source/ remarks	Implementation process: Regulatory factors	Tools for promoting ESCO	Financing
Japan (Murakoshi and Nakagami 2009)	Standard and regulation	Private sector facilities (industrial sector), public facilities (pilot), Japanese association of ESCOs	Shared savings contracts (90%), incentives financing mechanism for large scale customer
Australia (Vine 2005)	Government accredited, development of EPC facilitators, standard contract	Australasian energy performance contracting association limited (AEPCA), energy efficiency council	Multiyear budget, treasury funds for repaying the guaranteed saving loans, commercial leasing
Austria (Bertoldi et al. 2014)	Certification and accreditation scheme, mandatory of energy consultation, standardization contract	Small medium enterprises, professional Association for Energy Contracting	EPC, shared savings model, commercial banks financing
Belgium (Bertoldi et al. 2014)	Public ESCO establishment and funding	Public ESCOs and private firms (large and SMEs)	Public energy service contracts, third party financing (TPF), EPC and smart EPC
		Green building platform, information campaigns	
Germany (Marino et al. 2010; Seefeldt et al. 2013; Bertoldi et al. 2014; LLC 2013)	National and European legislation (key movers)	Numerous associations	Shared savings model, 80–85% energy supply contracting (ESC), 8–10% EPC
UK (Bertoldi et al. 2014)	Climate and energy conservation policy	Various trade associations	EPC (both shared savings and guaranteed savings)

The government has also provided loan guarantee fund for energy efficiency project. However, although there are several financing schemes, the energy efficiency marketing is still the major problem in Brazil. The implementation processes of ESCO are summarized in Annex 1. Table 26.1 shows the ESCO projects in developed countries and Table 26.2 are developing countries.

Table 26.2 The implementation of ESCO in developing countries

Country and source/remarks	Implementation process		
	Regulatory factors	Drivers for promoting ESCO	Financing
Korea (Vine 2005),(Lee et al. 2003)	Pilot project, multiyear procurement	Korean association of ESCOs (KAESCO)	Tax credit and long term and low interest
China (Murakoshi and Nakagami 2009; Lo 2014; Da-li 2009)	Founding energy conservation information and dissemination centre (NECIDC), energy conservation law revision	China energy conservation service industry association (EMCA), ESCO association and partial loan guarantee	Shared savings, guaranteed savings and outsourcing management contract, short payback period (< 2 years), loan guarantee system, subsidy
Thailand (Murakoshi and Nakagami 2009; (CCAP) 2012)	Providing incentives for ESCO, pilot project for industrial facilities	Electricity Generating Authority of Thailand (EGAT) as a role of ESCO promotion	Shared and guarantee savings, energy conservation promotion funds (ENCOND funds), revolving fund
Africa (Egypt, Kenya, South Africa) (Ellis 2009; (Vine 2005)	Egypt: training for energy audit, EE technologies and ESCO project evaluation and financing; Kenya: survey banks, hiring energy consultant to explore ESCO establishment	Egypt: credit guarantee, Egyptian energy service business association (EESBA); Kenya association of manufacturers industrial energy efficiency project	EPC Cooperation with an international organization to get a fund/credit
India (Murakoshi and Nakagami 2009)	Small project, involving international organization	Federation of Indian Chambers of Commerce and Industry, Indian Council for Promotion of energy efficiency business	EPC, decentralized investment of ESCO, low-interest financing
Brazil (Ellis 2009; Vine 2005)	Establishing National Electricity Conservation Programme to fund EE project	Brazilian Association of ESCOs (ABESCO)	Loan guarantee fund Creating fund and join cooperation with PROCEL

26.4 Barriers in Implementing ESCO in Promoting Energy Efficiency

Barriers to energy efficiency are multifold, stretching over issues concerning institutional and legal frameworks, financial and economic incentives and information, knowledge and technology gaps. There are several barriers faced in both develop and developing countries in implementing ESCO for energy efficiency. These barriers can be categorized by (i) institutional, (ii) financial and economic and (iii) knowledge and information as given below:

26.4.1 Institutional Barriers

- Lack of policies, legal and regulatory frameworks and enforcement
- Lack of government support
- Limited institutional capacity (both in public and private)
- Breakup in energy consumption amongst a diverse range of end-users and business models
- Lack standardization

26.4.2 Financial and Economic Barriers

- Energy prices
- Unfavourable market structures
- Lack of financial incentives

26.4.3 Knowledge and Information Barriers

- Lack of awareness about ESCO operation
- Low priority on energy efficiency programmes
- Lack of information, education and training
- Lack of energy efficiency technologies available
- Lack of infrastructure

26.5 Conclusion and Lessons Learned

Energy service company (ESCO) has been implemented widely in the world in both developed and developing countries. The development of ESCO is very demanding since the energy efficiency is needed for the most country in the world. There are potential factors that will affect the development of ESCO such as energy price, subsidy, environmental issue, international policy and market of ESCO. However, although the concept and model of ESCO in the world are quite similar, the actual implementation of ESCO in every country varies, resulting in different progress of development with various barriers. The barrier can be institutional, financial and knowledge and information. Although some developing countries successfully promoted ESCO, the lessons learned from developed countries would help ESCO promotion in developing countries. As discussed in Sect. 26.3, the policy for implementing ESCO depends on many factors such as culture, the acceptance of energy efficiency and the level of energy efficiency development. The same policy for ESCO implementation in several countries can cause a different result and impact. Thus, market participation adjustment is required to support ESCO implementation. The financing and funding are the two important issues in all ESCO project. Several lessons learned from developed countries are summarized below:

- Almost all of countries which implemented ESCO mostly used the shared savings model.
- Creating a market driver and identifying market is essential to design the appropriate ESCO policy. Not all developed countries have the large market such as Japan, Germany and the UK. Developed country such as Austria has an efficient market policy for small and medium enterprises.
- It is essential to eliminate institutional weakness by providing appropriate institutional organization such as creating ESCO association. The ESCO association contributes in promoting the development of ESCO, disseminates information and assists government for policy development.
- Implementing a pilot project through government support can boost the ESCO market.
- Enhancing standard and regulation and improving the EPC and accreditation of ESCO are important. The accreditation of ESCO is necessary for the reliable service in energy efficiency project. While, a standard contract will minimize uncertainty and problem for distinguishing the energy performance contract in each sector (industrial, commercial or public sector).
- In developing countries, the funding source is hard to obtain, and, therefore, the role of an international organization is necessary to promote the financial ESCO.
- Increasing the awareness of local bank on ESCO to improve the financial mechanism is needed to implement energy efficiency in developing countries.
- Loan guarantee policy also can be a solution in the financing mechanism in both developed countries and developing countries since it will ensure the security of financial institution to finance the project by minimizing credit risk as done in Australia and China.
- Providing incentives for a financial institution is important. The incentives can be given depending on the market situation.

From the above lessons learned, the barriers of promoting ESCO in developing countries can be removed. Due to the growing economy in many developing countries, there is a potential for development of ESCO for energy efficiency especially in industry sector. With proper policy regulation and financial incentives, the development of ESCO in developing countries can be improved.

References

(CCAP), C. f. C. A. P. (2012). Revolving and Esco funds for renewable energy and energy efficiency finance: Thailand. Funding the future: The growing might of financing boosts renewable energy and energy efficiency. Washington, DC 20002, Center for Clean Air Policy (CCAP).

Agency, I. E. (2010). Energy poverty: How to make energy access universal? Special early excerpt of the World Energy Outlook (WEO) 2010 for the UN general assembly on the millennium development goals. Paris, United Nations.

Bertoldi, P., et al. (2006). Energy service companies in European countries: Current status and a strategy to foster their development. *Energy Policy, 34*(2006), 1818–1832.

Bertoldi, P., et al. (2014). ESCO market report 2013. Luxembourg, European Commission.

Bobbino, S., et al. (2013). Budget-neutral financing to unlock energy savings potential: An analysis of the ESCO model in Barcelona. B. W. P. Series, Basque Centre for Climate Change.

Commission, E. (2016). *Energy performance contracting*. Retrieved 3 October 2016, 2016, from http://iet.jrc.ec.europa.eu/energyefficiency/european-energy-service-companies/energy-performance-contracting

Da-li, G. (2009). Energy service companies to improve energy efficiency in China: Barriers and removal measures. *Procedia Earth and Planetary Science, 2009*, 1695–1704.

Dornan, M. (2011). Solar-based rural electrification policy design: The Renewable Energy Service Company (RESCO) model in Fiji. *Renewable Energy, 36*(2010), 797–803.

Dreessen, T. (2003). *Advantages and disadvantages of the two dominant world ESCO models; shared savings and guaranteed savings*. Proceedings First Pan-European Conference on Energy Service Companies.

Ellis, J. (2009). *Energy service companies in developing countries: Potential and practice*. International Institute for Sustainable Development (IISD).

Geller, H., et al. (2006). Polices for increasing energy efficiency: Thirty years of experience in OECD countries. *Energy Policy, 34*(2006), 556–573.

Hansen, S. J. (2003). *Lesson learned around the world*. First European Conference on Energy Service Companies (ESCOs): "Creating the Market for the ESCOs Industry in Europe, European Communities.

Hansen, S. J. (2011). ESCOs around the world. *Strategic Planning for Energy and the Environment 30*(3), 9–15.

IEA. (2015). *World energy outlook 2015*. 75739 Paris Cedex 15, France, International Energy Agency.

Kenny, C. (2016). *Goal 1—end poverty in all its forms everywhere*. Retrieved 17 Jan 2017, 2017, from https://unchronicle.un.org/article/goal-1-end-poverty-all-its-forms-everywhere.

Lee, M.-K., et al. (2003). Promoting energy efficiency financing and ESCOs in developing countries: Experiences from Korean ESCO business. *Journal of Cleaner Production, 11* (2003), 651–657.

Liming, H. (2009). Financing rural renewable energy: A comparison between China and India. *Renewable and Sustainable Energy Reviews, 13*(2008).

Lipp, J. (2001). Micro-financing solar power the Sri Lankan SEEDS model. *Refocus, 2*(8), 18–21.

LLC, M. I. (2013). The energy service market 2020.

Lo, K. (2014). A critical review of China's rapidly developing renewable energy and energy efficiency policies. *Renewable and Sustainable Energy Reviews, 29*(2014), 508–516.

Marino, A., et al. (2010). *Energy service companies market in Europe*. Luxembourg: European Commission.

Morgado, D. (2014). *Energy service companies and financing*. Retrieved Oct 4, 2016, from https://www.iea.org/media/training/presentations/latinamerica2014/8A_Energy_Service_Companies_and_Financing.pdf

Murakoshi, C., & Nakagami, H. (2009). *Current state of ESCO activities in Asia: ESCO industry development programs and future tasks in Asian countries*.

Okay, N., & Akman, U. (2010). Analysis of ESCO activities using country indicators. *Renewable and Sustainable Energy Reviews, 14*(2010), 2760–2771.

Osborn, J., et al. (2002). Assessing U.S. ESCO industry: Results from the NAESCO database project. Berkeley, Ernest Orlando Lawrence Berkeley National Laboratory.

Seefeldt, F., et al. (2013). *Marktanalyse Und Marktbewertung Sowie Erstellung Eines Konzeptes Zur Marktbeobachtung Für Ausgewählte Dienstleistungen Im Bereich Energieeffizienz*. Berlin: Prognos.

Taylor, R. P., et al. (2008). *Financing energy efficiency: Lessons from Brazil, China, India, and beyond*. The International Bank for Reconstruction and Development, The World Bank, Washington DC.

Urmee, T., et al. (2009). Issues related to rural electrification using renewable energy in developing countries of Asia and Pacific. *Renewable Energy, 34*(2009), 354–357.

Vine, E. (2005). An international survey of the energy service company (ESCO) industry. *Energy Policy, 33*(2005), 691–704.

WorldBank. (2016). Fostering the development of eSCo markets for energy efficiency. Retrieved 1 Oct 2016, 2016, from http://documents.worldbank.org/curated/en/709221467753465653/pdf/103932-BRI-LW54-OKR-PUBLIC.pdf

Wu, J. (2015). *Goal 7—ensure access to affordable, reliable, sustainable and modern energy for all*. Retrieved 13 Jan 2017, 2016, from https://unchronicle.un.org/article/goal-7-ensure-access-affordable-reliable-sustainable-and-modern-energy-all

Chapter 27
The Development of a Performance Indicator for PV Power Generators

Saad Odeh

27.1 Introduction

Rooftop solar PV power generators become one of the major sources of solar energy electricity and play significant role in the renewable energy strategies of many governments worldwide. The installation of such type of systems is spreading vastly in residential and commercial sectors due its suitability for the different roof designs.

Long-term assessment of a residential rooftop solar PV system was conducted based on 4 years of operation data (Odeh 2016). Solar electricity delivered to grid was verified with the results from a computer simulation package (PVSYST) and high consistency between both values was reported. New method of benchmarking the PV system was introduced and compared with the currently used performance indicators. The effect of tilt and azimuth angle on the performance ratio and annual yield of a commercial PV rooftop system was carried out (Singh et al. 2016). The maximum estimated yield for this system was found when the tilt angle is equal to 15° and 45°. Energy simulation model was used to design and assess 110 kW stand-alone rooftop solar PV system for commercial application (Shukla et al. 2016). Three types of softwares, Sunny design, SAM, and Blue Sol, were used to simulate the system performance and cost analysis. The results showed that the difference between these three PV system performance simulators is between 3% and 5%. The output of an existing rooftop PV grid-connected system was recorded and analyzed by Ayompea et al. (2011) to evaluate the annual final yield, performance ratio, PV module efficiency, system efficiency, and system different losses. The results of the performance indicators evaluated at this site were compared with the results at

S. Odeh (✉)
Science and Engineering Program, Sydney Institute of Business and Technology,
Western Sydney University – City Campus, Sydney, NSW, Australia
e-mail: S.Odeh@city.westernsydney.edu.au

A. Sayigh (ed.), *Transition Towards 100% Renewable Energy*,
Innovative Renewable Energy, https://doi.org/10.1007/978-3-319-69844-1_27

different sites in Europe and show superiority among others only in reference yield where its value was found quite below the average. The impact of PV panel slope on energy fed to grid was studied for 11 cities in Queensland, Australia (Liu et al. 2012). The study showed that the optimum design of panel slope for all cities considered in that study is when slope angle is between 20 and 25°. Performance analysis based on capacity factor and performance ratio was carried for an existing 20 kW industrial PV rooftop system (Kumar et al. 2014). The study showed that there are factors other than clearness index affecting the power generation such as variation in ambient temperature, partial shading, and accumulation of dust on the PV module surface. The average performance ratio and capacity factor of the PV system of this study are found falling within the global range reported by Ayompea et al. (2011). The minute by minute output measurements of a 2.1 kW grid-connected PV system were conducted to show the effect of high-frequency irradiation data on the overall performance (Su 2015). The analysis showed that the output power and therefore PV system performance respond nonlinearly to minute by minute irradiation changes. When the high-frequency data are used to evaluate system performance, some of the results may not reflect the actual trend since the input energy to the PV panel used in these performance indicators is always at laboratory maximum condition and not the actual values.

This study is an extension to the work of Odeh (2016) to develop new performance indicator method for PV power systems. The developed method correlates system actual output with the maximum output that can be achieved by the same system when it operates in two-axis solar tracking mode. Although the tracking mode cannot be applied to some designs of PV systems such as the roof designs, it represents the actual operation condition to benchmark system performance. It is basically useful for the PV system designers to measure the compliance of the PV array design (e.g., orientation or inclination) with the standard design that allows maximum output when irradiation is normal to PV panel surface. The new method results are compared with the current performance measures available in the literature, and significant difference was reported. The impact of this method on system design and end-user system pricing were investigated at different sites and latitudes.

27.2 Rooftop PV Power Generator Model

A residential rooftop PV power generator is used to develop a model of a performance indicator. The system which its layout is shown in Fig. 27.1 consists basically of ten PV panels of total installed capacity equals to 1.75 kW, a DC/AC inverter, and a grid energy meter to export the generated AC power to the local energy provider. Two types of energy meters (gross and net meter) are currently used as a gateway to connect the rooftop PV system to the grid. In case of gross meter, all the generated electricity from the power generator is fed directly to the grid and imports the daily household demand independently from the grid. The net

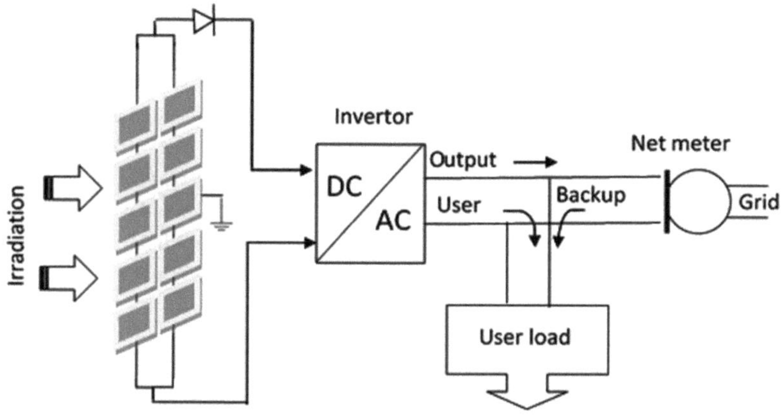

Fig. 27.1 Rooftop PV system layout with net meter

meter circuit allows the household to use the generated power and exports the surplus to the grid. In both cases the meter can import the required household load during nighttime or periods of low or zero solar electricity. The PVSYST package (Mermoud 2012) is used to develop a model of rooftop PV power generator which allows to select different brands and types of system components and weather data stations.

The model is validated against 4 years of actual system reported data, and high consistency was achieved (Odeh 2016). The system components and site latitude at major cities in Australia are presented in Table 27.1. The average daily load of a typical house in Sydney was concluded from energy provider record (Energy Australia) and was found in the range of 18 to 23 kWh/day.

27.3 Performance Indicators of PV Power Generators

The traditional method of assessing power generator performance is by estimating the ratio of system output (electricity) during a period of time (such as daily, monthly, or yearly) to the generator maximum capacity for the same period of time. Although this type of assessment method is applicable to the different types of power generator such as oil, gas, coal, renewable, etc., solar power generators require some other considerations. The intermittent solar irradiation and other weather conditions affect significantly the long-term performance of PV power generators (Odeh and Behnia 2009). The five major performance measures found in the literature can be summarized as follows:

Table 27.1 Major system specifications and site latitude used in this study

Roof tilt and azimuth angles	23° and 10°
Type of PV panel	175 W monocrystalline silicon
DC/AC inverter type	SMA, Sunny Boy SB1700
Average load of the house	
Spring	18 kWh/day
Summer	21 kWh/day
Autumn	23 kWh/day
Winter	21 kWh/day
Latitudes	
Darwin	12.2°
Alice Springs	23.5
Brisbane	27.5
Perth	31.6°
Sydney	33.5°
Adelaide	34.5°
Canberra	35.3°
Melbourne	37.5°
Hobart	42.5°

27.3.1 System Efficiency (η_{system})

It is the ratio of the output energy of the PV system to the energy of the incoming irradiation incident on the same PV panels' area and is given in the following form (The European Standard 1998; Ueda et al. 2008; Odeh and Abu-Mulaweh 2012):

$$\eta_{\text{system}} = \frac{\text{Generated energy from the PV system in kWh}}{\text{Solar irradiation incident on the PV array's area in kWh}} \tag{27.1}$$

The total PV system efficiency is quite low compared with other conventional generators and depends significantly on the PV panel efficiency which is in the range of 14–17% at standard test conditions and inverter efficiency which is in the range of 95–98% at actual operation condition (Photovoltaics Report 2016). The results of this study show that the system efficiency does not vary too much during the year, and it is in the range of 10.1–10.9% as shown in Fig. 27.2. This method is useful when it is required to compare the different designs and brands of PV systems. However, considering this method of assessment in marketing, the solar generators have a drawback if it is compared with conventional power units which have much higher energy conversion efficiency than the solar energy system.

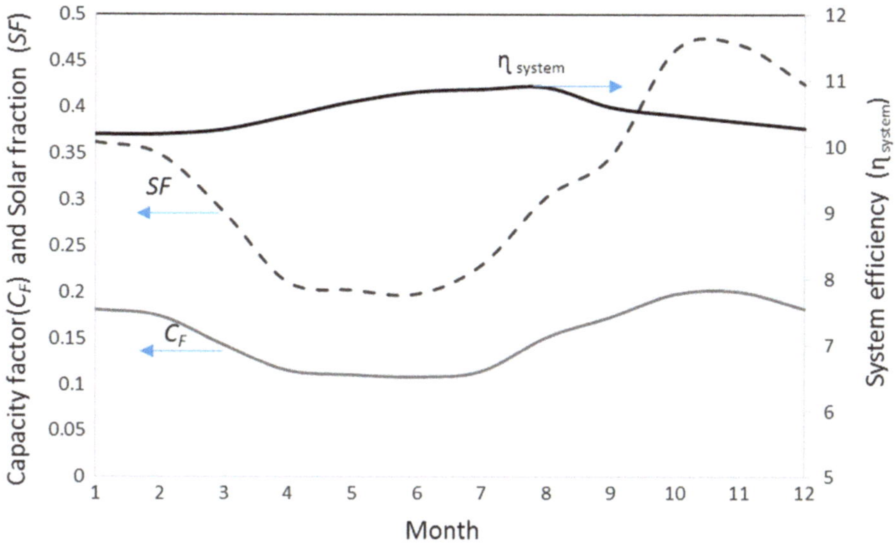

Fig. 27.2 System yearly performance using SF, C_F, and Y_f indicators

27.3.2 Solar Fraction (SF)

It is the amount of energy produced by the PV system to the amount of load required at the respective site (Fujisawa and Tani 2001, Odeh 2003). The value of solar fraction depends mainly on the PV system contribution to the site load and cannot be used to compare system performance with other similar systems at different sites. Solar fraction can be represented by the following equation:

$$SF = \frac{\text{Generated energy from the PV system in kWh}}{\text{Site energ load(heat and/or electricity)in kWh}} \tag{27.2}$$

There is no specific range for the value of solar fraction because it depends on the percentage of solar contribution to the site energy load. This is quite clear in Fig. 27.2 which shows the annual SF for the system adopted in this study. The generated energy from the rooftop system was capable to cover the energy demand of the house in the range of 20–47%. From the economic perspective, SF cannot approach 100% due to the requirement of the costly energy storage battery bank to cover the periods of low or zero irradiation.

27.3.3 Capacity Factor (C_F)

It is defined as the ratio of the actual annual energy output from the PV system to the energy generated by the PV system when it operates at its full rated power, i.e., 24 h and 7 days a week (Ayompe et al. 2011, Kumar et al. 2014). Based on this definition, the equation of C_F is given by:

$$C_F = \frac{\text{Generated energy from the PV system(kWh)/year}}{\text{PV array maximum capacity(kW)} \times \text{8760h/year}} \tag{27.3}$$

The definition of this performance indicator shows that the expected value of C_F cannot be high because the actual system capacity is bounded by the number of sunshine hours. Figure 27.2 shows the monthly variation of the capacity factor for the rooftop system in Sydney. It is clearly showm that the maximum C_F value occurs in summer which is around (0.2) while the lowest value takes place in winter (around 0.1). This is not far from the literature results such as the work of Ayompe et al. (2011) which showed that the average yearly C_F in Europe is only 0.1.

27.3.4 *Performance Ratio* (P_R)

It is the actual amount of energy produced by a PV system to the energy produced by the PV array when operating continuously at STC and same global irradiation (The European Standard 1998; Pless et al. 2005; Ayompe et al. 2011; Kumar et al. 2014; Mermoud 2012) therefore,

$$P_R = \frac{\text{Generated energy from the PV system(kWh)}}{\text{Incident global irradiation on PV array x } \eta_{STC} \text{ (kWh)}} \tag{27.4}$$

η_{STC} *is PV panel efficiency at standard test conditions.* P_R is independent of system size and is typically evaluated on a monthly or yearly basis by considering system total losses. Equation 27.4 shows that P_R does not change much with the type of the PV system and depends basically on the constant STC values. In large-scale commercial systems, P_R can be used to investigate the occurrences of component failures by calculating it for smaller intervals, such as weekly or daily,. The average value of the performance ratio is found in the literature within the range of 0.6–0.9 (Ayompea et al. 2011; Kumar et al. 2014; Mermoud 2012; Photovoltaics report 2016). The results found by this study quite similar and show that P_R monthly values do not change dramatically and vary between 0.75 and 0.8 (see Fig. 27.3).

27.3.5 *The Final PV System Yield or the PV System Specific Power Production* (Y_f)

It is the actual net energy output during a certain period of time (i.e., daily, monthly) divided by the maximum installed power capacity of the PV array and has the unit

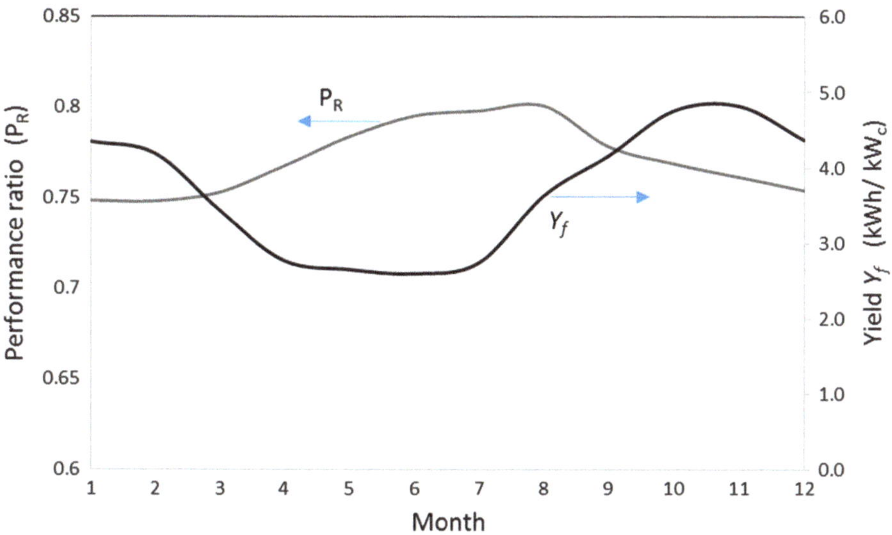

Fig. 27.3 System yearly performance using P_R and η_{system} indicators

(kWh/kW$_c$). This performance indicator can be presented by the following equation (The European Standard 1998; Ayompe et al. 2011; Kumar et al. 2014; Mermoud 2012):

$$Y_f = \frac{\text{Generated energy from the PV system in (kWh)}}{\text{PV array maximum capacity(kW)}} \tag{27.5}$$

System yield is a convenient way to compare the energy produced by the PV systems of different sizes and brands. The monthly average daily yield presented in Fig. 27.3 shows that the PV system of this study has Y_f in the range of 2.6–4.8 kWh/kW$_c$ which is considered high compared with the results found in the literature (Ayompe et al. 2011).

It is quite obvious from Figs. 27.2 and 27.3 that the performance indicator yearly profiles do not have the same trend and can be classified into two groups:

I. SF, C_F, and Y_f which show that the PV system has minimum performance in winter and maximum system performance in summer. This trend of performance is due to the actual PV system output is benchmarked to the system operation at ideal or arbitrary condition.

II. P_R and η_{system}: These two indicators show that the maximum system performance occurs in winter, while the minimum system performance occurs in summer. This trend of system performance is due to both indicators use input irradiation as a reference of comparison.

In Fig. 27.4 the different PV performance indicators were benchmarked considering the effect of latitude and weather conditions. It can be clearly noticed that the

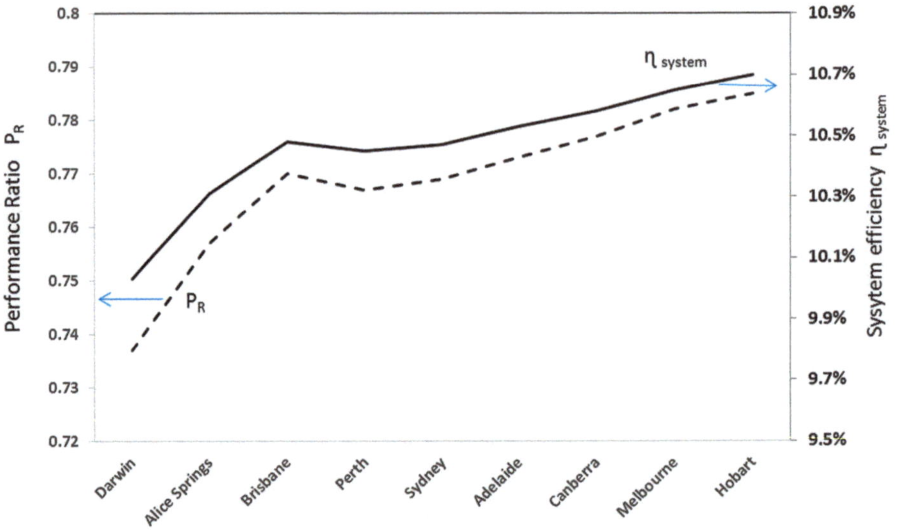

Fig. 27.4 The effect of latitude on P_{R} and η_{system}

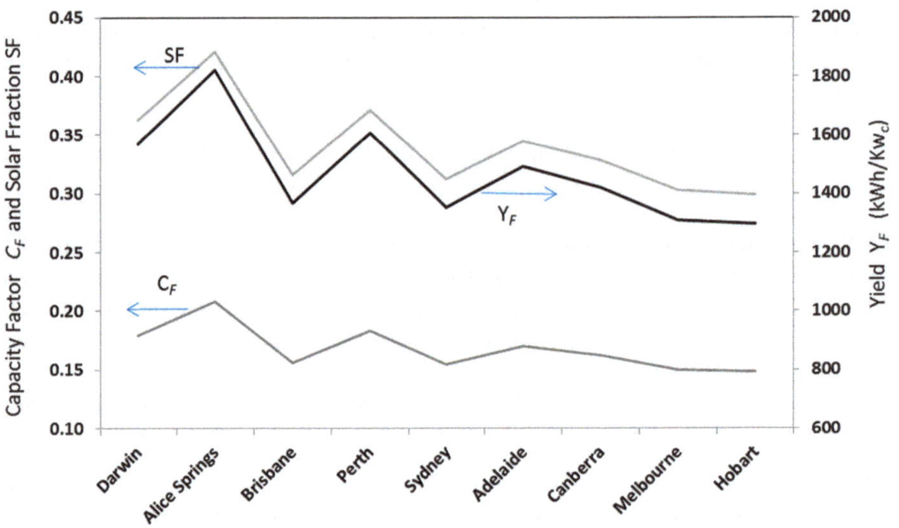

Fig. 27.5 The effect of latitude on SF, C_{F}, and Y_{F}

P_R and η_{system} indicators are correlating very well with site latitude, and the values of these two indicators are directly proportional to latitude. This result contradicts the result in Fig. 27.5 which shows that system performance given by the indicators SF, C_F, and Y_F is inversely proportional to latitude.

The fluctuation of performance values in Fig. 27.5 shows that it is not only the latitude that affects system performance but other factors such as level of irradiation and ambient temperature.

27.4 Development of New PV Performance Indicator

It can be concluded from the previous analysis that the different performance indicators (P_R, η_{system}, SF, $C_{F,}$ and Y_F) found in the literature do not provide the end user with a scale that describes the actual capacity of the PV generators, and two major drawbacks were found:

I. The first one is they benchmark the performance of the solar PV grid-connected system to the standard test condition (STC) given in Table 27.2. Five standard conditions (STC) are being used in testing the PV performance: zero incident angle, fixed solar irradiation, fixed ambient temperature, fixed air mass, and zero system losses. During actual operation of the solar PV system, the standard operation conditions (2–5) in Table 27.2 cannot be achieved constantly, i.e., they are laboratory virtual conditions. Considering these arbitrary conditions in the performance indicator of a solar PV system will underestimate the actual value of the system performance and give the end user false impression about the PV system capacity. Therefore, it was required to introduce an indicator that benchmark the compliance of the PV system design with its maximum capacity design at reachable operation conditions.

   The common PV design of almost all residential systems in Australia is the rooftop. This design requires the PV panel inclination to comply with the building code of a residential roof which sets roof slope angle between 20 and 24° (National Construction Code 2016). The building code allows also wide range of roof orientation angle from east to west (i.e., azimuth angle from 0 to 180).

Table 27.2 Standard test condition STC

	Condition type	STC	Real condition
1	Solar incident angle	Always zero; irradiation beam always normal to the PV panel	Variable and depends on time, date, and site latitude. In case of rooftop system, roof orientation and inclination governs system capacity
2	Solar irradiation	Always equal to 1,000 W/m 2	Variable and depends on the time, date, and site latitude. Limited sunshine hours bound system capacity
3	Ambient temperature	Always 25 °C,	Variable and depends on the time, date, weather condition, and site latitude. Higher ambient temperature degrades PV panel efficiency and reduces system output
4	Air mass coefficient (AM)	Always equal to 1.5	Variable and depends on the time, date, and site latitude. Higher AM with higher latitudes
5	System losses (e.g., wiring, inverter)	Always zero	Variable and depends on the design and location of PV panels, inverter, and grid meter

II. The second drawback found in the performance indicators is the time period used in Eq. 27.3 which does not describe the actual operation time of the PV system. This point can be well clarified by the following equation which represent the energy capacity of conventional power generators such as diesel, gas, and petrol engines;

$$E_c = P_c \times \text{number of operation hours} \tag{27.6}$$

where,

E_c – energy capacity (kWh)
P_c – power capacity (kW)

Equation 27.6 shows that energy capacity (E_c) may reach its maximum when the power generator operates full time (daily, monthly, or yearly). However, this is not possible in case of the solar PV system since its output is limited by two factors: the number of sunshine hours during the year and the PV array orientation angles. Therefore, it is not correct to specify the energy capacity of a PV panel by referring to Eq. 27.6 because these two factors frame its capacity and nothing can be generated beyond this limit. To clarify this further, a system of PV array size (1 kW) cannot produce 1 kWh constantly during the 24 h and the 365 days of the year.

Based on these two drawbacks and from the end-user perspective, PV system capacity must be readjusted for each site based on its individual operation condition so that the end user will not be penalized financially for the unreachable standard test conditions. The only STC of Table 27.2 that can be achieved in actual PV operation is the first condition by using solar tracking system that tracks the solar beam and keeps incident angle always zero. This finding led to develop an indicator that can benchmark the PV system performance to its maximum reachable capacity during actual operation condition. Condition 1 in Table 27.2 is the only STC that can be achieved during the real operation of the PV system by using two-axis solar tracker which keeps solar incident angle always zero (solar beam always normal to PV panel) to maximize system output. Therefore, the PV system output that considers only the first STC of Table 27.2 can be used as reference point to benchmark PV output at specific site. Consequently the performance ratio (P_R) and the capacity factor (C_F) given by Eqs. 27.3 and 27.4 can be readjusted and represented in terms of new performance indicator called optimum performance compliance ratio (P_{CR}) and given by the following equation:

$$P_{CR} = \frac{E_p}{E_{max}} \tag{27.7}$$

where,

E_P – energy produced by the actual PV system (kWh/year)

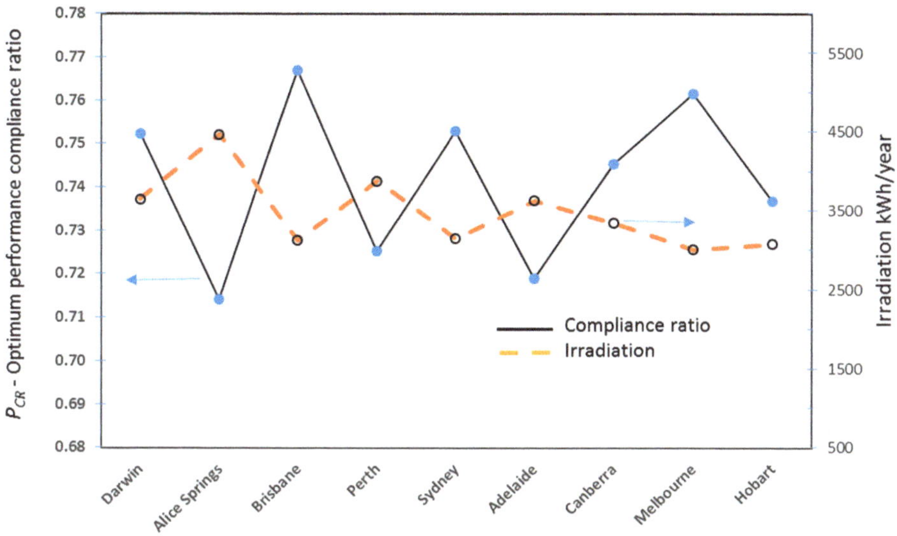

Fig. 27.6 Optimum performance compliance ratio and irradiation at different sites in Australia

E_{max} – maximum energy produced by the same PV system with two-axis solar tracking mode (kWh/year)

The optimum performance compliance ratio (P_{CR}) given by Eq. 27.7 is a measure of the compliance of the PV system output to the optimum design (solar tracking system) output. Therefore, if a system operates at $P_{CR} = 0.7$, then this means that there is 30% deviation from the maximum capacity design. Furthermore, if the PV system operates in two-axis tracking mode, then its $P_{CR} = 1$. This indicator provides the system designers, contractors, and energy providers with a tool to draw actual picture of the PV system price offered to the end users. For example, if a system of 1 kW is required to be installed in two different sites with different roof orientation, their cost price ($/kWh) must be adjusted to their P_{CR} value because this is the actual output that the end user can receive and not the PV panel output at STC condition. Figure 27.6 shows that the highest P_{CR} occurs in Brisbane because the rooftop system output at this latitude become closer to tracking mode system output. Although other sites like Alice Springs has higher irradiation, yet the optimum performance compliance ratio of its rooftop PV system is minimum compared with other sites in Australia. However, it can be observed from Fig. 27.6 that the maximum difference in P_{CR} among the different sites is only 6%.The effect of different roof angles on P_{CR} are presented in Fig. 27.7. Maximum P_{CR} (0.76) or 24% drop in maximum capacity occurs when the roof inclination is around 23°, and azimuth angle is 0 (toward north).

We can conclude from Fig. 27.7 that PV panel orientation has more influence on P_{CR} than site latitude. The worst case design of the rooftop PV system orientation (roof facing east or west) may cause 32% drop in its maximum capacity. The result

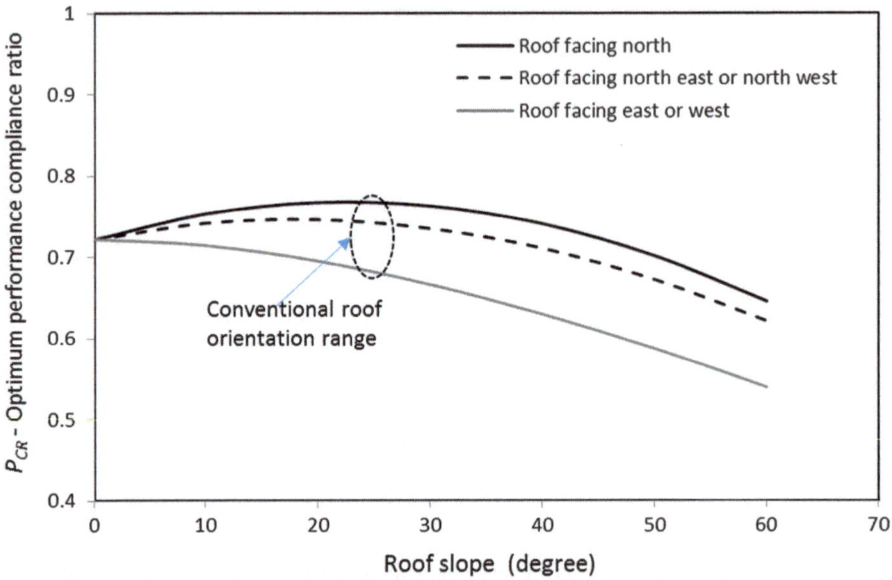

Fig. 27.7 Optimum performance compliance ratio for different rooftop system orientations at Brisbane

of Fig. 27.7 is for the site at Brisbane and may have slight change for other latitudes due to the change in solar beam incident angles.

27.5 Benchmark the PV System Price

The pricing of PV system in general identified by the PV panel capacity that measured at laboratory standard test condition. Such type of pricing charges the end user for unreachable PV array power capacity and do not consider the drop in energy production caused by the PV system design restrictions and weather condition. Traditionally the total cost of the solar power generator consists of hardware (e.g., PV panels, framing structure, inverter, wiring, installation, and land) and non-hardware or soft costs (e.g., marketing and profit). This system pricing is highly dependent on region, local retail electricity rate structures, local rebate, and government incentive structures (Fu et al. 2016). However, this system pricing does not consider the actual operation conditions of the PV systems at each site and benchmark system cost to one reference (generic) price, i.e., the hardware and soft costs. To have fair pricing from the end-user perspective, third component which is called "cost adjustment factor" is considered in the traditional pricing system. The developed method modifies the reference price by using the cost adjustment factor that considers the PV system P_{CR} at real operation condition. The reference cost value of tracking mode system can be estimated from the local market and then used to benchmark different site prices. To calculate the price of a PV system at a certain site, the following equation can be used:

$$\text{System cost} = \text{Reference cost} - \text{Reference cost} \times (1 - P_{\text{CR}}) \quad (27.8)$$

where,

Reference cost – (hardware cost + soft cost) of two-axis tracking mode PV system
$(1 - P_{\text{CR}})$ – cost adjustment factor

Equation (27.8) shows that for each site, there is special price of the PV system based on its actual output. The reference cost (maximum cost) is always reduced by the adjustment factor if the PV system design does not comply with the optimum design. When PV system has $P_{\text{CR}} = 1$ (i.e., system designed with two-axis tracking mode), then system cost will equal the reference cost.

Prices of rooftop PV system of 1.75 kW array at major cities in Australia were calculated using Eq. 27.8 and presented in Fig. 27.8. It can be concluded that system sale price is not fixed, and it can be adjusted and modified based on the operation condition and limitation in some design parameters. Therefore even for the same site, the rooftop solar power generators must have different prices if the P_{CR} value is not equal.

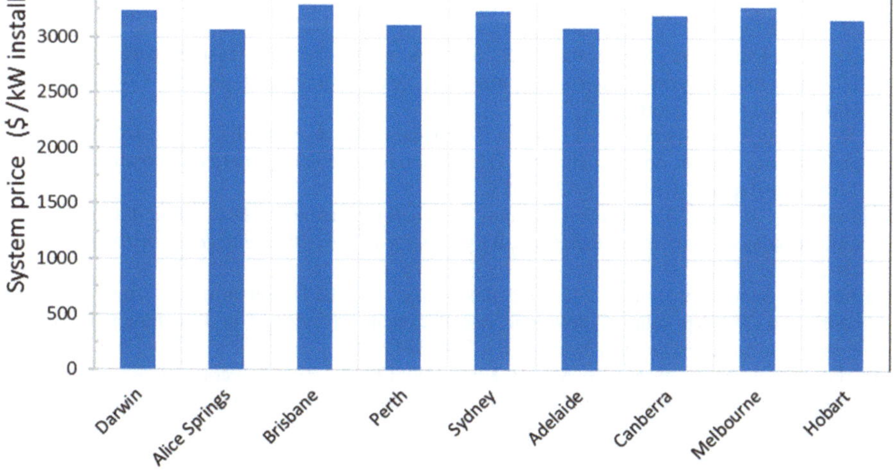

Fig. 27.8 Rooftop system prices at major cities in Australia

27.6 Conclusion

The analysis of the different performance indicators found in the literature (P_R, η_{system}, SF, C_F, and Y_F) showed that there are two drawbacks in applying these indicators from the end-user perspective: the first one is that they benchmark the performance of the solar PV grid-connected system to the standard test condition (STC) which cannot be achieved during real operation and the second drawback is the time period used in some of these indicators which does not describe the actual operation time of the PV system. New performance indicator called optimum performance compliance ratio (P_{CR}) is developed based on actual operation condition of the PV power generator. It is a measure of the compliance of the PV system output to the optimum design output when solar irradiation is normal to PV panel surface. The analysis conducted on the new performance indicator showed that P_{CR} cannot be correlated with the latitude as in the conventional method, and its value depends on the PV system design restrictions (e.g., PV panel orientation, or inclination) and weather conditions. The optimum performance compliance ratio was considered in developing new PV pricing component called "cost adjustment factor." The developed pricing method considers the price of the system with two-axis tracking mode as the reference price for other systems with lower P_{CR} value. It can be concluded that PV system sale price to the end user can be adjusted and modified based on the operation condition and design parameters. The proposed performance indicator is useful for the PV system designers to measure the compliance of the PV array design with the standard design that allows maximum output when irradiation is normal to PV panel surface.

References

Ayompe, L., Duffya, A., McCormackb, S., & Conlonc, M. (2011). Measured performance of a 1.72 kW rooftop grid connected photovoltaic system in Ireland. *Energy Conversion and Management, 52*(2), 816–825.

Fu, R., Chung, D., Lowder, T., Feldman, D., Ardani, K., & Margolis, R. (2016). U.S. Solar photovoltaic system cost benchmark: Q1, National Renewable Energy Laboratory (NREL), Technical Report NREL/TP-6A20-66532.

Fujisawa, T., & Tani, T. (2001). Optimum design for residential photovoltaic-thermal binary utilization system by minimizing auxiliary energy. *Electrical Engineering in Japan, 137*(1), 28–35.

Kumar, K., Sundareswaran, K., & Venkateswaranc, P. (2014). Performance study on a grid connected 20 kWp solar photovoltaic installation in an industry in Tiruchirappalli (India). *Energy for Sustainable Development, 23*, 294–304.

Liu, G., Rasul, M., Amanullah, M., & Khan, M. (2012). Techno-economic simulation and optimization of residential grid-connected PV system for the Queensland climate. *Renewable Energy, 45*, 146–155.

Mermoud, A. (2012). *PVSYST photovoltaic software*. Switzerland: The University of Geneva.

National Construction Code. (2016). *Building code of Australia class 2 to class 9 building*, Vol 1, ABCB.

Odeh, S. (2003). Unified model of solar thermal electric generation systems. *Renewable Energy, 28*, 755–767.

Odeh, S. (2016). Long-term assessment of a grid connected solar PV system in Sydney. *Journal of Energy and Power Engineering, 10*, 591–599.

Odeh, S., & Abu-Mulaweh, H. (2012). Design and development of experimental setup of hybrid PV/thermal collector. *Global Journal of Engineering Education, 14*(2), 170–176.

Odeh, S., & Behnia, M. (2009). Development of PV module efficiency using water cooling. *Heat Transfer Engineering Journal, 30*(6), 499–505.

Photovoltaics Report. (2016). Fraunhofer Institute for Solar Energy Systems, ISE with support of PSE AG, Freiburg, Germany.

Pless, S., Deru, M., Torcellini, P., & Hayter, S. (2005). Procedure for measuring and reporting the performance of photovoltaic systems in buildings, National Renewable Energy Laboratory (NREL), Technical Report NREL/TP-550-38603.

Shukla, A., Sudhakar, K., & Baredar, P. (2016). Design simulation and economic analysis of standalone roof top solar PV. *Solar Energy, 136*, 437–449.

Singh, H., Sirisamphanwong, C., & Santhi S. (2016). Effect of tilt and azimuth angle on the performance of PV rooftop system. *Applied Mechanics and Materials, 839*, 159–164.

Su, Y. (2015). A comparison analysis of the performance of a grid-connected photovoltaic system based on low- and high-frequency. *International Journal of Greemn Energy, 12*, 1206–1214.

The European Standard. (1998). Photovoltaic system performance monitoring – guidelines for measurement, data exchange and analysis, EN 61724, BSI, UK.

Ueda, Y., Kurokawa, K., Itou, T., Kitamura, K., Akanuma, K., Yokota, M., & Sugihara, H. (2008). Advanced analysis of grid-connected PV system's performance and effect of batteries. *Electrical Engineering in Japan, 164*(1), 247–258.

Chapter 28
Performance of Solar-Thermal Organic Rankine Cycle (STORC) Power Plant with a Parabolic Trough System

O.Y. Odufuwa, K. Kusakana, and B.P. Numbi

28.1 Introduction

There is a high demand for alternative modes of energy provision to reduce the potential for negative environmental impact. Addressing the problem of sustainable energy supply is one of the major engineering challenges of the twenty-first century (Katayama and Tamaura 2005). One potential means of overcoming this challenge lies in the development of renewable energy technology through a dedicated research effort. The effort includes, for example, exploration of biomass utilisation, waste heat energy recovery, wind energy and solar energy, among others. Among these alternative sources, renewable energy sources provide economical, safe and renewable energy technologies and create opportunities for the sustainability of power generation (Resch et al. 2014).

Moreover, renewable energy technologies have only recently achieved widespread adoption. Similar to conventional energy technologies, many of them are still to be tested and proven. One way of hastening the adoption of these technologies is to reduce the amount of field testing and field optimisation by using numerical models to optimise the technologies. Modelling of the specific renewable energy technology, such as the Organic Rankine Cycle (ORC) technology, reduces

O.Y. Odufuwa (✉)
Department of Mechanical and Mechatronics Engineering, Central University of Technology, Free State, Bloemfontein, South Africa
e-mail: oodufuwa@cut.ac.za

K. Kusakana
Department of Electrical, Electronic and Computer Engineering, Central University of Technology, Free State, Bloemfontein, South Africa

B.P. Numbi
Centre for the Development of Green Technology, Department of Electrical, Electronic and Computer Engineering, Mangosuthu University of Technology, Durban, South Africa

A. Sayigh (ed.), *Transition Towards 100% Renewable Energy*,
Innovative Renewable Energy, https://doi.org/10.1007/978-3-319-69844-1_28

the cost of testing and optimisation by providing tools for the evaluation and optimisation of existing and proposed ORC plants (Orosz et al. 2013).

Thermal power generation is a proven technology with several hundreds of plants in operation. Current large-scale systems rely on traditional steam-based Rankine cycles for power production. Most of these plants develop megawatts of electricity. On the other hand, ORC power plants are more compact and less costly than traditional steam cycle power plants and can better exploit lower temperature thermal resources. ORC develops kilowatts of electricity. The utilisation of ORC allows for solar-thermal power generation to become a more modular and versatile means of supplanting traditional fuels (Mendelsohn et al. 2012). The parabolic trough collector with the ORC power plant is currently being utilised for power generation and will generate up to 100 MW clean power supply to some areas in South Africa (Abengoasolar.com 2016).

The objective of the study was to provide a tool, in the form of a computer simulation and modelling capability, for optimising the performance of a solar ORC plant in the medium output category. Another objective of the study was to possibly compare the performance of components of an ORC solar-thermal power plant system into one numerical model that can be used to analyse entire system energy conversion processes for varying cases regarding the generation of electricity.

28.2 Methodology

Modelling the power plant considers the thermodynamic analysis of each component that transforms energy. The simulation programme was developed using Matlab Simulink® and Thermolib Simulink® library. The steady state hourly variations of the parameter were considered for analysis of the STORC power plant.

28.2.1 Solar Radiation Resource Model

Based on the study focus, which is on the South African weather, reliable data for the solar resource is available on a database of the Southern African Universities Radiometric Network (SAURAN) (Brooks et al. 2015). Weather data is also available from the South African Weather Service (SAWS) database, but there is no provision for any form of solar radiation data (Weathersa.co.za 2016).

The main requirement in this part of the resource model is to read the specific meteorological data at a particular hour of a given day for the location of interest. This variable at the instance of the hour is the output to the collector model. The resource Simulink block looks for the required value from the given set of data. This flow chart is given in Fig. 28.1.

The n-dimensional lookup block was adopted to determine the DNI, atmospheric temperature, atmospheric pressure, wind speed and relative humidity at a specific

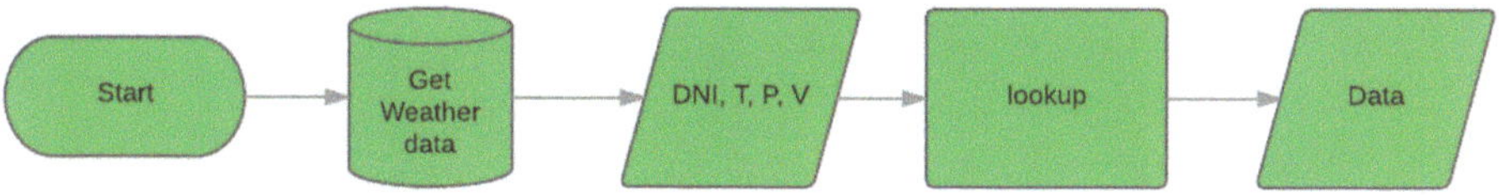

Fig. 28.1 Lookup hourly parameter flow chart

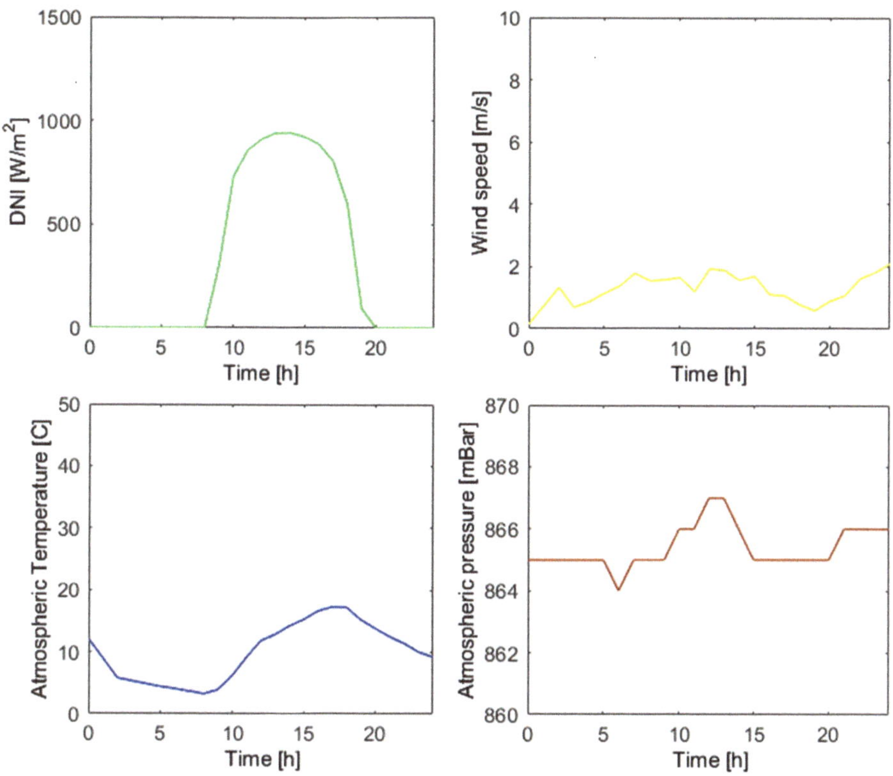

Fig. 28.2 Simulink plot of the meteorological data for Bloemfontein on 1 January

hour of a chosen location of interest. The model allows the change of data for different places and time periods. The lookup block made it easier to read the different meteorological data for any location for 24 h during the various seasons of the year. The user is only required to modify the lookup table data for the measured data of the site and periods and dates of the seasons of interest. Below is the Fig. 28.2, which shows Bloemfontein's first 24-h weather variations for January 1, 2015.

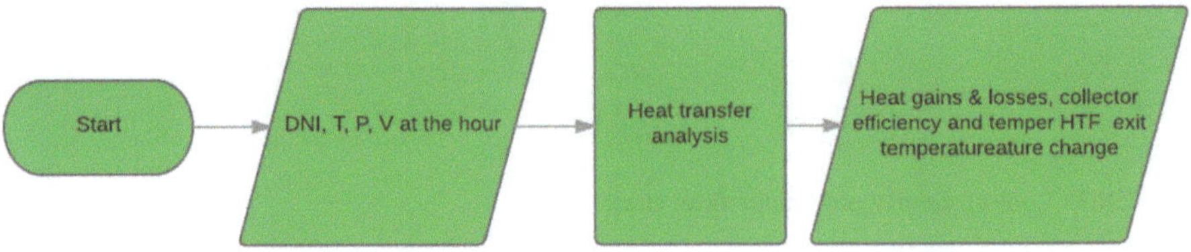

Fig. 28.3 Thermal collector flow diagram

28.2.2 Solar-Thermal Collectors

Utilisation of the energy from sunlight requires means to capture radiation from this source efficiently. The parabolic trough collector (PTC) was modelled based on the choice of the existing power plant type of collector. PTC modelling involves thermodynamic analysis. Such analysis allows for the consideration of the thermophysical properties, geometric parameters and other factors that determine the output of the collector unit of the power plant. The model applies to PTCs. The previous study was used for comparisons and validations. Flow chart of the model is given in Fig. 28.3.

The most important output parameters (collector efficiency, heat gain and HTF exit temperatures) are considered for three cases. The first instance will review the variation for these important output parameters over 24 h. The second instance is the change of the output with the dimension and HTF in the collector system. Lastly, the model investigated the variations of the output parameters with the location of interest. The model considers the thermal efficiency of the collector in some cases by assuming the optical efficiency ranges from 60% to 100%. Matlab Simulink® and Thermolib library were used to develop the models.

28.2.3 Solar-Thermal Storage

Storage of solar energy to stabilise the periodic fluctuations in the radiation is required for the solar plant system. Reasonable performance could be achieved in the system taking into consideration the relative function of its components. Load requirements, capacity and condition of the source of energy to the system determine the expectation of the system. Analysis of the capacity and rate of inputs and outputs provides a reasonable tool towards system optimisation. The storage was modelled using the Simulink ® Thermolib library. Considering the Thermolib, the mass and energy balance of the liquid storage tank input and output flows will be based on Eqs. 28.2 and 28.2 below. Equation 28.1 is used to calculate the mass balance using the molar fraction and the molar flow rate of the fluid in the system. The energy balance is done by considering the enthalpy rate at the inlet and outlet of the components. Hence, energy balance is expressed in Eq. 28.2.

$$\frac{\mathrm{d}n_i}{\mathrm{dt}} = n_{i,\mathrm{in}} - n_{i,\mathrm{out}} \tag{28.1}$$

$$\frac{\mathrm{d}H}{\mathrm{dt}} = H_{\mathrm{in}} - H_{\mathrm{out}} \tag{28.2}$$

28.2.4 Organic Rankine Cycle Outputs

The operation of the simple Rankine cycle still applies to an Organic Rankine Cycle. The difference is based on the type of transport fluid used in the system. The ORC uses a high molecular mass organic fluid which allows heat recovery from a low-temperature source such as industrial waste heat and solar-thermal collectors. The study is centred on an ORC type applicable to parabolic trough collectors (PTCs) (Fig. 28.4).

The thermodynamic analysis follows the first law of thermodynamic of non-flow and steady-flow processes for the components of the system (Fig. 28.5) (EUtech scientific engineering GmbH).

The model is composed of the evaporator to the turbines and a pump coupled to the WF storage tank. The block determines the power output and thermal efficiency of the plants at the endpoint. A double-stage turbines and air-cooled condenser were used in the system. On the other hand, the pump and evaporator are set up for pumping and counterflow heat exchange, respectively. Figure 28.6 shows the configuration of the ORC in Simulink®.

28.2.5 Heat Transfer Fluids and Working Fluids

Heat transfer fluid is the transport medium at the high-temperature zone. Glycol compound found in car antifreeze fluids and other HTF candidates were simulated. The working fluid is utilised at the turbine region for transport to work output areas. The choice of transport fluid is also relevant to optimisation of the system.

28.3 Results

28.3.1 Solar-Thermal Collectors Model Outputs

The collector efficiency for different types of HTF in the collector was plotted in Fig. 28.6. The heat gain at the various locations revealed that the heat gain varies with the site of interest as plotted in Fig. 28.7. Figure 28.8 shows the variations of heat gain with the length of the collector. The HTF exit temperature was also

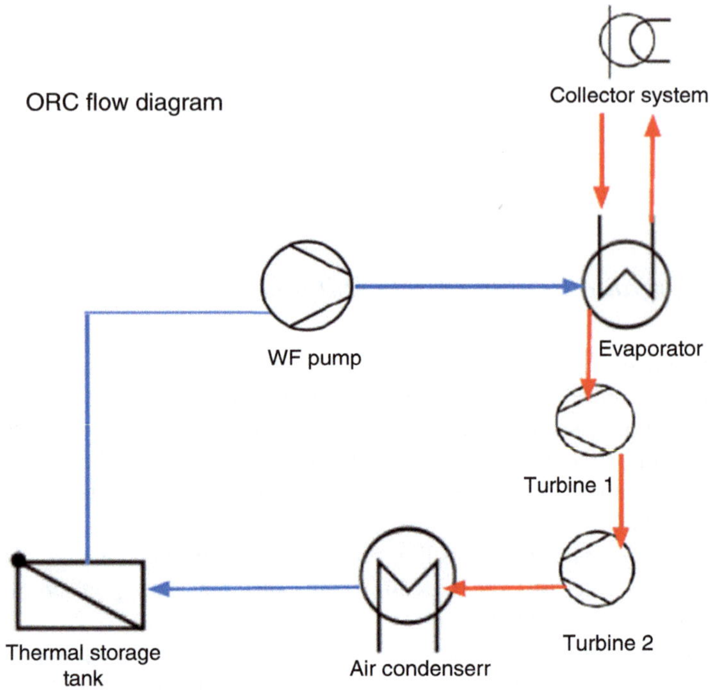

Fig. 28.4 ORC flow diagram

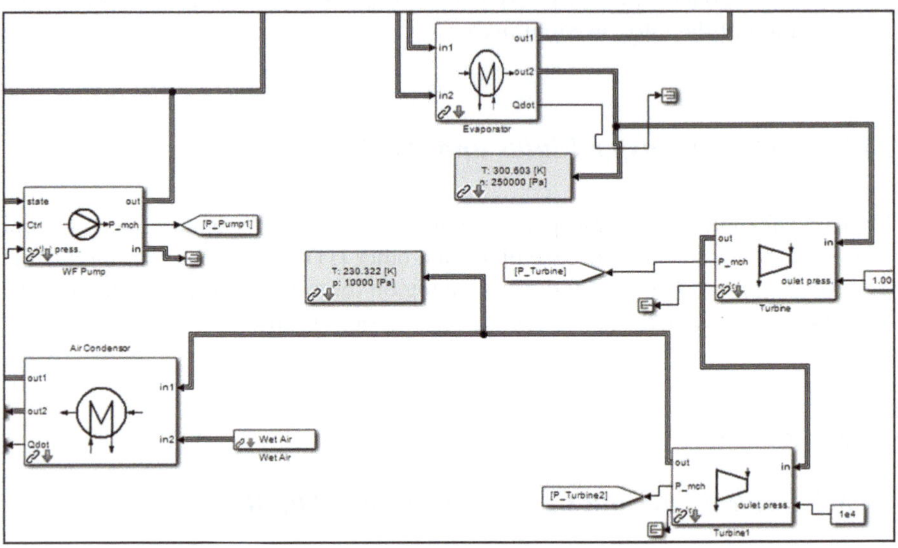

Fig. 28.5 ORC part of the STORC power plant

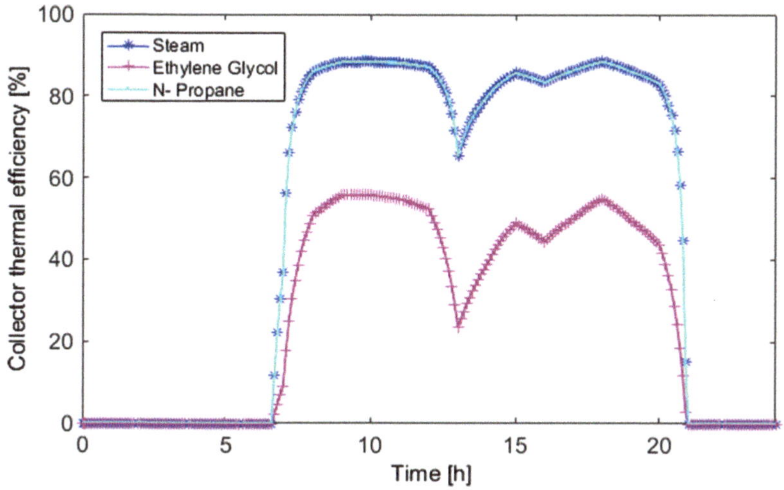

Fig. 28.6 Variation in collector efficiency over the same day with different HTF in the system

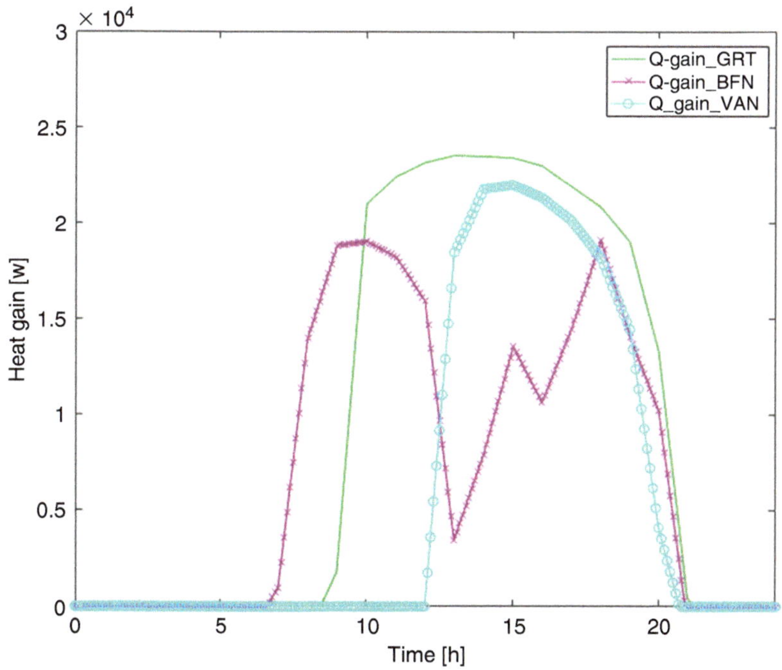

Fig. 28.7 The hourly heat gain for different locations

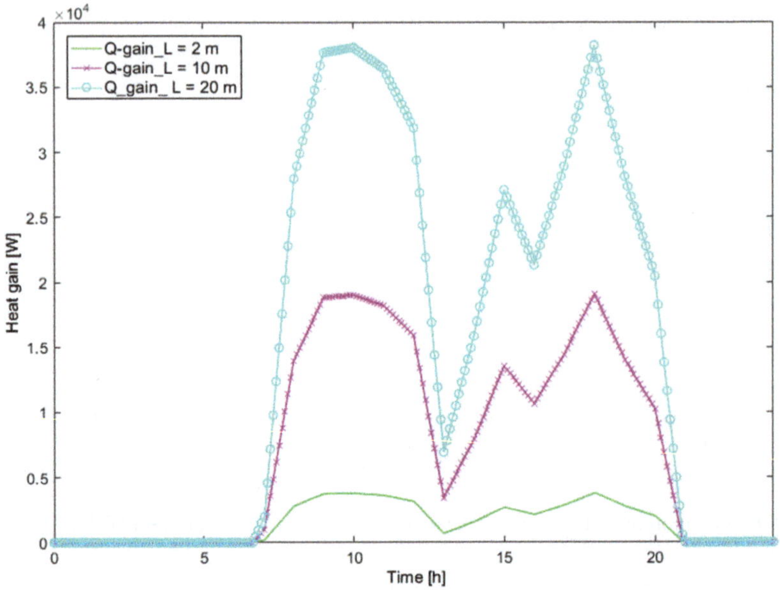

Fig. 28.8 The hourly heat gain for various lengths of the collector at a chosen day of the year

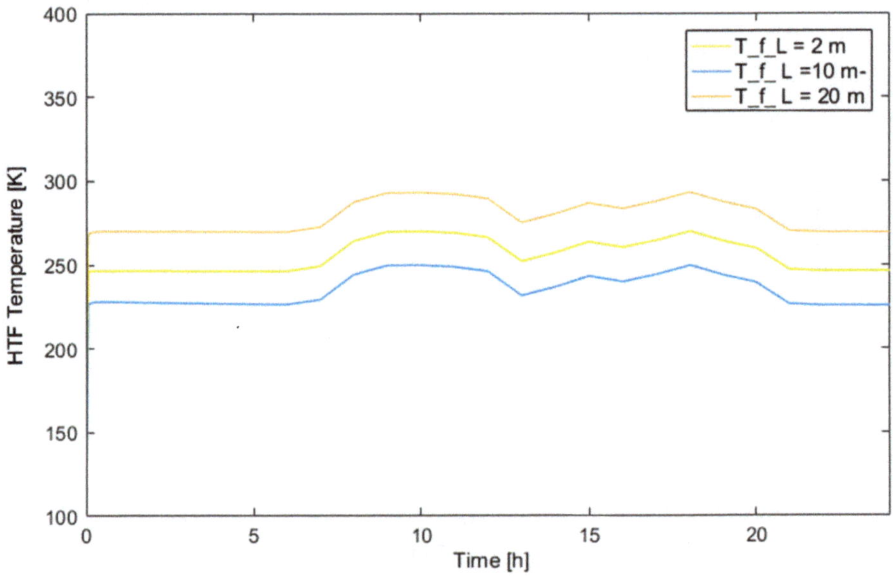

Fig. 28.9 Plot of fluid exit temperature for 24 h at different lengths of the collector

considered as shown in Fig. 28.9, and it was seen that the fluid exit temperature stays constant for different lengths of the collector.

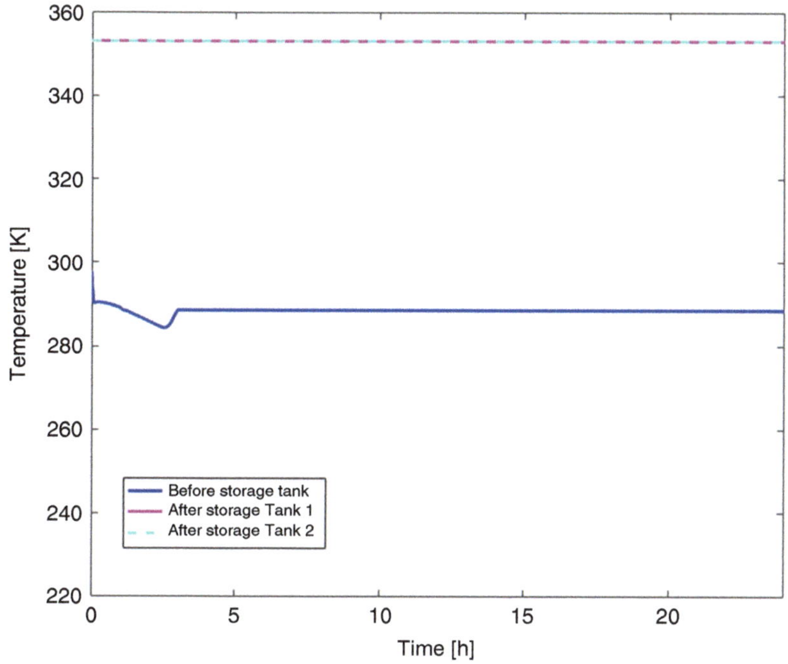

Fig. 28.10 Comparisons of the temperatures (K) before and after storage tank boundary in the system

28.3.2 Solar-Thermal Storage Model Output

Figure 28.10 shows the comparisons of the temperature before and after storage tank entrance and exit at a different point in the system over 24 h. It was observed that the storage tank helps to stabilise and accumulate the heat of the system. The model was done using the Thermolib library in Simulink®.

28.3.3 Organic Rankine Cycle Model Output

Figures 28.11 and 28.12 show the power output over a day and with a different variable that affects the output. Figure 28.12 presents the variation of power output with WF and HTF. The plot shows different gradients for the power output indicating the effect of fluid on the output of the system. The length of the collector has also affected the power output as seen in Fig. 28.12. The gradient of the power output curve is steeper more for the lower length than the higher length, indicating the rapid increase with a shorter or smaller system. Figure 28.13 shows the cycle efficiency plotted over 24-h time. The curve shows a constant efficiency after the system attained a steady state.

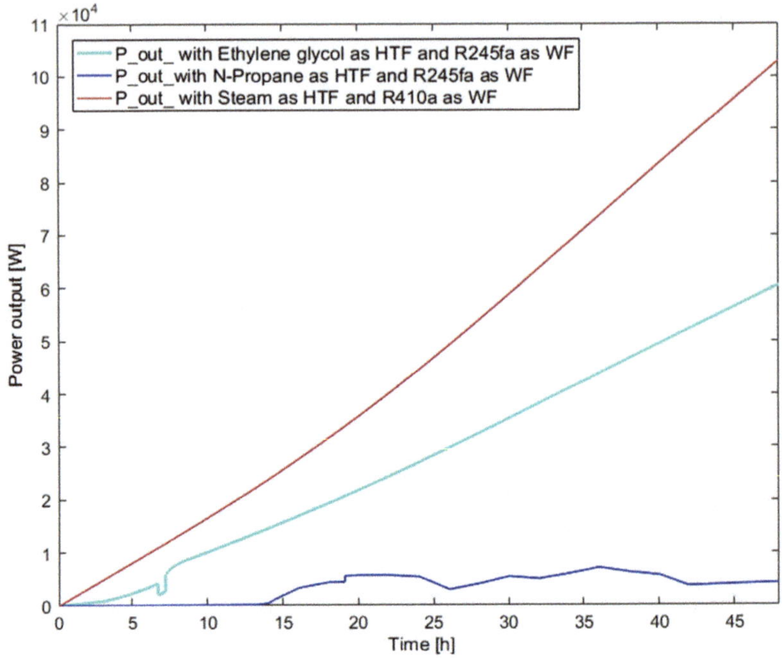

Fig. 28.11 Power output for different HTF and WF in the system

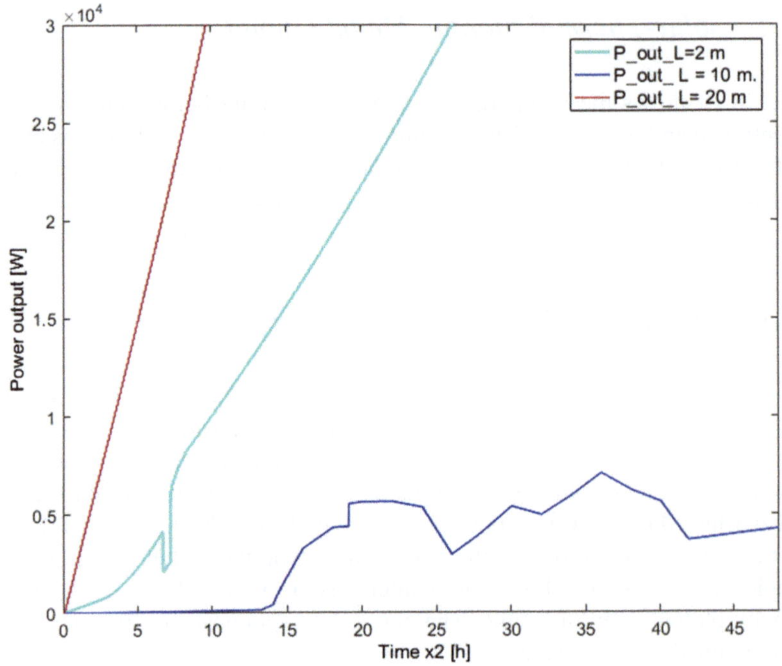

Fig. 28.12 Power output with different lengths of the collector

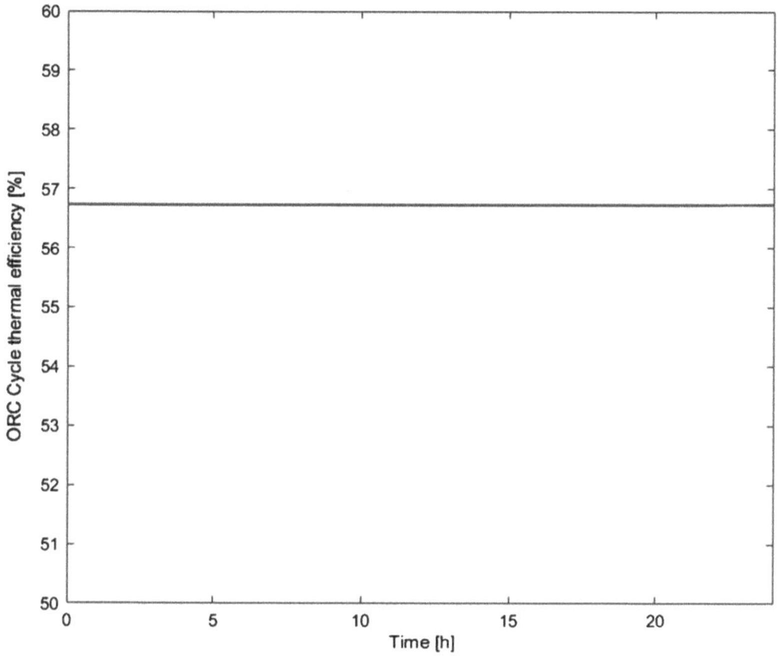

Fig. 28.13 Thermal efficiency of the system for the chosen day

28.3.4 Discussion of Results

It is essential to estimate the output and the input requirement of a power plant before investing in the technology. From the model developed above, the outputs of the components of the plant as well as that of the integrated unit were investigated. Knowledge of the optimum heat gain and temperature output of the collector, the capacity of the thermal storage and the power output of the ORC with regard to the size of the plant and the HTF used in the systems provided a reasonable basis for the design of the power plant.

The validity of the model was also determined by using parameters from different plants and comparing the results with the current model. For instance, the input parameter from the AZTRAK test was used on the present model, and the results were plotted for the different cases. The graph in Fig. 28.14 shows that the results in both cases differ less than 10% in terms of the graph. The same procedure was applied to check the other output parameters of the collector, and the maximum discrepancy was 15%. The outcomes from the validations of the outputs indicate the extent of model acceptability.

Furthermore, collector optimisation could be achieved by using the best HTF, building with the correct dimension of the aperture, suitable length and diameter of pipes and choosing a location that will generate the required power. For instance, Bloemfontein and its vicinity tend to provide more heat gain than the other two sites considered.

Fig. 28.14 Heat gain (in Watt on vertical axis) of the test model compared to heat gain of the present model over a day (in hours on horizontal axis)

28.4 Conclusions

From the study results and validations, the verified solar irradiance values sourced from the database could be used for correct input parameters for the collector model. The values from SAURAN were compared to other values from different weather databases and provided reasonable acceptance after statistical analysis. The solar-thermal collector of the model determines the useful energy gain, HTF outlet temperatures and collector thermal efficiencies and other useful parameters for analysis. It was observed that the larger the aperture and the longer the length of the collector, the higher the heat gain. Concerning storage, the correct storage capacity must be chosen to achieve the required performance. From the ORC engine model, the power output shows a realistic output could be generated from the plant. It was also found from a recent investigation that the PTCs technology is now in operation by Abengoa's in KaXu Solar One, located near the town of Pofadder in the Northern Cape Province of South Africa. Other plants are under construction and are also located in the Northern Cape Province. The plants are the Khi Solar One, which is expected to generate around 50 MW, and Xina, which has a total installed capacity of 100 MW.

The research will be useful for future maintenance and optimisation of the related plants.

References

Abengoasolar.com. (2016). *Abengoa solar : Products & services: Solar field components: Parabolic trough collectors: E2, Phoenix and Astr0.* [online] Available at: http://www.abengoasolar.com/web/en/Productos_y_Servicios/Componentes_campo_solar/Colectores_cilindroparabolicos/E2_Phoenix_Astr0/. Accessed 13 Sep 2016.

Brooks, M., du Clou, S., van Niekerk, W., Gauché, P., Leonard, C., Mouzouris, M., Meyer, R., van der Westhuizen, N., Evan Dyk, E., & Vorster, F. (2015). SAURAN: A new resource for solar radiometric data in Southern Africa. *Journal of Energy in Southern Africa, 26*(1), 2–10.

Katayama, Y., & Tamaura, Y. (2005). Development of new green – fuel production technology by combination of fossil fuel and renewable energy. *Energy, 30*(11), 2179–2185.

Mendelsohn, M., Lowder, T., & Canavan, B. (2012). *Utility-scale concentrating solar power and photovoltaic projects: A technology and market overview.* [online] NREL. Available at: http://www.nrel.gov/docs/fy12osti/51137.pdf. Accessed 7 Dec 2014.

Orosz, M., Mueller, A., Dechesne, B., & Hemond, H. (2013). Geometric design of scroll expanders optimized for small organic Rankine cycles. *Journal of Engineering for Gas Turbines and Power, 135*(4), 042303.

Resch, B., Sagl, G., Törnros, T., Bachmaier, A., Eggers, J., Herkel, S., Narmsara, S., & Gűndra, H. (2014). GIS-based planning and modelling for renewable energy: Challenges and future research avenues. *ISPRS International Journal of Geo-Information, 3*(2), 662–692.

Simulation toolbox for the Design and Development of Thermodynamic Systems in MATLAB®/Simulink®. (2011). EUtech scientific engineering GmbH.

Weathersa.co.za. (2016). *Forecast.* [online] Available at: http://www.weathersa.co.za/. Accessed 28 Apr 2016.

Chapter 29
Solar Short-Term Forecasts for Predictive Control of Battery Storage Capacities in Remote PV Diesel Networks

Dorothee Peters, Thilo Kilper, Martina Calais, Taskin Jamal, and Karsten von Maydell

29.1 Introduction

Increasing photovoltaic (PV) system penetration in small, isolated networks faces a number of technical, regulatory, and economic challenges but also opportunities. The replacement of costly and polluting diesel generation is highly desirable; however, a key technical issue is system stability, particularly on intermittent cloudy days. In this context, the step load response capability and spinning reserve (SR) provided by diesel generation are critical to cover for the sudden loss of PV power.

In previous simulation studies, we varied the PV capacity in a remote diesel network while concurrently observing the feasible benefit of short-term forecasts based on sky imagers. It was found that stable grid operation can be guaranteed with an installed PV capacity of up to 30% of peak network load (even on intermittent cloudy days) allowing for fuel savings of up to 8% (Peters et al. 2016). Reduction of SR according to highly probable irradiation derived from sky imager forecasts resulted in minor improvement of <1% of fuel savings. Similar findings of around 2% fuel reduction were made by Bourry et al. 2016 for the application of short-term forecasts at 33% PV penetration.

Building on previous work, this simulation study implemented in MATLAB introduces storage into the network to enable higher PV penetration and further benefits from sky imager forecasts. Malhotra et al. 2016 classified and prioritized applications of stationary battery technologies. In remote electricity systems, the

D. Peters (✉) • T. Kilper • K. von Maydell
Division Energy Systems & Storage, NEXT ENERGY · EWE Research Centre for Energy Technology at the University of Oldenburg, Oldenburg, Germany
e-mail: dorothee.peters@dlr.de

M. Calais • T. Jamal
School of Engineering and Information Technology, Murdoch University, Perth, Australia

A. Sayigh (ed.), *Transition Towards 100% Renewable Energy*,
Innovative Renewable Energy, https://doi.org/10.1007/978-3-319-69844-1_29

most important applications (following the definition of Battke and Schmidt 2015) are load following, renewable energy technology (RET) firming (i.e., storing excess RET production to be used at a later time and increasing its dispatchability), and RET smoothing and frequency regulation. This study includes ramp rate control for smoothing, provision of spinning reserve, and RET firming. A combination of storage capacities and short-term forecasts is introduced by combining pre-mentioned techniques (Method 1). For the second method (Method 2), the ideas of Fulcrum3D (2016) are reevaluated. With this control method, PV generation is curtailed according to the lowest forecasted value to decrease use of available storage in the network. Fulcrum3D (2016) states that the amount of curtailed energy is lower than the losses due to battery system efficiency.

The following sections introduce the network configuration and input data and define the simulation cases used (2), depict results of selected characteristics of all simulation cases (3), and discuss the results in the greater context (4).

29.2 Methodology

The analysis is based on a remote network with a peak load of 590 kW and an average load of approx. 270 kW which is covered by a diesel generation capacity of 5×140 kW. The irradiance and load data set for the studies has been derived from a scaled measured load time series of approx. 10 months in 15-min resolution, and solar data from one measurement station in Kalgoorlie in 1-min resolution is used (Australian Bureau of Meteorology 2015). Actual forecast data is not available, but a time series matching characteristics of sky imager-based short-term forecasts found by Schmidt et al. (2016) is modeled according to Eq. 29.1 by introducing normal distributed errors. The forecast deviates from the original time series based on the variability of the actual irradiance time series. Additionally, an error is introduced to the temporal position of data as contribution to temporal ramp detection problems. The RMSE of actual forecast performance can be retrieved from the synthesized time series.

$$G_{\text{forecast}}(t) = G(t + t_{\text{error}}) + \mathcal{N}(0,1) \cdot \text{Variability}_{\text{interval}} \cdot G_{\text{clearsky}} \tag{29.1}$$

In total, six control strategies are defined and compared to operation with diesel generation only (see Table 29.1). This approach was chosen to allow for the comparability of controls as results strongly depend on network configuration and load time series.

For each time step, residual load and SR for load as well as PV are calculated. Curtailment of PV (which is applied in cases 2–6) is determined from diesel generator loading. Then the need for battery ramp smoothing, SR, charging, or discharging is checked before diesel generation is reconsidered and energy balance of load and generation is verified. In all control schemes with battery, the battery can be charged from curtailed PV power, from diesel gensets overnight, and in case

Table 29.1 Overview of control strategies in this simulation study

#	Case name	Description of case
1	No PV	Base case, total fuel consumption is used for comparison with other cases. Diesel generators run with a minimal load of 30% and need 5 min for start-up. Gensets meet current load plus 20% of current load as SR
2	PV only	Spinning reserve is provided according to current PV power. Tested cases are 0%, 25%, 50%, 75%, and 100% of current PV power for SR
3	PV + short-term forecast	Spinning reserve is provided by diesel gensets according to the lowest forecasted PV power in the next 10 min
4	PV + small battery	(a) Battery is used to guarantee predefined ramp rates (RR) of PV. Maximum ramp rates are 6 min for ramp-up from 0% to 100% and 12 min for ramp-down from 100% to 0% (Horizon Power 2013). Results for this case include 25% of current PV power as spinning reserve provided by gensets
		(b) Additionally to control case 4a, the battery provides spinning reserve when generator higher or lower operation border is reached
5	PV + short-term forecast + small battery	1. Method 1: forecast is used as in case 3; battery is used for ramp rate control and spinning reserve (case 4a + b)
		2. Method 2: PV power is curtailed to the lowest forecasted PV power; when curtailed, higher than maximum allowed ramps are smoothed by the battery as described in case 4a (adaption from Fulcrum3D 2016)
6	PV + large battery	The battery is discharged in the peak load hours determined from load time series (20% of highest values of mean) at fixed power (30% of inverter capacity). It is only charged with diesel power at low SOC and charged to a higher threshold SOC from PV power. Additionally to RET firming (RF), methods 4a and 4b are used to increase benefits of battery for the integration of PV

of low SOC also during the day. The charging current is restricted to less than total power to maintain the possibility of further power input as grid service. For case 4 and 5 the battery capacity is set to 100 kWh at 40% PV penetration and 150 kWh at 60% PV penetration, while the necessary inverter power (85% of battery capacity) was derived from the simulation at 40% PV penetration. At 100 kWh capacity, the rating of 85 kW includes the 3-σ-range (99.7% of values) which covers ±70 kW for case 4b, ±66 kW for case 5.1, and ±72 kW for case 5.2. Figure 29.1 depicts an exemplary time frame with battery controls used in case 4 and 5.

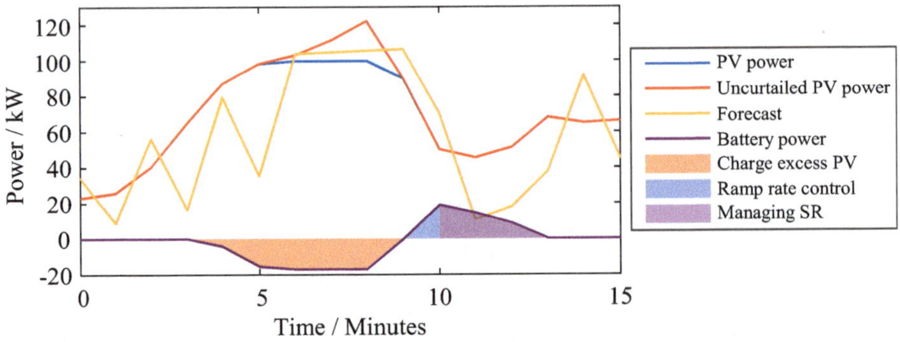

Fig. 29.1 Operation strategy of the battery showing ramp rate smoothing (light blue area), RET firming to manage SR (purple area), and charging with otherwise curtailed PV power (orange area)

29.3 Results

The effect of short-term forecasts on the remote PV diesel network is evaluated in terms of fuel savings, PV curtailment, and lack of spinning reserve which may cause stability problems. Further, battery throughput is analyzed as an indicator of battery usage among the different control strategies.

29.3.1 Fuel Savings

Considering PV penetrations of 40% and 60% for all cases defined in Table 29.1, the corresponding diesel fuel savings were determined in relation to the reference case 1 (no PV integration). The results are shown in Fig. 29.2 and indicate that the achieved fuel savings are in the range 14.7–27.7%. Without application of neither forecasting nor storage capacity, PV integration leads to fuel savings of 14.7% and 17.2%, respectively (case 2). At 40% PV penetration, the configuration case 4b performs best (19.4% fuel savings). However, at 60% PV penetration, case 6 leads to the largest fuel savings (27.7% at 1 MWh battery capacity). The effect of combining solar short-term forecast with battery storage capacity can be studied by comparing the cases 5.1 and 5.2 with case 3 (only forecast) and cases 4a and 4b (only battery). Without additional storage, the forecast application leads to an increase of fuel savings of 3.1% at 40% PV penetration and of 5.5% at 60% PV penetration (case 3) compared to PV only. With storage integration (case 4a+b), the resulting additional fuel savings are 4.7% and 6.7%. If forecasting and storage are both embedded into the network control, fuel savings of 3.9% and 5.6% are achieved (case 5.1). Halving the battery capacity (100 kWh → 50 kWh) in case 4b and 5.1 has no effect on fuel savings. The additional savings of case 5.2 are 0.6–1.3%.

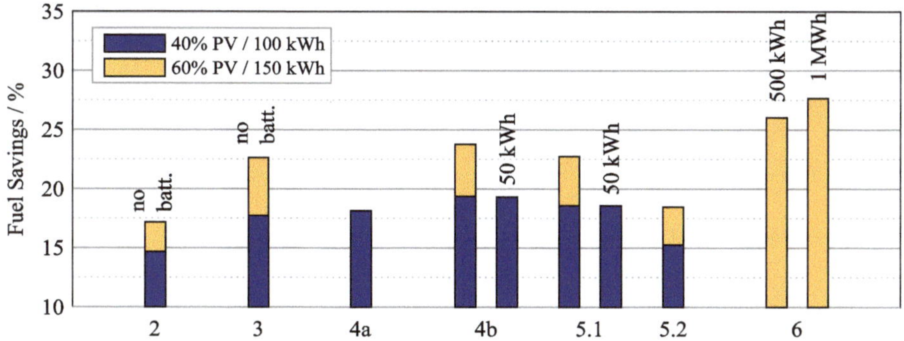

Fig. 29.2 Achieved diesel fuel savings at PV penetrations of 40% and 60% for simulated control strategies (cases 2–6; see Table 29.1 for definition)

29.3.2 PV Curtailment

With increasing PV penetration, PV needs to be curtailed. Curtailment increases in periods of low electricity demand at daytime due to minimal loading and SR provision of gensets. The results of cases 2–6 are shown in Fig. 29.3, and the arising PV curtailment is in the range of 4.4–36.4%. At 60% PV penetration, application of conservative spinning reserve policies (case 2) results in the peak value of 36% PV curtailment which raises LCOE of PV by approx. 50%. Similar high results are obtained by applying the control strategy of case 5.2. All other control strategies distinctly decrease PV curtailment. At 40% PV penetration, case 4b exhibits the smallest losses (approx. 7.5% at 100 kWh and 50 kWh storage capacity). Compared to case 2, 3, 5.1, and 5.2, the control strategy of case 4b also leads to the smallest losses (18.8%) at 60% PV penetration. However, considering a PV penetration of 60%, the control strategy of case 6 enables a further significant reduction of the curtailment losses (10.2% at 500 kWh and 4.4% at 1 MWh storage capacity). Another result is that halving the storage capacity (100 kWh → 50 kWh) in case 4b and 5.1 has again no negative effect concerning PV curtailment.

29.3.3 Network Stability

To evaluate the network stability, periods with insufficient spinning reserve are counted. A reduced spinning reserve enables higher PV shares, but simultaneously the network stability deteriorates. Figure 29.4 depicts the probability of a lack of spinning reserve per one million time steps. Compared to the conventional control strategy (SR = 100%), the occurrence of event lacking of spinning reserve is distinctly increased in case 3 due to uncertainties of the forecast and also in case 4a. However, control strategies that use the battery storage for SR provision (cases 4b, 5.1, and 5.2) have clearly reduced occurrences leading to higher network

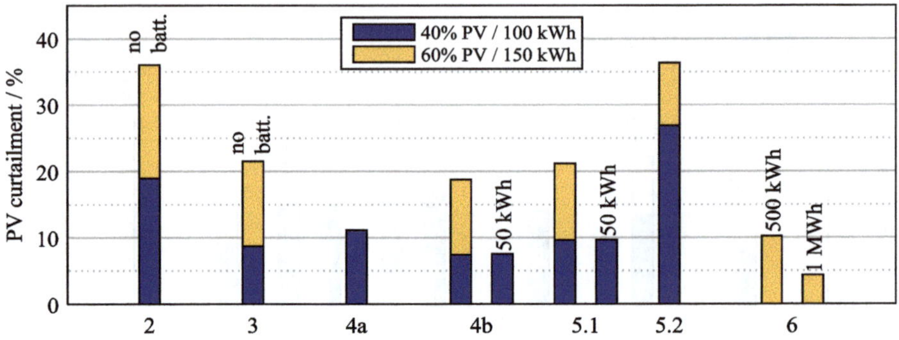

Fig. 29.3 Arising PV curtailment for cases 2–6

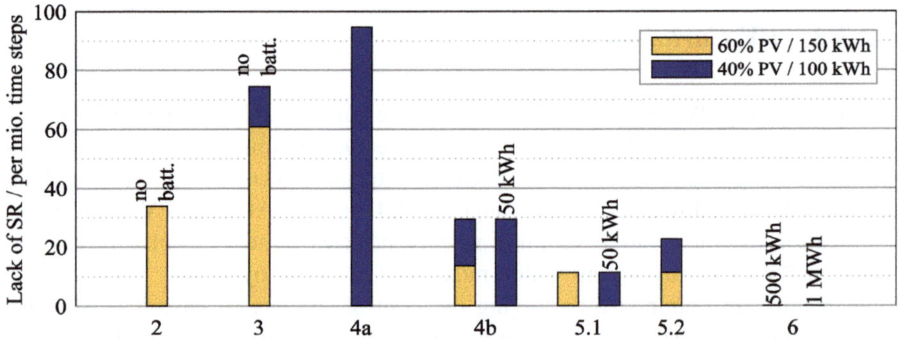

Fig. 29.4 Arising lack of spinning reserve for cases 2–6

stability. In case 6 the larger battery capacity can provide enough SR for all occurring operation situations.

29.3.4 Battery Throughput

The normalized energy throughput in kWh per installed kWh storage capacity over the whole simulation period is depicted in Fig. 29.5 and enables the assessment of battery usage. Case 4b and 6 with the smaller storage capacities of 50 kWh and 500 kWh, respectively, show the highest normalized battery throughput of each approx. 100 kWh/kWh_{cap}. The highest total battery throughput arises in case 6 with 50 MWh (500 kWh storage capacity) and 82 MWh (1 MWh storage capacity). The combination of forecast and battery storage (case 5) leads to a battery throughput reduction compared to battery-only operation (case 4). A reduced storage capacity (100 kWh → 50 kWh) in case 4b and 5.1 changes normalized throughput but does not change total throughput significantly, though reduced storage capacities increase the share of diesel power used for battery charging. The simulation results

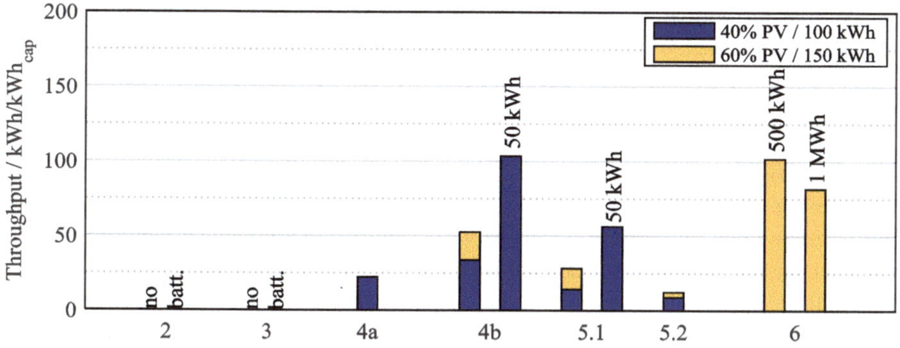

Fig. 29.5 Normalized battery throughput for cases 4–6

also revealed that the share of charged PV energy compared to total PV energy is below 1% in case 4 and 5 and 8.9–10.5% in case 6.

29.3.5 Discussion

The presented results illustrate that the compared control strategies (case 2–6) for 40% and 60% PV penetration have differing impacts on the evaluated criteria: fuel savings, PV curtailment, network stability, and battery throughput. For an overall assessment, the consideration of the following aspects could be helpful:

1. As long as capital cost for battery storage capacities remains on a relative high level, it will be favorable to aspire smaller battery storage capacities. Under these conditions the control strategies 5.1 and 5.2 are a promising approach. Case 5.2 (Method 2) exhibits the lowest battery throughput and increases the network stability. However, the fuel savings are only marginally decreased compared to PV only (0.6–1.3%). Case 5.1 (Method 1) still features a comparatively low battery throughput and a high network stability, while the increased fuel savings are now noticeable (3.9–5.6%).
2. Due to the increasing demand for renewable energy technologies like PV systems and electric vehicles, Hoffmann (2014) predicts that battery storage cost will decrease strongly due to economies of scale and technological innovations. In this case, large battery storage capacities become feasible and preferred due to high fuel savings. Furthermore the results of case 6 show the lowest curtailment of PV power and the highest network stability. The battery storage handles all power fluctuation, while the benefit of short-term forecast decreases. This perception might change if further parameters like charging and discharging rates, cycle numbers, and depth of discharge are considered for the overall assessment.

Comparing with literature, this study yields quite similar results for usage of forecast only (case 3). For example, Bourry et al. (2016) analyzed cases at 33%, 66%, and 100% penetration with resulting fuel savings of 2%, 2–5%, and 5–12%, respectively. At 60% PV penetration our corresponding values are in the range 3.1–5.5%. The results of Method 2 (case 5.2) adapted from Fulcrum3D (2016) applied to our data led to the envisaged low battery throughput but also to high PV curtailment and low fuel savings. The reliability of forecasted irradiance is of great importance when integrating it into network controls. Though accuracy of the used forecast time series was not determined, results from the literature are indicative. Schmidt et al. (2016) state that persistence forecasts have in total less RMSE than short-term sky imager forecasts, but the benefit of sky imager-based forecasts is a better performance in times with high irradiance variability. Most important for SR is the prediction of cloud events (or nonpersistence in clear sky conditions) at least 2 min ahead which is accurate in 90% of cases for single-point forecasts (Schmidt et al. 2017). When using single-point irradiance (as done in this study), it includes higher ramps than profiles of distributed or large-area PV plants. In terms of forecast accuracy, effects of non-detected small clouds are overexaggerated.

29.4 Conclusions

The results of the conducted simulation studies provide several useful inputs for benchmarking the different approaches of PV integration into remote diesel networks. A basic aspect was the additional benefit of solar short-term forecasts besides the integration of battery storage. It was found that storage capacities and short-term forecasts benefit from each other. Predicted SR is covered by the gensets and reduces battery usage, while storage provides further SR for lacking prediction accuracy. The combination of a comparatively small storage capacity and short-term forecast enables good results for all considered aspects. Applying 100 kWh storage capacity at 40% PV penetration, fuel savings are 18.6% (3.9% more than conventional control), curtailment losses are 9.6%, and lack of spinning reserve occurs 11.3 times per million time steps. At significant larger storage capacities (500 kWh or 1 MWh) and a higher PV penetration of 60%, these parameters further improve (26.0–27.7%, 4.4–10.2%, and no lack of SR). Our work will be continued with economical evaluations. For rating the cost-effectiveness of storage integration and short-term forecast systems in remote PV diesel networks, the investment and O&M cost have to be compared to fuel savings, reduced PV curtailment, and an improved network stability. Moreover, in the case of storage integration, it is necessary to consider further battery parameters (e.g., aging, charging and discharging losses, C-rate, cycles, and DOD) in order to improve the simulation precision.

References

Australian Bureau of Meteorology. (2015). *One minute solar data*. http://reg.bom.gov.au/climate/reg/oneminsolar

Battke, B., & Schmidt, T. S. (2015). Cost-efficient demand-pull policies for multi-purpose technologies – The case of stationary electricity storage. *Applied Energy, 155*, 334–348.

Bourry, F., Do, T. -P., Le Pivert, X., & Riou, M. (2016). Benefits from short-term photovoltaic power forecasts for hybrid system operation. In *International conference on Solar Technologies & Hybrid Mini Grids to improve energy access* (pp. 236–241), Bad Hersfeld, 21–23 Sept 2016.

Fulcrum3D Pty Ltd (Poole, M.). (2016). *Cloud detection and prediction for maximising solar PV utilisation in off-grid hybrid power systems*. Published via https://arena.gov.au/projects/

Hoffmann, W. (2014). PV as one of the major contributors to a future 100% renewably powered world – Importance and evidence for cost effective electricity storage. In *EUPVSEC*, Amsterdam, September 2014.

Horizon Power. (2013). Distribution design manual volume 1 – Quality of electric supply, Document Number: HPC-5DC-07-0001-2012. https://horizonpower.com.au/

Malhotra, A., Battke, B., Beuse, M., Stephan, A., & Schmidt, T. S. (2016). Use cases for stationary battery technologies: A review of the literature and existing projects. *Renewable and Sustainable Energy Reviews, 56*, 705.

Peters, D., Völker, R., Kilper, T., Calais, M., Schmidt, T., Carter, C., v. Maydell, K., & Agert, C. (2016). Model-based design and simulation of control strategies to maximize the PV hosting capacity in isolated diesel networks. In *32nd EUPVSEC*, Munich, June 2016.

Schmidt, T., Kalisch, J., Lorenz, E., & Heinemann, D. (2016). Evaluating the spatio-temporal performance of sky imager based solar irradiance analysis and forecasts. *Atmospheric Chemistry and Physics, 16*(5), 3399–3412.

Schmidt, T., Calais, M., Roy, E., Burton, A., Heinemann, E., Kilper, T., & Carter, C. (2017). Short-term solar forecasting based on sky images to enable higher PV generation in remote electricity networks. In *World renewable energy congress XVI*, Perth, 5–9 Feb 2017.

Chapter 30
Criteria for Sustainable Operation of Off-Grid Renewable Energy Services

Bharat Poudel and Kevin Parton

30.1 Introduction

Significant efforts have been made to achieve off-grid RE penetration globally. In Nepal, our case-study location, about 10% of the population (mostly in remote areas) is connected to electricity by isolated micro-hydro plants (MHPs) and other off-grid energy solutions. Thousands of similar plants are expected to come in operation in the coming years. However, the functionality of such off-grid plants is a major concern among the stakeholders, as many of them are found to be unsustainable in terms of performance, service delivery and/or revenue generation. This research will aim to thoroughly assess the existing practices of the off-grid RE-based electrification initiatives and identify key barriers to sustainable operation. More specifically, it will identify preconditions and criteria (institutional, technical, financial and social) for the off-grid RE services (distributed generators) to be operated in a sustainable way and to be integrated more with local economic activities.

The evidence and data collection phase of the research will be field based. The research will test for degree of association between project attributes and performance using a consistent data set and methods based on significant samples. Key performance indicators for five dimensions of sustainability (economic, social, environmental, technical and institutional) of off-grid RE projects will be defined and measured during the field survey. Overall evaluation and ranking of the sampled cases will be carried out using multivariate analysis techniques. Grounded theory will be employed to identify essential project attributes. Finally, multiple correspondence analysis (MCA) will be performed to estimate relationships

B. Poudel (✉) • K. Parton
Faculty of Business, Justice and Behavioural Sciences and Institute for Land, Water and Society, Charles Sturt University, Bathurst, NSW, Australia
e-mail: bhpoudel@csu.edu.au

A. Sayigh (ed.), *Transition Towards 100% Renewable Energy*,
Innovative Renewable Energy, https://doi.org/10.1007/978-3-319-69844-1_30

between project attributes and performance and to identify key variables which explain sustainable performance of off-grid RE services. The outcomes of the study will be instrumental in terms of designing off-grid RE plans/programmes and other supporting mechanisms. Additionally, the research will help to ensure proper scale and better quality of infrastructure, provision of effective after-sales services and integration with local economic activities.

As the project is at the planning stage, it is only possible to present suggestive methods and tools applicable to solve the aforementioned problems. The rest of the paper is structured as follows: the second section provides a sector overview and research background, while the third section deals with methodology. Some conclusions are presented in the final section.

30.2 Background

30.2.1 Energy and Economic Development

Access to electricity is a precondition for economic development and livelihood enhancement. A new, endogenous theory of economic growth explicitly recognises energy as one of the three factors, capital, labour and useful work (product of energy and efficiency), in a production function (Ayres et al. 2007). At every stage of the economic system, from extraction to finished goods, transportation, services and information/communication, some form of energy is required. The well-being of people in all countries is now measured in terms of the amount of electricity they consume. A 1% increase in per capita electricity consumption is associated with a 0.22% increase in Human Development Index (HDI) (Ouedraogo 2013). The United Nations' Millennium Development Goals (MDGs), the global development agenda for the twenty-first century, and the Sustainable Energy for All (SE4ALL) initiative have played an important role in placing universal access to modern energy services at the centre of the UN's sustainable development goals (Shyu 2014).

The world electricity market is dominated by fossil fuel generators. For example, out of a total of 23,815 kWh electricity produced in the year 2014, 66.7% was supplied by the fossil fuel generators. As a result, about 40% of the total anthropogenic greenhouse gas comes from those generators (IEA 2016a). On the other hand, almost 20% of the total world population (1.2 billion people), who happen to live mostly in rural areas of Africa, Asia and Pacific regions, still do not have any access to electricity (IRENA 2016). Therefore, a deliberate transition to a truly sustainable electricity supply is now considered obligatory because of the finite fossil fuel reserves, climate change impacts and development needs.

30.2.2 Development of Renewable Energy Technologies

The oil embargo in 1973 brought dramatic change to the global energy market. It emphasised the need to reduce dependence on fossil fuel and made more attractive, more sustainable and cleaner energy sources. Since then, various renewable energy (RE) technologies have come into the market. Out of world total primary energy supply (TPES) of 13,700 Mtoe in 2014, 13.8% was produced from renewable energy sources. The share of renewables excluding hydropower used to produce electricity has grown from 1.3% in 1990 to 6.0% in 2014 (IEA 2016c).

The OEDC countries, which represented 45.3% of the world electricity generation in 2014 and are equipped with energy security and climate change supportive policy instruments and financial incentive mechanisms, notably the feed-in tariff, strongly support grid-based renewable systems (IEA 2016b). The non-OEDC countries, which significantly use renewable energy sources (70.1%) for cooking and heating, promote decentralised off-grid renewable energy systems (IEA 2016a). This research primarily intends to deal with some prominent issues of the renewable energy services in the later part of the world.

Off-grid RE systems are generally promoted under a programme/project model and supported for a certain time period with specific donor support strategies. More often, these programmes are focused on their short-term physical targets and outputs rather than long-term sustainability, particularly overlooking after-sales services. Approximately 100 million people (25 million families) across the world presently benefit from one or more off-grid RE systems (IRENA 2015). In Nepal, where the field survey will be carried out, about 10% of the population (mostly in remote areas) is connected to electricity by off-grid micro-hydro plants (MHPs) and other RE systems (AEPC 2014). Thousands of similar plants are expected to come into operation in the coming years. Most of these technologies are believed to be sustainable, anticipating the circular economic effect of income generation through productive use of electricity. Once electricity services are established in a village, local income will be increased so that the paying capacity of users will be increased, lowering the non-payment risk and growing energy demand over time and ultimately ensuring project sustainability (Bhattacharyya and Palit 2016). However, the present functionality of such off-grid plants in Nepal and other parts of the world is a major concern among the stakeholders as many of them have been found to be unsustainable in terms of performance, service delivery and/or revenue generation. A thorough assessment of such efforts needs to be undertaken, and successful models need to be identified, analysed and expanded to benefit more people.

30.2.3 Research Objectives and Purpose of This Paper

This PhD research aims to assess the sustainability of off-grid RE systems and identify key project attributes that influence the performance of such systems. This paper highlights the research approach and methods to determine performance of off-grid RE systems as a measure of sustainability, identifies key project attributes that affect performance and outlines tests for degree of association. It is considered high time to bring this emerging issue of sustainability of off-grid RE systems into the broad picture as a novel approach within the transition towards 100% renewable energy – through the World Renewable Energy Congress XVI 2017 in Perth, Australia.

30.3 Methodology

30.3.1 Justification for Off-Grid RE Systems

In some parts of the world, with supportive policy instruments and technological change, grid-connected renewable energy technologies are commercially viable. As a result, countries like Germany, Australia, Denmark and the UK have been successful in making progress in achieving their renewable energy targets. But we should not forget those 1.2 billion people, mostly in Africa, Asia and Pacific regions, who do not have access to electricity and live in rural areas where extension of the grid is unlikely for many reasons. Many rural communities, often located in difficult geographical locations and possessing the lowest energy demand, are excluded from the mainstream electrification programme (van Ruijven et al. 2012). According to the IEA's recent projections, by 2040, the world electricity market will increase by 30%, but 500 million people will still remain unelectrified. This is a huge challenge for the UN initiative to provide universal access to sustainable energy by 2030. High penetration of sustainable off-grid RE systems may be the only option to close this gap.

30.3.2 Sustainability Assessment of Off-Grid RE Systems

In the context of ex ante evaluation and selection of off-grid RE systems, multi-criteria decision analysis has been extensively used. Various sustainability indicators and criteria have been employed (Mainali et al. 2014; Maxim 2014; Wang et al. 2009). In contrast, the ex post assessment of operational off-grid RE systems using these methods has received less attention in the literature. The latter involves project attributes, which are basically project features that influence the performance of the project, and separately those sustainability indicators that actually measure performance. For example, in the case of a micro-hydro project, type of design is one of the project attributes, whereas reliable supply of electricity is a sustainability indicator. Thus, the aim is to construct and measure project attributes (as categorical variables) and performance indicators (as dependent variables).

30.3.3 Research Approach/Strategies

The research employs both qualitative and quantitative methods. The research will test for degree of association between project attributes and performance using a consistent data set and methods based on significant samples. A final set of sustainability indicators, which represent five sustainability dimensions, will be selected through the predefined selection criteria, and their relative importance will be defined by employing the analytic hierarchy process (AHP). Similarly, the project attributes will be identified through qualitative research methods. Finally, after collection of data from the field, performance of each project will be measured, and a sustainability index of all sample projects will be calculated. Multiple correspondence analysis (MCA) will be performed on a multidimentional contengency table to estimate degree of association between the attributes (categorical variables) and sustainability indicators (dependent variables). The overall research approach and methods are presented in Fig. 30.1.

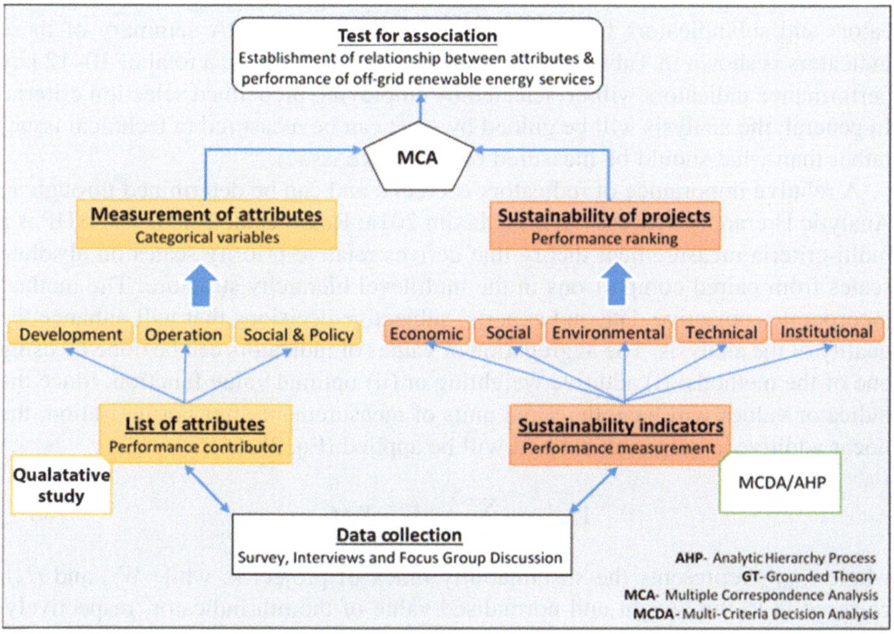

Fig. 30.1 Overall research approach and methods

Table 30.1 List of indicators for sustainability assessment

Economic	Social	Environmental	Technical	Institutional
Generation cost, revenue and profitability, local business and IG activities, appropriate tariff structure and presence of competition	New jobs, universal access, acceptance and willingness to pay, social services and drudgery reduction	Substitution of fossil fuels, local impact, land use, functional impact on landscape, aesthetic impact, noise pollution, other serious environmental impacts	Reliability of supply, plant factor, safety and technical compliances, load compatibility/ grid connection, repair and maintenance services	Local capacity, client relation, inclusiveness, governance and financial audit, accountability, tariff collection and defaulter rate, staff turnover and documentation

30.3.4 Research Methods

There is no standard method for sustainability assessment of renewable energy projects. However, many researchers have utilised multi-pillar models to evaluate the economic, social and environmental dimensions of energy projects. Recent literature emphasises the importance of using technical and institutional aspects of energy projects, particularly for decentralised renewable energy services (Mainali and Silveria 2015; Maxim 2014; Rosso et al. 2014; Singh et al. 2009). This research will consider all five sustainability dimensions of Fig. 30.1 and select subsequent indicators. The literature collectively suggests a large variety of indicators and subindicators for each sustainability dimension. A summary of these indicators is shown in Table 30.1. Considering the limitations, a total of 10–12 key performance indicators will be selected by employing predefined selection criteria. In general, the analysis will be guided by what can be measured (a technical issue) rather than what should be measured (a normative issue).

A relative importance of indicators is crucial and can be determined through an Analytic Hierarchy Process (AHP) (Maxim 2014; Rosso et al. 2014). The AHP is a multi-criteria measurement theory that derives relative priority scales on absolute scales from paired comparisons in the multilevel hierarchy structure. The method incorporates objective data and experts' subjective decisions that will enhance the quality of the analysis. The aggregation of values of indicators can be done by using one of the methods: (i) additive weighting or (ii) optimal value function. Since the indicator values will have the same units of measurement after normalisation, the linear additive aggregation method will be applied (Eq. 30.1).

$$\mathrm{I}_{\mathrm{sust},k} = \sum\nolimits_{n=1}^{i} W_{in}\, X\, (I_n)_k \tag{30.1}$$

where $I_{\mathrm{sust},k}$ represents the sustainability index of project k, while W_{in} and $(I_n)_k$ correspond to the weight and normalised value of the nth indicator, respectively. Once the final sustainability index value of each project is computed, all the projects will be ranked using their aggregated score.

Table 30.2 Major project attributes of off-grid RE systems

Project development phase	Project operation phase	Social and policy issues
Project investment cost and funding sources, system optimisation (capacity and coverage), local contribution (cash, labour), ownership type, social mobilisation and capacity building activities, construction contract type, involvement of local government and provision of maintenance fund	Power output and availability of resources, daily management, power supply schedule, routine maintenance schedule in place, standard set of guidelines (operation) in place and implemented, tariff collection rules and practices, local conflicts, electricity-based business type and monitoring practices	Policy supports (financial and technical), decision making process, leadership selection and transition, credit facilities, support infrastructures and expertise readily available, remoteness, possibility of grid extension, service life of technology, local economic status and operational and business plan of the RE system in place and followed

In this research, qualitative research methods will be employed to obtain key attributes from decision makers. This kind of exercise allows the capturing of a full range of project themes that are common to each case but with different variants. These themes (attributes) will be obtained through interaction with the community which owns and operates the off-grid RE systems. For demonstration purposes only, a list of attributes, collected from the literature and the authors' experiences, is summarised and presented in Table 30.2.

Finally, MCA will be performed on a multidimensional contingency table (the Burt matrix). All selected themes (attributes) are included as explanatory variables, while the dependent variables measure system performance and hence the degree of sustainability. This enables exploration of combinations of variables associated with the performance outcome of each project (Gross and Hagy 2017). Similarly, based on Chi-square statistics, MCA is able to elicit the interdependences among various project attributes and performance levels simultaneously (Wang and Nien 2016). The results will be used to determine whether projects with certain attributes or combinations of attributes can be related to different performance levels (sustainable index).

30.4 Conclusions

The sustainability of operational off-grid RE systems is an emerging issue. The proposed method will be helpful for ex post assessment of operational systems and for planning and designing of new systems. All the statistical methods and tools proposed for this research have been applied in previous studies. The main contribution of the research is to develop and test a method that will fill the current knowledge gap, particularly related to ex post sustainability assessment of off-grid RE systems. The sustainable operation of such systems is essential to environmentally sound power generation particularly in developing countries.

References

AEPC. (2014). A year in reviews, fiscal year 2012/2013, Alternative Energy Promotion Centre, Lalitpur, Nepal.

Ayres, R. U., Turton, H., & Casten, T. (2007). Energy efficiency, sustainability and economic growth. *Energy*, *32*(5), 634–648. https://doi.org/10.1016/j.energy.2006.06.005.

Bhattacharyya, S. C., & Palit, D. (2016). Mini-grid based off-grid electrification to enhance electricity access in developing countries: What policies may be required? *Energy Policy*, *94*, 166–178. https://doi.org/10.1016/j.enpol.2016.04.010.

Gross, C., & Hagy, J. D. (2017). Attributes of successful actions to restore lakes and estuaries degraded by nutrient pollution. *Journal of Environmental Management, 187*, 122–136. https://doi.org/10.1016/j.jenvman.2016.11.018.

IEA. (2016a). *Key electricity trends*. Paris: International Energy Agency.

IEA. (2016b). *World energy outlook*. Paris: International Energy Agency.

IEA. (2016c). *Key renewables trends*. Paris: International Energy Agency.

IRENA. (2015). *Off-grid renewable energy systems: Status and methodological issues*. Abu Dhabi: The International Renewable Energy Agency.

IRENA. (2016). *Renewable energy statistics*. Abu Dhabi: The International Renewable Energy Agency.

Mainali, B., & Silveira, S. (2015). Using a sustainability index to assess energy technologies for rural electrification. *Renewable and Sustainable Energy Reviews, 41*, 1351–1365. https://doi.org/10.1016/j.rser.2014.09.018.

Mainali, B., Pachauri, S., Rao, N. D., & Silveira, S. (2014). Assessing rural energy sustainability in developing countries. *Energy for Sustainable Development, 19*, 15–28. doi:https://doi.org/10.1016/j.esd.2014.01.008.

Maxim, A. (2014). Sustainability assessment of electricity generation technologies using weighted multi-criteria decision analysis. *Energy Policy*, *65*, 284–297. https://doi.org/10.1016/j.enpol.2013.09.059.

Ouedraogo, N. S. (2013). Energy consumption and human development: Evidence from a panel cointegration and error correction model. *Energy, 63*, 28–41. https://doi.org/10.1016/j.energy.2013.09.067.

Rosso, M., Bottero, M., Pomarico, S., La Ferlita, S., & Comino, E. (2014). Integrating multicriteria evaluation and stakeholders analysis for assessing hydropower projects. *Energy Policy, 67*, 870–881. https://doi.org/10.1016/j.enpol.2013.12.007.

van Ruijven, B. J., Schers, J., & van Vuuren, D. P. (2012). Model-based scenarios for rural electrification in developing countries. *Energy, 38*(1), 386–397. https://doi.org/10.1016/j.energy.2011.11.037.

Shyu, C.-W. (2014). Ensuring access to electricity and minimum basic electricity needs as a goal for the post-MDG development agenda after 2015. *Energy for Sustainable Development, 19*, 29–38. https://doi.org/10.1016/j.esd.2013.11.005.

Singh, R. K., Murty, H. R., Gupta, S. K., & Dikshit, A. K. (2009). An overview of sustainability assessment methodologies. *Ecological Indicators, 9*(2), 189–212. https://doi.org/10.1016/j.ecolind.2008.05.011.

Wang, C.-H., & Nien, S.-H. (2016). Combining multiple correspondence analysis with association rule mining to conduct user-driven product design of wearable devices. *Computer Standards and Interfaces, 45*, 37–44. https://doi.org/10.1016/j.csi.2015.11.007.

Wang, J.-J., Jing, Y.-Y., Zhang, C.-F., & Zhao, J.-H. (2009). Review on multi-criteria decision analysis aid in sustainable energy decision-making. *Renewable and Sustainable Energy Reviews, 13*(9), 2263–2278. https://doi.org/10.1016/j.rser.2009.06.021.

Chapter 31
Energy and Material Constraints in India's Economic Growth

Ravi Prakash

31.1 Introduction

The energy challenge for India is enormous. A large population with relatively low income levels, very low per capita energy consumption levels and the imperative to increase the rate of economic growth are all going to put pressure on the demand for energy. This pressure is building up on a system that is already under critical strain, functioning in a situation of shortages. This is discussed in the first part of the paper. A similar and significant challenge lies in materials availability (e.g. metals, minerals, biomass, etc.) to sustain the economic growth in India, which is elaborated in the later part of this paper.

India has witnessed a rapid growth in electricity consumption since its independence. From a meagre 1,362 MW at the time of independence in 1947, the current installed power generation capacity is nearly 308,834 MW (Ministry of Power 2016). For demand projection of energy, energy intensity of GDP is expressed in the form of energy-GDP elasticity, which is the ratio of the growth rates of the two. For India, the electricity-GDP elasticity was 3.0 till the mid-1960s; this reduced to 1.2 during 1991–2000 (Department of Atomic Energy 2016) and has further reduced to about 0.8, as observed through recent historical trends in the Indian economy (Economic Survey 2015–2016). Reasons for these electricity-economy elasticity changes are demographic shifts from rural to urban areas, structural economic changes towards lighter industry, impressive growth of services, increased use of energy-efficient devices, etc. Though electricity intensiveness of the economy has apparently decreased in the last seven decades, the electricity growth requirements will still be high in order to maintain high GDP growth rates.

R. Prakash (✉)
Department of Mechanical Engineering, Motilal Nehru National Institute of Technology, Allahabad, UP, India
e-mail: rprakash234@gmail.com

A. Sayigh (ed.), *Transition Towards 100% Renewable Energy*,
Innovative Renewable Energy, https://doi.org/10.1007/978-3-319-69844-1_31

With the current average elasticity being 0.8, if India's GDP growth is to be sustained at 8%, the electricity growth rate should be 6.4%. However, this needs to be achieved with a restraint in carbon emissions due to climate change obligations as well as supply vulnerabilities of fossil/nuclear fuels, leaving only limited options to the policymakers and the government.

Presently, renewable energy accounts for nearly 15% of India's total installed power generation capacity, i.e. about 46 GW (Ministry of Power 2016). With solar and wind power becoming commercially viable in comparison to marginal mainstream sources (particularly imported coal and nuclear-based generation), India's Intended Nationally Determined Contribution (INDC) aims to base 40% of the total installed power generation capacity on non-fossil fuel resources by 2030 with international support on technology transfer and financing. This includes the government of India's ambitious target of achieving 175 GW of renewable energy by the year 2022 that marks 75 years of India's independence. 100 GW of this is planned through solar energy, 60 GW through wind energy, 10 GW through small hydropower and 5 GW through biomass-based power projects. Of the 100 GW target for solar, 40 GW is expected to be achieved through deployment of decentralized rooftop projects, 40 GW through utility-scale solar plants and 20 GW through ultra-mega solar parks. It also aims to reduce the emission intensity of GDP by 33–35% from 2005 levels by 2030 (Report of the Expert Group on 175 GW RE by 2022, 2015).

31.2 Methodology

For future electricity generation on a large scale (e.g. from solar photovoltaic modules), one should take into account the manufacturing energy investments required for such systems. The net energy reaching the demand sector of the economy would be gross power generation minus the energy investments in the construction and commissioning of various power plants. This is illustrated through Fig. 31.1.

The gross energy G gets reduced by feedback energy F, which needs to be invested in the energy supply system. It is the net energy N which drives the national economic engine to deliver the economic output with an efficiency "f".

The static net energy analysis examines the energy benefits of a single energy conversion plant. It helps in the evaluation of the energy payback time of a power plant, which should be less than the life of the plant if it has to be of any benefit to the society. Similar to economic analysis, a smaller payback time represents a better plant.

The dynamic net energy analysis examines the energy profitability of a programme of several energy conversion plants with different growth rates. It may be pointed out here that the higher the growth rate of an energy programme, the higher would be the energy investments. Such a growth rate would reach a critical value, when accumulated energy investment exceeds the cumulative energy

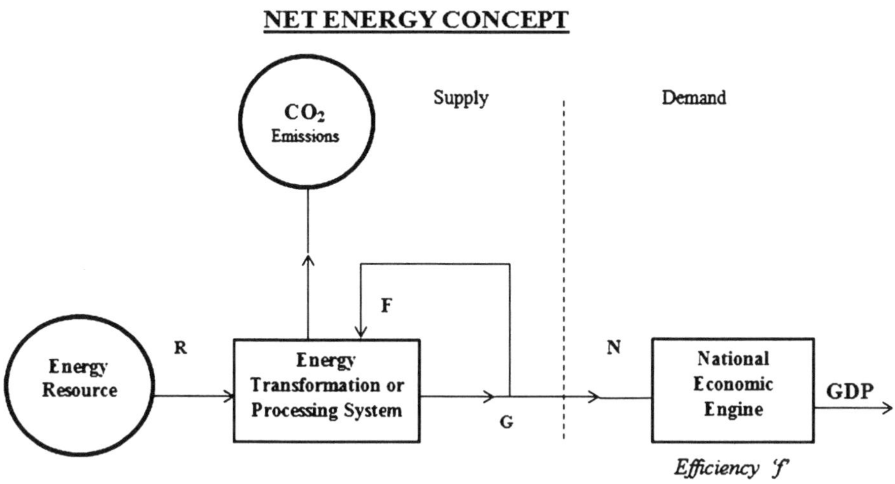

Fig. 31.1 Relation of net energy and national GDP

output, thus rendering the energy programme an energy sink. This would cause an overall reduction in energy availability, which would be detrimental to economic growth.

31.3 Results from Net Energy Analysis

The static and dynamic net energy analysis conducted by the author earlier (Prakash and Bansal 1995) for SPV module production in India provided the following results:

A. The energy payback time for monocrystalline Si module production in India is nearly 4 years.
B. That very high growth rates (beyond 20%) would make the SPV programme an energy sink, as the energy investments in such a programme would be more than the energy yield, thereby negatively impacting the GDP growth.

Hence, with government's particular focus on "Make in India", a very ambitious SPV programme would be counter-productive for India's GDP growth. If the SPV modules are imported, the energy embodied with these modules may apparently be considered an energy gain. However, such imports need to be matched with corresponding exports, whose embodied energy should also be taken into account.

Similar studies may be conducted for other renewable energy programmes, e.g. wind power, which is another important focus area of the government. In view of the reduced net energy benefits at high power generation growth rates, a restraint needs to be exercised for future energy investments in the country. Also, the focus on energy supply should not divert the focus from energy efficiency so as to reduce the demand.

31.4 Drivers of Demand for Materials in India

The structural changes faced by India as an emerging economy – rising population, rising industrial and service-related production, rising middle class, rising income, rising urbanization – will change the resource demand significantly in the country. The need for food, water, fossil fuels, minerals and metals will clearly increase in the coming decades. Following the medium growth scenario of 8%, it is estimated that India would require around 2.7 billion tonnes of biomass, 6.5 billion tonnes of minerals, 4.2 billion tonnes of fossil fuels and 0.8 billion tonnes of metals in 2030; per capita consumption would reach around 9.6 tonnes in that year, which is nearly the current global average. Thus, material consumption in India will triple by 2030 compared to its consumption in 2009 as shown in Fig. 31.2 (IGEP 2013).

Based on the accumulated analyses (IGEP 2013) of the material requirements until 2030 with the base year as 2009, the historic and projected material consumption is given in Table 31.1.

In the light of increased material consumption in India and also globally, a key question arises: from where should these materials be sourced? Materials come either from domestic resources or can be imported as raw materials or as finished goods. Like most countries, India meets most of its demand for resources domestically. Currently, around 97% of all materials consumed are extracted within India, while only 3% are net imports. Thus, India is, on the whole, self-sufficient (but not

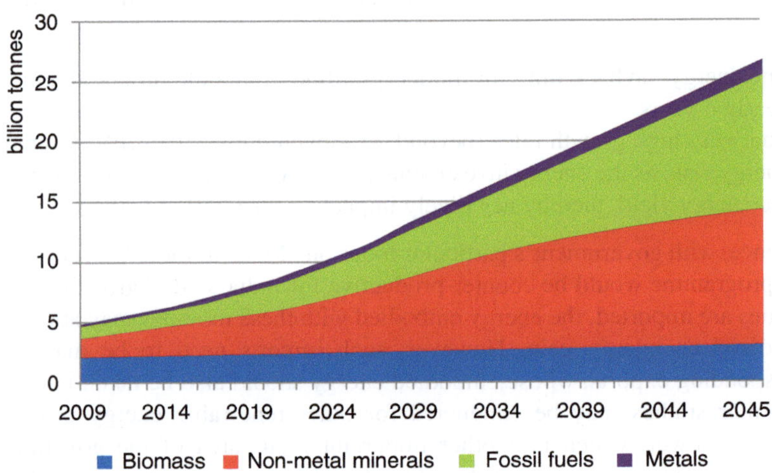

Fig. 31.2 Future material consumption by material categories

Table 31.1 Historic and projected material consumption in India

Material type	Material consumption during 1990–2009	Material consumption during 2010–2030
Biomass	37 billion tonnes	51 billion tonnes
Non-metal minerals	18 billion tonnes	81.6 billion tonnes
Fossil fuels	8 billion tonnes	45 billion tonnes
Metals	2.6 billion tonnes	10.5 billion tonnes
Total	66 billion tonnes	188 billion tonnes

with regard to all materials). The net imports in physical terms equalled a net trade value of US$161 billion in 2011.

While India is, on the whole, still self-sufficient in renewable materials, it is already import dependent in non-renewable materials, in particular with regard to petroleum and specific minerals and metals. For example, India imports 70% of its petroleum and 95% of its copper. With regard to several minerals, India is completely dependent on imports, e.g. nickel, cobalt, antimony, molybdenum and magnesite.

For the three strategic sectors of automobiles, housing and renewable energy, the observations made by the Indo-German study (IGEP 2013) are as follows.

31.5 Automotive/Mobility Sector

Currently, India is the fifth largest auto producer worldwide. Continuing the strong growth of the industry in the past, automotive demand is expected to grow at an average annual rate of 12% until 2025. The number of registered cars in India is estimated to reach 100,000,000 by 2029, along with a corresponding growing material demand for car production, such as steel, aluminium, copper, lead, zinc, chromium and nickel. Consequently, the annual steel demand for cars will account for 10% of total steel production in India by 2025.

Energy use and supply during the usage phase of a compact car constitute the largest amount of extracted primary materials during the entire life cycle of the car. Thus, fuel efficiency, which is closely linked to the weight and design of a car, is the predominant factor in efforts aimed at saving raw materials. The study identifies resource-efficient measures that hold significant material-saving potentials:

- Lightweight compact cars save up to 10% of accumulated raw material demand.
- Full steel recycling reduces primary material demand by 23%, especially chromites and molybdenum ores.
- Public transport has a material efficiency that is five times higher than that of a conventional car, as it has higher occupancy rates.

31.6 Housing/Construction Sector

India's construction sector has been growing at an average annual growth rate of 10% over the past 12 years and is predicted to continue growing rapidly over the next several years. Based on the prospects of housing demand, India's housing stock is estimated to increase from 330 million housing units in 2011 to 770 million housing units in 2030.

The construction and the operational phases of a building have significant raw material footprints, including the consumption of minerals (sand, gravel), cement, steel, bricks and aggregates and energy use – depending on the size and design of a housing unit. For instance, it is estimated that India's proven limestone reserves will be available only for about next 50 years taking into account future prospects for cement demand. The study identifies resource-efficient measures that hold significant material-saving potentials:

- The type of cement used, resource-efficient methods of cement production and efficient usage of cement reduce limestone demand. A promising new material for this purpose is Limestone Calcined Clay Cement (LC^3). Typically LC^3 consists of 50% clinker, 30% calcined clay, 15% limestone and 5% gypsum. The clinker content is about half as much as in Ordinary Portland Cement (OPC), which has 95% clinker. LC^3 technology has several advantages (Development Alternatives Newsletter 2016) over OPC and Portland Pozzolanic Cement (PPC) such as:
  - Reduced emission of CO_2 up to 30% compared to OPC
  - Lower energy consumption even with respect to PPC in some cases
  - Comparable strength with OPC
  - Improved durability
- The choice of building materials determines the material demand during the life cycle of a building. For instance, primary raw material savings of more than 40%, especially in the construction and retrofitting, can be achieved by deploying alternative wall materials (such as aerated cement concrete blocks, fly ash bricks and natural insulation) per unit housing space per person per year. To meet India's housing demand by 2030, the resource-efficient option assessed can save up to 50 billion tonnes of required raw materials compared to the material demand for constructing existing basic buildings, under the assumption that living space remains the same.

31.7 Wind Energy/Renewable Energy Sector

India is the third largest energy consumer worldwide. The renewable energy potential (from different renewable energy sources) in India is estimated to be 245,880 MW. The market for renewable energy is growing at an average annual rate of 15%. Wind energy, which forms the majority of India's renewable energy supply, requires certain key metals such as copper, cobalt and rare earth metals. Despite the high demand for the input materials needed, the potential resource savings using wind energy over coal-based power plants are still much higher. The design of a wind turbine with respect to the tower, magnet and drive system holds potential for raw material savings if smart and efficient design options are adapted.

Similarly, material-saving possibilities with regard to solar cell manufacturing can also be explored through techniques such as thin-film, optical concentrating and multi-junction solar cells.

31.8 Conclusions and Policy Implications

Given a GDP growth scenario of 7–8%, India's energy and material demand will grow significantly. In the light of greater global competition for resources, supply vulnerabilities and rising commodity prices, India will have to find ways to meet this significant growth in demand. In order to reduce India's dependence on imports, and to reduce its environmental and climate burden, the study builds the case for resource efficiency. Resource efficiency comprises all kinds of activities that are aimed at improving the input-output relation of material and energy-consuming or energy-transforming processes, while contributing to the mitigation of impacts on the environment caused by these processes. Such efforts in efficiency improvement will facilitate the decoupling of resource consumption with economic growth.

References

Department of Atomic Energy. (2016). Government of India. http://www.dae.nic.in/?q=node/128

Development Alternatives Newsletter (2016). Development Alternatives. New Delhi. http://www.devalt.org/newsletter/sep16/of_1.htm

Economic Survey. (2015–16). Vol.2. Chapter 1. Table 0.1: Key indicators, Government of India, Ministry of Finance. p. 2.

IGEP. (2013). *India's future needs for resources: Dimensions, challenges and possible solutions.* New Delhi: Indo-German Environmental Partnership. (www.igep.in).

Ministry of Power. (2016). Government of India. http://powermin.nic.in/

Prakash, R., & Bansal, N. K. (1995). Energy analysis of solar photovoltaic module production in India. *Energy Sources (Taylor & Francis), 17*(6), 605–613.

Report of the Expert Group on 175 GW RE by 2022. (2015). National Institution for Transforming India, Government of India. pp. 12–13.

Chapter 32
Magnetic Monitoring of the Dieng Geothermal Area

Dewi Maria Rahayu, Imam Supriyadi, Hilmi El Hafidz Fatahillah, Nugroho Adi Sasongko, Amarulla Octavian, and Yanif Dwi Kuntjoro

32.1 Introduction

Indonesia is a huge archipelago located around the Pacific Ring of Fire. The Pacific Ring of Fire is a major horseshoe-shaped area where numerous earthquakes and volcanic eruptions occur in the basin of the Pacific Ocean. This phenomenon is associated with tectonic activities as subduction at active margin plate boundaries (i.e., Pacific, Eurasia, and Indo-Australia plates). Volcanoes in Indonesia are predicted to be the most active of the Pacific Ring of Fire, with approximately 130 active volcanoes. The eruption of an active volcano poses a natural hazard because of the ash, gasses, and lava. However, volcanoes fertilize the soil with nutrients, attract tourists, and provide geothermal energy.

With numerous active volcanoes in Indonesia, the Ministry of Energy and Natural Resources of Indonesia (2015) has calculated the geothermal potential. According to the latest data from 2014, there are 28,910 GWe of geothermal potential for both resources and reserve scattered from Sumatera Island to Papua

D.M. Rahayu (✉) • I. Supriyadi • A. Octavian • Y.D. Kuntjoro
Energy Security, Faculty of Defense Management, Indonesia Defense University, Bogor, Indonesia
e-mail: dewi.rahayu@idu.ac.id; imamsup@gmail.com

H. El Hafidz Fatahillah
Geophysics, Faculty of Matemathic and Natural Science, Universitas Gadjah Mada, Yogyakarta, Indonesia
e-mail: hilmi.el.h@gmail.com

N.A. Sasongko
Energy Security, Faculty of Defense Management, Indonesia Defense University, Bogor, Indonesia

The Agency for Assessment and Application of Technology (BPPT), Central Jakarta, Indonesia
e-mail: nugroho.adi@bppt.go.id

A. Sayigh (ed.), *Transition Towards 100% Renewable Energy*,
Innovative Renewable Energy, https://doi.org/10.1007/978-3-319-69844-1_32

Island. Installed power plants produce 1403.5 MWe, which only covers about 4.9% of the geothermal potential already developed in Indonesia. Geothermal energy has the potential to meet the national target of 35 GWe by 2025.

One interesting study of geothermal development in Indonesia was a project by Geo Dipa Energy in Central Java. Geo Dipa Energy has successfully obtained 22 MW of 60 MW installed capacity from Dieng Unit 1. Based on the work of Pambudi et al. (2013), a single-flash system was adopted from power generation in the Dieng geothermal power plant. The Dieng geothermal area (Fig. 32.1) is part of the Dieng-Batur volcanic complex located in Central Java, Indonesia. The complex is situated on a broad plateau comprising several volcanic peaks with elevations of 2200–2565 m above sea level, enclosing an upland area (1600–2100 m) of moderate topographic relief (Zen 1971; Harijoko et al. 2010). According to Komarudin et al. (1992), there are three main structures from tectonic deformation that control the Dieng geothermal system; these are, from oldest to youngest:

- East–west trending faults that created the Dieng Plateau
- Northwest–southeast trending faults from the Merdada-Pangonan volcanic complex
- Northeast–southwest trending faults

Geophysical studies have been carried out at the Dieng-Batur complex using gravity, magneto-tellurics, and electrical geophysical methods. Boedihardi et al. (1991) interpreted gravity, magneto-telluric, and well data and showed that the Dieng geothermal field in Central Java can be divided into the Sileri area (northwestern), Sikidang-Merdada area (central), and Pakuwaja area (southeastern). The low permeability and low temperature that show in well data correspond with high gravity and low conductivity regions in the geophysics data. These regions are predicted to act as a barrier that separates Sikidang-Merdada from Sileri and Pakuwaja.

Electrical geophysical measurements were done in 1970 by Group 7 to delineate the boundaries of a subsurface hot water reservoir and its association with the boundaries of conductive areas. The results showed diversified areas with resistivity as contour boundaries. The area with thermal activity was enclosed with a 2.5 ohm-meter (Ω m) contour, whereas the 5 Ω m contour bounded an area with extensive thermal activity. Three areas were indicated within the thermal belt: one is centered around Pagerkandang fumarolic fields, one around the Sikidang crater, and a third south of Dieng Kulon (Zent 1970).

Furthermore, Boediardi et al. (1991) related that drilling has been done in the Sikidang-Merdada area, 4 of 13 wells being productive with capacities of 5–10 MW, produced at 1500–2000 mVD (meters of vertical depth). Another three wells produced steam for a total of about 10 MW from conductive zones between 1500 and 2000 m. Sikidang-Merdada is an interesting location to study as it has the highest potential reserves. In this study, a magnetic method was used to monitor the prospect geothermal area and update geophysics exploration data with another perspective of the physical properties of the Dieng-Batur geothermal area. Corresponding with previous research, magnetic monitoring was conducted in the

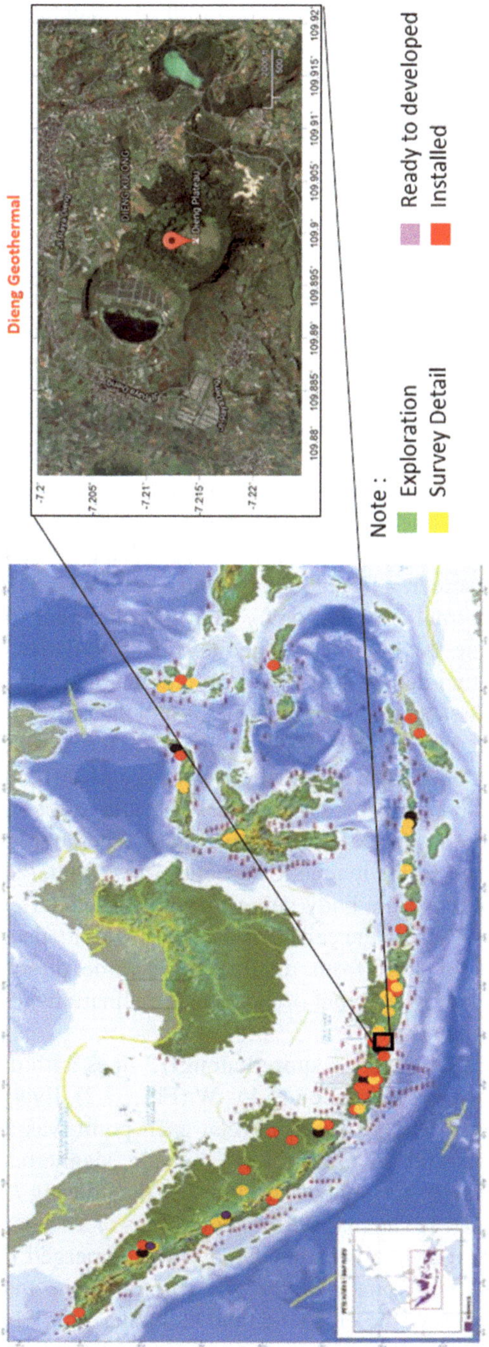

Fig. 32.1 Geothermal potential of Indonesia

Sikidang-Merdada area, around Telaga Balekambang-Kali Tulis, as marked by points in the geological maps in Fig. 32.2. From the geological maps, it can be seen that the research area is dominated by volcanic eruption product in the form of andesite (Qhpa) and that the area is crossed by faults with northwest–southeast and northeast–southwest trends.

32.2 Methodology

Research was carried out using a magnetic method to monitor the geothermal prospecting area. The Dieng-Batur geothermal area was selected as an ideal place for the research, based on a literature review of both geology and geophysics. Magnetic measurement is a geophysics method that is widely used to investigate evidence for geothermal exploration. According to Mariita (2010), variations in the magnetic field can be used to delineate geologic features that relate to hydrogeology (e.g., bedrock lineation, intrusive geologic contacts, heat sources, etc.). Magnetic measurement is not only one of the most cost-effective techniques for screening an area but it is also time-effective. The equipment is handy and simple to operate, making it possible to screen a limited area rapidly.

Evidence of geothermal activity comes from measurement of various physical properties to determine the presence of geophysical anomalies. A magnetometer measures total magnetic intensity (TMI) or resultant magnetic field (Res) as the sum of the main magnetic fields or geomagnetic fields and the effect of magnetization or the source's field at the surface (Fig. 32.3). In Fig. 32.3, the magnetizable body has a spherical form. The effect of magnetization can vary at different points of measurement because of local magnetic anomalies (i.e., the effect of variation in the magnetic mineral content of nearby rock structures). The degree of magnetization of a body is determined by its magnetic susceptibility. Higher susceptibility indicates a higher possibility of the body being magnetized, and vice versa. Measurement along the traverse across the center of the source might also vary, as shown in Fig. 32.3. Variations in TMI from relatively the same direction and point of measurement are caused by different orientations (inclinations) of the geomagnetic field (Dentith and Mudge 2014).

Magnetic data for Dieng were acquired along 12 lines with direction N 225° E and cut cross the graben with direction N 45° W (Fig. 32.2). Every line consisted of 41 measurement points. The spacing between each point was 50 m and spacing between each line was 100 m. The Proton Precision Magnetometer (PPM) Geometric G-856AX was used for collecting magnetic data during *Fisika Gunung Api* Labwork, in January 2015. The magnetometer consists of a sensor and console. The sensor (Fig. 32.4a) contains coils of insulated wire submerged in a hydrogen-rich fluid (Geometrics 2013). Atoms of hydrogen (proton) spin and become aligned parallel with the direction of the magnetic field along the coils while it is electrified. If the electric current is stopped, the protons return to alignment with the geomagnetic field. These changes can be influenced by an external field. Shifting of a

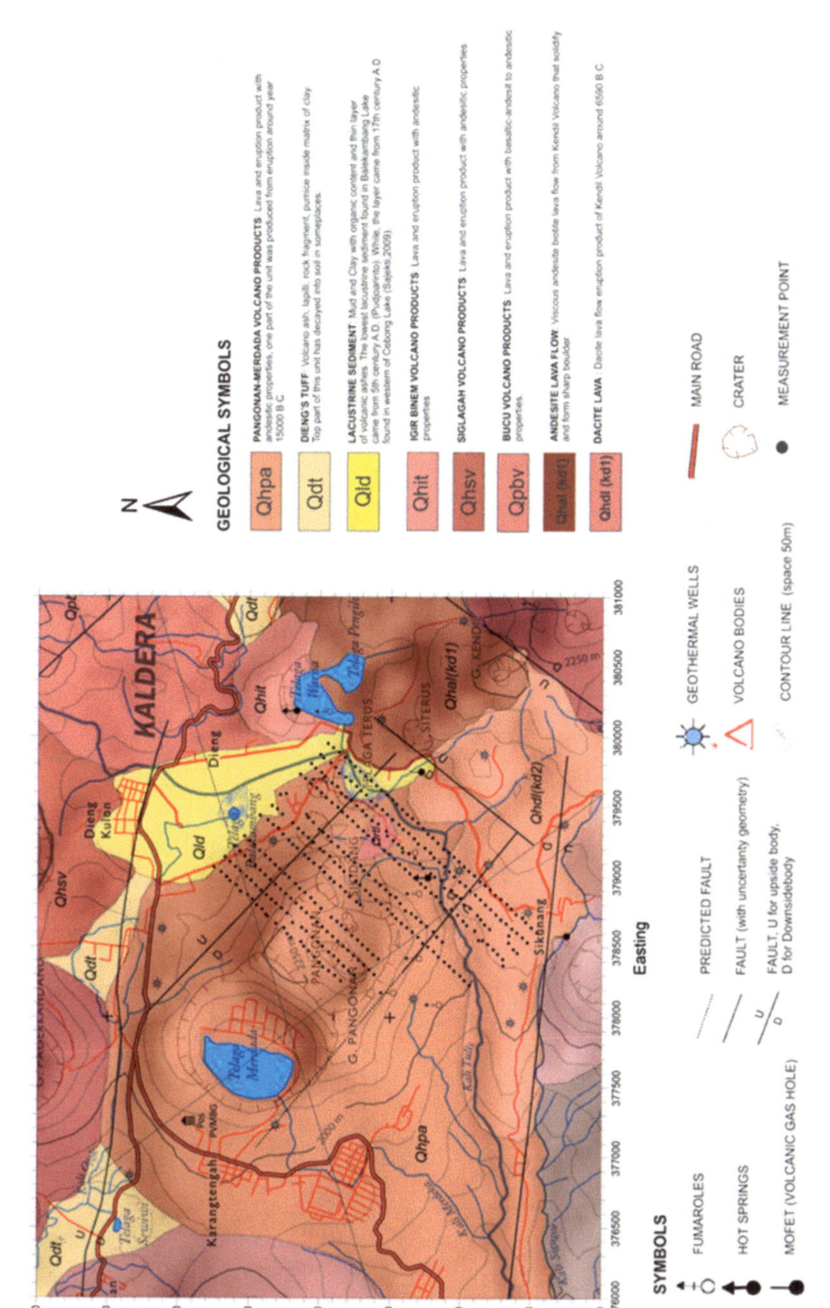

Fig. 32.2 Measurement point positions on the modified geological map of Dieng

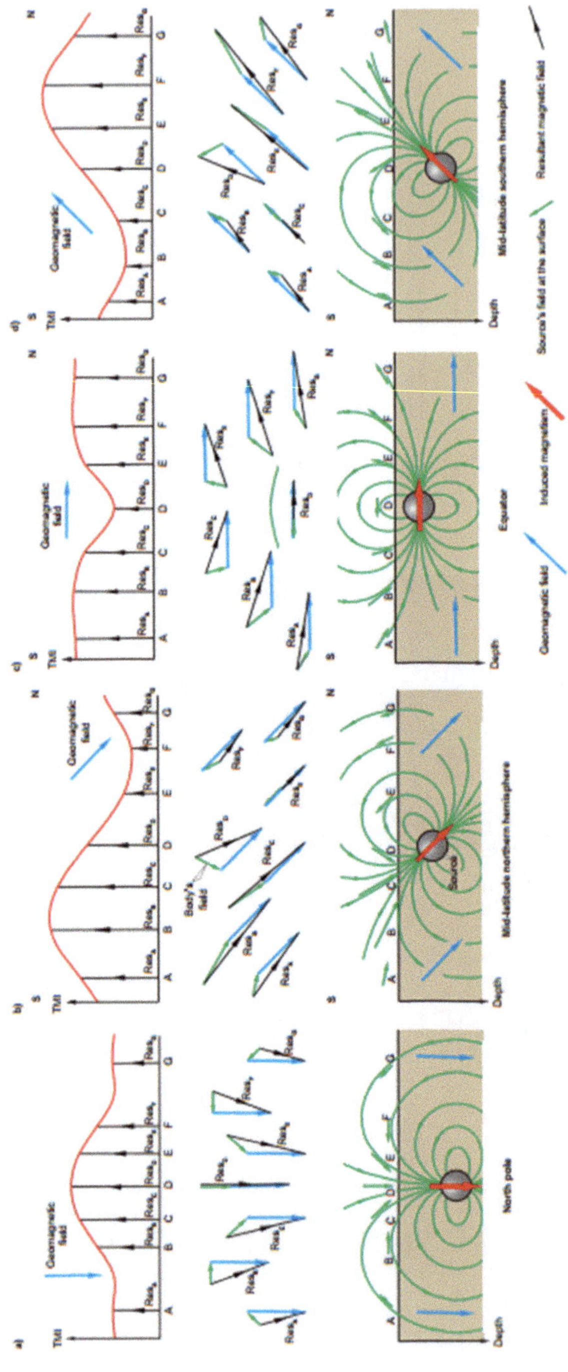

Fig. 32.3 The result of measurement of total magnetic intensity (TMI) using the magnetic method, with variation of point measurement and inclination (Dentith and Mudge 2014)

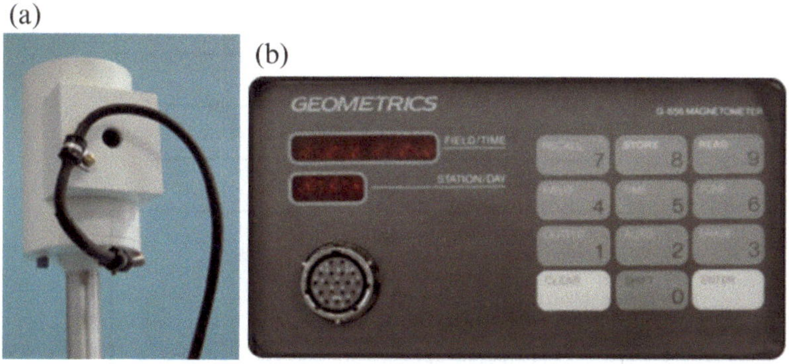

Fig. 32.4 Magnetometer geometric G-856AX (**a**) sensor and (**b**) console (Geometric 2013)

proton's rotary axis direction from an external field to the geomagnetic field is analyzed as a signal of TMI at the measurement point. When measuring a point, the north arrow of the sensor should be directed toward the magnetic north so that the axis of the internal coils in the sensor are perpendicular to the Earth's magnetic field; this produces the optimal signal. The console is used to control and display the magnetic data measured at each measurement point.

Magnetic observation data collected at each measurement point (Fig. 32.2) are the main data needed for further processing. The data are corrected according to the International Geomagnetic Reference Frame (IGRF) and diurnal magnetic data collected at the base point. Processing of magnetic data was done by applying the reduce to pole (TR) and upward continuation methods; a 2.5D geology model was then created. After processing, the magnetic data and 2.5D geology model were interpreted. The interpretation step, using a base assumption of magnetic anomalies, was correlated with geology properties. The potential of a geothermal area could be calculated using the size of area or prospect that corresponded with a magnetic anomaly and its reservoir temperature. The following equation based on Boedihardi et al. (1991) was used to calculate the potential developed energy from a geothermal area:

$$Q = C \times A \times \Delta T \tag{32.1}$$

where Q is potential (MWe), C is a conversion factor (MWe/°C km^2), A is area (km^2), and ΔT is the difference between reservoir temperature and cut-off temperature.

The scope and limitations of this study are illustrated in Fig. 32.5.

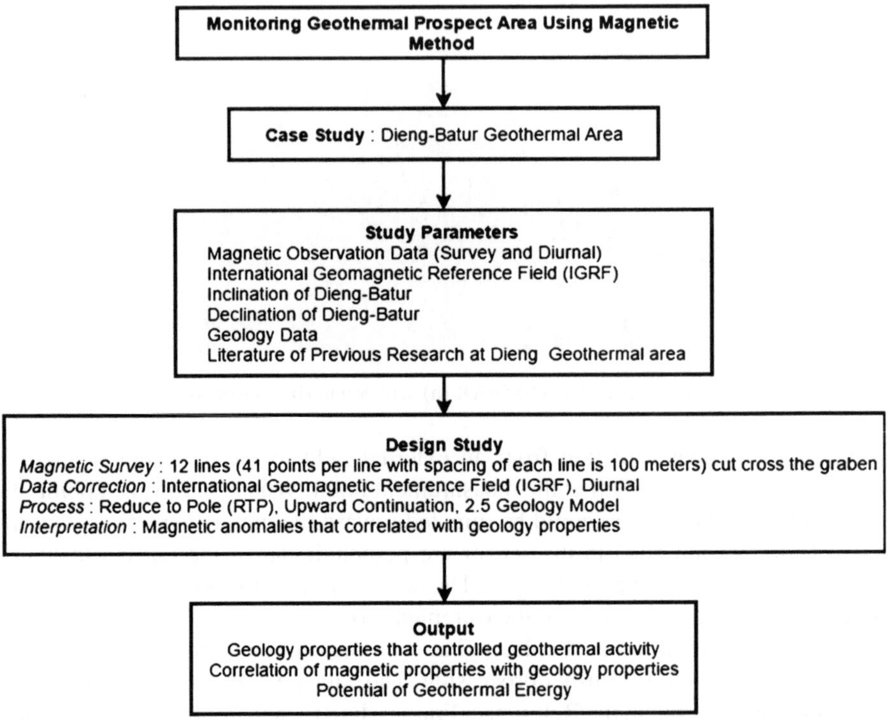

Fig. 32.5 Scope and limitation of the study

32.3 Results and Discussion

The IGRF is a model that is widely used to correct variations in the Earth's main magnetic field given by its spatial point (longitude and latitude) on the Earth's surface. The diurnal correction is related to diurnal variation caused by electric currents induced by Earth's external magnetic field and can be of the order of 20–30 nT with a period of about 24 h. By applying IGRF and diural corrections, anomalies in the magnetic field can be calculated and mapped as a total-field map (Fig. 32.6). For Dieng, the IGRF value is 44,778,1 nT.

The RTP filter was applied to data corrected for IGRF and diurnal variation. This filter requires the magnetic declination and inclination of a field as input data. Dieng area has inclination of $-31.7252°$ and declination of $0.9632°$. In this study, the upward continuation process was applied to the total-field map after the RTP filter. The value of continuation is based on trial and error up to the contour of the total-field map considered stable (100 m). The effect of applying the RTP filter can be seen on the total anomaly-field map after the upward continuation process. The comparison between upward continuation with and without applying RTP is shown in Fig. 32.7.

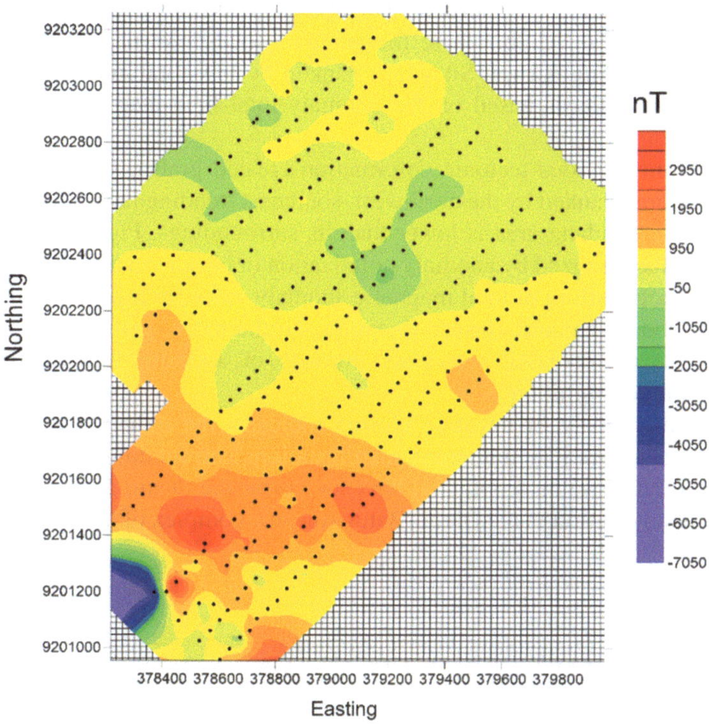

Fig. 32.6 Anomaly total-field map of Dieng

The total anomaly-field map of Dieng (Fig. 32.6) reveals that there are two closures, negative and positive. These closures indicate a pair of magnetic fields or dipoles in the Dieng area. The pair closures in this map are created by the effects of regional and local anomalies. To interpret the regional anomaly, a low-pass filter or upward continuation is applied. The result is a total anomaly-field map generated by objects located below the surface (Fig. 32.6).

Both continuation maps (Fig. 32.7a, b) show two main closures. The total-field anomaly map with RTP processing (Fig. 32.7a) showed low-total anomaly flanked by high-total anomaly. By contrast, the continuation map with RTP processing (Fig. 32.7b) showed high-total anomaly flanked by low-total anomaly. This difference is caused by the effect of inclination and declination of the magnetic poles. Geological information is needed to define which total anomaly-field map best represents the subsurface from both geological and physical analysis.

Based on the geological map of Dieng (Fig. 32.2), the dominant volcanic eruption product is andesitic (Qhpa) in the measurement area. This indicates that high-total anomaly was caused by high susceptibility and ferromagnetic properties. Numerous manifestations such as fumarole and hot springs surrounding Kawah Sikidang resulted in low-total anomaly caused by demagnetization and alteration of the andesitic. Fault structures in this area could result in closure and tight contours of the high-total anomaly in north and south-east areas. However, this presumption is not shown in Fig. 32.8b, where the anomaly total-field map shows high-total anomaly surrounding Kawah Sikidang, flanked by low-total anomaly. Therefore, further interpretation is based on the anomaly total-field map with applied RTP (Fig. 32.7a).

Faults from previous tectonic deformation created graben structures or subsidence in the area flanked by the northwest–southeast trending fault. The topography of the Kawah Sikidang area is lower than its surroundings. Figure 32.7a shows a fault effect characterized by anomaly polarization or closure in the north and south-east, significant shifting of total magnetic anomaly value, or tight contour anomaly at the fault boundary and closure lineation. Furthermore, graben is a permeable zone because of increased secondary porosity of andesitic and can become a fairway for fluid migration.

Low total-anomaly or negative closure indicates the influence of geothermal activity, as evidenced by numerous manifestations surrounding Kawah Sikidang. Volcanic activity can be a potential heat source for the Dieng geothermal system, considering its position in the Dieng-Batur volcanic complex of the late Quaternary.

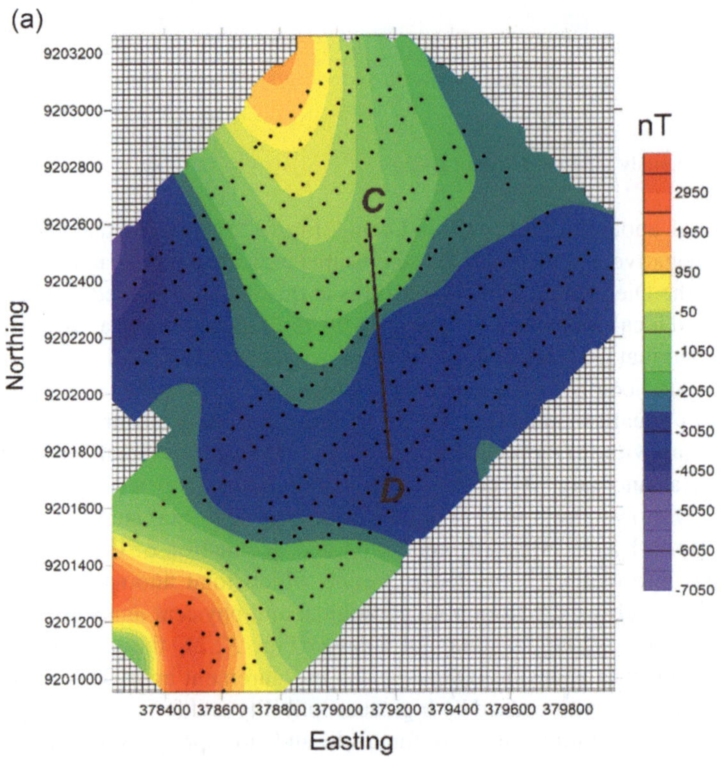

Fig. 32.7 Total anomaly-field applied upward continuation 100 m (**a**) RTP; (**b**) non RTP

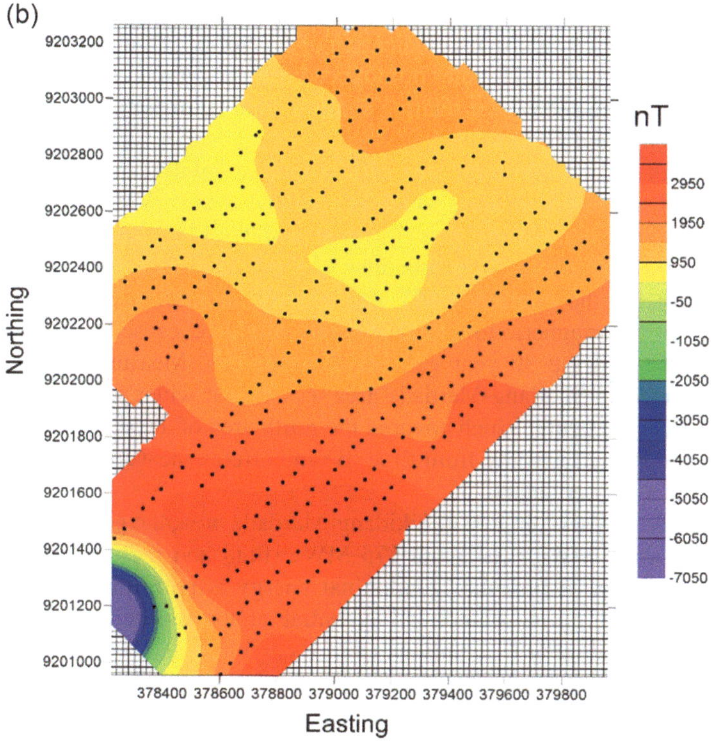

Fig. 32.7 (continued)

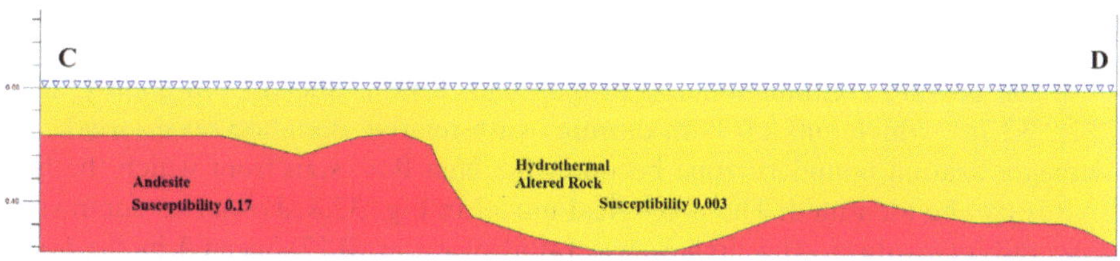

Fig. 32.8 2.5D geological model

The complex consists of numerous extrusions (Zen 1971) consentient with geochemical assessment (Purnomo and Pichler 2014) of Dieng hot springs having high HCO^{3-} and a high Mg^{2+}/Na^{2+} ratio, classified as volcanic and fed by liquid-dominated geothermal systems. Sirait et al. (2015) related heat source to magmatic body/bodies responsible for recent volcanic activity of the Dieng plateau, considering the age of volcanic product in the central and south-east areas to be less than 0.5 million years. The very young age and sub-active nature of the magmatic chamber is the origin of these products. Volcanism has been active in the Dieng plateau since the middle Pleistocene. The large volume of erupted volcanism within the concession area is an indirect indication of the volume of magmatic chamber/s. Because of the location and rather pronounced differentiation of volcanic products, it is possible to hypothesize a relative shallow heat source, probably located at a depth between 5 and 10 km.

On the other hand, the total area of low-total anomaly can be used for rough prediction of the reservoir area, which is around 1.2 km^2. Validation of reservoir size needs to be done using geo-electrical and well-log data. The size of the magnetic low-total anomaly is less than the 6.5 km^2 measured by Boedihardi et al. (1991) for the combined MT and gravity anomalies. The different sizes measured for the magnetic low-total anomaly might be caused by the size of the area surveyed and the 30 year gap between surveys. Measurement of MT and gravity around the Sikidang-Merdada area was over a wider survey area than the magnetic measurements, which covered the north area of Telaga Merdada. Fluctuations in geothermal activity from time to time affects measurement data that are not simultaneously collected.

For a 6.5 km^2 reservoir area at Sikidang-Merdada area, the geothermal potential could develop 90 MWe (fluid) and 175 MWe (fluid and rock) (Boedihardi et al. 1991). Calculation of potential geothermal energy is performed using Eq. 32.1. Furthermore, the optimum electrical potential of the measurement area can be calculated by combining assumptions from Pambudi et al. (2013) and Boedihardi et al. (1991). Prasetio et al. (2010) in Pambudi et al. (2013) state that the range of Dieng reservoir temperatures is 240–333 °C. However, Boedihardi et al. (1991) used 180 °C as cut-off temperature, and the constant C was equivalent to 0.1 (fluid) or 0.19 (fluid and rock). Using Eq. 32.1, the optimum electrical potential of the measurement area is 7.2–18.36 MWe (fluid) and 13.68–34.884 MWe (fluid and rock). The content of gas and steam in the reservoir after 30 years development is estimated at 1–5 %wt and 100 kg/s (Boedihardi et al. 1991).

If the upward continuation 100 m map with RTP is sliced from point C to D (Fig. 32.7a), which cuts off both closures with relative direction north–south, the demagnetization boundary zone becomes the heat flow sign from which the heat source can be identified. The geological model in Fig. 32.8 shows two contrasting susceptibility values. The higher susceptibility value (0.17) located in the lower body of rock is interpreted as andesite, a product of volcanic eruption. The lower susceptibility value (0.003) located in lower body of rock is interpreted as hydrothermally altered rock. This is caused by the heat from a source flowing conductively to the underlying rock. The flowing heat also increases the environmental temperature to significantly below the Curie point. As a reaction to its environment, a demagnetization process occurs in the underlying rocks, decreasing their magnetic properties (susceptibility).

The secondary porosity of andesite is essential for the transfer of hydrothermal heat to an upper layer and probably to the surface (a process known as manifestation). The alteration probably also occurs during hydrothermal interaction with hot rock. Alteration is a complex process that involves changes in the mineral, structural, and chemical composition of the rock. Both alteration and demagnetization cause a decrease in the magnetic properties of hot rock, showing as low-total anomaly. The alteration process is also evidence of a geothermal up-flow zone. This correlates with surface alterations in the Kawah Sikidang area and manifestation as fumarole and hot springs that occur in negative closures.

32.4 Conclusions

1. The total anomaly-field with RTP gives better maps to show both geological and physical features in the Dieng field. Two main closures indicate subsurface structure (fault and graben) and the presence of demagnetization and alteration processes. High-total anomaly or positive closure is caused by high susceptibility lithology such as andesite. Low-total anomaly or negative closure is caused by demagnetization and alteration processes around manifestation.
2. Geothermal activity is clearly visible from cloistered bodies in the geological model, distinguished by their susceptibility. The susceptibility of the lower body is predicted to be about 0.17 as a non-altered body, which is higher than the susceptibility of 0.003 for the upper body, as an altered body.
3. The size of the low-total anomaly indicates that the size of reservoir is around 1.2 km^2, with optimum electricity production of 7.2–18.36 MWe for fluid and 13.68–34.884 MWe for fluid and rock.
4. The geothermal system controlling by heat source as magmatic activity in the Dieng-Batur volcanic complex. Andesite as hot rock and the secondary porosity of andesite as a permeable zone for transfer of hydrothermal fluids in an up-flow zone are shown by manifestation in the surrounding Kawah Sikidang area.

Acknowledgements We would like to thank the Energy Security Department of Indonesia Defense University: Lieutenant General I Wayan Midhio, M.Phil, and Energy Security Co-5 supported, helped, and encouraged us during the preparing and writing process. We thank Mr. Imam Suyanto, M. Irfan Agus Faizal and assistants of the geophysics subdepartment of Universitas Gadjah Mada, who supported and helped us learn about the acquisition and interpretation processes. We also thank Alutsyah Luthfian for help preparing the references and composing the geological and topography maps, and GEOFISIKA 2011 for being the best mates during these 5 years.

References

Boedihardi, M., Suranto, & Sudarman, S. (1991). Evaluation of the Dieng geothermal field; review of development strategy. In *Proceedings Indonesian Petroleum Association, twentieth annual convention.*

Branov, V. (1957). A new method for interpretation of aeromagnetic maps: Pseudogravimetric anomalies. *Geophysics, 22*, 359–383.

Dentith, M., & Mudge, T. S. (2014). *Geophysics for the mineral exploration geoscientist.* Cambridge: Cambridge University Press.

Geometric. (2013). *Operation manual G-856AX memory-mag™ proton precession magnetometer*. San Jose: Geometrics.

Grant, F. S., & West, G. F. (1965). *Interpretation theory in applied geophysics*. New York: McGraw-Hill.

Harijoko, A., Uruma, R., Wibowo, E. H., Setijadji, D. L., Imai, K., & Watanabe, K. (2010). Long-term volcanic evolution surrounding Dieng geothermal area, Indonesia. In *Proceedings world geothermal congress*.

Kementerian Energi dan Sumber Daya Mineral. (2010). *World geothermal potential equivalents to 40,000 GW*. Retrieved October 10, 2016, from Kementerian Energi dan Sumber Daya Mineral: http://esdm.go.id/index.php/post/view/potensi-geothermal-dunia-setara-40000-gw

Komarudin, U., Sunaryo, D., Prayitno, R., & Hansen, S. (1992). Evaluation of geothermal igneous reservoirs. In *Proceeding Indonesian Petroleum Association, twenty-first annual convention* (pp. 607–630).

Mariita, O. N. (2010). *Strengths and weaknesses of gravity and magnetics as exploration tools for geothermal energy*.

Ministry of Energy and Natural Resources of Indonesia. (2015). *Statistic of new and renewable energy 2015*. Jakarta: Directorate of New and Renewable Energy.

Pambudi, A.N., Itoi, R., Jalilinasrabaddy, S., & Khasani (2013). Performance evaluation of double-flash geothermal power plant at Dieng using second law of thermodynamics. In: *Proceedings thirty-eighth workshop on geothermal reservoir engineering*.

Purnomo, J. B., & Pitcher, T. (2014). Geothermal system on the Island of Java, Indonesia. *Journal of Vulcanology and Geothermal Research, 285*(2014), 47–59.

Sirait, P., Ridwan, H. R., & Battistelli, A. (2015). Reservoir modeling for development capacity of Dieng geothermal field, Indonesia. In *Proceedings, fortieth workshop on geothermal reservoir engineering*.

Spector, A., & Grant, F. S. (1985). Statistical models for interpreting aeromagnetic data. *Geophysics, 50*, 1951–1960.

Telford, W. M., Geldart, L. P., & Sheriff, R. E. (1990). *Applied geophysics* (2nd ed.). Cambridge: Cambridge University Press.

Zen, M.T. (1971). Geothermal system of the Dieng-Batur volcanic complex. Proceedings ITB 6 (1): 23–37. http://journal.itb.ac.id/index.php?li=article_detail&id=611

Chapter 33
Sustainability Analysis of Net Zero Emission Smart Renewable Hybrid System Solution in Bangladesh Rural Context

Md. Nafeez Rahman, Md. Rayhan Sharif, Md. Hafezur Rahman Chowdhury, Khondakar Sami Ahamad, and Md. Asaduzzaman Shoeb

33.1 Introduction

Energy plays a vital role for both rural and urban development. Besides, at present time, uses of energy have been increased widely, and among various energy resources, renewable energy utilization is emerging and under development at present time. Moreover, electricity demand is increasing day by day due to urbanization and industrialization. In developing countries, rural areas are still facing problems like electricity outage. Thus, different kinds of power-generating technologies are developing to meet up adequate load demand especially in rural villages. Currently, solar and biomass power is the most in use among all renewable energy resources and is utilizing actively in electricity production and transportation (plug-in hybrid vehicle and electric vehicles) powering. Besides, biofuel production brings an added advantage for both current and future transportation sector, and it will replace the dependency on conventional fossil fuel like oil, gas, and coal. For rural development, biogas utilization will be increased not only for cooking facility but also for electricity production, and it is a better solution along with solar power which will work as active distribution network to meet up rural

Md. N. Rahman
Dalarna University, Falun, Sweden

Md. R. Sharif (✉)
Tampere University of Technology, Tampere, Finland
e-mail: sharif.mdrayhan@gmail.com

Md. H.R. Chowdhury
American International University Bangladesh, Dhaka, Bangladesh

K.S. Ahamad
Chittagong University of Engineering and Technology, Chittagong, Bangladesh

Md. A. Shoeb
Murdoch University, Perth, Australia

A. Sayigh (ed.), *Transition Towards 100% Renewable Energy*,
Innovative Renewable Energy, https://doi.org/10.1007/978-3-319-69844-1_33

and urban electricity demand. Due to high solar intensity and various biomass resources available in rural areas of developing countries like Bangladesh, power-generating infrastructure (power plants based on renewable energy and smart grid system) should be developed to secure electricity and keep rural areas free from power outage problem. This paper has been written to propose a desirable and efficient solution in terms of rural energy development, and it will help to get rid of power outage problem especially in the rural areas of Bangladesh.

33.2 Background

33.2.1 Field Study in Gokulnagar Village, Savar, Bangladesh (Gobar gas – Biogas and Renewable Energy n.d.)

The village has only one main wide road and several side roads in between. Houses are mainly built with galvanized tin. Others have mixed with tin and brick. There are some two or three floor buildings also present there. It has one main market and two small markets. NGO offices and high school are just beside the market. The land is quite low in height from the water level. Many of the houses have foundation which starts from a certain high level on the ground to prevent water access. During rainy season, the spaces get shorter due to lowland. However, in winter and summer land becomes dry but cultivable. There are some private kindergarten schools which are mainly placed inside of living houses. Shops are situated mostly inside the local market, whereas some are placed randomly with necessary goods for households around the village. The comprehensive analysis of the survey suggests that most of the villagers are middle-aged and young people. One could barely find an old man inside the village. The number of teenage people and infants is moderately high. The village is located just near the big town. Thus, they used to get all kind of medical supports and educational supports. However, a quite large number of villagers have come from outside the village nowadays. In addition, few outsiders also come to rent cheap houses for few days or permanently instead of residing in town as the living cost is comparatively high there. The interesting thing that was observed during the survey was that almost all the rickshaw pullers and battery-operated auto rickshaw drivers come from outside the village, while the local village people used to work in nearby town garments to earn more money in a short period of time. Surprisingly the whole of the village area is under LPG gas supply line for cooking, whereas electricity supply and demand are facing totally opposite scenario. People rarely get electricity supply during the peak hours or even whole day. During survey, some questions came up regarding these issues. According to local people's interpretation, the reason was high constant demand in loads of nearby garment factories, big offices, and warehouses in the adjacent town. Rural electrification board of Bangladesh has also agreed with the claim of mass people, but they are unable to take steps due to electricity production shortage. One of the corresponding REB engineers was asked about this issue and he said:

REB is liable to provide constant supply to the customers who're having higher load consumptions. Like, Hospitals, Factories, Multiplexes though there would be any emergency issues. On the other hands, village people use to have comparatively lower load consumption. That's why to balance the supply and demand it's essential to occur load shading in those areas. However, he also claimed that many auto rickshaw pullers are stealing electricity from the phase line to charge up their dc batteries several times in day which exceeds the load requirement of the provided feeder from the distributing line, causes load shading. It's hard to identify or prevent such disturbance, he added.

The essential two things were identified clearly during survey observation. People badly need pure drinking water and continuous electricity no matter what prices are going to be asked for those services. Almost all villagers have own water pumps in their houses regardless their economic situations. But they could barely fill up their tanks fully in a day due to load shading. Currently they used to run the pumps during 2–3 h in a day to fill up their reservoirs two times in a day. The water pumps are in between 600 and 746 W which draws the greater portion of electricity for a house. People have a lot of critics regarding electricity bill to the local utility, like most of the time they use to pay bills more than expectations. But the local utility answered differently when the question was asked. They said, "It's always identified several occurrences of fault in meter boxes. Some dishonest consumers use to tamper the meter boxes to have less amount of charges for the electricity connections. That's why we need to impose strict supervision every now and then. However, electricity bills are reasonable he added at last."

While sitting in a local shop, the shopkeeper said, "The supply of electricity could be said as no electricity at all. I had to repair two of my fridges couple of times in a month due to the discontinues electricity and voltage drops in line. It affects my business and there is no solution for that." It seems the people and the authority have no balanced solution to solve these problems at all. They are just complaining about each other's faults. As in the village, there are different types of communities living together; they have different lifestyles and habits of using energy. They used to throw their waste material such as cow dung outside in their unused lands to make it cultivable for daily vegetables need of their own. Sometimes they dry it and put it in the cooking clay pot instead of wood as a burning material to cook foods. Other waste materials like poultry wastes, leftovers of vegetables, rice husk, maze, etc. are used as food for fishes, or sometimes they use to sell it or just throw it outside as a garbage.

33.3 Methodology

33.3.1 Biomass Feedstock Requirement

In rural areas, there are a lot of resources of biomass which can be utilized by processing biofuel for sustainable development of energy systems. For proposed biogas plant, various kinds of biomass feedstocks have been considered for

Table 33.1 Biomass feedstock requirement for producing electricity

Biomass feedstocks	Quantity (Kg)	Digester size(m^3)	Calorific value (KWh/m^3)	Electrical power (KW)	Total electricity production (in KWh)
Cow manure	4,000–4,500	200–225	6	400–450	1,200–1,350
Food waste/ energy crop/ others	300–1,500	100–500	6	200–1,000	600–3,000

generating biofuel, and it will be utilized completely for power generation. So, feedstock requirement for biofuel is discussed below (Table 33.1) (Biogas Electricity (Small-scale)|SSWM n.d.; Designing of an optimized building integrated hybrid energy generation system – IEEE Conference Publication 2017; Electricity From Cow Dung, Power Generation from Biogas, Gobar Gas n.d.).

Generally, two main types of biomass feedstocks are cow manure and food waste which are available in most of the rural villages. Besides, cow manure will be collected from around 300 or more cows, and each cow is capable of producing 13 kg of manure per day (on average). Moreover, around 300–1,500 kg of food waste will be stored per day to reserve facility of biomass feedstocks, and energy crop will also be reserved to the facility in case there is a shortage of food waste. Necessary quantity of manure will be transported from nearest available places (outside village area) if there is any shortage of cow manure.

33.3.2 Electrical Load Analysis for Biogas-CHP Plant

Different electrical loads are considered and analyzed for figuring out total electrical load demand for the entire rural village. Generally, the loads have been analyzed for houses, village market and shops, mosque, and village schools. Typical loads are ceiling fan, fluorescent light, energy-saving bulb, refrigerator, frost refrigerator, CRT television, LED television, computer, laptop, and water pump for houses, mosques, and village schools. Besides, some heavy load like welding machine has been considered for village markets along with light loads. Below a load table has been provided for a period of 24 h (Table 33.2) (Uddin et al. 2015).

From these loads around 60–70% load demand will be fulfilled by biogas-CHP plant, and the rest will be supported by solar PV power.

33.3.3 Compositions of Chemical Substrates in Biogas

Biomass resources are natural sources of energy. By processing biogas from resources like cow manure, food waste, and energy crops, we will be able to get the quantity of chemical substrates (in terms of %). Generally, chemical substrates

Table 33.2 Hourly load demand of rural village

Hours	Electrical load (KW)	Hours	Electrical load (KW)	Hours	Electrical load (KW)
00.00–1.00	106.746	8.00–9.00	49.616	16.00–17.00	84.69
1.00–2.00	106.746	9.00–10.00	56.535	17.00–18.00	75.03
2.00–3.00	106.746	10.00–11.00	165.947	18.00–19.00	165.002
3.00–4.00	106.746	11.00–12.00	84.747	19.00–20.00	139.971
4.00–5.00	113.22	12.00–13.00	136.82	20.00–21.00	190.717
5.00–6.00	49.146	13.00–14.00	119.962	21.00–22.00	188.881
6.00–7.00	48.93	14.00–15.00	115.314	22.00–23.00	88.406
7.00–8.00	48.87	15.00–16.00	118.84	23.00–24.00	88.406

Table 33.3 Composition of biogas

Biogas composition of cow manure (%)	Biogas composition of food waste (%)	Biogas composition of energy crops (%)
Methane(CH_4), 60–70%	Methane(CH_4), 60%	Methane(CH_4), 68%
Carbon dioxide (CO_2), 25–35%	Carbon dioxide (CO_2), 33%	Carbon dioxide (CO_2), 26%
Hydrogen sulfide (H_2S), 0.1%	Oxygen (O_2), 0%	Oxygen (O_2), 0%
Carbon monoxide(CO), 4.7%	Water (H_2O), 6%	Water (H_2O), 5%
Nitrogen(N_2), 1%	Nitrogen(N_2), 1%	Nitrogen(N_2), 1%
Others, 0–10%		

are composed of methane (CH_4), carbon dioxide (CO_2), nitrogen (N_2), oxygen (O_2), water (H_2O), carbon monoxide (CO), hydrogen sulfide (H_2S), etc. Biogas composition of different chemical substrates for cow manure, food waste, and energy crops is given below (Table 33.3) (Biogas composition n.d.; Electricity From Vegetable, Fruit & MS – Waste, Electricity from Bio-Wastes n.d.).

33.3.4 Proposed Biogas-CHP Plant Design (Figs. 33.1 and 33.2)

This is the design prototype of biogas-CHP power plant for a rural village. Biogas-CHP plant is composed of different components like biomass storage facility, biomass processing units, mixing tanks, water tanks, biodigesters, slurry storage tank, fertilizer processing unit, biogas holder, and CHP plant. For biodigesters, vertical mixing technology will be used, and integrated gas filtering technology will be used for the biogas-CHP plant. Modern fertilizer processing unit will be used for processing fertilized from waste slurry. Biomass resources will be collected from village area, and it will be stored per day to the storage facility of the plant. CHP plant will be located near to the biogas plant. The entire biogas-CHP plant will be

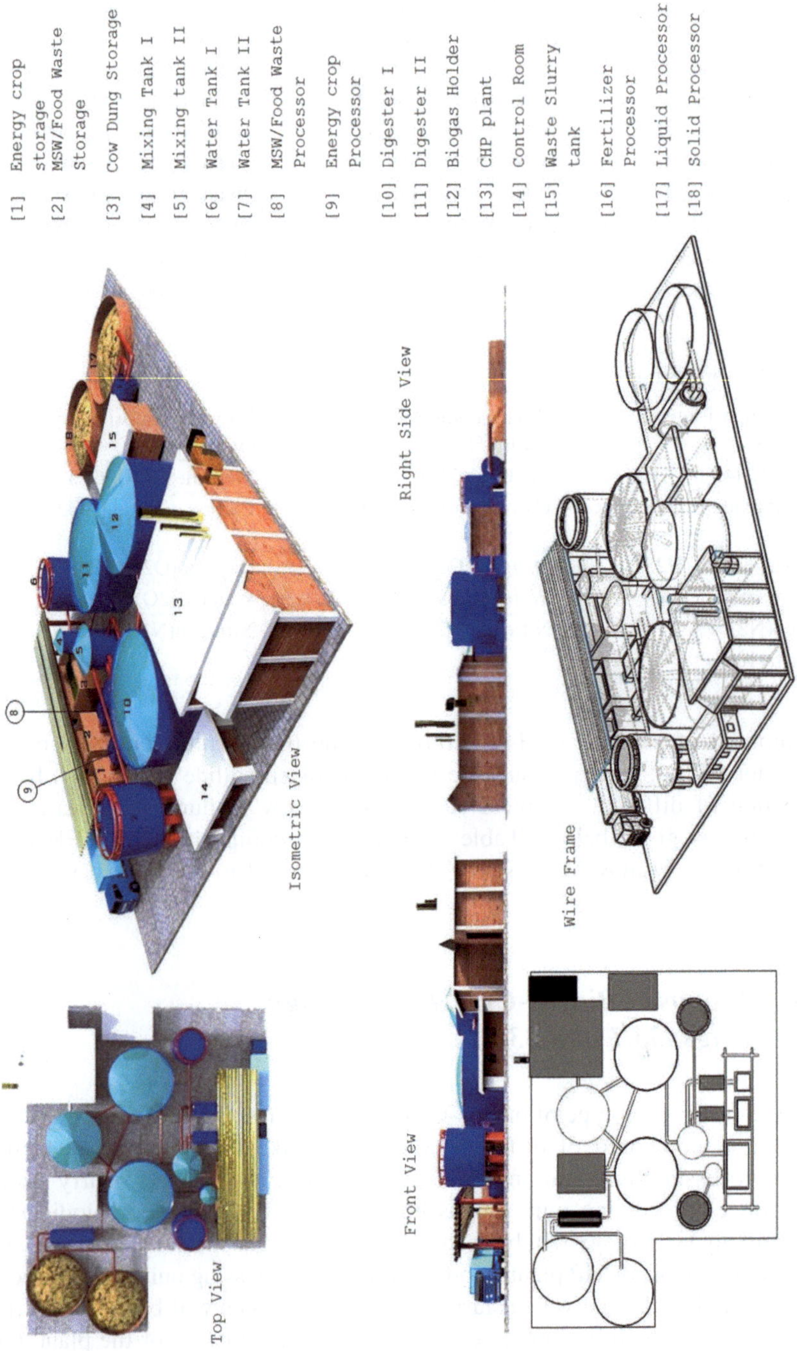

Fig. 33.1 Proposed model of biogas-CHP power plant

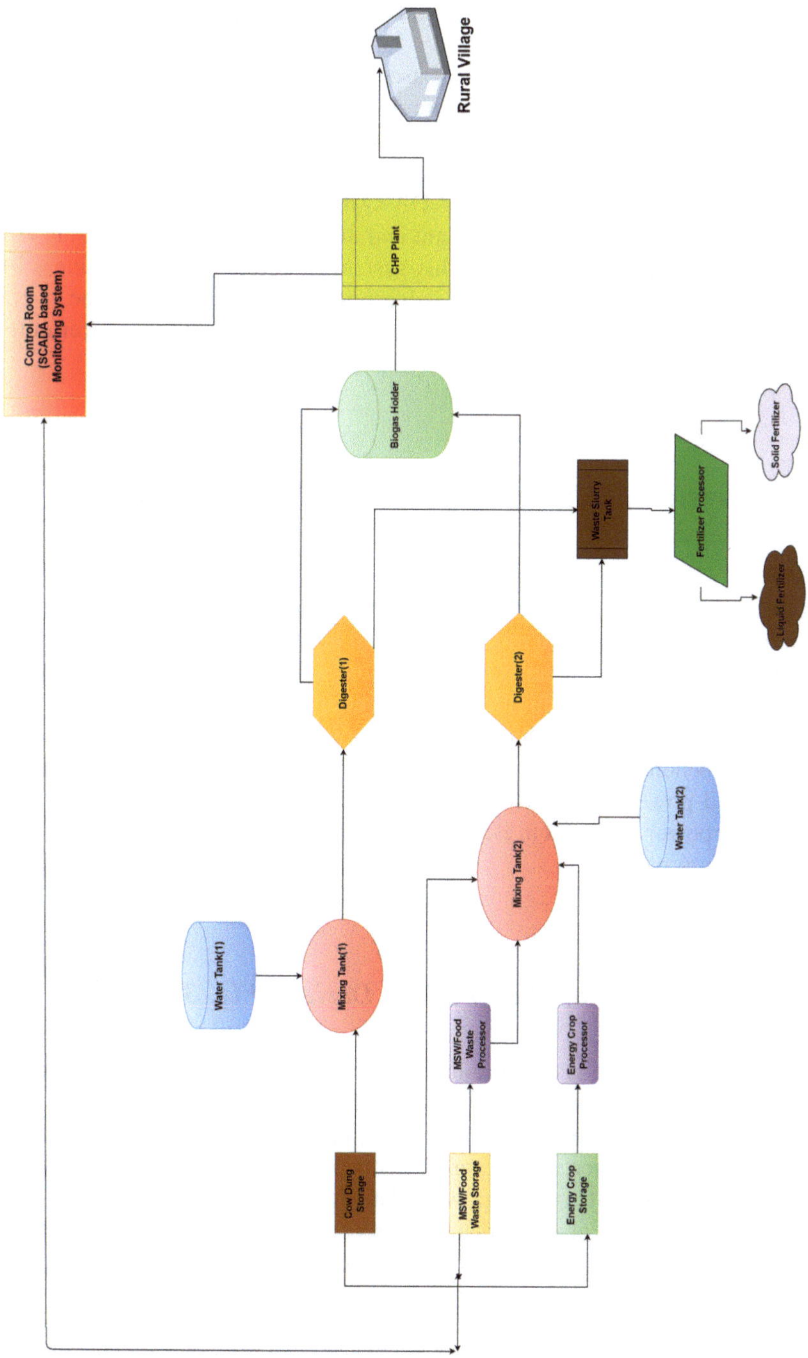

Fig. 33.2 Block diagram of biogas-CHP power plant with SCADA-based monitoring system

located to a suitable place of the rural village. Economic matters and land area requirements are needed to be analyzed before establishing a plant in an allocated area.

For this biogas-CHP plant, SCADA-based monitoring system will be used to monitor the entire process from biomass storage to electrical power output. SCADA (supervisory control and data acquisition) is a unified monitoring system used to control a complete system in industries such as thermal power plant, glass production plant, and nuclear power plant. SCADA system collects the information from plant and transfers the information to plant and display the information using HMI (human-machine interface) program. This system monitors some parameters like gas flow, gas pressure, tank level, and valve position, and it controls the overall system like start/stop by local and remote. It also monitors electrical power output from CHP plant (Abilash et al. 2016).

33.3.5 Overview of Combined Heat and Power (CHP) Plant (Fig. 33.3)

The figure shows conventional biogas-CHP unit for a biogas fuel-operated power plant. Generally, CHP units are more efficient in terms of generating output power than other power-generating units. Efficiency of a CHP plant is more than 80% which is better for utilizing a fuel for getting better output. Benefits of a CHP unit can be defined in four ways which include efficiency, environmental, economic, and reliability benefits. Nowadays, CHP units are installed by a large number of biogas-operated power plants which bring reliability and sustainability as a new technology for power generation. Besides, this type of power generation unit is also utilized by natural gas.

33.3.6 Proposed Solar Power Plant Design (Fig. 33.4)

The above figure shows the conventional solar power plant model for rural village, and it will jointly supply electrical power with biogas-CHP plant to the village loads. Solar power plant consists of several equipment which include PV panels, pyranometer, energy meter, converter system, battery banks, and control units. Generally, control unit of a solar power plant is designed based on SCADA to monitor the performance of solar plant and records data for different parameters like solar power generation, conversion, and energy consumption. Energy meters and other compatible devices have been integrated to the plant system with modern power plant infrastructure.

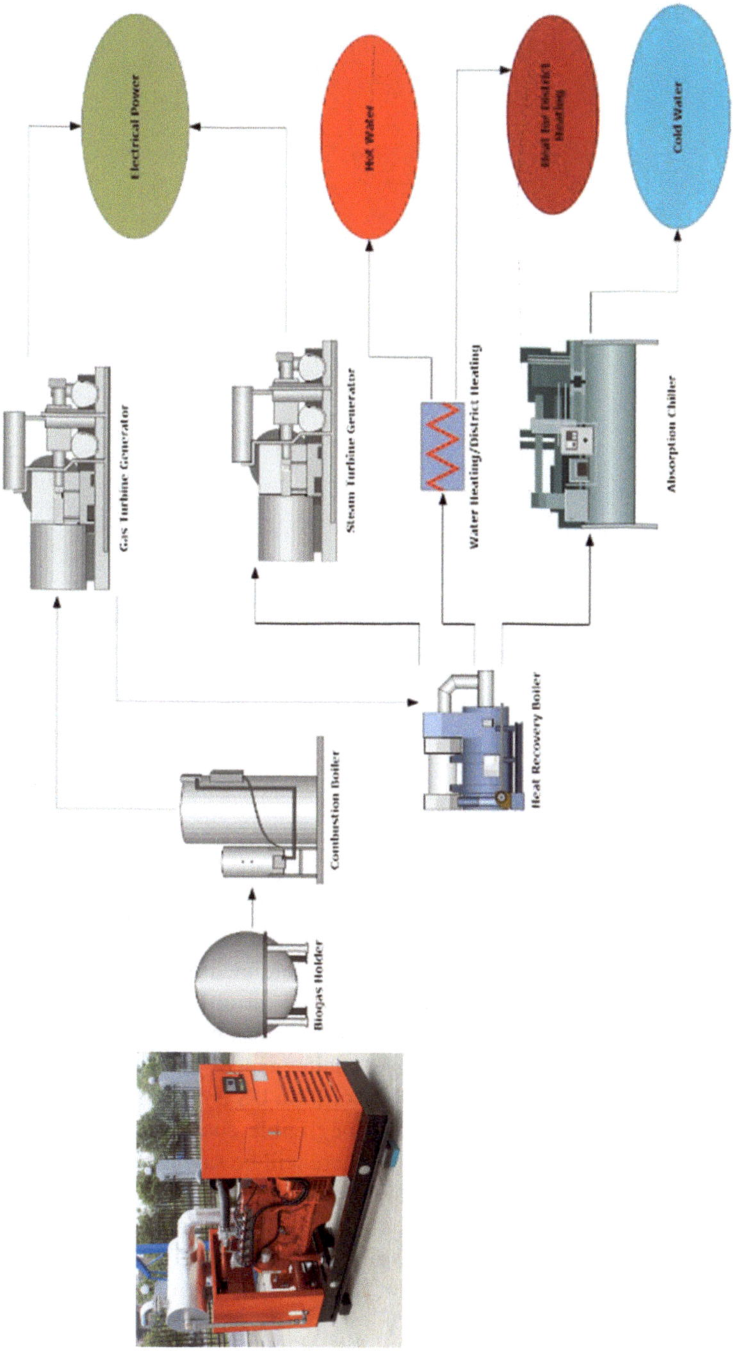

Fig. 33.3 20–500 KW CHP unit (Courtesy: Cummins Generators) and block diagram of CHP plant

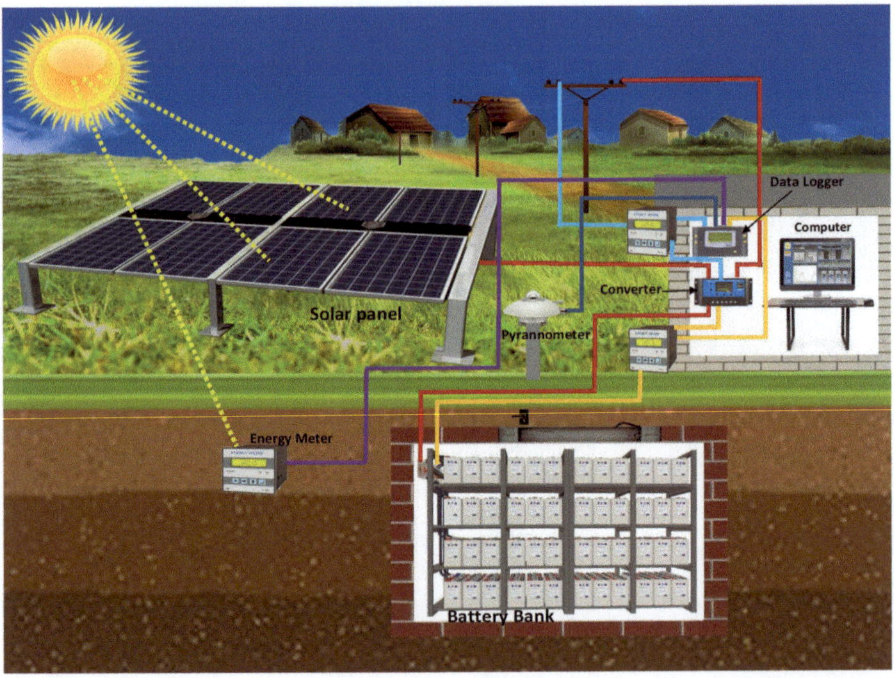

Fig. 33.4 Proposed model of solar power plant

33.4 Results

33.4.1 Simulation Analysis

The system is being designed for 25 years of lifetime. Extensive simulations in HOMER software tool show that the goal of reaching 60–70% from biogas electricity production and 30–40% from solar PV electricity production could be made as expected. It's very convenient to have a concrete idea of net present cost of the whole system and also the cost of operation per year at the first place. The result shows it would take almost 0.2 million US dollars as a net present cost and yearly 4.5 thousand US dollars for operation and maintenances and 0.017 USD/Kwh for COE (cost of electricity). After around 0.14 million US dollars of capital expense in the beginning, the system will run almost negligible cost of others (replacement, fuel, operating). At the end of the lifetime, there will be salvage value of the system as a benefit. The replacement cost of Vision 6FM200D converter takes large portion in capital cost. However, all components are available in local territory which ensures quick maintenance measure while it is needed. Total electrical energy production will be 1,018,509 Kwh/year, and the demand of electrical energy is around 889,143 Kwh/year.

33.4.2 System Architecture (Table 33.4 and Fig. 33.5)

33.4.3 Electrical Production and Consumption, Cost Summary (Tables 33.5, 33.6, and 33.7, Figs. 33.6 and 33.7)

The measure of internal rate of return (IRR) from nominal cash flow indicates negative value which means the sum of the post-investment cash flows are less than initial investment. That is quite relevant for self-sustaining off grid system. Annual savings for the system can be obtained by comparing the yearly amount of cost reduction from the cash flow graph which are very large quantities. So, it may conclude that the sustainability of the system is high due to large sum of annual savings although there are no initial paybacks during system lifetime.

Table 33.4 System configuration

PV array	180 kW
Biogas-CHP	120 kW
Battery	200 vision 6FM200D
Inverter	80 kW
Rectifier	80 kW
Dispatch strategy	Cycle charging

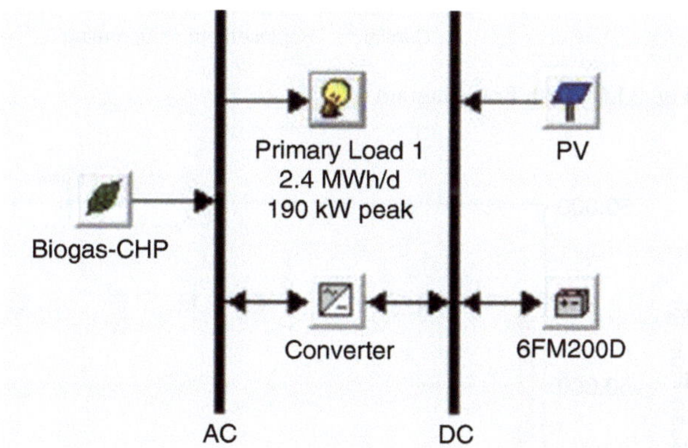

Fig. 33.5 HOMER representation

Table 33.5 Electrical production

Load	Consumption (kWh/year)	Fraction (%)
AC primary load	889,143	100
Total	889,143	100

Table 33.6 Electrical consumption

Component	Production (kWh/year)	Fraction (%)
PV array	324,815	32
Biogas-CHP	693,693	68
Total	1,018,509	100

Table 33.7 Cost summary

Total net present cost	$198,757
Levelized cost of energy	$0.017/kWh
Operating cost	$4,440/year

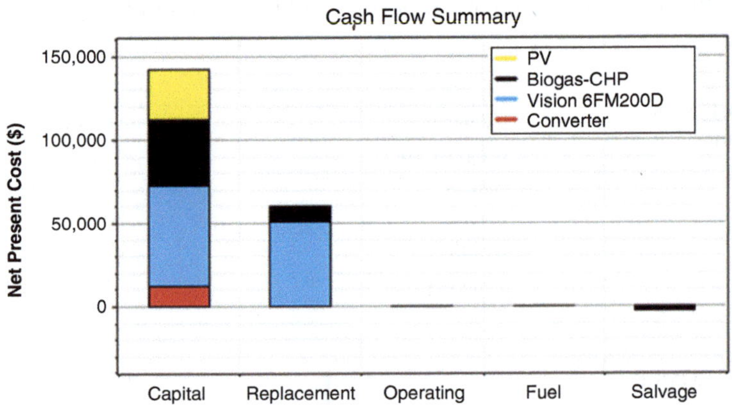

Fig. 33.6 Cash flow diagram

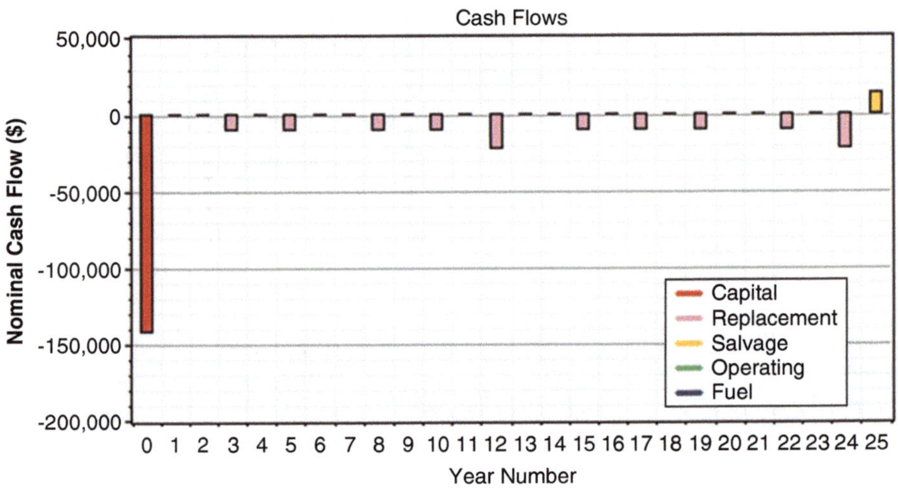

Fig. 33.7 Net present cost

The final factor is the estimation of risk factor which is considered to be quite low as the local surveys had made regarding social, political, energy issues quite elaborately before the starting of the system design. The reason was to ensure local people opinions and expectations involvement since the system is going to provide service to those community. The whole planning of designing part was then initiated from the real demand of that village community in Bangladesh. However, analyzing the occupation status and financial condition, the system was organized as such those people could involve in with their doings directly. Thus, the system itself will be sustainable with no risk at all as it includes more local business people to the system organizing network.

33.5 Conclusion

Nowadays, solar and biomass is one of the best sustainable sources of energy, and these have diverse applications not only in power generation purpose but also in transportation sector. Besides, these kinds of power plants would be a best solution for rural areas to meet up electricity demand, transportation charging facility, and advanced irrigation systems. Moreover, these kinds of power generation unit will be highly useable as distribution generating (DG) unit for smart grid system not only in developed countries but also in developing countries. So, as a reliable and sustainable source of power-generating system, biogas-CHP and solar PV system will play an enormous role not only in grid power but also for electric vehicle (EV) charging station. So, finally we can say that solar and biomass power are the efficient contributors of energy technology to meet up future energy demand and will bring one of the best solutions for rural areas to meet up power outage problem and help to bring reliability and sustainability as environment-friendly power generation system.

References

Abilash, D. A., Kayalvizhi, P., Rakesh, R., & Balamurugan, S. (2016). Automation in biomethanation plant using PLC and SCADA. *International Journal of Bio-Science and Bio-Technology, 8*(2), 171–180. 10.14257/ijbsbt.2016.8.2.16.

Biogas composition. (n.d.). Retrieved from http://www.biogas-renewable-energy.info/biogas_composition.html

Biogas Electricity (Small-scale)|SSWM. (n.d.). Retrieved from http://www.sswm.info/content/biogas-electricity-small-scale

Designing of an optimized building integrated hybrid energy generation system – IEEE Conference Publication. (2017). Designing of an optimized building integrated hybrid energy generation system – IEEE Conference Publication. [ONLINE] Available at: http://ieeexplore.ieee.org/document/7421498/

Electricity From Cow Dung, Power Generation from Biogas, Gobar Gas. (n.d.). Retrieved from http://www.perfectbiowastepower.in/electricty-from-cow-dung.htm

Electricity From Vegetable, Fruit & MS – Waste, Electricity from Bio-Wastes. (n.d.). Retrieved from http://www.perfectbiowastepower.in/electricity-from-vegetable-and-fruit-waste.htm

Gobar gas – Biogas and Renewable Energy. (n.d.). Retrieved from http://www.care-india.com/gobar-gas.html

Uddin, M., Rashed, M., Rahaman, M., Islam, M., Mobin, M. (2015). A survey study on eco-friendly Gokulnagar Village of Savar, Dhaka. *Journal of Environmental Science and Natural Resources, 7*(1). doi:https://doi.org/10.3329/jesnr.v7i1.22177

Chapter 34
New Highly Thermally Conductive Thermal Storage Media

Samuel Reed, Heber Sugo, and Erich Kisi

34.1 Introduction

The implementation of renewable energy systems such as solar PV and wind energy across the globe is providing electricity through an environmentally safer alternative and as such is decreasing carbon-based emissions from fossil fuel-fired power stations. Given the intermittency of the most renewable sources of energy, the need for large-scale energy storage is the largest impediment to baseload renewable energy. One of the more promising solutions is thermal energy storage (TES) within concentrated solar power (CSP) plants. TES systems enable heat to be accessed during intermittent periods and also when no energy is being generated due to the diurnal cycle. Currently many materials have been tested for TES, including concrete, phase change materials (PCMs), rock beds and, currently the most common TES material, molten $NaNO_3$-KNO_3 salts. These and other TES materials have been extensively reviewed (Xu et al. 2015; Rathod and Bannerjee 2013; Laing et al. 2012; Khare et al. 2012; Kenisarin 2010) and their performance evaluated (Khan et al. 2016).

Molten salt storage is the industry standard and has been implemented within a number of CSP plants around the world. The implementation of molten salt TES has allowed for thermal energy to be stored during daylight hours and accessed during the evening peak electricity load and as a result has directly progressed TES CSP plants. Molten salt TES uses sensible heat over a wide temperature range. The relatively low thermal conductivity of the salt is overcome by using very high pumping rates and high contact area heat exchangers leading to considerable plant complexity and elevated installation and maintenance costs. The considerable

S. Reed (✉) • H. Sugo • E. Kisi
Structure of Advanced Materials Group, School of Engineering, University of Newcastle, Callaghan, NSW, Australia
e-mail: samuel.reed@uon.edu.au

A. Sayigh (ed.), *Transition Towards 100% Renewable Energy*,
Innovative Renewable Energy, https://doi.org/10.1007/978-3-319-69844-1_34

latent heat of fusion is not used in this type of plant. In addition, there is additional complexity and parasitic energy loss of up to 24% due to the need to maintain the salt in the molten state (Relebohele and Dinter 2015). These characteristics suggest that an alternative approach based around solid thermal storage utilising some latent heat of fusion may provide high-performance, lower-risk TES within CSP plants.

Recent studies at the University of Newcastle have concerned an alternate method to store thermal energy through the combination of certain metals so as to exploit the immiscible areas of their binary phase diagrams (Sugo et al. 2013). Miscibility gap alloys (MGA) are two-phase combinations. One phase is a high thermal conductivity, thermodynamically stable matrix which encapsulates a lower melting temperature metal. This inverse microstructure relationship is achieved through simple powder metallurgy techniques. Unlike normal TES media, MGA have the ability to store energy as both sensible and latent heat. Latent heat storage is accomplished by the dispersion of the active phase (low melting point metal) throughout the matrix. The active phase undergoes a melting phase transition upon heating. The ability to store energy in this manner has many advantages including a narrower band of operating temperature (say ±50°C) around the melting temperature of the active phase. This is valuable for controlling steam quality and maintaining turbine efficiency. It also provides the ability to manipulate the materials used so as to choose to store/deliver thermal energy anywhere in the range 123–1100 °C, which allows custom implementation into already existing infrastructure.

Miscibility gap alloys are manufactured using powder metallurgy. Metal (or semimetal) powders are mixed to a desired volume ratio, set in a die and compressed using a hydraulic ram. This method forms a solid component which is referred to as a green compact. To increase the strength and durability of the pressed sample, it is sintered in an inert atmosphere at the desired temperature. The sintering temperature is dependent on the components used within the MGA. The sintering process is crucial as it allows diffusion to occur within the sample, which strengthens the component and allows the matrix to support any thermal expansion that occurs due to the melting of the active phase. Recently, several carbon-based MGA systems have been identified as possible candidates. As carbon has an extremely high melting temperature, the true sintering temperature is approximately 2000 °C. To facilitate lower-temperature manufacture, binders are used (Copus et al. 2017). To stabilise the microstructure, most MGA are sintered just below the melting temperature of the active phase.

Since the identification of the first MGA (Sugo et al. 2013), extensive studies have been undertaken to identify alternate combinations of materials that exhibit the same properties in a cost-effective way. It has been established that the C-Zn system is viable with an operating temperature band around 419 °C (Jackson 2015). In order to increase potential working fluid temperature and turbine efficiency, three systems, C-Al, C-Mg and C-Cu, operating at >600 °C were studied in this work.

34.2 Methodology

The systems investigated, C-Al, C-Mg and C-Cu, all have C in the form of graphite as the matrix phase due to its good thermal stability and high thermal conductivity. These combinations were selected for several reasons. The primary reason is the immiscibility that all of these active phases exhibit within carbon meaning that upon mixing and sintering, both materials keep their individual properties and neither form an intermetallic compound nor a solid solution. This is demonstrated for the C-Cu system in Fig. 34.1a where the solubility of C in Cu is shown to be ~0.0078 wt% at equilibrium. The remainder of the diagram below 1085 °C contains a simple mixture of C and Cu and so any ratio of the two elements is suitable. In order to avoid percolation, a 50:50 volume ratio was chosen for the C-Cu system and also for the C-Mg system which behaves similarly. The C-Al system, shown in Fig. 34.1b, is somewhat different as at equilibrium it *can* form the carbide phase Al_4C_3 and therefore has only a metastable miscibility gap. The carbide is undesirable as a constituent of a C-Al MGA thermal storage material as it locks up the Al latent heat phase in a high melting point compound. Fortunately, at the operational temperature of such an MGA (600–720 °C), the kinetics of carbide formation are quite slow. There are suggestions in our preliminary studies that this arises because the first carbide to form on the outside of the Al particles acts as a diffusion barrier preventing further reaction at low temperature. Due to this possibility, the compositional ratio was adjusted to 80:20 as an initial trial.

Other promising properties of these systems are high thermal conductivity, large latent heat of fusion and high volumetric heat capacity which enlarges the temperature range in which thermal energy can be stored. The predicted thermal properties are shown in Table 34.1 compared with the industry standard molten nitrate salt storage material.

Raw materials used in the manufacture of all MGA samples were graphite powder (<45 μm, Alloys International), Cu powder (<85 μm, Alloys International) and Al powder (<75 μm, Alloys International). In order to maintain consistency when manufacturing, all samples weighed 10 g. This standardisation allowed easy comparison of properties and minimal material wastage. Many samples were produced for each system to refine the manufacturing process and assess whether samples were being manufactured reliably and reproducibly. Table 34.2 documents the preparation details for the final suite of samples.

All samples were prepared by hand mixing with 20 vol% of a sodium silicate binder followed by cold uniaxial pressing in a 19 mm diameter die to form green pellets. The pressing pressure was kept constant and is documented in Table 34.2. Binder dry-out was initially at 80 °C for 24 h followed by weighing. As there was no weight loss following a second 24 h drying period, it was concluded that all free water had left the samples after the initial 24 h. Sintering was conducted on the Al foil-wrapped samples under an argon atmosphere under the conditions given in Table 34.2.

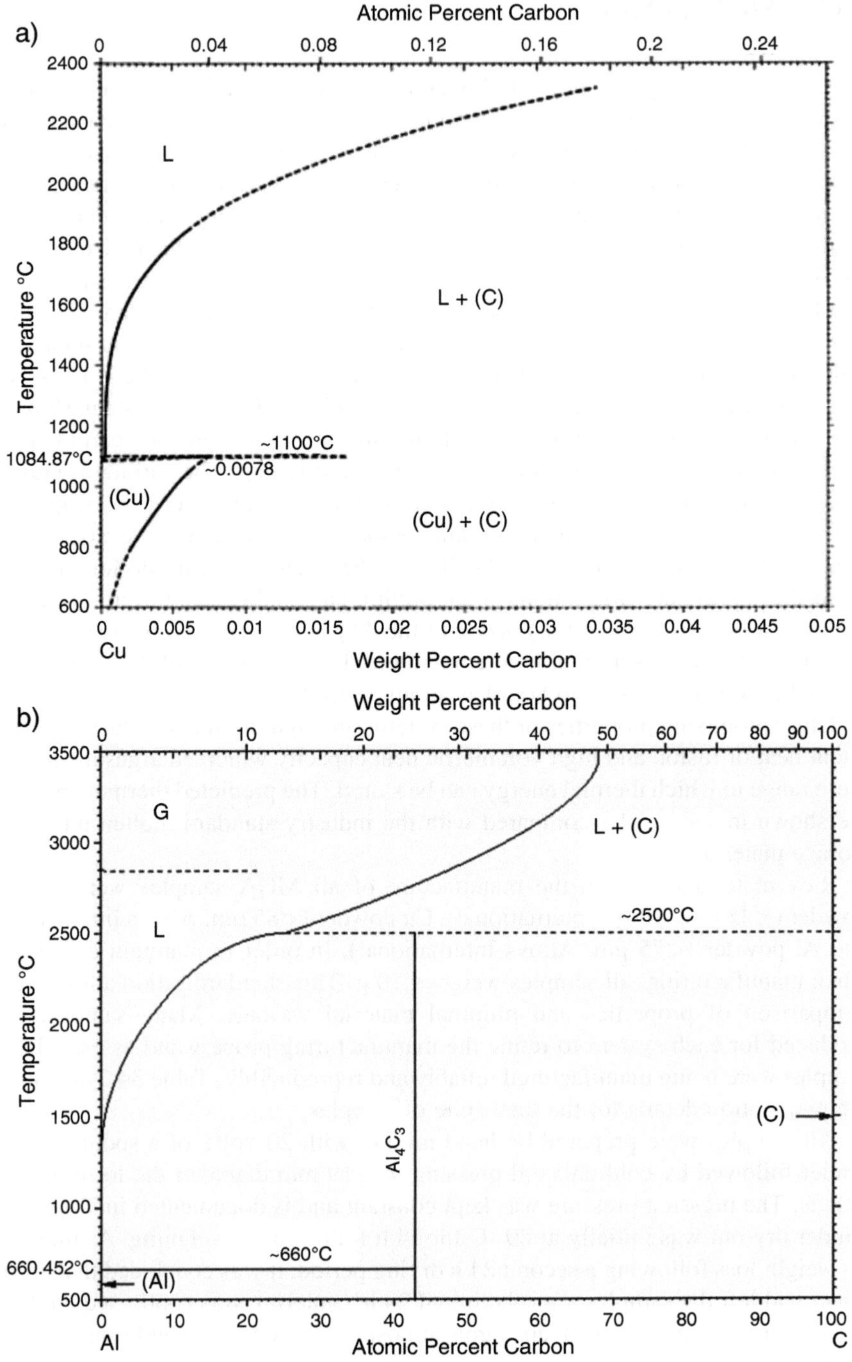

Fig. 34.1 Binary phase diagrams for (**a**) C-Cu and (**b**) C-Al systems (Massalski 1990)

Table 34.1 Thermal storage properties for new MGA systems (Rawson et al. 2014)

System	Phase change (°C)	Heat of fusion (kJ/L)	Energy density (MJ/L)	Thermal conductivity (W/mK)
C-Zn	420	843	0.65	100
C-Mg sensible heat	650	640	0.51	125
C-Al	660	1075	0.75	158
C-Cu	1085	1837	1.26	120–165
Molten salt	N/A	161	0.204	<1

MGA energy density data are for 50 vol% of the melting phase plus 100 °C of sensible heat. Heats of fusion are given for the pure melting phase. Molten salt data are for a temperature range of 100 °C

Table 34.2 MGA sample manufacturing parameters

Constituents	Volume ratio	Mass ratio	Pressing pressure (MPa)	Sintering temperature (°C)	Hold time (h)
C-Cu	50:50	20:80	175	500	2
C-Mg	50:50	57:43	175	500	2
C-Al	17:83	20:80	175	600	2

In order to identify if the manufactured MGA were suitable for thermal storage, they were examined using metallographic techniques and investigated using X-ray diffraction (XRD). All samples were mounted in Bakelite and polished to a 1 μm finish for metallographic examination. The same samples were also used for XRD.

X-ray diffraction was used to determine the crystallinity and phase composition of the system using a PANalytical X'Pert Pro diffractometer with Cu-k_α radiation within a scanning angle of $2\theta = 10–90°$. The crystal structure of the MGA is important as it allows prediction of how the systems will behave thermally as well as demonstrating the immiscibility of the two materials which is extremely important. XRD data allowed phase identification using the International Centre of Diffraction Database (ICDD).

Along with phase identification, an understanding of the microstructure of the potential MGA is important, as a key requirement is that the active phase be fully encapsulated within the thermally conductive matrix phase. The potential for percolation leading to egress of the molten metal during use can be assessed from the size and relative placement of the particles within the microstructure. Microstructural imaging was conducted using a Zeiss Sigma VP FE-SEM. Both secondary electron and backscattered electron images were taken, generally with a 15 kV accelerating voltage.

34.3 Results

The C-Cu system was the first to be manufactured and investigated. A representative X-ray diffraction pattern is shown as Fig. 34.2. It shows both carbon and copper have maintained their respective crystal structures without forming compounds.

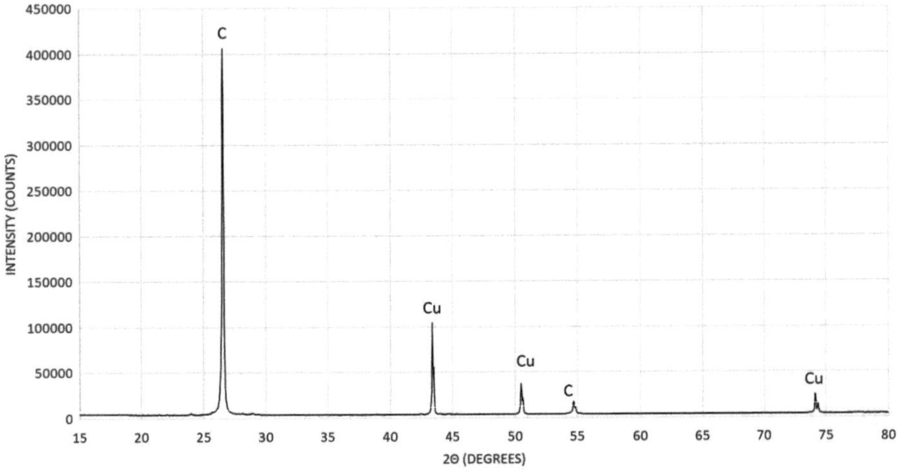

Fig. 34.2 X-ray diffraction pattern from a C-Cu sample. Significant diffraction peaks are identified

The figure shows carbon (graphite) peaks at $2\theta = 26.5°$ and $54.5°$ from the (0002) and (0004) planes, respectively. Other diffraction peaks from graphite are suppressed as this pattern was recorded perpendicular to the pressing direction and the graphite crystals have oriented during pressing. Also visible are strong copper peaks at $2\theta = 43.5°$, $50.5°$ and $74.2°$ from the (111), (200) and (220) planes, respectively, indicating no preferred orientation of the Cu. In addition, no oxide peaks were detected within the diffraction pattern, indicating that no oxygen has penetrated the sample throughout the manufacturing process. This type of diffraction pattern represents an ideal outcome for an MGA – essentially complete immiscibility coupled with no degradation due to external factors.

An example backscattered electron image of a C-Cu same sample is shown in Fig. 34.3. The bright 80 micron copper particles are clearly identifiable encapsulated within the dark carbon matrix. The distribution of the copper balls is consistent throughout the sample. A majority of the copper particles do not touch and are completely surrounded by the carbon. This means that upon the phase transformation, the molten copper will stay encapsulated and discreet throughout the sample, i.e. percolation leading to draining of Cu from the material in service cannot occur.

A representative XRD pattern for the C-Al system is shown in Fig. 34.4. This system exhibited the same trends as the C-Cu MGA. Both the C and the Al maintain their original crystal structures as evidenced by the peaks from hexagonal graphite at $2\theta = 26.5°$ and 54.5 and from face-centred cubic aluminium at $2\theta = 38.5°$, $45°$, $65°$ and $78.5°$. This is particularly significant in the C-Al system as the miscibility gap is only metastable. Clearly with the low-temperature treatment given to these samples, no observable quantity of Al_4C_3 has formed which bodes well for the suitability of this system to use in MGA thermal storage blocks.

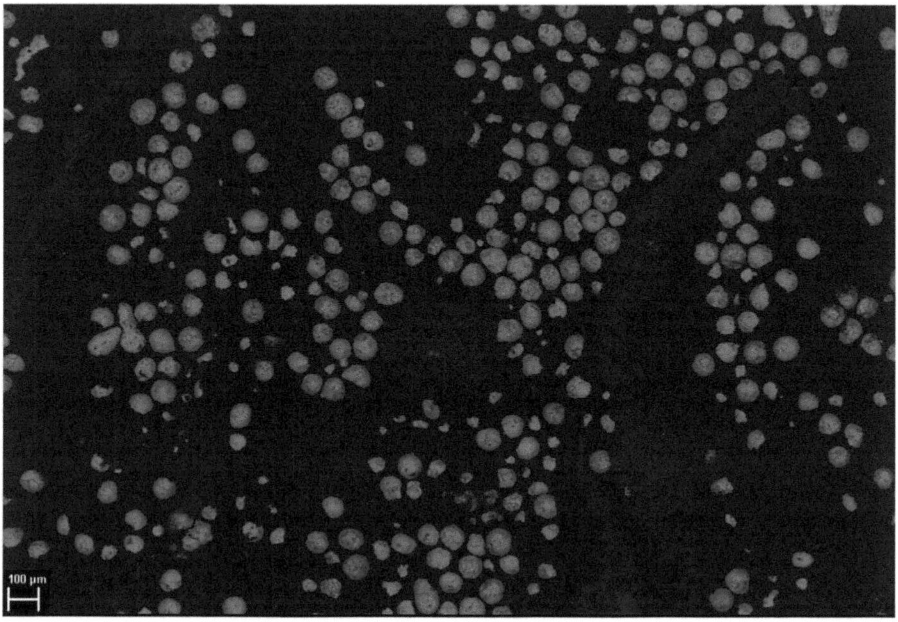

Fig. 34.3 Backscattered electron image of C-Cu MGA

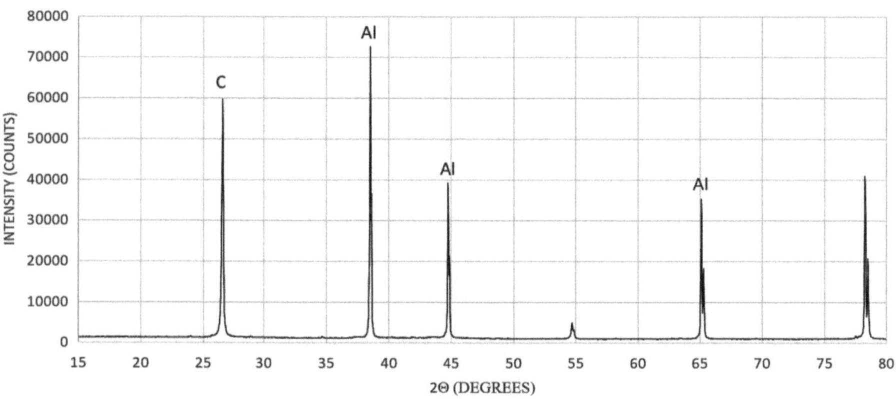

Fig. 34.4 XRD pattern from a C-Al sample. Significant diffraction peaks are identified

Shown in Fig. 34.5 is a backscattered electron image for the C-Al system. Aluminium is visible as the brighter phase with mean particle size around 50 µm in agreement with supplier data. As depicted within the image, the aluminium is quite prominent, occupying the majority of the space. This is largely due to the 80:20 volume ratio that was used. The Al distribution is slightly inhomogeneous with some regions densely packed and others Al-free which indicates that some agglomeration of C occurred during manufacture. Fortuitously, there are sections

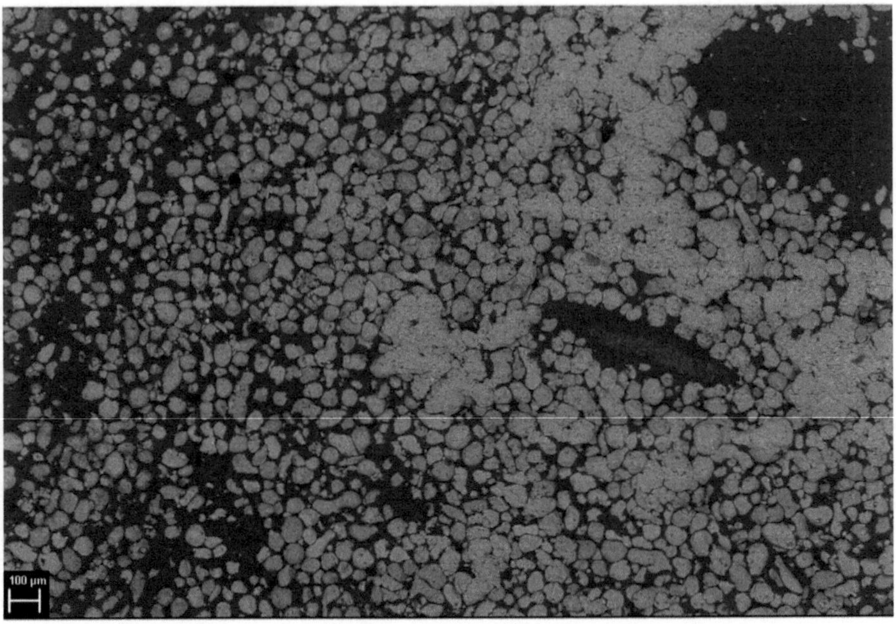

Fig. 34.5 Backscattered electron image of C-Al

such as the left-hand side of Fig. 34.5, which have Al perfectly encapsulated by carbon, along with no visible signs of oxide. From this, it is postulated that if the ratio of aluminium was decreased, the distribution of the aluminium particles would be more satisfactory and there are good prospects that this system is suitable as a MGA thermal storage material.

The XRD pattern shown in Fig. 34.6 shows the graphite peaks as before, as well as magnesium peaks at $2\theta = 32°$, 34.5°, 36.5° and also 63°. No peaks from unwanted phases were observed. The carbon and magnesium have remained immiscible and there has been no significant oxidation. Therefore, in terms of the XRD results, C-Mg is consistent with the other two systems.

Figure 34.7 shows a representative backscattered electron image from the C-Mg system. In this case, the (bright) Mg particles are approximately 200 µm in size, somewhat larger than the active particles in the previous systems. The distribution of the magnesium within the sample was not as consistent as in the other two systems largely due to some regions of agglomerated sodium silicate binder. Nonetheless, the carbon has successfully encapsulated the slightly irregularly shaped magnesium particles which on close examination were found to not be touching and therefore not at risk of draining from the MGA when molten. Elemental examination using EDS within the electron microscope found that minimal oxygen was present within the sample. Therefore, the XRD results and general form of the microstructure in Fig. 34.7 make this system a promising one for use as an MGA thermal storage material.

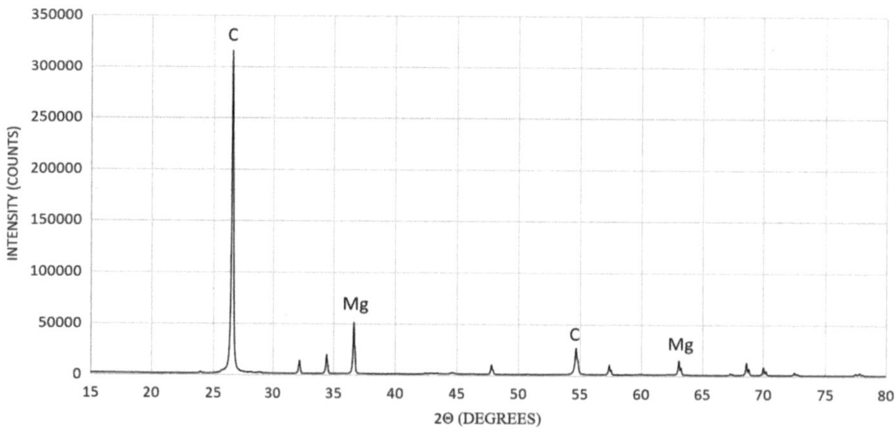

Fig. 34.6 XRD pattern from a C-Mg sample. Significant diffraction peaks are identified

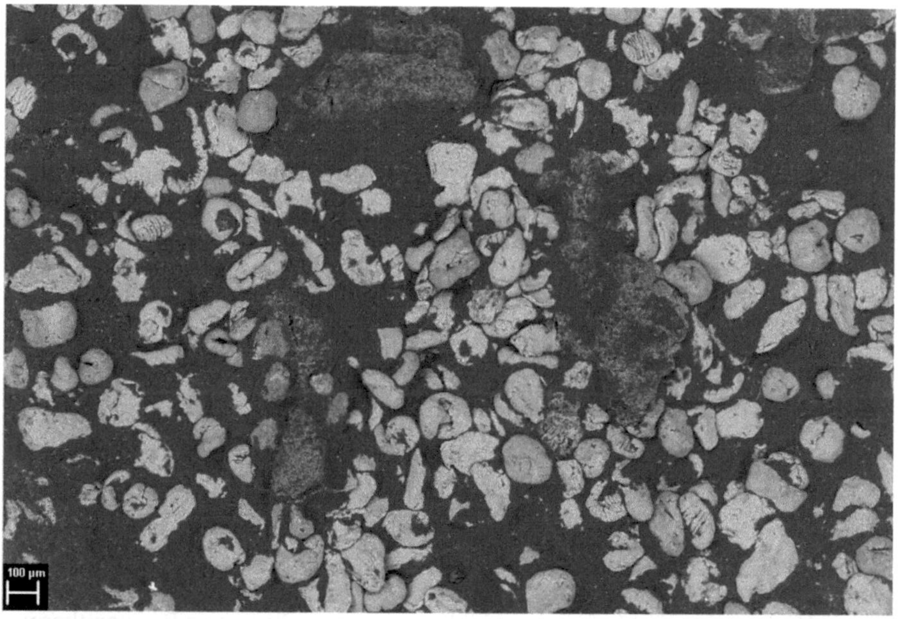

Fig. 34.7 Backscattered electron image of C-Al

34.4 Discussion

The purpose of this research was to investigate the suitability of three new systems with properties deemed promising for the manufacture of MGA thermal storage materials. Success in terms of MGA manufacture comprises three key elements:

(i) The two primary components remain immiscible.
(ii) No compounds between the constituents or with external elements such as oxygen are formed.
(iii) An inverse microstructure is formed, with the lower melting point phase fully encapsulated within the higher melting point phase.

The XRD investigations have demonstrated unequivocally that under the conditions used, the C-Cu and C-Mg systems show the expected immiscibility in agreement with their respective binary phase diagrams. In addition, it was demonstrated that although the C-Al binary phase diagram shows an intermediate compound, Al_4C_3, the use of low-temperature processing has avoided its formation during sample manufacture. In addition, the XRD results did not show any signs of oxidation or other deleterious interactions between the samples and their environment during manufacture. Therefore, according to the XRD data, all three systems satisfy the first two criteria.

Metallographic examination in the scanning electron microscope has confirmed that manufacturing left the active (low melting point) phase completely individual with little to no penetration of oxygen. The backscattered images indicate that the highly thermally conductive carbon matrix phase completely encapsulates the lower melting temperature active phase in the C-Cu and C-Mg cases. This encapsulation of the active phase allows perfect isolation so that when the low melting point component (Cu or Mg) undergoes its melting transformation, as the molten material is stored in pockets, completely surrounded by the carbon matrix, there is no prospect of it leaking out. The C-Al samples were somewhat different as they were mixed at a much higher concentration of the low melting point phase (Al). Nonetheless, in areas of the samples where the local Al concentration was approximately 50 vol.%, the correct inverse microstructure was formed as may be seen at the left of Fig. 34.5. Therefore, these samples have also locally met the third criterion, and it may be concluded that all three systems are suitable for further investigation regarding their manufacture into MGA thermal storage blocks.

Two of the three potential MGA systems investigated (C-Al and C-Mg) have predicted thermal properties (Table 34.1) that make them attractive for use in concentrated solar thermal power (CSP) using conventional Rankine cycle steam turbine-generator technology due to the melting temperature of the latent heat phase which is approximately 100 °C above the required subcritical steam temperature conventionally used. The third system, C-Cu, with a latent heat delivery temperature of 1085 °C, is suitable for use in high-temperature Stirling or Brayton thermodynamic cycles. The assessment of the long-term suitability of these three systems for such applications will require accelerated thermal cycling tests and manufacturing scale-up tests; however, the prospects of success appear good.

References

Copus, M., Bradley, J., Kisi, E., Sugo, H., Reed, S. (2017). *Scaling up miscibility gap alloys for thermal storage*. Proceedings World Renewable Energy Congress (WREC XVI), Perth, Australia, 5–9 January 2017, pp. XXX–XXX.

Jackson, M. (2015). *Design and construction of miscibility gap alloy high energy density thermal storage demonstration apparatus*, Bachelor of Engineering Honours Thesis, University of Newcastle, Newcastle Australia.

Kenisarin, M. M. (2010). High-temperature phase change materials for thermal energy storage. *Renewable and Sustainable Energy Reviews, 14*, 955–970.

Khan, Z., Khan, Z., & Ghafoor, A. (2016). A review of performance enhancement of PCM based latent heat storage system within the context of materials, thermal stability and compatibility. *Energy Conversion and Management, 115*, 132–158.

Khare, S., Dell'Amico, M., Knight, C., & McGarry, S. (2012). Selection of materials for high temperature latent heat energy storage. *Solar Energy Materails & Solar Cells, 107*, 20–27.

Laing, D., Bahl, C., Bauer, T., Fiss, M., Breidenbach, N., & Hempel, M. (2012). High-temperature solid media thermal energy storage for solar thermal plants. *Proceedings of the IEEE, 100*(2), 516–524.

Massalski, T. (1990). Alloy phase diagram volume 3. ASM International, Materials Park Ohio USA.

Rathod, M. K., & Bannerjee, J. (2013). Thermal stability of phase change materials used in latent heat energy storage systems: A review. *Renewable and Sustainable Energy Reviews, 18*, 246–258.

Rawson, A., Kisi, E., & Sugo, H. (2014). *Thermal properties of miscibility gap alloys for thermal storage applications*. Melbourne: Australian Solar Council.

Relebohele, J.R., Dinter, F. (2015). Evaluation of parasitic consumption for a CSP plant. Proceedings Solar PACES 2015, AIP Conf. Proceedings 1734, document 070027, pp. 1–8.

Sugo, H., Cuskelly, D., Kisi, E. (2013). Miscibility gap alloys with inverse microstructure and high thermal conductivity for high energy density thermal storage applications. *Applied Thermal Engineering, 51*, 1345–1350.

Xu, B., Li, P. W., & Chan, C. (2015). Application of phase change materials for thermal energy storage in concentrated solar thermal power plants: A review to recent developments. *Applied Energy, 160*, 286–307.

Chapter 35
Electricity Demand and Implications of Electric Vehicle and Battery Storage Adoption

Paul Ryan

35.1 Introduction

The introduction of new, higher performance models of electric vehicles to the Australian market has produced a ground swell of public interest in adopting electric cars, in part due to their potential reduced environmental impacts and lower operational costs. Forecasts for sales of electric vehicles range from 10% to 30% of new car sales from the mid-2020s. However, any large scale adoption of electric vehicles will impact on demand on Australia's electricity networks and could have significant impacts on demand management, network investment and augmentation, and generation requirements.

The decreasing availability of high-value, solar feed-in tariffs as well as an increasing range of battery storage systems targeting the residential sector has raised consumer expectations that in the future, many households will be less reliant on the electricity grid or independent. Battery storage systems are related to electric vehicles due to their sharing of common technologies of low-cost, high-capacity batteries. This paper explores this link by examining the potential electric energy and demand impacts of different uptake scenarios for electric vehicles (EV) and battery storage systems (BSS) in the residential sector.

Recent projections in studies for the Australian Energy Market Operator (AEMO 2016a, b) suggest mild impacts over the next 20 years, with EVs adding almost 4% to 2035–2036 projections of total electricity use in Australia and approximately 2800 MWh of battery storage system capacity in the residential sector by 2030. By applying similar technical and usage characteristics used by AEMO of EVs and BSS, this paper models projected power and energy demand. This modelling shows that projected impacts were highly sensitive to EV take-up assumptions. When

P. Ryan (✉)
EnergyConsult Pty Ltd, Jindivick, VIC, Australia
e-mail: paul@energyconsult.com.au

A. Sayigh (ed.), *Transition Towards 100% Renewable Energy*,
Innovative Renewable Energy, https://doi.org/10.1007/978-3-319-69844-1_35

related only to the residential sector, residential electricity use may stabilise or reverse the original projected declining trend by 2030 shown in the Residential Baseline Study (RBS) (DIS 2015). The modelling also suggested that the take-up of BSS is underestimated as the BSS retrofit market is not included in the recent AEMO (2016a) study, only new installations of photovoltaic (PV) and integrated BSS.

35.2 Methodology

Projections of power and energy demand from EV and BSS were modelled using a highly detailed residential energy model recently used to produce the third Australian Residential Baseline Study (RBS) of national residential energy use (DIS 2015). This model calculates the energy and power demand impact of more than 120 different products, including PV generation, for the residential sector by each Australian state, using detailed data researched for each product, and was designed to also model the impact of new products, such as the introduction of EV or BSS.

A businesses-as-usual (BAU) (or neutral) scenario and alternative scenarios were explored using the residential energy end-use model developed for the RBS. The residential energy end-use model is classified as a bottom-up engineering model. It involved calculating the energy end-use consumption at the individual level and aggregating these figures to estimate the total locality or network consumption. A full explanation of the modelling methodology is available in the study documentation (DIS 2015). At the heart of this approach is the calculation of total energy consumption or generation for each product (Eq. 35.1):

$$\text{Total Energy Consumed} = \text{Stock Numbers} \times \text{Unit Energy Consumption (UEC)} \tag{35.1}$$

Determining the stock number of energy end-use equipment was undertaken by stock models which are effectively databases that keep a running tally of the number of equipment installed on a year by year basis. The stock in any year equals the sum of all past stock sales, less retirements of equipment.

The second aspect of the energy modelling determined the value of the unit energy consumption (UEC) for each product used in the residential energy end-use model. At its most basic level, UEC is determined by (Eq. 35.2):

$$\text{UEC} = \text{Hours of usage} \times \text{Unit Power Input} \tag{35.2}$$

This paper uses the baseline RBS projections from 2015 to 2030 and adds the impacts of EV and BSS for the residential sector; the first time such modelling has been completed. These baseline forecasts utilise the latest projections of EV and BSS take-up from the Australian Energy Market Operator (AEMO) undertaken in

2016, which have been adjusted to ensure that only the residential sector was included (i.e. excluding commercial/industrial uses of EV and BSS) (AEMO 2016a, b). An alternative scenario of higher uptake of EV and BSS is also reported here, based on likely decreases in battery prices and higher consumer purchasing assumptions. Therefore the baseline forecasts are in line with the latest studies by AEMO. The advantage of utilising the RBS model is to understand the impacts of battery technologies in a consistent model of the residential sector energy use.

35.2.1 Electric Vehicle Assumptions

The impact of EVs on the energy system is highly sensitive to the projected sales of EVs over the projection period. The projected sales of EVs related to the residential sector were obtained from the "neutral" and "strong" scenarios modelled be AEMO (2016b) as shown in Fig. 35.1. Sales of new passenger cars were 924,154 in 2015 (ABS 2016). The average annual growth rate of new car sales for the period till 2030 is assumed to 1% (Trading Economics 2017).

Other major assumptions and inputs were based on published data regarding the average distance travelled by passenger cars, energy efficiency of EV, and specifications of EV currently on the market, as shown in Table 35.1.

The rapid rise in EV share of new car sales under the strong and neutral scenario is due to the assumption in the AEMO (2016b) study that the Australian government will introduce international best practice emission standards by 2022 (strong) and 2026 (neutral).

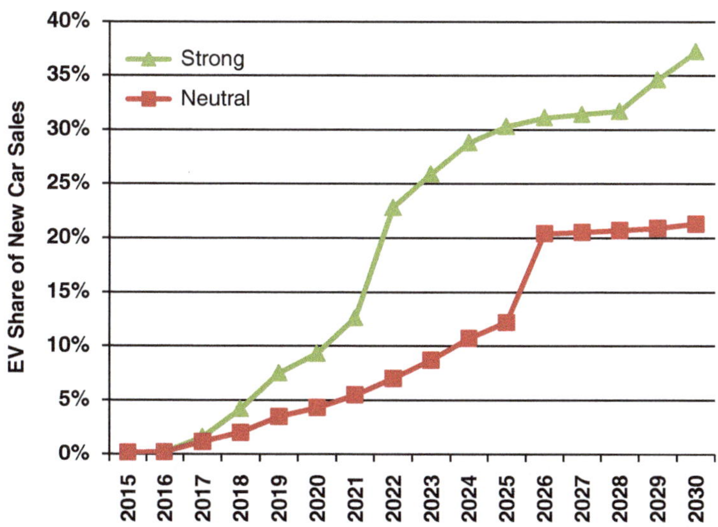

Fig. 35.1 Projected sales of EV as a share of the passenger car market (AEMO 2016b)

Table 35.1 Key EV modelling inputs

Key input	Value	Key input	Value
EV charging requirement	3.6 kW	Average EV efficiency	173 Wh/km
Average km travelled	14,000 pa	Coincidence factor (max demand)	0.05
Average annual load	2424 kWh	Coincidence factor (total demand)	0.5

Sources: www.greenvehicleguide.gov.au, ABS, AEMO (2016b), EnergyConsult calculations

Table 35.2 Key BSS modelling inputs

Key input	Value	Key input	Value
Average BSS capacity	6 kWh	Average BSS available capacity	4.8 kWh
Cycles pa	365 pa	BSS losses	10%
Average inverter capacity	2.4 kW		

35.2.2 Residential Battery Storage System Assumptions

The two most sensitive inputs relating to residential battery storage systems are installation take-up rate and the size of a system. For the installation of PV with integrated BSS, the AEMO (2016a) has made projections which are applied in this paper (the scope of the AEMO study did not include BSS retrofit to an existing PV household). The take-up of residential BSS that are retrofitted with an existing solar PV system were based on the number of households with PV systems which are receiving 5–7 c/kWh feed-in tariffs for exported electricity. Evidence from discussions with suppliers of BSS during 2016, suggests that economic benefits alone were not the only consideration in the installation of a BSS. These suppliers reported that the vast majority of their enquiries and early sales related to householders wanting to increase the proportion of their usage of solar energy and reduce their grid imported power. In addition, BSS market prices are decreasing rapidly with the recent introduction of the Tesla Powerwall 2, priced at $8000 plus installation (Tesla 2017). This paper makes the assumption that between 0.4% and 2% pa of existing householders with PV will instal BSS over the period 2017–2030 (up to a total 22% of households with PV will retrofit BSS). The key technical modelling inputs for BSS are shown in Table 35.2.

35.3 Results

35.3.1 Electric Vehicles

The impact of EV in terms of energy use over the period to 2030 is shown in Fig. 35.2, with the two scenarios, *EV – Neutral* increasing consumption by 3800 GWh in 2030 and *EV – Strong* increasing consumption by 7350 GWh in 2030. The *EV – Strong* scenario is equivalent to more than the combined electricity

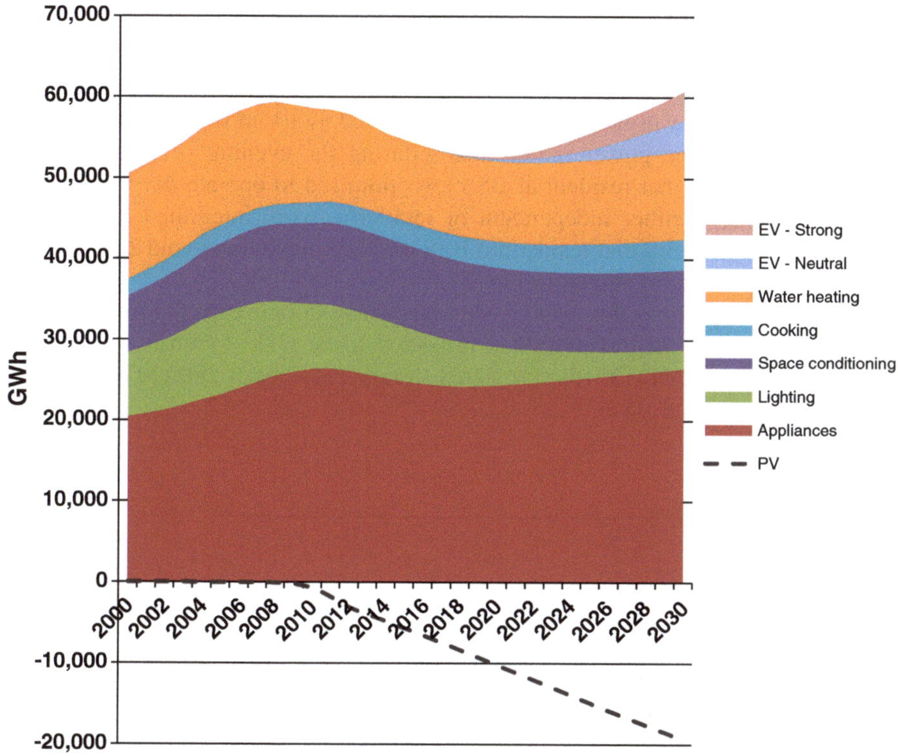

Fig. 35.2 Projected impact of EV energy use compared with residential electricity energy end-use in Australia (DISS 2015)

consumption of lighting and cooking in Australia in 2030. It is interesting to note that projected gross residential PV energy generation in 2030 is 19,000 GWh, well in excess of the projected EV energy consumption. Total residential electricity consumption is projected to rise by 2030 under both scenarios. Even with the highest EV consumption and taking into account residential PV generation, total net residential energy use is projected to be 15% lower than 2015.

Residential electricity peak demand is not projected to increase substantially over the 13-year period due to the contribution of EV, as it assumed that pricing and network incentives will ensure that most EV charging is undertaken in the off-peak night period. A small fraction of EVs (5%) were assumed to be charging during the residential peak due to unforeseen consumer requirements, and this is projected to increase peak demand by approximately 200–500 MW. However the electricity demand in the late evening/night period is projected to increase by 3–5 GW. This is equivalent to 25% of the total maximum residential summary demand, although this EV demand is assumed to be non-coincident with the summer peak (typically between 5 and 8 pm).

35.3.2 Residential Battery Storage

Residential BSS has the potential to reduce peak demand and provide benefits on the electricity grid by utilising the energy generated by PV to supply the household with some or all the power requirements during the evening peak period. The modelling assumes that residential BSS are optimised to operate during the early evening period by either independent or retailer/network integrated control systems. The model shows that residential BSS has the potential to avoid 2700 MW of peak electricity load, as shown in Fig. 35.3.

The modelling shows that total residential energy storage capacity for retrofitted BSS and PV with integrated BSS is projected to be 6000 MWh by 2030. This compares to the AEMO (2016a) projected storage capacity of 2800 MWh by 2030 for integrated PV and BSS.

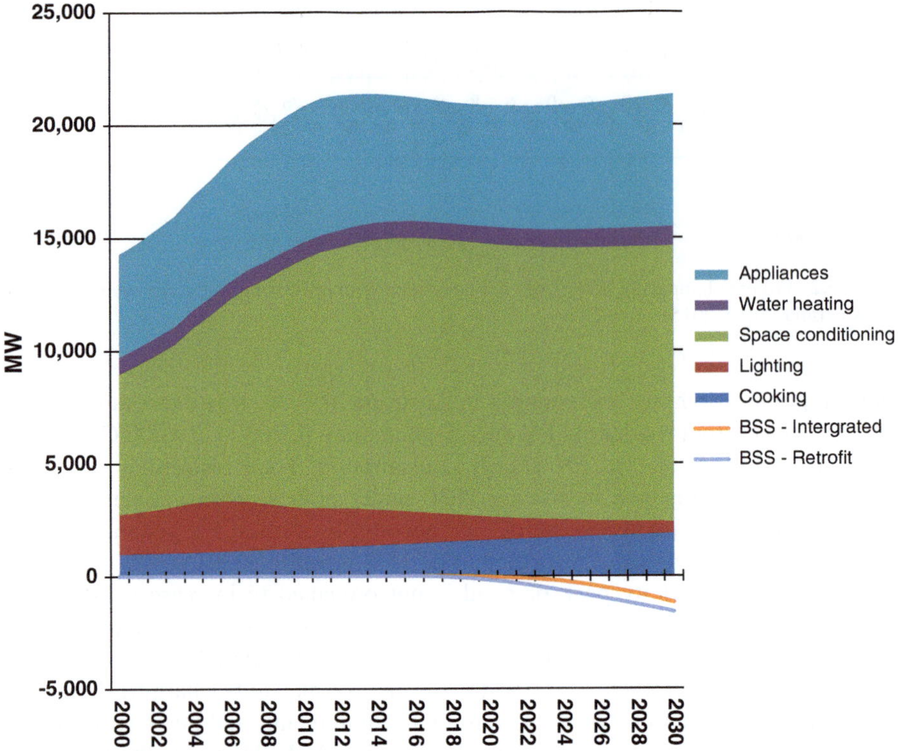

Fig. 35.3 Projected potential reduced electricity demand by BSS compared with Australian residential electricity end-use demand (DISS 2015)

35.4 Discussion

The impact of EV and BSS on residential sector electricity energy use and demand is highly dependent on the projections of EV sales and installations of BSS. The AEMO studies (2016a, b) are modelled in this paper as the base case. However, additional scenarios are also presented which are supported by alternative assumptions.

In the case of EVs, the penetration of EVs in Europe is forecast to be 20% of all stock by 2030 under their mid (base) scenario (Öko-Institut e.V 2016). In Australia the *neutral* scenario published by AEMO shows a 10% penetration by 2030 (2016b). The AEMO study models a *strong* scenario that is likely to eventuate in a penetration of 20% by 2030. This outcome is suggested to be more likely due to recent evidence showing lower cost EVs entering the market with battery costs lower than the costs assumed in the *strong* scenario by AEMO. General Motors has released its Chevrolet Bolt EV for US$37,495, and the estimated battery cost is US $150/kWh (The Guardian 2017). In addition, by 2020, several major car manufacturers are expecting sales of EVs to represent 15–20% of new sales, (Automotive News 2017). The AEMO *strong* scenario assumes 9% of sales in Australia are EVs in 2020. Given this latest information, the sales of EVs may be higher in Australia than currently estimated and so have a greater impact on the electricity demand.

To reduce the impacts on local distribution networks, EV charging will need to be controlled by smart charging systems that optimise the time of charging in accordance with network load. This is very important, considering the projections in this paper suggest that 3–5 GW of additional load will need to be addressed. The local electricity distribution systems feeding residential areas may need to be re-examined to ensure that these additional loads can be accommodated safely. Network demand charges and energy pricing that encourages charging strategies to mitigate local supply constraints will need to be developed. Daytime EV charging strategies may also address some of the issues in some states where high PV generation is creating minimum or low electricity demand in the middle of the day. This may be difficult to encourage as most vehicles are used during the day; however, some households will be able to take advantage of this charging profile depending on their EV usage patterns.

In the case of BSS, the major driver to the uptake of BSS is the ability of householders to utilise more self-generation (solar PV) during the periods of higher electricity use, which are typically in the evening when the PV is not producing outputs. The BSS enables the householder to reduce their costs, especially if they are receiving between 5 and 7 c/kWh during the day for power exported to the grid. The difference in price between the exported power and the evening prices ranges from 15 to 30 c/kWh for those customers not receiving state government feed-in tariffs. From 1 January 2017, over 275,000 customers with solar PV in NSW, SA, and Victoria have had their feed-in tariff reduced to between 5 and 7 c/kWh. This could become a major driver in the installation of BSS for these households, along with the estimated 300,000 customers who have installed PV since the state government feed-in tariff schemes closed to new entrants over the period 2010–2013. The AEMO study (2016a) does not include the impacts of BSS retrofit to existing residential PV; however, this is estimated to more than double the impacts according to the projections in this paper.

BSS systems can provide significant potential benefits to the electricity network distribution businesses and retailers, as well as householders. These benefits to the network business are mostly in the form of avoided costs to augment or upgrade the electricity network to meet peak demands when the BSS is used to meet some or all of the evening loads of householders. Retailers, network businesses and other energy market business can also obtain benefits if the PV generation is utilised locally (by the BSS) and not exported (when some distribution systems are experiencing voltage control issues). The use of smart control strategies and communications platforms that enable the sharing of network and energy system benefits with householder with BSS will increase the uptake of BSS. Some preliminary programmes providing these services are already underway in Victoria (ATA 2016) and can be evaluated to guide the development of future programmes.

References

Alternative Technology Association. (2016). *Mini-grid in Mooroolbark* (pp. 34–35). Renew. Issue 137. published by Alternative Technology Association, Melbourne, October 2016.

Australian Bureau of Statistics. (2016). *93140 Sales of New Motor Vehicles*, Australia, December 2016.

Australian Energy Market Operator. (2016a). *Projections of uptake of small-scale systems*. Prepared by Jacobs, for the AEMO, Melbourne, June 2016.

Australian Energy Market Operator. (2016b). *AEMO insights, electric vehicles*. Melbourne: AEMO and Energeia. August 2016.

Automotive News. (2017). Article *"Is the industry zooming toward a battery shortage?"* http://www.autonews.com/article/20170109/OEM05/301099969/is-the-industry-zooming-toward-a-battery-shortage. Visited 12/01/2017.

Department of the Industry and Science. (2015). *Residential energy baseline study: Australia*. Canberra: Department of the Industry and Science. August 2015, Prepared by EnergyConsult, Available from http://www.energyrating.gov.au/document/report-residential-baseline-study-australia-2000-2030.

Öko-Institut e.V. (2016). *Electric mobility in Europe – future impact on the emissions and the energy systems*. Berlin: Öko-Institut e.V. September 2016.

Tesla. (2017). Price of the Tesla Powerwall 2. https://www.tesla.com/en_AU/powerwall. Visited 12/01/2017.

The Guardian. (2017). Article *"GM delivers first Chevrolet Bolts, sparking electric car price race"*. www.theguardian.com/environment/2016/dec/14/gm-delivers-first-chevrolet-bolts-sparking-electric-car-price-race. Visited 11/01/17.

Trading Economics. (2017). *Australian new car sales 2017 to 2020*. http://www.tradingeconomics.com/australia/total-vehicle-sales/forecast. Visited 11/01/2017.

Chapter 36
Development of Green Concrete from Agricultural and Construction Waste

Abdul Aziz Abdul Samad, Josef Hadipramana, Noridah Mohamad, Ahmad Zurisman Mohd Ali, Noorwirdawati Ali, Goh Wan Inn, and Kong Fah Tee

36.1 Introduction

In the 11th Malaysia Plan (Economic Planning Unit 2015), the Malaysian construction industry has been urged to change from the conventional construction method to Industrialised Building System (IBS) to attain better construction quality and productivity. The use of IBS has been made compulsory in the construction of public buildings, and the adoption of this alternative construction system was fully supported by the government through programmes, incentives and encouragement policies stipulated under the IBS Roadmap 2011–2015 (CIDB 2010). Through the recently launched Construction Industry Transformation Plan 2016–2020 (CIDB 2015), see Fig. 36.1, the government, together with the Construction Industry Development Board (CIDB), will be emphasizing on a construction system which is environmentally sustainable, in line with the requirements of green construction and the reduction of carbon emission of CO_2.

It has been well documented that the global expansion of the greenhouse effect has been partly due to the production of cement. It was approximated that 2.7 billion cubic metres of concrete were consumed with production spread unevenly among more than 150 countries. This equals more than 0.4 m^3 of concrete consumed per person annually. The production of Portland cement, an essential constituent of concrete, leads to the release of a significant amount of CO_2 and other gases leading to the effect on global warming. Literature has proven that the manufacturing of 1 ton of Portland cement produces about 1 ton of CO_2. Since the production of Portland cement is not particularly environmentally friendly and the

A.A.A. Samad (✉) • J. Hadipramana • N. Mohamad • A.Z.M. Ali • N. Ali • G.W. Inn
Jamilus Research Centre, Universiti Tun Hussein Onn Malaysia, Batu Pahat, Johor, Malaysia
e-mail: azizs@uthm.edu.my

K.F. Tee
Department of Engineering Science, University of Greenwich, Kent, UK

A. Sayigh (ed.), *Transition Towards 100% Renewable Energy*,
Innovative Renewable Energy, https://doi.org/10.1007/978-3-319-69844-1_36

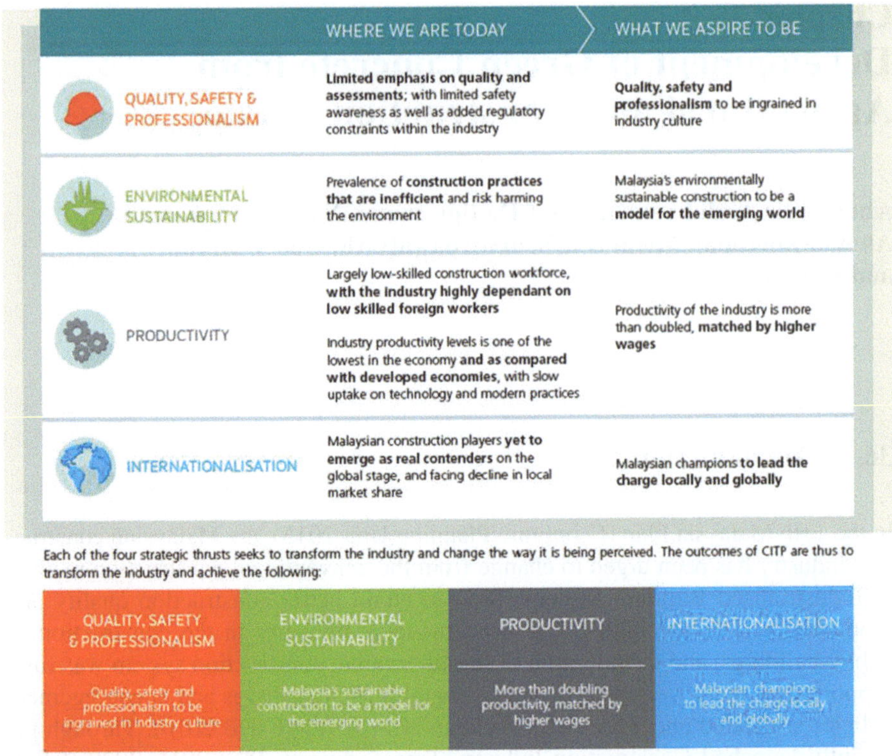

Fig. 36.1 Construction Industry Transformation Plan (CITP) 2016–2020 (CIDB 2015)

impact it causes to the environment due to carbon emission (CO_2) calls for the development of the innovative green concrete which utilizes agricultural and construction waste as cement replacement and the reduction of natural resources (Vlastimir et al. 2013; Megat Johari et al. 2011; Karim et al. 2011). The challenge for green concrete is therefore to use as little Portland cement as possible and replace the Portland cement with supplementary cementitious materials such as agricultural waste and waste from construction such as recycled concrete aggregates (Meyer 2009). The consumption of agricultural and construction waste in green concrete reduces the dependency of earth natural resources, reduces waste and improves the environment and quality of life. Hence, the design of green concrete as a sustainable building material is an innovative idea and has attracted the attention of many researchers and construction players around the world.

However, literature has shown lack of evidence on investigations related to the combination of the four waste materials (POFA, RHA, POF and RA) into one mix design. This causes a deficiency in the material and mechanical properties of the proposed green concrete comprising of the four components from agricultural and construction waste. Due to the high composition of waste material in the green concrete mix, it is expected that a severe non-homogeneous and anisotropic material would be produced and an effect to its strength and mechanical properties of the proposed green concrete is anticipated. Nevertheless, the paper attempts to understand and clarify the influence and performance of concrete with waste materials and its additives. Hence, a discussion on past and present development of green concrete including its characteristic strength properties compared to normal concrete is presented.

36.2 Green Concrete

Malaysia is one of the largest producers of palm oil fuel ash (POFA) in the world. This is apparent when 1000 tons of POFA were produced annually by 200 numbers of palm oil mills in operation. Figure 36.2 shows the production of palm oil producers around the world as recorded by Zeyad et al. (2017). These POFA if not utilized will be dumped into ponds as agricultural waste products. It has also been documented that approximately 110 million tons of rice husk and 16–22 million tons of rice husk ash (RHA) are generated worldwide. Figure 36.3 shows an example of a palm oil tree, its fruit bunches and fibres (from palm oil shells). Until today, no proper utilization of POFA and RHA has been developed and adopted in the construction industry around the world (Payam et al. 2014). However, as issues on sustainable construction gain more prominence, research on green concrete using agricultural by-product waste such as POFA and RHA and construction waste (recycled concrete aggregate) have been conducted and were given serious considerable work by researchers from Malaysia and other parts of the world.

The outcome of the research shows that the addition of agricultural waste or construction waste with additives for green concrete shows positive and satisfactory strength when compared to normal concrete. Palm oil fuel ash (POFA) and rice husk ash (RHA) are an agricultural waste containing large components of silica and have been accepted as a pozzolanic material in concrete for the past 20 years. The utilization of POFA or RHA as a pozzolanic material to partially replace Portland cement has been extensively investigated. However, the authors observe that a

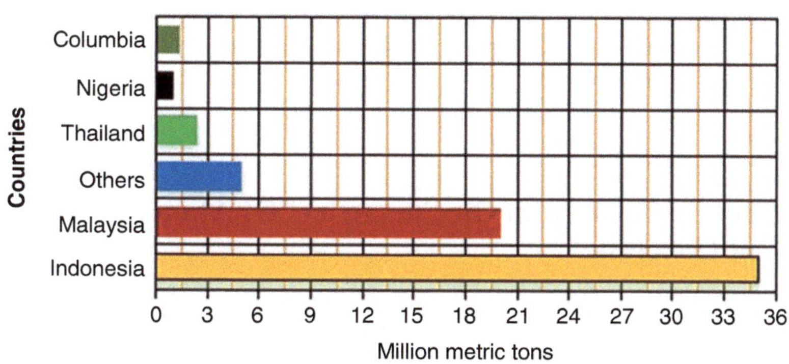

Fig. 36.2 Palm oil producer countries (Zeyad et al. 2016)

Fig. 36.3 Palm oil tree, fruit bunch and fibrous strands (Mohamed and Huzaifa 2008)

combination of both POFA and RHA in concrete has not been fully investigated and has yet to be fully accepted and officiated by the construction industry in Malaysia and neighbouring countries.

Likewise, research have shown that using natural palm oil fibre (POF) at 0.1–0.5% by percentage weight of cement enhances the strength of concrete by more than 40% from normal concrete. It is interesting to note that from research observation, the palm oil fibres tend to create an uneven distribution causing a balling-like effect to occur in the concrete.

Concrete also consumes large volume of natural aggregates (sand and gravel) as a key component of its material matrix. This raises concerns about the preservation of the earth natural resources. Hence, researchers have studied the cause and effect of this issue and recommended ways to balance the shortfall of future natural resources. One of the solutions to this is the recycling of demolished concrete structures to produce new structural concrete elements. Studies have shown that recycled concrete aggregate (RCA) is a viable option for a construction's material in terms of its mechanical properties and structural behaviour. Therefore, the successful use of agricultural and construction waste contributes to energy saving and conservation and preservation of natural resources and reduces rising construction cost. Producing green concrete is sustainable as it solves the disposal of waste and protects the environment.

36.3 Influence of Palm Oil Fuel Ash (POFA) on Strength of Concrete

Research on utilizing POFA as cement replacement has been conducted for the past two decades. POFA essentially contains siliceous compositions and reacts as pozzolans to produce a cement-like reaction to form a concrete. Table 36.1 shows

Table 36.1 Research on green concrete with POFA

Concrete properties	Researchers								
	Altwair et al. (2012) [1]	Altwair et al. (2012) [2]	Altwair et al. (2012) [3]	Tangchirapat et al. (2009)	Islam et al. (2016) [1]	Islam et al. (2016) [2]	Awal and Abubakar (2011) [1]	Awal and Abubakar (2011) [2]	Awal and Abubakar (2011) [3]
Ratio water/binder	0.30	0.36	0.38	0.32	0.30	0.40	0.48	0.42	0.38
Median particle size of POFA d_{50} (μm)	Ground 2.87	Ground 2.87	Ground 2.87	10.10	17.10	17.10	14.58	14.58	14.58
Fibre	Polyvinyl alcohol fibre 0.02			–	–	–	–	–	–
Superplasticizer	0.001	0.001	0.001	0.064–0.116	0.34–0.062	0.00–0.062	–	0.02	0.02

( ) Types of concrete mix design

a summary of various concrete mix designs with POFA as a cement replacement from various researchers (Islam et al. 2016; Altwair et al. 2012; Awal and Abubakar 2011; Tangchirapat et al. 2009). Figure 36.4 clearly illustrates the differences in compressive strength of concrete from various ratios of POFA as cement (ordinary Portland cement) replacement. Interestingly, the results show that the concrete compressive strength reduces as ratio of POFA increases. Tangchirapat et al. (2009) recorded the highest result of compressive strength ranging from 59 to 61 MPa compared to others. In his experimental work, POFA was initially grounded, and high fineness of POFA was produced with a median particle size d_{50} of 10.1 μm; this size was more subtle than the typical OPC type I cement which measures at approximately 14.6 μm. The degree of fineness of POFA shows significant influence to concrete strength properties (Tangchirapat and Jaturapitakkul 2010). The specific gravity of POFA and its shapes has also been observed to affect its water requirement further influencing the workability and strength (Chindaprasirt et al. 2007). However, the particle size of POFA by Altwair et al. (2012) was the most subtle (see Table 36.1; Fig. 36.4), but results clearly show lower strength of concrete was achieved compared to Tangchirapat et al. (2009). Other variable that may have influenced the strength of concrete was the use of superplasticizer which affected its workability and eventually its strength.

This observation highlights the influence of POFA in concrete which shows an obvious increase in its strength. Research by Altwair et al. (2012) replaces cement by 40% POFA (POFA/OPC = 0.4) by mass which recorded an increment in strength of 10.3% from normal concrete. POFA/OPC = 0.2 or 20% POFA replacement was conducted by Tangchirapat et al. (2009) which resulted in a 24% increase in strength compared to normal concrete. Meanwhile, research by Islam et al. (2016) with 10% POFA as OPC replacement observed an increase in concrete strength by 11%. However, when POFA exceeded 40% as cement replacement,

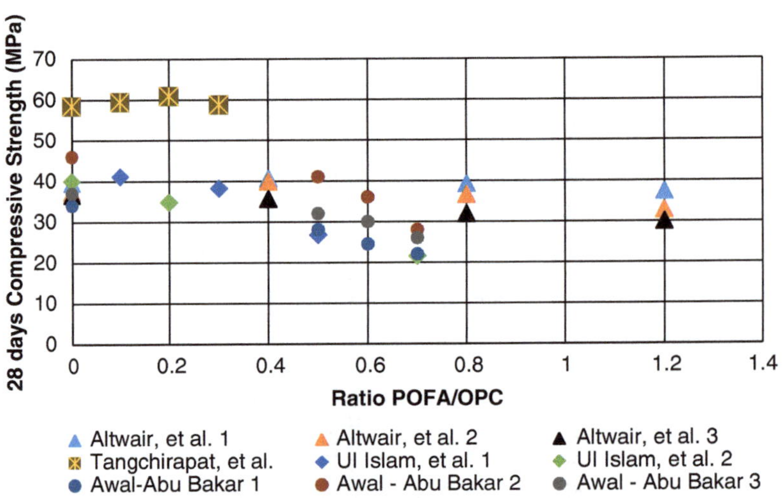

Fig. 36.4 Compressive strength of green concrete with different ratios of POFA/OPC

Awal and Abubakar (2011) show a substantial decrement to the strength of concrete.

36.4 Influence of Rice Husk Ash (RHA) on Strength of Concrete

In the 1970s, production of RHA was not a popular study for research when the uncontrolled combustion of RHA was found to have poor pozzolanic properties. However, Mehta (2001) described an investigation on the pyro-processing parameters on the pozzolanic reactivity of RHA, and the high reactive RHA was then obtained. Since then, RHA has been used as an alternative for active silica production to be used as cement concrete replacement in the construction and building industry (Della et al. 2002; Ganesan et al. 2008).

Figure 36.5 shows the concrete compressive strength distribution with different ratios of RHA as cement replacement. Unfortunately, it has been generally observed that by adding RHA as cement replacement, a reduction in its strength was evident. Kishore et al. (2011) conducted test on RHA from 0%, 5%, 10% and 15% cement replacement with two target concrete strengths at M40 and M50. Kishore observed that adding RHA (as cement replacement) reduces its compressive strength, but a higher target strength of 3% and 8% for grades M40 and M50 was recorded, respectively. Madandoust et al. (2011) tested concrete with up to

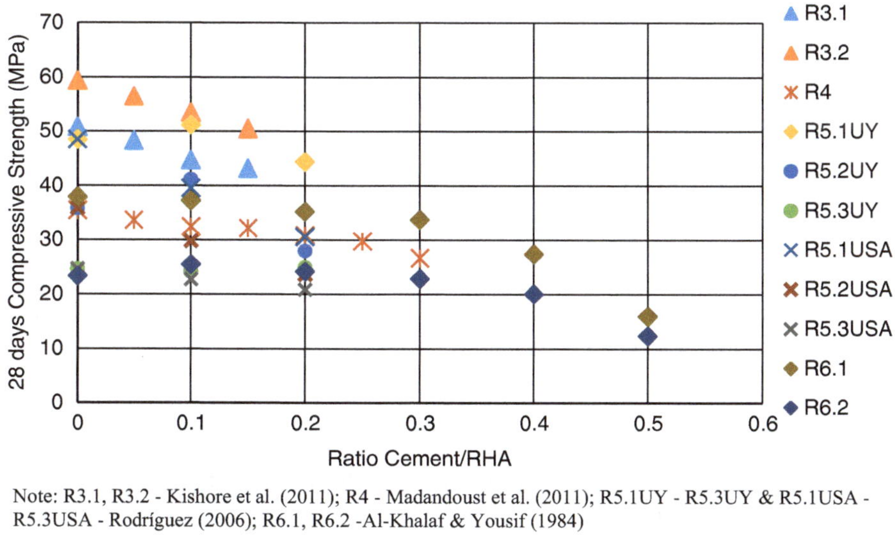

Fig. 36.5 Compressive strength of green concrete containing various percentages of RHA. Note: R3.1 and R3.2, Kishore et al. (2011); R4, Madandoust et al. (2011); R5.1UY–R5.3UY and R5.1USA–R5.3USA, Rodríguez (2006); R6.1 and R6.2, Al-Khalaf and Yousif (1984)

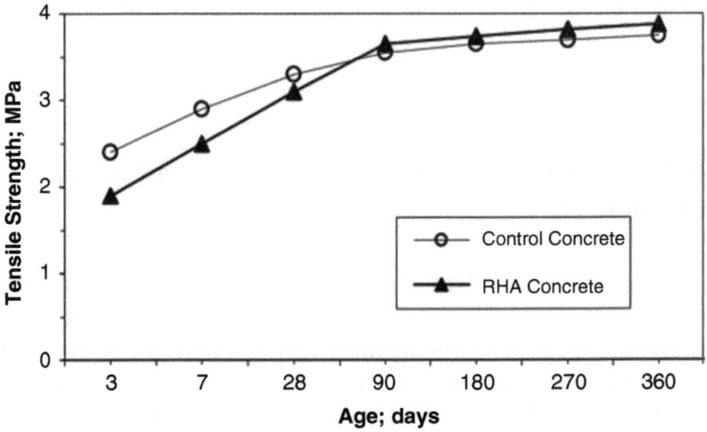

Fig. 36.6 Tensile strength with age (Madandoust et al. 2011)

30% RHA as cement replacement, and a reduction in strength was again observed from 38 MPa at 0% RHA reducing to 28 MPa at 30% RHA. By further observation of Madandoust's work, a 5% RHA cement replacement is shown to be ideal.

Rodriguez (2006) also conducted research on concrete with RHA at 0%, 10% and 20% cement replacement. However, the research observed that replacing cement with 10% RHA recorded higher compressive strength by 4%. Yet, similar trends (as observed by Madandoust et al. 2011) occur when higher percentage of RHA shows a reduction of strength by 6% compared to normal concrete. A high volume of RHA as cement replacement of up to 50% was conducted by Al-Khalaf and Yousif (1984). Results show that the strength of concrete containing 50% RHA produces the lowest strength with a reduction of 60% from normal concrete. However, concrete containing 10% and 20% of RHA achieved higher strength by 9% and 3% compared to normal concrete.

A long-term study on the tensile strength of concrete was conducted by Madandoust et al. (2011). Figure 36.6 shows clearly the tensile behaviour of normal concrete with respect to concrete with RHA. The figure shows that when concrete with RHA reached 90 days, higher tensile strength values were achieved than normal concrete. This trend indicates that the hardening process of concrete with RHA was still progressing beyond 90 days (similar to normal concrete). This hardening process sustained its behaviour even at 360 days.

36.5 Influence of Palm Oil Fibre (POF) to Strength of Concrete

Ahmad et al. (2010) reported that the addition of palm oil fibres will enhance the mechanical properties of concrete especially its fatigue and tensile stresses as well as cracking and impact resistance. It has also been reported that adding POF also

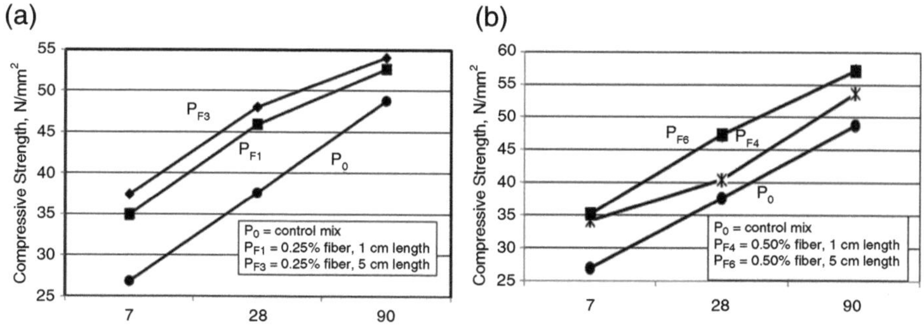

Fig. 36.7 Compressive strength of concrete with different percentages and lengths of POF (Mohamed et al. 2008). (**a**) 0.25% POF with 1 cm and 5 cm length. (**b**) 0.5% POF with 1 cm and 5 cm length

increases its flexural strength. POF with length from 1 cm up to 5 cm, and initially immersed in water, was observed to be applicable for concrete. However, as the length of POF exceeds 5 cm, a balling effect in the concrete will occur to the POF which causes the fibres to be less active. The effective fibre content used was observed to fall within 0.1–0.2% of cement by weight; however, by the addition of 1% of fibre, the concrete will not achieve its desired strength. Interestingly, research by Mohamed et al. (2008) observed that concrete with 0.25% fibre content and 5 cm fibre length shows an enhanced compressive strength compared to normal concrete (see Fig. 36.7).

36.6 Influence of Recycled Concrete Aggregate (RCA) to Strength of Concrete

The major benefits of using recycled concrete aggregate in new construction include lower environmental pollution, reduction in the need for valuable landfill space and conservation of resources for natural aggregates. Furthermore, the application of recycled concrete aggregate (RCA) in concrete is expected to contribute economically and is environmentally viable. However, RCA obtained from crushing from old concrete can exhibit inconsistent properties depending on the composition, particularly the water-to-cement (W/C) ratio and cement content of the original concrete. The quality of RCA is generally inferior to that of normal aggregate. Alengaram et al. (2011) have used 0%, 30%, 50%, 70% and 100% RCA in their study. The researchers observed that the RCA replacement can produce high-workability concrete with higher compressive strength for 50% RCA. A compressive strength reduction was observed for percentage exceeding 50%. Figure 36.8 shows the compressive strength and flexural strength results of concrete using different ratios of RCA. However, the flexural strength of concrete shows a slight reduction (a maximum of just 5%) with different ratios of RCA.

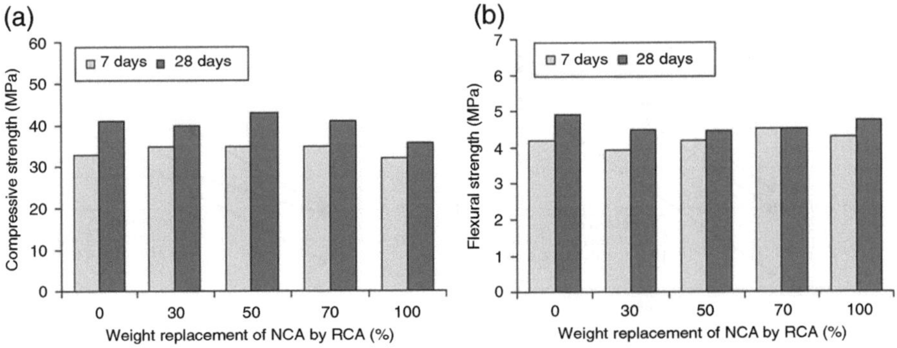

Fig. 36.8 Compressive strength and flexural strength of green concrete with different percentages of RCA (Alengaram et al. 2011). (**a**) Compressive strength. (**b**) Flexural strength

36.7 Conclusions

The need for going green is an essential element for producing an environmentally sustainable construction. Malaysia through its Construction Industry Transformation Plan (CITP) 2016–202 has laid its strategies in producing a construction system which is eco-friendly and environmentally sustainable. To ensure the success of this strategic plan, Jamilus Research Centre of Universiti Tun Hussein Onn Malaysia together with the University of Greenwich will be engaging in a research on green concrete utilizing agricultural and construction waste materials. Literature has shown that research on green concrete has been conducted for the past two decades worldwide. The utilization of agricultural waste with pozzolanic properties and construction waste shows an enhancement in the compressive and tensile strength of concrete. Further addition of palm oil fibres as binders essentially upgrades the concrete flexural strength and cracking properties. From observation of past and present research work, replacing cement with POFA and RHA by 20% and 10% enhances the concrete strength by up to 10%. Adding 0.1–0.2% of POF has also been recorded to have improved its strength and cracking properties of concrete with fibres not exceeding 5 cm long. Likewise, replacing natural aggregates with 50% RCA shows an improvement to its concrete compressive strength, but slight reduction in its flexural strength was observed.

Acknowledgement The authors acknowledge funding from the Fundamental Research Grant Scheme (FRGS) Vot. No. 1573, received from the Ministry of Higher Education of Malaysia. The first author also acknowledges the support received from Jamilus Research Centre, Universiti Tun Hussein Onn Malaysia, for the success of this research work.

References

Ahmad, Z., Saman, H. M., & Tahir, P. M. (2010). Oil palm trunk fiber as a bio-waste resource for concrete reinforcement. *International Journal of Mechanical and Materials Engineering, 5*(2), 199–207.

Alengaram, U. J., Salam, A., Jumaat, M. Z., Jaafar, F. F., & Saad, H. B. (2011). Properties of high-workability concrete with recycled concrete aggregate. *Materials Research, 14*(2), 248–255.

Al-Khalaf, M. N., & Yousif, H. A. (1984). Use of rice husk ash in concrete. *International Journal of Cement Composites and Lightweight Concrete, 6*(4), 241–248.

Altwair, N. M., Megat Johari, M. A., & Saiyid Hashim, S. F. (2012). Flexural performance of green engineered cementitious composites containing high volume of palm oil fuel ash. *Construction and Building Materials, 37*, 518–525.

Awal, A. S. M. A., & Abubakar, S. I. (2011). Properties of concrete containing high volume palm oil fuel ash: A short – term investigation. *Malaysian Journal of Civil Engineering, 23*(2), 54–66.

Chindaprasirt, P., Homwuttiwong, S., & Jaturapitakkul, C. (2007). Strength and water permeability of concrete containing palm oil fuel ash and rice husk–bark ash. *Construction and Building Materials, 21*(7), 1492–1499.

CIDB. (2010). *IBS road map 2011–2015*. Kuala Lumpur: Construction Industry Development Board.

CIDB. (2015). *Construction industry transformation plan (CITP) 2016–2020*. Kuala Lumpur: Construction Industry Development Board.

Della, V. P., Kühn, I., & Hotza, D. (2002). Rice husk ash as an alternate source for active silica production. *Materials Letters, 57*(4), 818–821.

Economic Planning Unit. (2015). *Eleventh Malaysia plan 2016–2020*. Putrajaya: Prime Minister's Department.

Ganesan, K., Rajagopal, K., & Thangavel, K. (2008). Rice husk ash blended cement: Assessment of optimal level of replacement for strength and permeability properties of concrete. *Construction and Building Materials, 22*(8), 1675–1683.

Islam, M. M. U. I., Mo, K. H., Alengaram, U. J., & Jumaat, M. Z. (2016). Durability properties of sustainable concrete containing high volume palm oil waste materials. *Journal of Cleaner Production, 137*, 167–177.

Karim, M. R., Zain, M. F. M., Jamil, M., Lai, F. C., & Islam, M. N. (2011). Use of wastes in construction industries as an energy saving approach. *Energy Procedia, 12*, 915–919.

Kishore, R., Bhikshma, V., & Prakash, P. J. (2011). Study on strength characteristics of high strength rice husk ash concrete. *Procedia Engineering, 14*, 2666–2672.

Madandoust, R., Ranjbar, M. M., Moghadam, H. A., & Mousavi, S. Y. (2011). Mechanical properties and durability assessment of rice husk ash concrete. *Biosystems Engineering, 110* (2), 144–152.

Megat Johari, M. A., Brooks, J. J., Kabir, S., & Rivard, P. (2011). Influence of supplementary cementitious materials on engineering properties of high strength concrete. *Journal of Construction and Building Materials, 25*, 2639–2648.

Mehta, P.K. (2001). Reducing the environment impact of concrete. *Concrete International, 23* (10), 61–66.

Meyer, C. (2009). The greening of the concrete industry. *Journal of Construction and Building Materials, 31*, 601–605.

Mohamed, A. I., & Huzaifa, H. (2008). Palm oil fiber concrete. The 3rd ACF International Conference ACF/VCA 2008, A.40, pp. 409–416.

Payam, S., Hilmi, M., Jumaat, M. Z., & Majid, Z. (2014). Agricultural wastes as aggregate in concrete mixtures – a review. *Journal of Construction and Building Materials, 53*, 110–117.

Rodríguez de Sensale, G. (2006). Strength development of concrete with rice-husk ash. *Cement and Concrete Composites, 28*(2), 158–160.

Tangchirapat, W., Jaturapitakkul, C., & Chindaprasirt, P. (2009). Use of palm oil fuel ash as a supplementary cementitious material for producing high-strength concrete. *Construction and Building Materials, 23*(7), 2641–2646.

Tangchirapat, W., & Jaturapitakkul, C. (2010). Strength, drying shrinkage and water permeability of concrete incorporating ground palm oil fuel ash. *Cement and Concrete Composites, 32*(10), 767–774.

Vlastimir, R., Mirjana, M., Snezana, M., & Ali, E. S. A. M. (2013). Green cycled aggregate concrete. *Journal of Construction and Building Materials, 47*, 1503–1511.

Zeyad, A. M., Megat Johari, M. A., Tayeh, B. A., & Yusuf, M. O. (2017). Pozzolanic reactivity of ultrafine palm oil fuel ash waste on strength and durability performances of high strength concrete. *Journal of Cleaner Production, 144*, 511–522.

Chapter 37
Study on an Economic Feasibility of Wind-Diesel Hybrid System

Case Study in Baragoi, Samburu County, Kenya

Khisa Sirengo, Ryohei Ebihara, Hannington Gochi, and Tsutomu Dei

37.1 Introduction

In the design of hybrid systems, there was need to consider the technological feasibility vis-à-vis economic feasibility. The research discussed in this paper analyzes the local wind resource and evaluates the economic feasibility of supplementing the current diesel-energy generation system in Baragoi village in Kenya with a wind turbine generator. This paper will cover the status of the existing energy system, the data collection, and analysis procedures for the prefeasibility portraying the resource availability. Finally the conceptual design considering the cost analysis of the hybrid wind-diesel generation system will be presented.

In the promotion of renewable energy, seeking solutions to the energy challenge using appropriate technology is important while considering the locations, environment, culture, and social background. In Baragoi majority of the population live a nomadic lifestyle. However, a number of permanent residential houses are being developed. Therefore, there is potential for growth in the near future.

Generally, for non-oil-mining countries, the cost of power production by diesel generation is relatively high compared to the relative power cost from the main grid (price per kWh energy). Therefore, the price of power produced by diesel generators is considerably higher especially when considering the constantly increasing petroleum oil prices and the related fuel transportation costs.

K. Sirengo (✉) • R. Ebihara • T. Dei
Dei LAB, Department of Renewable Energy, Ashikaga Institute of Technology, Ashikaga, Japan
e-mail: ksirengo@gmail.com

H. Gochi
Rural Electrification Authority, Nairobi, Kenya

A. Sayigh (ed.), *Transition Towards 100% Renewable Energy*,
Innovative Renewable Energy, https://doi.org/10.1007/978-3-319-69844-1_37

37.1.1 Location and Topography

Baragoi is located in Samburu County in the Great Rift Valley. It is the most densely populated area of Samburu County with a population of 4,694 and 924 households (2009 census), covering an area of 2.2 Km^2 (2,102 persons/ Km^2). The main economic activity in Baragoi is livestock farming and small-scale trading (JICA 2013). Figure 37.1 shows the location of Baragoi in the Kenyan map.

37.2 Review of Existing Power System

The system under study already has a 300 kVA diesel generator installed. The network as shown in Fig. 37.2 comprises of one diesel generator connected to a small grid supplying mainly two schools, a hospital, and the administration offices.

Fig. 37.1 Location of Baragoi

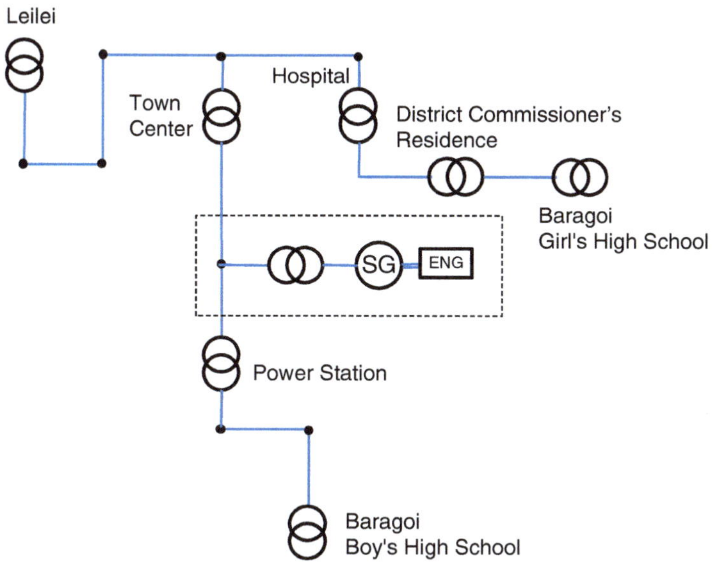

Fig. 37.2 Distribution network (11kVA) in Baragoi

In the network there are five step-down transformers at a girl's high school, boy's high school, district commissioner's residence, hospital, and power station.

There are also two step-down transformers at the town center and at Leilei. The consumption rate of diesel fuel was about 3.382 kWh/L.

37.2.1 Present Status of the System

From the site, a one-day data was collected, and the results are shown in the graph on Fig. 37.3. It can be seen that the peak consumption was recorded at around midday. From this data diesel generator, consumption was about 350 l per day (JICA 2013). However, due to insufficient previous data on the fuel consumption rates, an assumption of 25% of the full load rate was used.

37.3 Resource Assessment

37.3.1 Wind Analysis

Wind monitoring and analysis of obtained data are so important for implementing any wind project. As a result, proper sizing can be done for the wind turbine. Overestimating the wind speed will result in less wind energy being available than

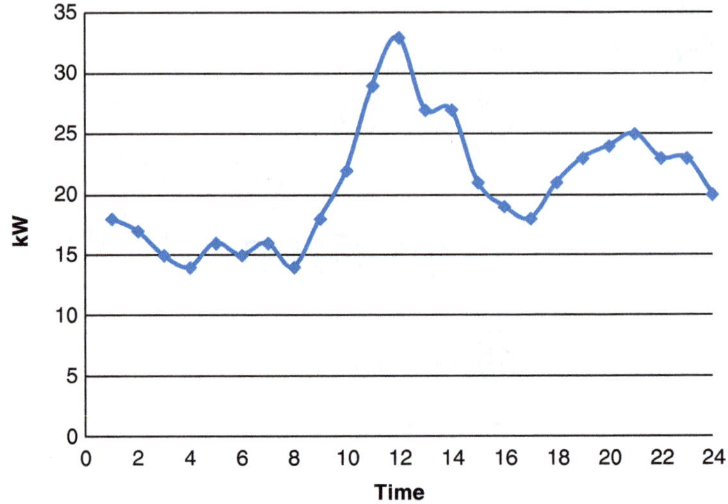

Fig. 37.3 Power demand curve (Baragoi, 13 December 2013)

Table 37.1 Baragoi average wind/max wind speed

Year	Month	Average wind speed (m/s)		max. wind speed (m/s)	
		20 m a.g.l.	40 m a.g.l.	20 m a.g.l.	40 m a.g.l
2010	Sept	4.4	5.1	14.2	14.8
	Oct	4.7	5.6	15.9	17.3
	Nov	3.4	5.1	16.7	17.6
	Dec	5.1	6.0	17.3	20.6
2011	Jan	5.3	6.1	17.4	17.8
	Feb	5.0	5.8	17.2	18.7
	Mar	5.1	6.1	20.1	22.0
	Apr	5.0	6.1	26.2	26.4
	May	4.0	5.0	18.1	20.0
	Jun	3.6	4.2	12.8	14.3
	Jul	3.8	4.4	13.6	14.2
	Aug	3.8	4.4	13.5	15.2
Average		4.4	5.3	16.9	18.2

expected, which may be uneconomical for the project. Underestimates are less dangerous in the economic sense but may mean that the wind potential of the area has not been fully exploited, with possible overall saving cost. The monitoring of wind was done for 12 months, and the data was used to study the economic feasibility of a wind-diesel hybrid system as shown in Table 37.1.

The annual average wind speed at 20 m above ground level was 4.4 m/s and that of 40 m above ground level was 5.3 m/s. On the basis of the summarized monthly wind data, monthly average wind speed was high between December 2010 and April 2011 and low between June and August 2011.

37.4 Wind-Diesel Hybrid

Several literatures have studied the economic feasibility of hybrid energy systems, and the use of a hybrid wind-diesel system is a promising solution to reduce the fuel cost and apparently the energy cost compared to only a diesel generator. Further literature also shows that generation capacity is determined to best match the power demand by minimizing the difference between total power generation and load demand over a period of 24 h (Azmy and Erlich 2005).

Based on economic savings, benefits are maximum with high-penetration wind-diesel hybrid systems. However, this means that the energy produced from the wind turbine at any time must be a little more than the load supplied to the system. As a result, high and medium penetration systems need to have a means of energy storage. Accordingly, if the system was designed in such a way that the diesel engine has to run full time, then it can be classified as low or medium penetration system. As is well known, batteries have a high life cycle cost [2]; therefore, for low wind penetration, storage can be eliminated from the system design. As a result of this, the penetration factor must be kept low due to the constraints related to the operation of the diesel generator. Hence, the power from wind has to be made less than 50% peak instantaneous penetration.

Since wind fluctuates greatly during the day, it is difficult to determine the spinning reserve of the diesel engine in the low and medium penetration systems, and therefore care must be taken as this can cause an overload to the diesel engine (Ko 2008). The diesel can be operated in idle mode when there is enough energy from the wind turbine. However, beyond a certain penetration, running the diesel engine at 20–30% of its nominal output power is very inefficient.

37.4.1 *Proposed Wind-Diesel Hybrid System*

In this prefeasibility study, low wind penetration system was recommended for Baragoi diesel station also to accumulate experience of operation of wind-diesel hybrid system. Therefore, target annual energy penetration should be less than 20% of the total energy demand.

$$\text{AEP}\ (\%) = \frac{\text{AEO (kWh)}}{\text{APD (kWh)}} \times 100(\%)$$

Where:

AEP: annual energy penetration
AEO: annual wind energy output
APD: annual primary energy demand

The following calculation assumed appropriate wind capacity to meet with wind penetration. In this simple prefeasibility study, target wind penetration was decided as 15% for low wind penetration system.

Annual primary energy demand can be estimated as follows:

$$\begin{aligned}\text{Annual Primary Energy Demand} = &\ 500\,\text{kWh/day} \times 365\ \ (\text{day/year}) \\ &= 182{,}500\,\text{kWh/year}\end{aligned}$$

Therefore, WG (kW) for 15% wind penetration becomes as follows:

$$\text{WG(kW)} = 0.15 \times 182{,}500/1752 = 15.63$$

Therefore, in this simple prefeasibility study, total capacity of 15 kW wind turbine was selected.

37.4.2 Cost Estimate

Cost for installation works and the transportation to project site are depending on the contractor and selected project sites. Table 37.2 shows a sample of installation costs for the wind turbine installation.

37.4.3 Economic and Financial Evaluation

The cost of installation of wind turbine is simply the cost of the wind turbine, tower, wiring installation, taxes, and so on. Maintenance costs are expenses for servicing or repairing the wind system.

Table 37.2 Costs of wind turbine installation

	Unit	Unit cost (USD)	Units	Cost(USD)	Cost(kSh)
Wind turbine	kW	5,000	15	75,000	6,600,000
Inverter	kVA	700	15	10,500	924,000
Battery	kAh	3,600	4	14,400	1,267,200
Control house	(1/system)	2,000	1	2,000	176,000
Installation materials	kW	600	15	9,000	792,000
Installation work	kW	500	15	7,500	660,000
Transportation	kg	3	1,000	3,000	264,000
Total investment cost	–	–	–	121,400	10,683,200

Table 37.3 Estimated installation cost

Generation capacity of the wind turbine	15	kW
Capacity factor (wind)	16.0%	
Operation hour per day	24	Hours
Operation day per year	360	Days
Annual power generation (wind)	20,736	kWh/Year
Life time of wind turbine	20	Years
OM cost of the generation system (% of the investment)	2.0%	
Diesel fuel cost per litter	105.0	kSh/Litter
Fuel consumption	0.67	Litter/kWh
Fuel cost per kWh	70.35	kSh/kWh
Benefit (replaced diesel cost)	1,458,778 kSh/Year	

One of the advantages using wind energy over generating electricity by conventional means is the free fuel. Once wind turbines are installed, the energy produced costs little over the remaining life of wind turbines. Table 37.3 shows a sample benefit from an installation of wind turbine with 15 kW. The benefit is the diesel cost which was replaced by power supplied from the wind turbine.

Table 37.4 shows EIRR of the installation of wind turbines at an existing diesel power plant. The costs for the rehabilitation of the existing diesel power plant are not included. According to the Central Bank of Kenya, the discount rate was 7% as of 31 December 2010 in Kenya. In April 2014 the World Bank released its commodity forecast, which predicts that world crude oil price will decrease from $104/barrel in 2013 to $97/barrel by 2020. In Table 37.4, fuel price escalation rate for 20 years was estimated as 2%.

37.5 Conclusion

The conceptual design of hybrid system with existing diesel power station was studied. The methodology of qualifying the hybrid system considered economic feasibility and technical performances too. Then using life cycle cost modelling, the optimum configuration of a hybrid system was obtained based on economic internal rate of return (EIRR).

The system under study is suitable for areas with a small energy demand and where connection to the national grid is economically not feasible due to distance from the national grid transmission line. The goal of designing the hybrid system is mainly to reduce fuel consumption rate of the diesel generator. From the analysis, the EIRR was obtained as 10%. From the research findings, the wind generator size was calculated as 15 kW, which proved that a low penetration system is suitable for Baragoi, considering also that low penetration is useful for acquiring experience of a wind-diesel hybrid system.

Table 37.4 EIRR wind turbine installation

Year	Benefit saved fuel cost (kSh)	Cost						Cumulative
		Initial investment	O&M	Battery	Inverter	Controller	Total	Revenue
0		−10,683,200					−10,683,200	−10,683,200
1	1,458,778		−132,000				−132,000	1,326,778
2	1,487,953		−134,640				−134,640	1,353,313
3	1,517,712		−137,333				−137,333	1,380,379
4	1,548,066		−140,079				−140,079	1,407,987
5	1,579,028		−142,881				−142,881	1,436,147
6	1,610,608		−145,739	−1,267,200			−1,412,939	197,670
7	1,642,821		−148,653				−148,653	1,494,167
8	1,675,677		−151,627		−924,000	−330,000	1,405,627	270,050
9	1,709,190		−154,659				−154,659	1,554,531
10	1,743,374		−157,752				−157,752	1,585,622
11	1,778,242		−160,907				−160,907	1,617,334
12	1,813,807		−164,125	−1,267,200			−1,431,325	382,481
13	1,850,083		−167,408				−167,408	1,682,675
14	1,887,084		−170,756				−170,756	1,716,328
15	1,924,826		−174,171				−174,171	1,750,655
16	1,963,323		−177,655		−924,000	−330,000	−1,431,655	531,668
17	2,002,589		−181,208				−181,208	1,821,381
18	2,042,641		−184,832	−1,267,200			−1,452,032	590,609
19	2,083,494		−188,529				−188,529	1,894,965
20	2,125,164		−192,299				−192,299	1,932,864
	35,444,459		−3,207,253	−3,801,600	−1,848,000	−660,000	−20,200,053	15,244,406
EIRR 10.0%								
Project life	20		Years					
Number of target households	150		HHS	Fuel price escalation rate		2%		
Exchange rate (as of June 2014)	102.0 Yen/US$			Discount rate		7% (31 Dec. 2010, Central Bank)		

References

Azmy, A. M., & Erlich, I. (2005). Impact of distributed generation on the stability of electrical power systems. *IEEE Power Engineering Society General Meeting, 2*, 1056–1063.

Japan International Corporation Agency. *Attachment M-2*, (2013). Simple pre-feasibility study on wind power development in Baragoi.

Ko, H. S., Lee, K. Y., Kang, M. J., & Kim, H. C. (2008). Power quality control of an autonomous wind–diesel power system based on hybrid intelligent controller. *Neural Networks, 21*(10), 1439–1446.

References

Chapter 38
Prospects of Renewable Energy Development Within Remote or Rural Areas in Indonesia

Sumarsono, Koesmawan, Harry Santoso, and Iskandar Andi Nuhung

38.1 Introduction

National Energy Policy (REN) aims for the independence and national security of energy to support sustainable national development, with objectives that (i) energy resources are not used as merely export commodities, but as national development capital; (ii) technological ability, the energy industry, and domestic energy services are developed to be independent and have improved human resources; (iii) jobs are created; and (iv) environmental functions are preserved (government regulation no. 79, 2014).

Indonesia's energy security is needed for the welfare of society and is determined by sustainability, environmental safety, and economic considerations. Two energy resources exist, namely, (i) non-renewable resources, which include oil, gas, coal, and nuclear power; and (ii) renewable resources, which include micro-/mini-hydro, geothermal, and solar power, biofuels, and biomass, wind, and ocean energy.

Sumarsono (✉)
Ffaireness Indonesia Daya (FID), State Oil Company Republic of Indonesia, Jakarta, Indonesia
e-mail: sumarsono318@gmail.com

Koesmawan (✉)
Ffaireness Indonesia Daya (FID), Ahmad Dahlan School of Economics, Jakarta, Indonesia
e-mail: mkoesmawanas@gmail.com

H. Santoso (✉)
Ffaireness Indonesia Daya (FID), Ministry of Forestry,
Republic of Indonesia, Jakarta, Indonesia
e-mail: harrysantoso52@gmail.com

I.A. Nuhung (✉)
Ffaireness Indonesia Daya (FID), Ministry of Agriculture,
Republic of Indonesia, Jakarta, Indonesia
e-mail: andi_nuhung@yahoo.co.id

A. Sayigh (ed.), *Transition Towards 100% Renewable Energy*,
Innovative Renewable Energy, https://doi.org/10.1007/978-3-319-69844-1_38

The content of this chapter is derived from research results and a summary of a seminar on "Acceleration of Indonesia Electricity Program Year 2019 by Renewable Energy and Environmental Friendly: The Debottlenecking of Electricity Provision 35.000 MW," which was held in Jakarta in November 2016.

38.1.1 The Importance of Energy for Human Life, Especially Economies

Energy supports human beings and civilizations. The existence of energy accompanies human life. Energy possibly already existed in the world before the first human was born. Human beings need energy to stay alive and continue life to the next generation. The more people are in the world, the more energy must be prepared and provided. Therefore, energy must be maintained and continuous for human life and cultures. Moreover, living organisms also requires energy to stay alive; humans, for example, require energy from food and to metabolize oxygen. Civilizations require a supply of energy for various activities. Energy sources such as fossil fuels are an important topic in economics and politics.

Climate change and ecosystems are affected by the energy obtained from solar radiation and from geothermal energy derived from the earth. Therefore energy is very important for human life – both the development of life and the development of economics. Economic developments should be supported by development in energy; when Indonesia requires economic growth of 6%/year, then it must provide energy growth of 12%/year (Minister of Financial 2016).

38.1.2 National Energy Conditions and Challenges

Energy management for national interests in Indonesia creates major challenges: (i) energy governance currently does not add optimal economic value; (ii) energy use in various sectors is still not efficient; (iii) dependence on fossil energy is still high and is not offset by the supply within the country, and non–fossil fuel energy use is still low; (iv) gaps exist in supplies of energy to users as a result of difficulties in distribution and a lack of infrastructure; (v) low investment in energy because of high risk; (vi) delays in the development of renewable energy because energy prices are not economical and the application of subsidies in some types of energy is less precise; (vii) dependence on imported energy technology; (viii) lack of public access to electric energy and gas, especially in eastern Indonesia; (ix) renewable energy development is moving slowly because of budget limitations; and (x) renewable energy research has not been integrated well, with the result that many energy research results are not useful for the development of renewable energy policy (Ministery of Energy 2016).

38.1.3 Forms of Energy

Energy takes various forms, but all types of energy must fulfill various conditions: they can be converted into other forms of energy, they follow the law of conservation of energy, and they cause changes in body mass upon exposure. Common forms of energy include kinetic energy of a moving object; radiation energy from light and electromagnetic radiation; potential energy stored in an object because of its position within a gravitational field, electric field, or magnetic field; and heat energy, which consists of potential energy and microscopic from movements of irregular particles. Fossil fuels are running out; therefore, several countries, such as the Netherlands, have disrupted supplies of fossil energy and are replacing them with other alternative energy, namely, wind and wave power. Oil will soon run out in Indonesia; the country's oil reserves were only about 3.5 billion barrels. If the average production is 800,000 barrels/day, then the reserves are enough for only 11 years (Ministry of Energy and Mineral Resource 2015).

38.1.4 Commitment to Economic Development: Low Emissions

Indonesia ratified the Paris COP 21 Agreement to reduce the rate of global temperature increase of 1.5–2 °C, with the following commitment: (i) assign nationally determined contributions gradually, during the first phase to reduce greenhouse gas emissions by 29% for business as usual or 41% with international cooperation by 2030, through the use of energy sectors including transportation (16.87%), forestry (7.22%), agriculture (1.21%), industry (0.71%), and waste (2.99%); (ii) develop a green economy and sustainable development, namely, economic development of low emissions through the use of mixed energies by the year 2025, in which renewable (nonfossil) energy will be increased to 23%, while fossil fuels will be decreased (coal 30%, oil 25% and gas 22%.), and expand peatland conservation efforts (National Committee of Energy and Industry 2016).

38.1.5 Rules and Regulations

The Indonesian government does have some rules and regulations in energy processing:

(i) The Constitution of 1945, article 33, is the main reference for our law for all activities. Energy resources are a natural wealth; they must be controlled by the state and must be used fully for the welfare of the Indonesian people.

(ii) Law No. 30 of 2007 of Energy.

Provision of energy by the Government and/or regional governments take [s] precedence in underdeveloped regions, remote and rural areas using local energy sources, especially renewable energy sources. Provision of new energy and renewable energy shall be enhanced by the Government and local governments in accordance with their authority. Provision of energy from new energy sources and renewable energy sources conducted by enterprises, permanent establishments, and individuals may obtain facilities and/or incentives from the Government and/or regional government in accordance with its authority for a certain period until the economic value is reached.

38.1.6 Potential, Conditions of Use, and New and Renewable Energy

38.1.6.1 Potential

The fossil energy resource potential versus the production reserve ratio is 7.73 Million barrels (MB) for petroleum (18 years), 152.89 TCF for natural gas (61 years), and 103 metric tons for coal (147 years). Moreover, the potential versus nonfossil energy resources (percentage of energy used) are as follows: marine power (ocean waves, tides, currents, and sea temperature difference) 60 GW (0.000458%), geothermal 29 GW (4%), solar 4,8 KWh/m^2 or 11 GW (0.00001%), wind 9.9 GW (0.005%), micro hydro 769,7 GW (0.1%), biofuel 9 GW (ethanol, biodiesel, biogas, biomass). Oil, gas, and coal comprise 95.3% of fossil energy consumption (41.8% oil, 23.8% gas, 29.7% coal) (Ministry of Energy and Mineral Resource 2015).

38.1.6.2 Used Condition of Energy in Indonesia

Energy can be used for many purposes: industry, transportation, households, and others. According to the most recent data, 37.2% of energy is used in industry, 33.9% in transportation, 10.1% in domestic use, 6.5% commercially, and 12.3% from raw materials. About 777,000 bpd of oil is being produced, and 1.6 million bpd of fuel is being consumed, with growth of ~5–8%. In addition, fuel consumption in 2025 is estimated to be 2.5 million bpd.

38.1.6.3 New and Renewable Energy

According to Rencana Umum Energy Nasional (National Energy General Planning) (RUEN), the target for the primary energy supply of new and renewable energy by 2025 are 400 MTOE, which includes 22% gas, 30% coal, 25% oil, and 23% new and renewable energies. The target of new and renewable energy, at 92.2 MTOE, consists of 45 GW electricity from new and renewable energy; 69.2 MTOE from geothermal, hydro, biofuels, solar, wind, and others; and the remaining 23 MTOE from biofuel, biomass, biogas, and Carbon Per Millimeter (CPM). A pilot nuclear power plan in Indonesia must be stopped. We set nuclir energy as the last alternatives.

38.1.7 Prospects for Renewable Energy in the Region

This chapter focuses on future prospects or renewable energy at the level of villages in Indonesia – villages that have been developed and remote villages far from major cities, particularly those far from the capital city of the province.

Indonesia is an archipelago located near the equator. One positive aspect is that the state always received solar light each day and has much wind and many natural rivers. The country also has a position as a state of a continent, receiving a portion of the huge profits associated with energy supply for the future. Simply put, energy can be prepared by the government in rural areas across the country, include (i) sunlight available year-round; (ii) wind, although only some wind are consistent; (iii) hydropower by utilizing the flow of rivers and waterfalls, as well as motion (waves) and temperature differences in the ocean; and (iv) bioenergy, where raw materials can be obtained from biomass waste and energy crops. Waste biomass can be used to generate gas and electricity, especially for households. Energy crops can be used to create (i) bioethanol as a substitute for gasoline, from crop species such as sugar cane (*Saccharum officinarum*), palm sugar(*Arenga pinnata*), cassava (*Manihot esculenta*), corn (*Zea mays*), and sorghum (*Sorghum* spp); (ii) biodiesel as a substitute for diesel fuel, which comes from palm oil (*Elaeis guineensis*), coconut (*Cocos nucifera*), jarakpagar (*Jatropha curcas*), kemirisunan (*Reutealis trisperma*), and nyamplung (*Calophyllum inophyllum*); and (iii) integrated power generation and biofuels.

Availability and suitability of land, cultivation technology, production/productivity, processing machines, and business partnerships between the private sector and rural communities need to be prepared further in order to develop energy crops.

38.1.8 Efforts of the Government

The government's policy has been formulated and needs to be realized, namely, (i) maximizing renewable energy, (ii) minimizing the use of petroleum, (iii) optimizing the use of natural gas and new energy, (iv) using coal as the mainstay of the national energy supply, and (v) using nuclear energy as a last option.

Conservation strategies for energy efficiency have been proposed to be continued and maximized: (i) a mandatory minimum energy performance standards program and labeling of household appliances; (ii) human resources development for managers and energy auditors for industries, buildings, and others; (iii) implementation of Presidential Instruction no. 13 (2011) on energy and water savings in buildings and buildings owned by the government; (iv) the use of light-emitting diodes in public facilities, such as street lighting; and (v) a campaign to cut the budget for energy efficiency by 10% (Ministry of Energy and Mineral Resource 2016).

To accelerate future economic growth, we hope for no more restrictions on electricity consumption for individuals and business communities and industry. Therefore, it is necessary to prepare electricity that is available in sufficient quantities and is inexpensive. Because it is no longer possible to rely on fossil fuel–based electricity, the government has requested that an electrification program be accelerated, and that renewable energies be sovereign based on environmentally friendly concepts.

The government will formulate a regulatory body and an independent supply of energy so that energy policy is no longer a rivalry among various sectors. The government must maximize the utilization of renewable energy. Involving the public and private sectors will reduce the government's financial burden. Provision of energy should benefit all parties – the government, private companies, and communities.

38.1.9 Research Purposes

The main objective of this research is to explore the possibility of better prospects for renewable energy in rural areas by maximizing local wisdom so that the energy can be properly managed, and there is no other word to say, that the future energy will not depend on fossil energy of universe (nature), it is true, but without the man who seeks to provide alternative friendly to substitute for fossil energy environments. Otherwise there will be no more life on earth since life is depended on energy. Given that country comprises thousands of islands, the focus on energy development must be initiated from either the boundary areas between islands and between countries, or the periphery.

38.2 Theoretical Framework and Methodology

38.2.1 Theoretical Framework: Some Future Alternative Energies

Renewable energy is briefly explained below.

(i) Renewable energy is energy derived from natural processes which are sustainable, such as solar power, wind power, current air-process biology, geothermal energy, and biomass. The concept of renewable energy was introduced in the 1970s in an effort to offset the development of nuclear energy and fossil fuels. The most common definition is an energy source that can be quickly restored naturally, in an ongoing process. Based on this definition, nuclear and fossil fuels are not included. A nonrenewable energy is eventually discarded.

(ii) All renewable energy is certainly also sustainable energy, because it is always available in nature for a relatively long time, so there is no need to worry that its source will be depleted. Non-nuclear energy does not include nuclear power as part of sustainable energy, because there are natural limits (say, hundreds of years) to the supply of uranium-235. However, nuclear activists have argued that nuclear energy, including sustainable energy if it is used as fuel in a fast breeder reactor for nuclear fuel reserve, could be "created" hundreds or thousands of times. Again, in Indonesia, nuclear reactors are the last option, if they are still in use.

(iii) Geothermal Energy.

Geothermal energy comes from radioactive decay in the center of Earth, which creates heat, and from the heat of the sun, both of which heat the Earth's surface. Geothermal energy can be used in three ways: For power plants and used in the form of electricity. As a heat source is utilized directly using a pipe into the bowels of the Earth. As a heat pump, which pumped straight from the bowels of the Earth.

(iv) Solar Energy.

The solar panel (photovoltaic arrays) on a small yacht at sea can charge 12 V to 9 amperes under conditions of full and direct sunlight. Because most renewable energy comes from "solar energy," this term can be a bit confusing. Solar power can be used to generate electricity using solar cells or a solar tower; then that energy is able to heat buildings directly or through a heat pump, to reheat food using a solar oven, or to heat water through solar-powered water heaters. Of course, the sun does not provide constant energy to any point on Earth, so its use is limited. Solar cells are often used to charge a battery at noon, and the battery power is used at night, when sunlight is not available.

(v) Wind Power.

The temperature difference at two different places produces different air pressures, resulting in wind. Wind is the movement of matter (air) and has been known for years. It is used to drive turbines, which in turn are used to generate kinetic and electrical energy.

(vi) Hydropower.

Water is used as energy because it has mass and flows. Water has a density 800 times that of air. Even slow movement of water is able to be converted into other forms of energy. Water turbines are designed to get energy from different types of reservoirs, calculated from the amount of water mass, altitude, and speed of water. Water energy is used from (i) power plants built into dams (the largest is the Three Gorges Dam in China); (ii) micro-hydro is built to generate electricity up to a scale of 100 kW; (iii) mini-hydro is commonly used in remote areas that have water sources and are built to harness the kinetic energy of the water flow without the need for a large water reservoir.

(vii) Biomass.

This kind of energy has several advantages: (i) biomass is the world's oldest and most efficient solar battery; (ii) biomass means healthy forests; (iii) biomass is abundant and one of the best available sources on Earth; (iv) biomass allows net-neutral carbon management and is a greenhouse gas control mechanism; (v) biomass creates a new energy economy.

Because fuel from biomass or wood is used in rural areas of Indonesia as a substitute for fossil fuels, forests become an important resource. The tree species used for firewood are those that grow quickly, are readily available, can be grown on marginal lands, are easily cultivated, quickly sprout after being cut, have a high calorific value, and create little smoke or poison gas. About 70 tree species seem to have potential for energy purposes where the calorific value is more than 100 GJ/ha/year.

Plants use photosynthesis to store solar energy, air, and carbon dioxide. Biofuels are fuels derived from biomass – organisms or products of animal metabolism, such as feces from cows. Biomass is also a source of renewable energy. Typically, biomass is burned to release the chemical energy stored in it; the exception is when biofuels are used to fuel a fuel cell (e.g., direct methanol or ethanol fuel cells). Biomass can be used directly as fuel or to produce other types of fuels, such as biodiesel, bioethanol, and biogas, depending on its source. Biomass takes three forms: solid, liquid, and gas. In general, biomass is produced through two methods: growing the organism producing biomass and clicking.

(viii) Liquid Biofuels.

Liquid biofuels are usually shaped bioalcohols such as methanol, ethanol, and biodiesel. Biodiesel can be used in modern diesel vehicles with little or no modification, and can be obtained from vegetable waste and animal oils and fats. Depending on the potential of each area, corn, sugar beets, sugar cane, and several types of grass can be cultured to produce bioethanol. Biodiesel is produced from crops or crops that contain oil (palm oil, copra, castor seeds, algae) and goes through various processes such as esterification.

(ix) Solid Biomass.

Solid biomass is usually directly used in the form of combustible solids, either wood or flammable plants. Plants can be grown specifically for combustion or may be used for other purposes, such as processing in specific industries and waste processing results, which can be burned as fuel. Biomass briquets can also use solid biomass, where the raw material can be a piece(s) of raw solid biomass or put through certain processes such as pyrolysis to increase the percentage of carbon and reduce the water content. Solid biomass can also be processed by means of gasification to produce gas.

(x) Biogas.

Various organic materials, biologically by fermentation or in physicochemical gasification, can release flammable gases. Biogas can easily be produced from a variety of industrial wastes that exist today, such as paper production, sugar production, and animal husbandry. Various waste streams must be diluted with water and allowed to ferment naturally, producing methane gas. The residue from the fermentation is a fertilizer rich in nitrogen, carbon, and minerals.

(xi) Small-scale Energy Sources.

(a) Piezoelectricity, an electric charge, results from the application of mechanical stress on a solid object. This object converts mechanical energy into electrical energy.
(b) Track time automatically through an automatic or self-winding clock or watch that is driven by stored mechanical energy, which is generated from the movement of the user's hands. The mechanical energy is stored within the spring mechanism.
(c) An electrokinetic platform (e.g., an electrokinetic road ramp) is a method of generating electrical energy by harnessing the kinetic energy of a moving car on a foundation built on a street. One such foundation has been installed in the parking lot of a Sainsbury's supermarket in Gloucester, United Kingdom, where the electricity generated is used to power the cash registers.
(d) Catch electromagnetic radiation that is not used and convert it into electrical energy using a rectifying antenna. This is one method of harvesting energy.

38.2.2 Matters of Renewable Energy Development

38.2.2.1 Design and Hazards

(i) The aesthetics of wind turbines or issues of nature conservation when solar panels are installed in rural areas. They must be well designed.
(ii) Wind turbines can be set up to avoid flying birds, and hydroelectric dam power plant cannot create barriers to migrating fish. Renewable energy should be safe for these animals.
(iii) The technologies must take advantage of existing infrastructure, for example, use solar collectors as a noise barrier along a highway, combine them as to create shade in the sun, put them on a roof that is already available or have them replace the roof entirely.
(iv) The carbon released into the atmosphere can be properly reabsorbed if the biomass-producing organisms are continuously cultivated.
(v) Another problem with many renewable energies, particularly biomass and biofuels, is the large amount of land needed for cultivation efforts. Fortunately, in much of Indonesia, it is still possible to develop land to create more energy.

38.2.2.2 Concentration

(i) Renewable energy is not merely concentrated in one particular area, but rather is set up in selected locations.
(ii) Wind energy is the most difficult to focus; thus the solution should require large turbines to capture as much wind energy as possible. This will be no problem for Indonesia, where much of the area has stable wind conditions.
(iii) Water energy utilization methods depend on the location and characteristics of the source water, so water turbine designs must be well prepared.
(iv) Solar energy can be used in various ways, but the issue is getting the energy to areas in need. For comparison, under standard conditions of testing in the United States, 1 m^2 of solar cells have an efficiency of 20% and will produce 200 W. The testing standard conditions in question are an air temperature of 20 °C and irradiance of 1,000 W/m^2.

38.2.2.3 Distance to the Electrical Energy Receiver

Renewable energy can solve the problem of geographical diversity because it can be distributed in several locations throughout the country (national grid). This is also a significant problem, as renewable energy sources such as geothermal, hydro, and wind can be located far away from the receiver of electric energy: geothermal in the mountains, water upstream from receivers, and wind energy offshore or on a plateau. The solution to the environment-related problem of utilization of these resources on a large scale is likely to require considerable investment in transmission and distribution networks, as well as in the technology itself. Renewable energy is well established close to people in specific locations.

38.2.2.4 Availability in Nature

Renewable energy is available throughout the country, and it must be intensified for the benefit of local/rural people. However, one significant shortcoming is the universal availability of renewable energy; some locations have it only occasionally or not at all, because (i) sunlight is available only during the daytime (the solution is to prepare a reservoir of solar energy that powers over time); (ii) water energy cannot be used when the river is dry (the solution must prepare a reservoir in another place); or (iii) biomass has problems with climate and pests.

The abovementioned three problems currently make those kind of renewable energies still infeasible. Therefore we have to make a feasibility study a priority in the future.

38.2.3 Research Design

This chapter describes a desk study, a qualitative paradigm, and discussion with relevant parties. Renewable energy will remain the preferred energy if all three parties benefit – namely, (i) the community, (ii) the energy company, and (iii) the government. They all must receive great benefits from the renewable energy created.

38.3 Results

The general conclusion of this research is that the prospects for energy generation in rural areas in Indonesia is very attractive, and this renewable energy sector has received much investment. Therefore, for the rural sector – especially renewable energy – to continue to grow, this chapter comes to the following conclusions:

1. Given the position of Indonesia as an archipelago, making a national grid difficult to achieve, the conditions of energy independence in each region must algin with local wisdom and should continue to be pursued. the goal is to form independent areas of energy that are environmentally friendly.
2. Indonesia, located at the equator, has some advantages: sunlight on a daily basis, a lot of wind, many natural rivers, and the potentially to develop bioenergy (energy crops). These conditions can create more renewable energy alternatives and more choices for local people.
3. At present, we are in an energy crisis; if likened to a medical emergency, Indonesia's energy is in the emergency department. Fossil fuels will soon run out, so attention on environmentally friendly renewable energy needs to be enhanced.
4. Indonesia has a great potential ability to develop all kinds of energies: geothermal energy, solar energy, wind power, hydropower (waves/sea currents, water flow), biomass, biofuels (biodiesel, bioethanol), and biogas.
5. Renewable energy must become a national priority because of its high association with national security now and in the future. The concepts of diversification, conservation, intensification, and indexation of energy need to be developed and their potential and sustainability taken into account.
6. Indonesia has ratified the Paris COP-21 Agreement with a commitment to the first phase of reducing greenhouse gas emissions by 29–41% by 2030, and developing a green economy with low emissions, less fossil energy (coal 30%, oil 25%, and gas 22%), and more renewable energy (up to 23% by the year 2025).

38.4 Discussion and Implications

The Indonesian government is constantly working to improve investment in new and renewable energies in Indonesia through various incentives. It is currently undertaking a mix of energies to reduce dependence on the fossil fuel sector and to increase non–fossil fuel energy.

Now we have an energy road map and a national energy general plan to develop the energy sector. In addition, with reference to the comparative advantages of Indonesia (e.g., an archipelago, located at the equator), the prospects of renewable energy in rural areas will be better in the future.

The magnitude of the investment in renewable energy, is currently more expensive than fossil fuel–based energy; however, in the future it will be cheaper.

Renewable energy and fossil no need to be disputed. The development of renewable energy has become a trend in many parts of the world. The development of renewable energy is a must. The Indonesian government continues to strive to optimize the development of renewable energy. As a consequence, Indonesia needs a special institution that administers or manages renewable energy, because the present situation, in which the management of renewable energy is spread out over various sectors, requires each sector to possess their own policies. Moreover, Indonesia needs special laws and regulations concerning renewable energy in order to accelerate the benefits of renewable energy.

International cooperation needs to be developed to exchange experience and supply energy. With regard to the World Renewable Energy Congress XVI in 2017, we propose the establishment of an International Forum for Rural Renewable Energy.

References

ASTM G 173-03. (2013). *Standard tables for reference solar spectral irradiances: Direct normal and hemispherical on 37 tilted surface*. New York: ASTM International.

Demirbas, A. (2009). Political, economic and environmental impacts of biofuels: A review. *Applied Energy, 86*, S108–S117. London.

European Wind Energy Association. (2011). *EWEA executive summary, Analysis of Win Energy in the EU-25*. Brussels: European Wind Energy Association Accessed.

Ministry of Energy and Mineral Resources. (2015). *Statistics of energy and mineral resources*. Jakarta: ESDM Publisher.

Chapter 39
Present Status and Target of Japanese Wind Power Generation

Izumi Ushiyama

39.1 General

Japanese wind power market is still sleeping, due to the complicated cumbersome EIA procedure which was applied for all wind farms over 10 MW since Oct. 2012, and it needs about 4 years to finish this procedure. Only 203 MW with 10 projects have won the EIA permission, and the other 6226 MW with 88 projects are still in the EIA process at the end of Oct. 2014. The dawn light will shine in 2016, when most of these projects complete the EIA and start operation gradually. A lot of discussions and efforts are paid for cutting red tapes in Japan, especially for the EIA, grid access, and land-use restrictions now.

39.2 Market

At the end of 2015, 3120 MW of wind capacity had been installed in Japan, representing 0.5% of the total power supply in 2015. It includes 52.6 MW of offshore wind power (4 MW of floating, 4.4 MW of fixed foundation, and 44.2 MW of semi-offshore wind turbines). 130.43 MW (119.357 MW in net) with 64 new turbines were installed in 2014. It increased 2.7 times larger than in 2013, but it was still a very low level for the last 10 years. There were no additional offshore wind turbines in 2014. One 3 MW semi-offshore wind turbine starts operation in Feb. 2015 at Akita Port, and one 7 MW floating offshore wind turbine is to start operation in summer 2015 by Fukushima FORWARD project (Fig. 39.1).

I. Ushiyama (✉)
Ashikaga Institute of Technology, Ashikaga, Japan
e-mail: ushiyama@ashitech.ac.jp

A. Sayigh (ed.), *Transition Towards 100% Renewable Energy*,
Innovative Renewable Energy, https://doi.org/10.1007/978-3-319-69844-1_39

Fig. 39.1 Kaminokuni wind farm, in Hokkaido, 28 MW, MHI 2.4 MW × 12 units, ©J Power

39.3 Feed-In Tariff

The feed-in tariff (FIT) for onshore wind has maintained JPY 22/kWh (EUR 0.164/USD 0.185). And the Japanese Ministry of Economy, Trade and Industry (METI) had newly offered JPY36/kWh (EUR 0.27/USD 0.30) for offshore wind which use jack-up ship for turbine installation (i.e., it excludes semi-offshore wind turbines constructed from seashore). This tariff for offshore wind is set 1.64 times higher than onshore wind.

The tariff is to be reestimated every year according to the latest experience in Japan. On the other hand, the qualification of FIT can be gotten when the project almost finished the EIA procedure. It forces Japanese developers to take risk of forecasting FIT price at about 4 years in the future, when they start new projects. Only few developers who have strong financing ability can afford such uncertainty. Therefore, the Japan Wind Power Association (JWPA) requests Japanese government to shift the qualification timing a little earlier so as to make Japanese wind power development bankable.

39.4 Deregulation

The Ministry of Environment (MOE) and the METI have tried to shorten the EIA process period to within 2 years. The reestimation for the EIA contents was discussed. And the MOE started new subsidies supporting 50% cost for pre-EIA investigation, and it was applied for about 20 sites in FY 2014.

The strict rule for land use especially for farm was one of the big red tapes in Japan. The Ministry of Agriculture, Forest and Fisheries (MAFF) made a new law named "Act for the Promotion of Renewable Energy in Rural Districts (APRERD)" on 15 Nov. 2013, and APRERD went in operation on 1 May 2014. It will help the change of land-use purpose from "farm/agriculture" to "wind power/industry." It means the big increase of potential land area for wind power in Japan.

39.5 Grid Restriction

Solar power developers have rushed to propose FIT approval by Mar. 2014 (about 27 GW new proposal for Mar.), responding the announcement of FIT price reduction (from JPY 36/kWh to JPY 32/kWh) from April by the METI. The accumulated proposed photovoltaic capacity in Japan by the end of April had reached 68.43 GW which was about one fourth of Japanese total electric power plant capacity.

The requested solar FIT in Kyushu district was over 12GW, which is about 1.5 times larger than the maximum electricity demand of 8 GW at that district in low-demand holiday season. Therefore, Kyushu Electric Power Company had denied to new grid connection for all renewable energies including wind power. Other six electric companies (Hokkaido, Tohoku, Hokuriku, Chugoku, Shikoku, Okinawa) have followed to Kyushu. A big discussion had broken up in Japan about this "no" declaration against renewable energies. The METI had gathered all the stake holders urgently, made discussion, and announced the new grid connection rule in Oct. 2014. This new rule expands acceptable capacity by introducing the wider output control for renewable energies by power supply reasons. The new acceptable capacity for wind power at seven electric companies was 5.73 GW in total. 2.1 GW has been already in operation in these seven regions; therefore, 3.63 GW was available for new connection. The central three regions (Tokyo, Chubu, Kansai) have huge electric demands and are free from this kind of restriction.

The acceptable capacity may be reestimated for expanding in the future, because the above new grid connection rule did not consider the interregional grid operation. The electric power system reform was conducted by the Japanese government now. The new Organization for Cross-regional Coordination of Transmission Operators (OCCTO) who is in charge of dealing Japanese electric grid lines integrally shall start its work from April 2016. If Japanese grid lines are operated more favorably for renewable energies, the above grid restriction shall be eased dramatically in the future.

Most of Japanese wind resources exist in the northern rural regions called Hokkaido and Tohoku. Low population needs only small grid lines. The METI had started to build new grid lines for wind power at northern rural area in Japan (Hokkaido and Tohoku). The METI subsidizes about 50% of the building cost (25 billion JPY for every year). The grid building consortium for Hokkaido (for about 3 GW) had decided last year, and new two consortiums for Tohoku (for 600 MW at Akita and for 900 MW at Aomori) have also decided in 2014.

NEDO started a new national project to build up nationwide wind power output forecast system for 5 years. The fund for the first year was 4 billion JPY. The JWPA cooperates with this project.

39.6 Industry/Manufacturer

Three Japanese wind turbine manufacturers keep more than 60% domestic market share for several years. Their flagship new wind turbines start operation from 2013 to 2015. MHI's MWT167/7.0 7 MW turbine is to be applied for Fukushima FORWARD floating offshore wind power demonstration project in summer 2015. Hitachi's HTW5.0–126 5 MW turbine, as shown in Fig. 39.2, is to be used for several offshore wind power projects in Japan (Table 39.1).

39.7 Cooperation with Foreign Companies

The oversea experience and investment are useful to promoting wind power development in Japan, especially for offshore wind power.

Hitachi Zosen Corporation made technology alliance for floating offshore wind power with Norwegian StatoilHydro in Nov. 2012.

Maeda Co. who develops 60 MW Yasuoka wind farm in Yamaguchi pref. made technology alliance with Australian Macquarie group in Oct. 2013.

39.8 Offshore Wind Power Development

Japan has 52.6 MW of offshore wind power at the end of Feb. 2015, including two 2 MW floating wind turbines. Fukushima's floating turbine is located 20 km offshore from seashore. Further 914.5MW of 13 offshore wind power projects are under planning, and 504 MW of 6 projects within it are under the EIA process and preparing for the construction by around 2020 as shown in Table 39.2.

Fig. 39.2 Hitachi's HTW5.0-126 5 MW turbine at Kashima port side ©Hitachi

Table 39.1 New wind turbines developed by Japanese manufacturers

Company	Wind turbine	Rated output	Start operation	Type
MHI	MWT167/7.0	7.0 MW	Feb. 2015 at the UK	Hydraulic drive
Hitachi	HTW5.0-126	5.0 MW	Mar. 2015	Downwind
	HTW2.0-86	2.0 MW	Mar. 2014	Downwind
JSW	J100-3.0	3.0 MW	Sep. 2013 (2.7 MW)	Gearless PMSG

Fukushima FORWARD is a national project and the others are commercial projects. Kashima, Ishikari-Shinko, and Mutsuogawara projects are located in a designated area (port area). Fukushima and Yasuoka projects are located in an undesignated area. To get permission, social acceptance and compensation for

Table 39.2 Coming offshore wind power projects in Japan

Project name	Prefecture	Output	Wind turbines
Fukushima FORWARD	Fukushima	14 MW	5–7 MW
Kashima port no. 1	Ibaraki	125 MW	5 MW
Kashima port no. 2	Ibaraki	125 MW	5 MW
Yasuoka	Shimonoseki	60 MW	4 MW
Ishikari-Shinko	Hokkaido	100 MW	2.5–5 MW
Mutsuogawara	Aomori	80 MW	2.5 MW

Fig. 39.3 Fukushima FORWARD 7 MW and floater set sailed from Nagasaki Port

fishermen are much easier at a designated area, because almost all rights are controlled by the port chief in the Ministry of Land, Infrastructure, Transport and Tourism (MLIT) who is positive to offshore wind power.

As for undesignated area, there is no law or regulation for offshore wind power development in Japan now. The JWPA started holding study meeting with voluntary lawmakers in Dai-Ichi Tokyo Bar Association to fulfil the lack of necessary regulations.

The world's biggest class offshore wind turbine named "Fukushima Shinpu" was installed at 20 km offshore of Fukushima in July 2015 as shown in Fig. 39.3.

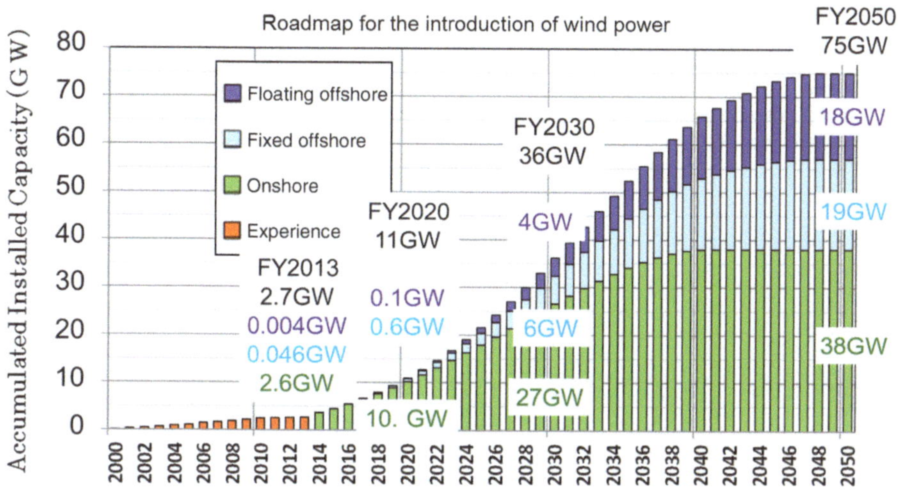

Fig. 39.4 Road map for the introduction of wind power in Japan

39.9 Wind Road Map 2050

One of the major issues in Japanese wind industry is a lack of long-term installation goal and road map. Without them, effective planning of subsidy by the government as well as of capital investment by manufacturers is not possible. Then, JWPA proposed wind road map 2050 based on wind potential in Japan. The Ministry of Environment published onshore and offshore wind potential of several wind speeds on March 2010, with consideration to almost all the social limitations including development regulation zones and distance from local residence.

The wind power target at 2050 was set as 50 GW so that it would produce 10% of the national electricity demand by wind. The installation targets for onshore and offshore wind at each utility zone were then calculated based on the wind potential. The road map 2050, which shows annual installation of onshore and offshore wind, was then estimated based on a well-known theory of "diffusion of innovation" that describes penetration of new technology in time follows S-shaped curve as shown in Fig. 39.4.

39.10 Conclusion

Japanese wind power market is waiting for the coming dawn. But, we are trying to cut many red tapes against the wind power development in Japan. Japan will grow up as the world's leading wind power market after 2016. We keep on making every effort to realize this dream in the future.

Reference

Ushiyama, I., Nagai, H., Saito, T., & Watanabe, F. (2010). Japan's onshore and offshore wind energy potential as well as long-term installation goal and its roadmap by the year 2050. *Wind Engineering, 34*(6), 701–720.

Chapter 40
Levelized Cost of Solar Thermal System for Process Heating Applications in the Tropics

Arifeen Wahed, Monika Bieri, Tse K. Kui, and Thomas Reindl

40.1 Introduction

In the ASEAN region, industrial energy demand accounts for more than a third of total energy consumption (Remap ASEAN 2015). According to the International Energy Agency (IEA), about one-third of industrial energy is required for heating applications (Renewable Heating & Cooling 2012). In general electricity and gases, which are the most common sources of energy, are utilized for commercial (e.g., hospitals, hotels, etc.) and industrial (e.g., preheating, pasteurization, evaporating, drying, distillation, etc.) heating of a fluid stream (e.g., hot water) and/or some reservoirs (e.g., liquid baths) in the temperature range of 60–150 °C. Due to the year-round availability of solar energy in the selected countries, an effective utilization of the solar thermal system can provide an alternative to conventional heating systems for industrial applications in the food, beverage, pharmaceutical, chemical, textile, machinery, and pulp and paper industry (Vannoni et al. 2008).

Even though this region has a good potential for the adaptation of industrial solar heating applications, very few installations have been reported except in Thailand (UNEP 2010) with a total installed capacity of about 450 KWh_{th}. Key challenges to integrate solar heating for industries is the lack of knowledge on technical knowhow and on economic viability of implementing this technology. To demonstrate the competitiveness of such technology, it is worthwhile to compare the solar heating investment with the long-term fossil fuel price trends. Thus, this paper studies an economic assessment for implementing solar heating systems for industrial applications in the ASEAN region.

A. Wahed (✉) • M. Bieri (✉) • T.K. Kui • T. Reindl
Solar Energy Research Institute of Singapore, Singapore, Singapore
e-mail: arifeen.wahed@nus.edu.sg; monika.bieri@nus.edu.sg

A. Sayigh (ed.), *Transition Towards 100% Renewable Energy*,
Innovative Renewable Energy, https://doi.org/10.1007/978-3-319-69844-1_40

40.2 Solar Heating System

For industrial heating applications, a variety of heat demand profiles has been observed. Integrating a solar thermal system into the industrial process heating requires to identify the most favorable integration concept from technical and economical viewpoints. An exemplary case of a solar thermal system integrated to an industrial heating application will be discussed in this section.

40.2.1 Solar Thermal System

The design baseline for integrating a solar thermal system for industrial heating is to ensure that the industrial heating demand matches either completely or partially by the heat supplied from the solar thermal system. The main components of a solar thermal system are solar thermal collectors, heat storage and discharge system, a hydraulic system, and a control system. As shown in Fig. 40.1, the solar thermal heat is harvested by the fluid (water) stream via solar thermal collectors in a closed-loop system. The absorbed heat is stored in a water storage tank. A discharging water loop is connected to the heat storage water tank where heat is transferred through a heat exchanger.

The solar thermal collector performance of irradiation-to-heat conversion depends on the collector technology, the operating mean temperature of collector fluid with respect to ambient temperature and incident irradiation. A variety of solar thermal collectors with different technologies, such as flat plate, evacuated tube, or concentrated collector, are available in the market. For tropical industrial heating applications in the temperature range of 60–150 °C, the evacuated tube collector is found to be the most appropriate technology. Three different collectors (type and manufacturer), as presented in Table 40.1, are considered for this analysis. A 2-h buffering heat storage (specifications presented in Table 40.1) is incorporated in order to level out the intraday irradiation fluctuation. Other systems of the solar thermal systems are the hydraulic systems including piping, fittings and valves, pumps, and insulation and the control systems including sensors, electrical, data

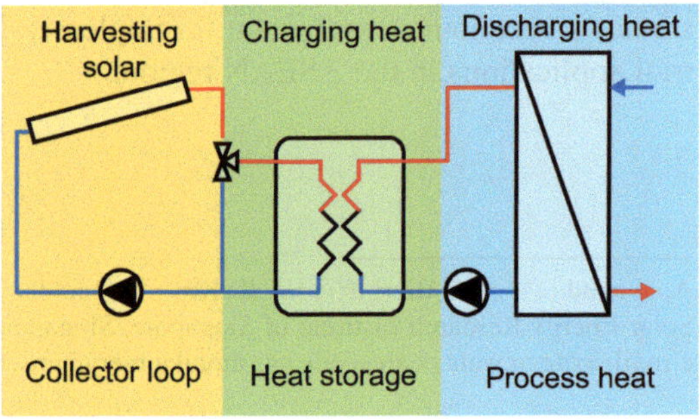

Fig. 40.1 Schematic diagram of a solar thermal system for process heat applications

Table 40.1 Solar thermal collector specification (KEYMARK 2016)

Parameters	Specifications		
Collectors			
Collector type, evacuated	A (water tube)	B (heat pipe)	C (heat pipe)
Aperture area, m^2	4.50	2.24	1.42
Efficiency parameters η_o, a, $W/(m^2K)$, b, $W/(m^2K^2)$	0.688, 0.583, 0.003	0.618, 1.376, 0.018	0.574, 1.917, 0.012
Water storage tank			
Tank capacity (m^3)	0.43		
Type	Commercial		
Total units (no.)	4		
Gas heater			
Fuel type	Natural gas		
Heating capacity (kWh)	117		
Tank capacity (m^3)	0.38		
Total units include backup (no.)	4		

acquisition, and monitoring devices. Costs of these subsystems are also included in the financial analysis.

In this study, a gas heater is chosen as a reference case to calculate the financial viability of a solar thermal system. Thus, a gas-fueled heating system is specified with the same heating capacity of the solar thermal system for comparison. A well-known gas heater, as specified in Table 40.1, is selected from a wide range of gas-fueled heating systems available in the market.

40.2.2 Solar Thermal Collector Model

The specifications of a solar thermal system for industrial heating depend on the specific heating load demand. In this analysis, an industrial heat demand profile is defined, as shown in Fig. 40.2, with assumptions – (i) the required heating temperature is in the range of 60–80 °C, and (ii) daily heat demand is during the day time from 7 a.m. to 6 p.m. for 52 weeks of a year and 4 weeks for maintenance.

A thermal system simulation program, TRNSYS 17, has been utilized to analyze the solar thermal collector performance at an hourly time step. The analysis assumes that the solar thermal system is capable to provide the defined heat load, as presented in Fig. 40.2. The collector thermal yield is considered as a basis for the comparison on different weather conditions for this analysis. For each collector performance evaluation, collector input data such as aperture area, efficiency parameters, incident angle modifiers, etc. were obtained from the database

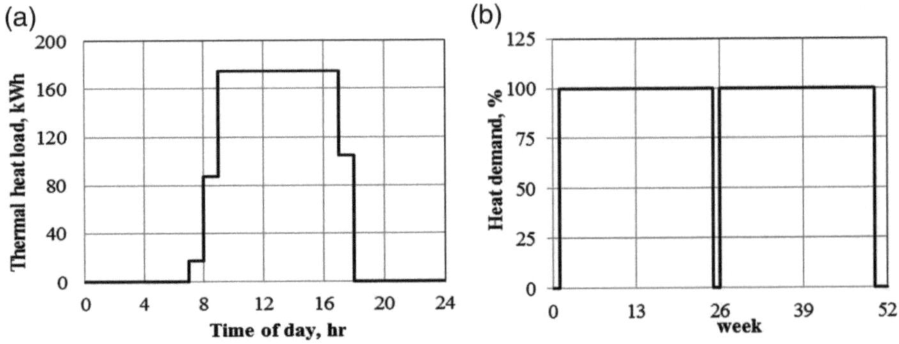

Fig. 40.2 Exemplary heat load demand profile of an industrial process heat application – (**a**) daily heat demand profile, (**b**) annual heat demand profile on weekly basis

(KEYMAR 2016). Hourly weather data were obtained from the TMY2 database for different countries – Indonesia (Medan Polonia), Malaysia (Kuala Lumpur), Singapore, and Thailand (Bangkok). The defined collector model was able to evaluate thermal yield of each evacuated collector type for those regions. Since there are variations of collector price and collector thermal yield performance, mean value of those matrices was considered in the financial analysis.

40.3 Financial Model

Based on the thermal yield of the collectors, a financial model was developed to estimate the life cycle cost of the thermal energy ($LCOE_{th}$) for both the solar thermal and the gas heating system. Figure 40.3 shows the flow diagram of the financial model.

40.3.1 Input Parameters

This work gives an update on the analysis done previously on a similar system configuration with the financial parameters being aligned to Singaporean market standards (Wahed et al. 2015). These parameters are updated here within and also extended to the other selected neighboring countries. Table 40.2 outlines the discount rates used for each country, based on country-specific risk-free rates (the 10-year bond yield was used), market risk premium, inflation, and tax rates. For all of the four markets, it was assumed that the investment's risk characteristics, i.e., investing in a solar or a gas-fueled thermal system, are in line with the overall market (beta = 1.0) with equal usage of financial leverage (i.e., 10-year debt

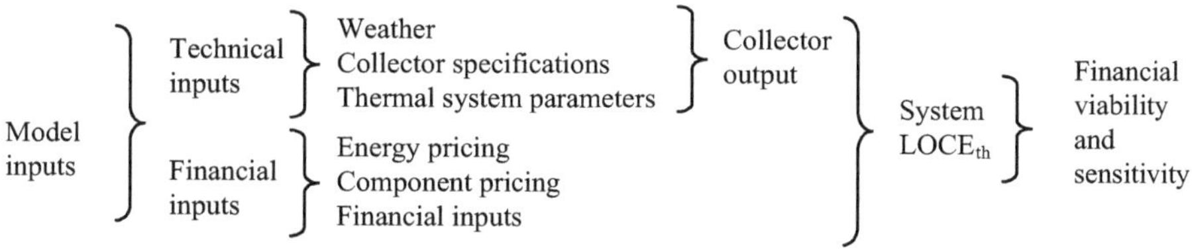

Fig. 40.3 Flow diagram of levelized cost modeling and analysis for solar thermal systems

Table 40.2 Financial assumptions for the base value calculation

Parameter	Singapore	Thailand	Indonesia	Malaysia
Risk-free rate (%)	2.5	2.7	7.7	4.3
Market risk premium (%)	6.1	8.4	8.0	6.5
Beta	1.0	1.0	1.0	1.0
Cost of equity (%)	8.6	11.1	15.7	10.8
Risk-free rate (%)	2.5	2.7	7.7	4.3
Debt premium (%)	3.0	3.0	3.0	3.0
Cost of debt (%)	5.5	5.7	10.7	7.3
Equity ratio (%)	40	40	40	40
Debt ratio (%)	60	60	60	60
Corporate income tax rate (%)	17	20	25	24
Nominal WACC (%) = discount rate	6.2	7.2	11.1	7.6
Annual inflation (%)	1.0	1.1	3.0	1.8

financing 60% of the upfront investment) at the same debt premium over local risk-free rates. No particular tax treatments are taken into account (e.g., accelerated depreciation schemes), and the operational life has been assumed to be 20 years.

Collector prices per square meter vary slightly among the countries due to the different aperture area required to fulfill the given fixed heat demand (see Table 40.3). The other system cost items include the hydraulic system (49% of the total, in the Singaporean context), the control system (16%), the 2-h water storage system (23%), and the installation (12%). Installation cost varies based on local labor cost. Labor cost contribution to overall system cost is hence quite different, being the highest proportion for Singapore (12%) to the smallest in the case of Indonesia (1%). No potential differences in labor productivity have been included. The replacement investment includes the electric pump replacement in the tenth year, and it differs on year 10 due to local inflation rate. The operating maintenance and insurance costs are assumed to be 1.0% and 0.5% of the initial investment, respectively, and inflated for the subsequent years. The electricity price is aligned to each country's industrial price (see Fig. 40.6) in 2016. For Singapore, the future progression of electricity price follows the SERIS contestable price

Table 40.3 Technical assumptions for the base value for the solar thermal system

Parameter	Singapore	Thailand	Indonesia	Malaysia
Collector price (USD/m^2)	278	276	274	275
Other system cost (USD/m^2)	170	155	151	154
Total system cost (USD/m^2)	480	461	468	455
Total system cost (USD)	311,925	276,395	297,670	299,272
Aperture area (m^2)	650	599	637	657
Collector efficiency [a](%)	69	70	69	71
Horizontal irradiance (kWh/m^2/year)	1,623	1,741	1,665	1,569
Specific heat yield (kWh/m^2/year)	1,123	1,217	1,145	1,109
Replacement investment (USD/year 10)	4,013	4,084	4,905	4,362
Operating and maintenance (USD, first year)	3,119	2,764	2,977	2,993
Insurance cost (USD, first year)	1,560	1,382	1,488	1,496
Pump electricity cost (USD, first year)	2,597	3,037	2,266	2,575
Degradation per annum (%)	1.0	1.0	1.0	1.0

[a]Based on average values of three different collector types – collector A, collector B, and collector C

Table 40.4 Technical assumptions for the base value for the gas-fueled heating system

Parameter	Singapore	Thailand	Indonesia	Malaysia
Gas heater price (USD/per piece)	13,380	13,380	13,380	13,380
Other system cost (USD)	110,645	93,114	95,884	101,360
Total system cost (USD)	175,640	156,882	164,326	164,156
Gas heater efficiency (%)	80	80	80	80
Replacement investment (USD/year 10)	59,997	62,317	74,848	66,571
Operating and maintenance (USD, first year)	2,108	1,883	1,972	1,970
Insurance cost (USD, first year)	878	784	822	821
Pump electricity cost (USD first year)	1,558	1,822	1,360	1,545
Gas fuel cost (USD, first year)	133,551	77,006	75,544	23,560
Degradation per annum (%)	1.0	1.0	1.0	1.0

scenario (SERIS-NSR 2017). For the other three countries, it was assumed that 60% of the electricity prices will follow the annual changes of the 30 December 2016 Brent oil forward curve until year 2024 with a flat progression assumed thereafter (i.e., the spot oil price was at 56.2 with the December 2024 price at 60.2 USD/barrel). No changes in the electricity landscape nor subsidy-related price politics have been taken into considerations.

Degradation has been assumed to be equal at 1.0% across the different countries as all of the regions are under tropical hot and humid climatic conditions. The assumptions for the gas-fueled heating system are summarized in Table 40.4. The other system cost has been aligned to the same value needed to install and monitor a solar thermal system. The replacement investment includes the gas heater and the electric pump replacements in the tenth year. The gas fuel cost is evaluated based on an 80% fuel efficiency and the local industrial gas prices. Future price progressions were fully pegged to the aforementioned Brent oil price forward curve. It is observed that the annual gas cost is considerably lower in Malaysia, where the industries' gas price is more than 80% lower than in Singapore. No residual value has been taken into considerations.

40.4 Results and Discussion

40.4.1 Collector Thermal Yield

As shown in Fig. 40.4, the predicted year-round collector thermal yield is in the range of 90–120 KWh_{th} per m^2 of collector area for each country – Indonesia (Medan Polonia), Malaysia (Kuala Lumpur), Singapore, and Thailand (Bangkok). It is also observed that the collector efficiencies vary in the range of 60–75% in those regions. Thus, the simulation studies confirm that solar heat can be utilized for industrial heating applications for those regions.

Figure 40.5 illustrates the breakdown of the different life cycle cost calculations. It can be seen that the fuel cost contributes to the main cost item (84% in the context of Singapore with the highest gas prices) for the gas-fueled heating system, while the capital cost is the main contributor for the solar thermal system. The tax cost includes the corporate income tax payment in the respective country, where an investment into a solar thermal system compared to a gas-fueled heating system is economically superior. The consideration of the income tax implication makes certain that the benefits are considered on an "after-tax" basis. No tax implication is included for Indonesia where the gas fuel savings are too low to compensate the higher upfront investment of the solar thermal system.

The base value $LCOE_{th}$ calculations per country are shown in Table 40.5 with a range illustrated in Fig. 40.6. The range is based on either (i) $\pm 10\%$ change in the underlying gas price for the gas-fueled heating system or (ii) $\pm 10\%$ change in the upfront capital investment for the solar thermal system. These two parameters are the most important drivers for each type of heating system's life cycle cost, respectively, as presented in Fig. 40.5. It can be shown that the life cycle cost is lower for solar thermal systems when compared to gas-fueled heating systems in all of the countries being analyzed. However, when it comes to the benefit calculation, i.e., gas fuel saved, thanks to a solar thermal system, the gas prices in Malaysia are currently too low to make the investment financial viable. Otherwise, the potential EIRR returns and DPBPs are at attractive levels (see Table 40.5).

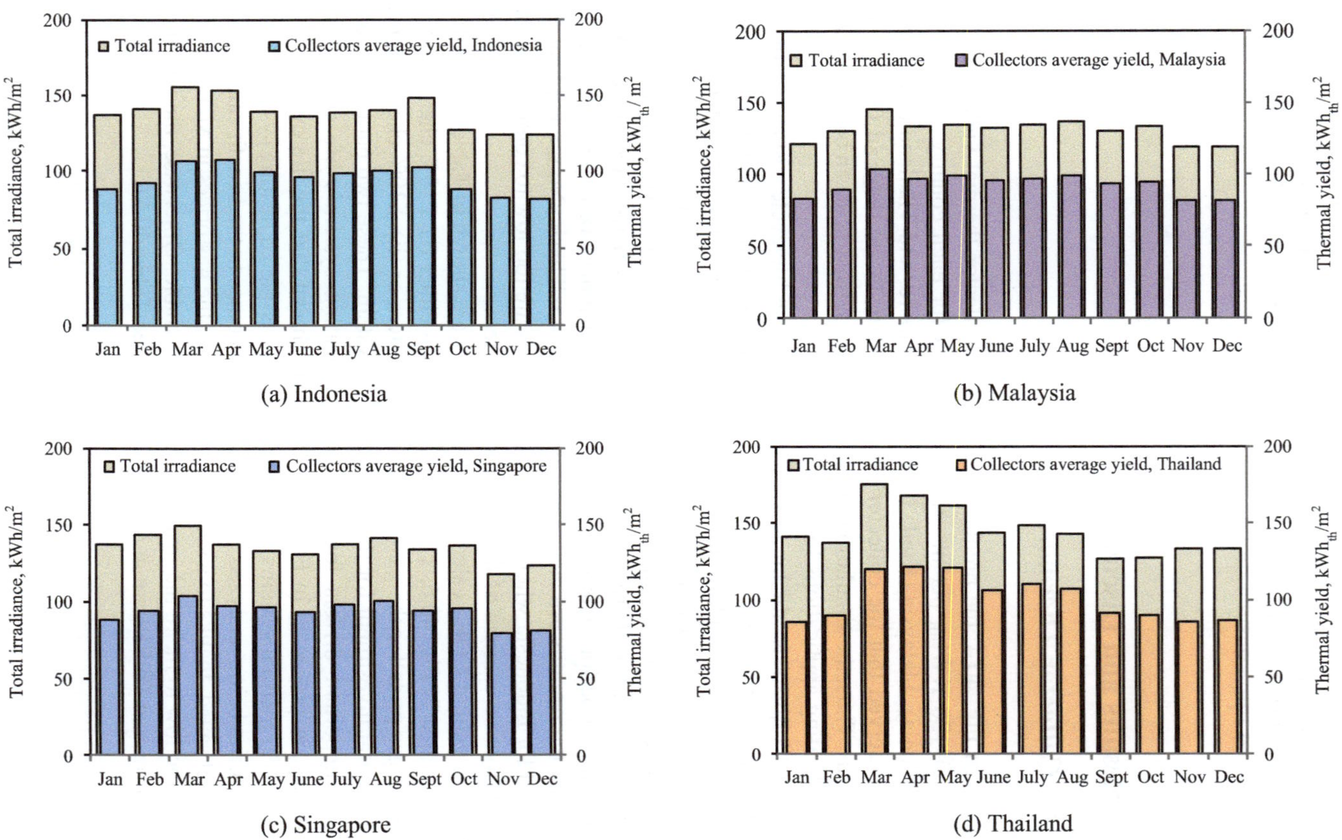

Fig. 40.4 Irradiance and collector yields per country. (**a**) Indonesia. (**b**) Malaysia. (**c**) Singapore. (**d**) Thailand

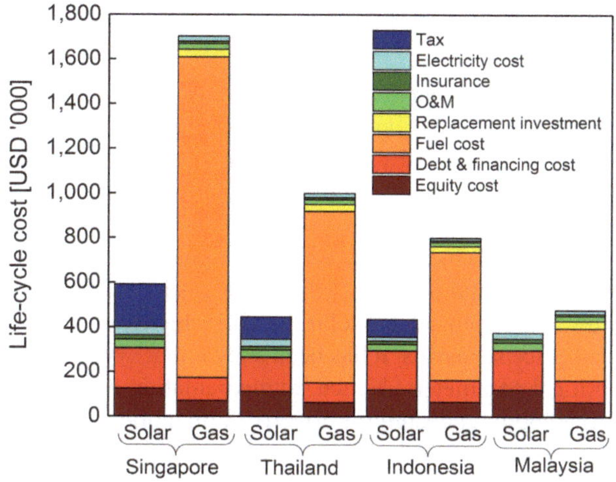

Fig. 40.5 Life cycle cost comparison between countries and heating system types

Table 40.5 Base value $LCOE_{th}$ and financial returns of investing in solar thermal heating systems

Financial result	Singapore	Thailand	Indonesia	Malaysia
Base $LCOE_{th}$ for solar thermal system (USD cents/kWh_{th})	7.7	6.3	8.1	5.5
Base $LCOE_{th}$ for gas-fueled system (USD cents/kWh_{th})	22.2	14.1	14.8	6.9
Net present value (NPV, USD)	846,046	321,426	137,047	−142,382
Discounted payback period (DPBP, years)	2	4	6	>20
Equity internal rate of return (EIRR, %)	67%	34%	25%	Negative

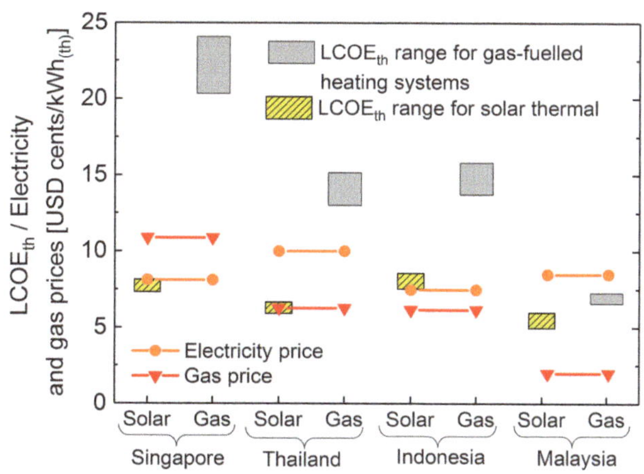

Fig. 40.6 $LCOE_{th}$ ranges compared to the gas and electricity price basis used for 2016

40.5 Conclusions

Solar thermal heat can be an alternative energy source for several industrial heating applications in the countries covered in this analysis. The levelized cost of thermal energy generated by a conventional gas-fueled system was determined between 7 and 23 USD cents per kWh_{th} for the example cases. Instead the levelized cost of thermal energy generated by a solar thermal system ranges only from 5 to 9 USD cents per kWh_{th}. The discounted payback period of a solar thermal system ranges between 2 and 6 years. Only in Malaysia the industrial gas prices are currently too low to compensate for solar thermal's higher upfront investment. Besides the future gas price scenarios, it has been shown that system price and collector efficiency are critical parameters. Hence, implementation of a solar thermal system must ensure (i) an optimal design for effective utilization of solar radiation, and (ii) the integration is economically viable for the industrial process heat application.

References

Remap 2030. (2015). *Renewable energy outlook for ASEAN, IRENA*. Available: http://www.irena.org. Accessed 10/01/2017.

Renewable Heating & Cooling. (2012). *Strategic research priorities for solar thermal technology*. Brussels. Available: http://www.rhc-platform.org. Accessed 10/01/2017.

Solar Energy Research Institute of Singapore (SERIS) National Solar Repository (NSR). *Solar economics handbook, future electricity price scenarios*. Available: http://www.solar-repository.sg/future-electricity-price-scenarios. Accessed 01/09/2017.

SolarkeyMark Certificate. Available: http://www.estif.org/solarkeymarknew. Accessed 10/01/2017.

UNEP (United Nations Environment Programme). (2010). *Solar heat for industrial processes (SHIP): IEA*. Available: http://ship-plants.info/solar-thermal-plants. Accessed 10/01/2017.

Vannoni, C., Battisti, R., & Drigo, S. (2008). *Potential for solar heat in industrial processes*. Rome: CIEMAT, Madrid.

Wahed, M. A., Bieri, M., & Hellwig, R.T. (2015). *Potential of solar thermal system for industrial process heat applications in the tropics*. SusTED'15. Kuala Lumpur.

Chapter 41
Selection of Adsorbents for Double-Effect Adsorption Refrigeration Cycle Combined Compressor

Fumi Watanabe and Atsushi Akisawa

41.1 Introduction

A system which is driven by low-grade heat around 60–90 °C is expected to enhance the use of the solar heat and waste heat. The adsorption refrigeration is capable of operating at low-grade heat while the challenge is how to increase the efficiency of current implementations of adsorption refrigerators. To accomplish this, research into systems that incorporate cascading adsorption refrigeration cycles or heat recovery functionality is underway. Akisawa (2015) performed a statistic response analysis of both adsorption heat recovery and condensation heat recovery for a double-effect adsorption refrigeration cycle with a heat source of 120 °C. In that study, the cycle which recovered adsorption heat had the higher COP of 1.1. At temperatures below 80 °C, temperature variation between the adsorption and desorption processes is small, and the adsorbent concentration range is too limited, meaning the cycle cannot operate. Hybrid designs which include compressors within their cycles have already been demonstrated for both absorption and adsorption refrigerators. These cycles feed a small amount of electricity to their compressors and enable the refrigerator to be driven by a low-temperature heat source. In order to improve the efficiency of the cycle, this study proposes a double-effect adsorption refrigeration cycle equipped with a compressor. The objective of this study is to investigate the performance of double-effect adsorption cycles with various installation position of a compressor and various adsorbent pairs. Cycle efficiency was calculated using a static analysis based on a state of equilibrium.

F. Watanabe (✉) • A. Akisawa
Graduate School of Bio-Applications and System Engineering, Tokyo University of Agriculture and Technology, Tokyo, Japan
e-mail: s146256s@st.go.tuat.ac.jp

A. Sayigh (ed.), *Transition Towards 100% Renewable Energy*,
Innovative Renewable Energy, https://doi.org/10.1007/978-3-319-69844-1_41

41.2 Double-Effect Adsorption Refrigeration Cycle Equipped with a Compressor

The double-effect cycle with compressor consists of a high-temperature adsorber, a low-temperature adsorber, a compressor, an evaporator, and a condenser. Conventional designs transport refrigerant vapor using differences in adsorption equilibrium pressure based on temperature variation, but this cycle creates a pressure difference using the compressor and temperature variation. With its high-temperature cycle (HTC) and a low-temperature cycle (LTC), the double-effect adsorption refrigeration cycle repeatedly carries out the processes of precooling, adsorption, preheating, and desorption in sequence. That behavior is indicated in the Dühring diagram, the regulation high-pressure side (Type H) as Fig. 41.1a while the regulation low-pressure side (Type N) as Fig. 41.1b. The horizontal axis represents the temperature in the adsorber, while the vertical axis represents the pressure of the refrigerant vapor; the diagonal line is the adsorption isotherm. During the precooling process of the HTC, the adsorber is disconnected from the evaporator and condenser, and a steady adsorbent concentration is maintained while pressure decreases in accordance with the lowering of temperature. The adsorption process begins once the pressure inside the adsorber reaches that of the evaporator, at which point the two are connected and the adsorbent concentration rises due to the state of pressure in the evaporator. The heat that is generated when the refrigerant is adsorbed by the adsorbent in the HTC is reused as the heat source for the preheating and desorption processes of the LTC. During the preheating process, the evaporator is disconnected from the condenser, and a steady adsorbent concentration is maintained while the heating causes temperature and pressure to rise. Once the pressure inside the adsorber reaches that of the condenser, the two are connected, and the adsorbent concentration decreases due to the state of pressure in the condenser. The desorption process ends at the point when the

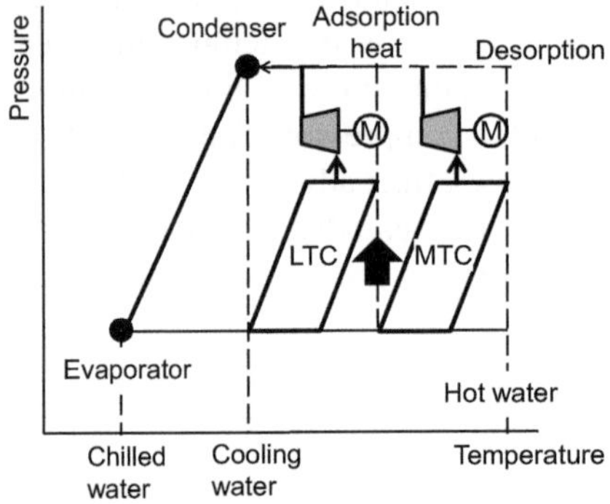

(a) Regulation high-pressure side(Type-H)

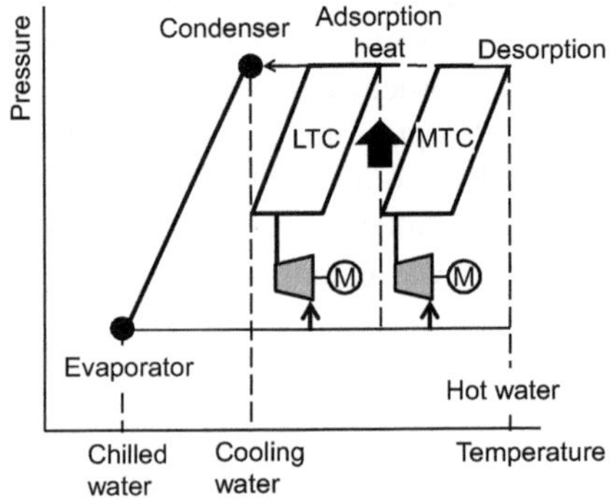

(b) Regulation low-pressure side(Type-L)

Fig. 41.1 Dühring diagram of double-effect adsorption refrigeration cycle with compressor. (**a**) regulation high-pressure side (Type H). (**b**) Regulation low-pressure side (Type L) adsorbent matches the temperature of the heat source. The LTC undergoes the same process, except its adsorption processes ends once it reaches the temperature of the cooling water and its desorption process once it reaches the intermediate temperature for the LTC. The compressor allows for regulation of adsorption or desorption pressure. Because the double-effect cycle is divided into HTC and LTC, the compressor is able to regulate the pressure in 2 cycles, and the adsorption and desorption processes of each cycle are performed with the variations in adsorbent concentration taking place when the adsorbent is dense.

41.3 Static Analysis

In order to study the efficiency of the adsorption refrigeration cycle, a static analysis was conducted based on the theoretical values of the Dühring diagram. Using the temperature and pressure of the state of equilibrium on the Dühring diagram, the ideal adsorption capacity and cooling capacity were calculated by the following assumptions: The upper limit of the HTC's adsorber is equivalent to the water temperature of the heat source. The lower limit of the HTC's adsorber is equivalent to the intermediate temperature. The upper limit of the LTC's adsorber is equivalent to the intermediate temperature. The lower limit of the LTC's adsorber is equivalent to the temperature of the cooling water. The precooling process ends when the pressure matches that inside the evaporator. The preheating process ends when the pressure matches that inside the condenser. For cooling capacity, the refrigerant's effective adsorption capacity was used as the evaporation amount, and the sensible heat returning from the condenser when an identical amount was desorbed from the adsorbent was excluded. The cooling capacity of the cycle was considered to be the total energy output of the HTC and LTC. The heat input into the cycle is the latent heat of desorption as well as the sensible heat created by temperature variations in the adsorbent and refrigerant from the adsorption process through the desorption process. For the quantity of heat required to desorb the refrigerant, the heat of adsorption was used to express the heat of desorption. Also, the adsorbent concentration of the refrigerant fluctuates during desorption, but here the average of the concentrations was used at the start and end of desorption. The heat of desorption and the heating by sensible heat are indicated using the respective formulas below. To preserve the balance of energy, the heat of adsorption from the higher temperature cycle is equivalent to the amount of heat supplied to the lower temperature cycle. The amount of electricity supplied to the compressor is determined using the following formula, which includes shaft efficiency and the enthalpy before and after the compressor. COP, SCE, and exergy efficiency were used to evaluate the effect of the compressor on the efficiency of the refrigeration cycle. COP is defined as the cooling capacity achieved within a single cycle divided by the sum of supplied electricity and heat energy. SCE is the value arrived at by dividing the cooling capacity by the total mass of adsorbent contained in all heat exchangers. Exergy efficiency (Bejan et al. 1996) is the ratio of exergy output by the evaporator to the exergy supplied by the compressor and the desorption process. This puts the Carnot cycle to work using an amount of heat with temperature and corresponds to the maximum effective work done when temperature T_0 is released to the external environment. T_0 is equivalent to the temperature of the cooling water.

$$Q_{\text{cool}} = M(x_{\text{ads}} - x_{\text{des}})[L - c_r(T_{\text{cond}} - T_{\text{eva}})] \tag{41.1}$$

$$Q_{\text{cool,all}} = Q_{\text{cool},h} + Q_{\text{cool},l} \tag{41.2}$$

$$Q_{\text{des}} = M(x_{\text{ads}} - x_{\text{des}})q_{\text{st}} \tag{41.3}$$

$$\begin{aligned} Q_{\text{sen}} &= M(c_{\text{ads}} + c_r x_{\text{ads}})(T_{\text{preh}} - T_{\text{cw}}) \\ &\quad + M\left(c_{\text{ads}} + c_r \frac{x_{\text{ads}} + x_{\text{des}}}{2}\right)(T_{\text{cw}} - T_{\text{preh}}) \end{aligned} \tag{41.4}$$

$$Q_{\text{heat,all}} = Q_{\text{des},h} + Q_{\text{sen},h} \tag{41.5}$$

$$Q_{\text{des},h} = Q_{\text{heat},l} \tag{41.6}$$

$$Q_{\text{comp}} = M(x_{\text{ads}} - x_{\text{des}})(h_{\text{con}} - h_{\text{mid}}) \tag{41.7}$$

$$Q_{\text{elec}} = Q_{\text{comp},h}\eta_h + Q_{\text{comp},l}\eta_l \tag{41.8}$$

$$\text{COP} = \frac{Q_{\text{cool,all}}}{Q_{\text{heat,all}} + Q_{\text{elec,all}}} \tag{41.9}$$

$$\text{SCE} = \frac{Q_{\text{cool,all}}}{M_{\text{all}}} \tag{41.10}$$

$$\eta_{\text{ex}} = \frac{\left|Q_{\text{cool,all}}\left(1 - \frac{T_0}{T_{\text{eva}}}\right)\right|}{Q_{\text{heat}}\left(1 - \frac{T_0}{T_{\text{des}}}\right) + Q_{\text{elec}}} \tag{41.11}$$

where c is specific heat (kJ (kg K)$^{-1}$), COP is coefficient of performance, H is enthalpy (kJ kg^{-1}), L is latent heat of vaporization (kJ kg^{-1}), M is mass of adsorbent (kg), P is pressure (kPa), q_{st} is adsorption heat (kJ kg^{-1}), SCE is specific cooling energy (kJ kg^{-1}), T is temperature (K), x is adsorbent concentration (kg kg^{-1}), and η_{ex} is exergy efficiency. Subscribe: ads is adsorption process or adsorbent, all is total, cond is condenser, cool is cooling, comp is compressor, cw is cooling water, des is desorption process, eva is evaporator, elec is supplied electricity, h is high-temperature side, heat is heating, l is low-temperature side, preh is preheating process, r is refrigerant, and s is sensible. Water is used as the refrigerant and zeolites (FAM Z01, FAM Z02, and FAM Z05) are used as the adsorbents. These adsorbents have high adsorptivity within a narrow range of relative pressures. The cycle's adsorption capacity at equilibrium is expressed by the following formula. A regression analysis was performed on Y_1, Y_2, Y_3, Y_4, and Y_5 which were set to approximate literature values (Kakiuchi et al. 2005a, b; Shimooka et al. 2007) with respect to temperature; the generalized function is expressed by the following equation:

$$x = \frac{Y_1}{1 + \exp\{-(Y_2 - Y_3 P_{\text{r}})\}} + Y_4 P_{\text{r}} + Y_5 \tag{41.12}$$

Table 41.1 Conditions of temperature

Item	Value	Unit
Heat source	60–90	°C
Cooling water	30	°C
Chilled water	10	°C

Table 41.2 Conditions of adsorbents and compressor

Item	LTC	HTC	Compressor
1	Z01	Z02	High-pressure side
2	Z01	Z02	High-pressure side
3	Z05	Z02	Low-pressure side
4	Z05	Z02	Low-pressure side

where P_r is relative pressure. The proposed cycle was simulated using the basic conditions listed in Tables 41.1 and 41.2. The intermediate temperature was set in such a way as to maximize COP. The HTC was loaded with 1 kg of adsorbent. Equation (41.6) was then used to determine load amounts for the LTC which would balance their heat of adsorption and supplied heat. The cycle efficiency has performed a simulation on the effect that heat source temperature has on COP, SCE, and exergy efficiency. The efficiencies of the single-effect, double-effect, and double-effect with compressor cycles are compared. The adsorbent selection was as follows: Z02 for the single-effect cycle, Z01 for the LTC, and Z02 for the HTC of the double-effect cycle.

41.4 Simulation Results

COP, SCE, and exergy efficiency are indicated in Figs. 41.2, 41.3, and 41.4. As seen in Fig. 41.2, COP numbers were highest with the Z01-Z02 Type H and Type L and Z05-Z02 Type H reaching about 1.2 or close to two times that of the single-effect cycle. At 70 °C, the temperature range between the HTC and LTC of the double-effect cycle was too narrow, eliminating the concentration range and making operation impossible. Since the double-effect cycle uses a compressor to increase adsorbent pressure, it is able to increase concentration range and thus operate even at low temperatures. Increasing the adsorbent concentration strengthened the relative effect of the output's cooling output and the input's heat of desorption while weakening the relative effect of the input's heating by sensible heat during the preheating process. This resulted in an increase in COP, which is a ratio of input to output. When increasing concentration range to a certain extent, COP reaches a nearly constant value since the influence of cooling capacity and desorption heat is high. At temperatures below 50 °C, double-effect cycle with a compressor is not able to operate. COP of Z01-Z02 Type H is highest in the 60 °C, and COP is not changed in the 60–90 °C. As indicated in Fig. 41.3, SCE of the single-effect cycle is almost three times greater than that of the double-effect cycle in the 80–90 °C

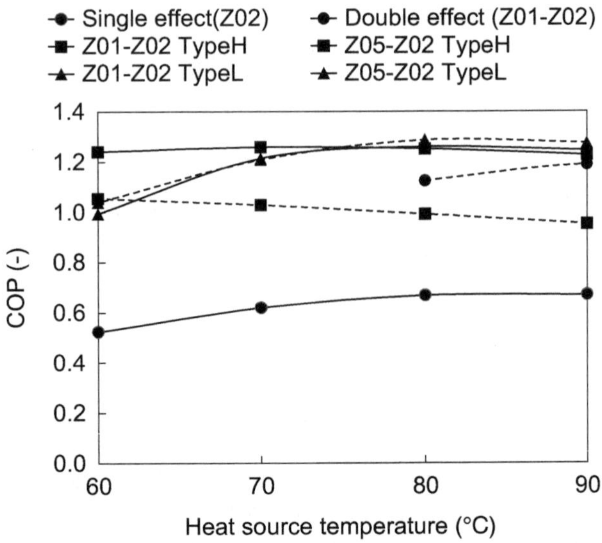

Fig. 41.2 COP

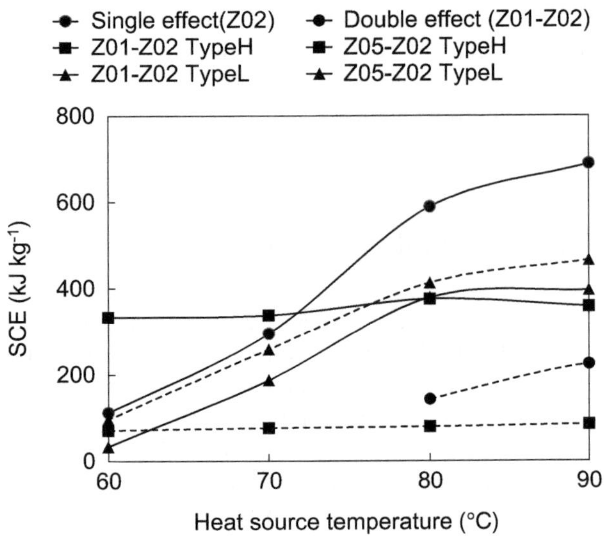

Fig. 41.3 SCE

temperature range. The maximum SCE value was 194 kJ/kg at 90 °C. Since the double-effect cycle has 2 cycles, the adsorbent concentration range per cycle is smaller than that of the single-effect cycle. The double-effect cycle with compressor is two times greater than that of the double-effect cycle due to the fact that it increases adsorbent concentration range using its compressor. SCE of Z05-Z02 Type L is highest in the 80–90 °C temperature range, while that of Z01-Z02 Type H is highest in the 60–70 °C. Figure 41.4 indicates that exergy efficiency numbers were highest with the double-effect cycle in the 80–90°C. Because exergy efficiency places a high value on electricity, it did not display the same superiority as

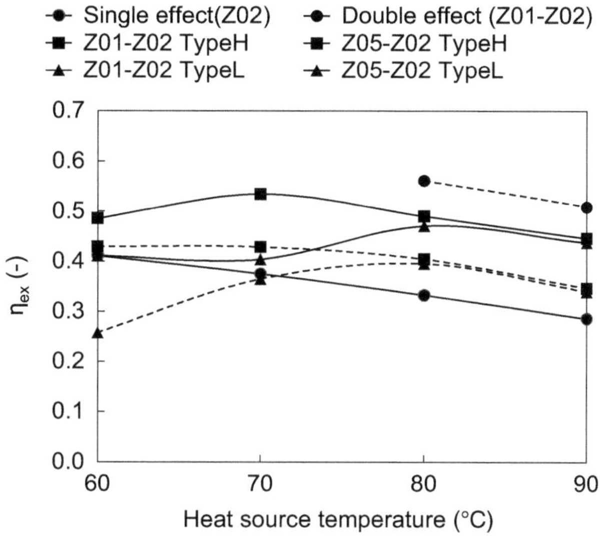

Fig. 41.4 Exergy efficiency

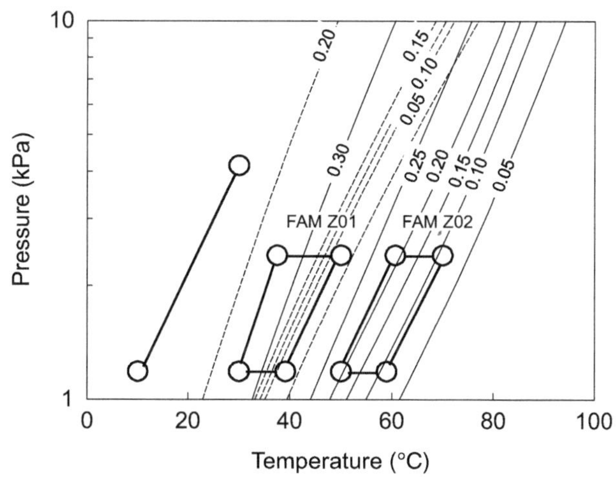

Fig. 41.5 Dühring diagram of double-effect adsorption refrigeration cycle with compressor (Z01-Z02 Type H)

COP. Exergy efficiency of Z01-Z02 Type H is highest and not almost changed in the 60–90 °C. For COP, SCE, and exergy efficiency, the double-effect with compressor of Z01-Z02 Type H obtained higher values. Those respective states are indicated on the Dühring diagram at heat source temperature of 70 °C (Fig. 41.5). The Dühring diagram shows that regulating the pressure of each cycle allows the cycle to operate in the area with the largest variation in adsorbent concentration. In addition, COP and exergy efficiency showed superior results because there was little increase in pressure from the desorption pressure to the condensation pressure. It is, therefore, possible to operate the double-effect cycle more efficiently with a small amount of electricity supply by regulating the pressure of each cycle.

41.5 Conclusion

The objective of this study is to investigate the COP, SCE, and exergy efficiency of double-effect adsorption cycles with various installation position of a compressor and various adsorbent pairs FAM Z01, Z02, and Z05. A static analysis was conducted based on a state of equilibrium to calculate the effect that adsorption pressure has on COP, SCE, and exergy efficiency and arrive at the following results. For heat sources in the temperature range of 60–90 °C, the method is superior when the compressor is installed on the high-pressure side and adsorbent pair is Z02-Z01. Utilizing the highly efficient double-effect adsorption refrigeration cycle with a compressor should contribute to promoting the use of renewable heat. This was a static analysis, however, which does not take heat loss into consideration. Since COP is expected to decline in a dynamic response analysis, the next challenge is to perform cycle benchmarks that take heat transfer performance and heat capacity into account.

References

Akisawa, A. (2015). Selection of adsorbents for double effect adsorption refrigeration cycle. In *The 20st National Symposium Power and Energy System, Miyagi* (pp. 127–130). (In Japanese).

Kakiuchi, H., Shimooka, S., Iwade, M., Oshima, K., Yamazaki, M., Terada, S., Watanabe, H., & Takewaki, T. (2005a). Water vapor adsorbent FAM-Z02 and its applicability to adsorption heat pump. *The Society of Chemical Engineers, 35*(4), 273–277.

Kakiuchi, H., Shimooka, S., Iwade, M., Oshima, K., Yamazaki, M., Terada, S., Watanabe, H., & Takewaki, T. (2005b). Water vapor adsorbent FAM-Z01 and its applicability to adsorption heat Pump. *The Society of Chemical Engineers, 35*(5), 361–364.

Shimooka, S., Oshima, K., Hidake, H., Takewaki, T., Kakiuchi, H., Kodama, A., Kubota, M., & Matsuda, H. (2007). The evaluation of direct cooling and heating desiccant device coating with FAM. *Journal of Chemical Engineering of Japan, 40*(13), 1330–1334.

Ventas, R., Lecuona, A., Zacarias, A., & Venegas, M. (2010). Ammonia-lithium nitrate absorption chiller with an integrated low-pressure compression booster cycle for low driving temperatures. *Applied Thermal Engineering, 30*(11), 1351–1359.

Chapter 42
Optimal Organizational Forms for Local Renewable Energy Projects

Yoshihiro Yamamoto

42.1 Introduction

Increasing attention has been paid to renewable energy (RE) development in a region, in particular, to generating and supplying electricity from RE sources within the region. This may contribute to the revitalization and regeneration of that region as well as the alleviation of environmental and energy issues.

RE development projects are undertaken through various forms of organizations. They include private companies, local governments, joint ventures, cooperatives, and so forth. One of the well-known forms may be community-based RE cooperatives (Schreuer and Weismeier-Sammerand 2010; DGRV 2014; Tarhan 2015). This organizational form of RE projects is observed in, among others, Denmark, Germany, the Netherlands, and the UK (Schreuer and Weismeier-Sammerand 2010; Viardot 2013). In Denmark and Germany, RE cooperatives have been widely established (Huybrechts and Mertens 2014). According to DGRV (2014), the number of energy cooperatives within the DGRV (German Cooperative and Raiffeisen Confederation) has increased from 8 in 2006 to 718 in 2013. The prevalence of this organizational form has historical background (Huybrechts and Mertens 2014).

On the other hand, recently in Japan, an increasing number of companies have partnered with the public sector to set up local RE projects. The public sector, usually at the municipal level, plays a central role in such projects but does not undertake them directly as public works. Instead, it establishes a company that undertakes the project, often jointly with other private companies. The first case of this arrangement appeared in 2013, and to date there are at least 14 cases in Japan. In this study, this kind of company is called a "municipal RE company."

Y. Yamamoto (✉)
Department of Economics, Takasaki City University of Economics, Takasaki, Japan
e-mail: ysyama@tcue.ac.jp

A. Sayigh (ed.), *Transition Towards 100% Renewable Energy*,
Innovative Renewable Energy, https://doi.org/10.1007/978-3-319-69844-1_42

The purpose of this study is to investigate the effectiveness of a municipal RE company as an organizational form for local RE projects. Supposing that an RE cooperative is one of the suitable organizational forms for local RE development, we may judge whether a municipal RE company will be successful or not by investigating the factor that we identify as being crucial for a successful RE cooperative.

The rest of the paper is organized as follows: Sect. 42.2 describes the methods to identify the crucial factor that makes an RE cooperative successful and examines its validity for a municipal RE company. The methodology consists of a literature review and interviews. Section 42.3 presents the results, and Sect. 42.4 concludes the study.

42.2 Methodology

This paper uses a literature review and interviews. The literature review consists of two steps. In the first step, the focus is on the characteristics of an RE cooperative to identify the crucial factor that makes it a successful organizational form for local RE projects. Following the results obtained in the first step, the literature review examines whether this crucial factor plays a role in the public sector.

Next, to verify the results of the literature review, an interview is conducted with a key figure at a municipal RE company. The municipal RE company selected for the interview, anonymously called Company X in this article, is in Municipality A in Japan. Company X is well known to some extent in Japan as a municipal RE company.

Let us describe Municipality A in brief. It is rural; more than 80% of the municipality is covered with forest. The population is composed of nearly 17,000 people living in 6800 households and has been decreasing over at least 30 years, leaving aside the merger with an adjacent village. Those aged 65 and over constitute more than one-third of the population. Forestry and agriculture had been the main industries but are currently on the decline, and no other industry has been substituted for them.

42.3 Results

This section has three subsections. In the first two subsections, the results of the literature review are presented, and Sect. 42.3.3 presents the interview results. The crucial factor that makes an RE cooperative successful is identified in Sect. 42.3.1, where the discussion is extended into community involvement. Then, based on the factor identified in Sect. 42.3.1, public service motivation will be the focus of the literature review in Sect. 42.3.2. The interviews with the manager of Company X to explore the findings from Sect. 42.3.2 are described in Sect. 42.3.3.

Before presenting the results, let us clarify the definition and principles of a cooperative. The International Co-operative Alliance (ICA), an independent, nongovernmental organization aiming to unite, represent, and serve cooperatives worldwide, has seven cooperative principles: voluntary and open membership; democratic member control; member economic participation; autonomy and independence; education, training, and information; cooperation among cooperatives; and concern for community (ICA 2017). Among them, democratic member control and concern for community seem relevant to successful RE projects. These features may be summed up by the notion of community involvement. Let us first verify that community involvement plays an important role in local RE projects.

42.3.1 Literature Review on Renewable Energy Cooperatives

Numerous studies deal with various characteristics of RE cooperatives (Schreuer and Weismeier-Sammerand 2010; Viardot 2013; Yildiz 2014; Yildiz et al. 2015). Many studies emphasize democratic decision-making in a cooperative. Viardot (2013) argued that cooperatives are democratically managed in that every member has an equal vote. Similarly, Yildiz et al. (2015) pointed out that the one member, one vote principle is one of the critical features of cooperatives. This decision-making system can encourage the members of a cooperative to play an active role in pursuing objectives (Yildiz 2014).

This characteristic plays a role in tackling the so-called NIMBY (not-in-my-backyard) problem. Schreuer and Weismeier-Sammerand (2010) argued that local participation and ownership may reduce local opposition to the installation of wind farms; the installation of wind farms is often refused by local citizens due to noise creation, shadow cast, impact on the landscape, and so forth (Musall and Kuik 2011). Viardot (2013) pointed out that the democratic decision-making process of a cooperative alleviates in an equitable way the concern over where a system will be installed.

There are also studies concerned with another aspect of cooperatives: community ownership of renewable energy facilities (Walker 2008; Warren and McFadyen 2010; Musall and Kuik 2011). According to Walker (2008), there is some evidence that if RE projects are owned or partly owned by the community, they will have fewer problems in obtaining permission and they will be more locally acceptable. There are questionnaire-based surveys that support the view that community ownership enhances public acceptance of local RE projects. For example, Warren and McFadyen (2010) showed that the wind farms owned by the community are associated with more positive local attitudes than those owned by commercial companies. Similarly, Musall and Kuik (2011) showed that community co-ownership of wind energy leads to higher levels of acceptance than ownership by a commercial company.

However, it should be appropriate to consider community ownership as part of community involvement if we are concerned with the conditions necessary for local

RE projects to be successful. For example, Walker and Devine-Wright (2008) examined the policy and practice of community RE in the UK by constructing a database of projects connected to the community, interviewing policymakers and managers involved in community projects, and conducting case studies. Their findings were that more direct and substantial involvement of local people contributes to greater acceptance and support for a project.

Community involvement in an RE project seems to be deeply related to the identity of the community itself (Walker et al. 2010; Bomberg and McEwen 2012; Seyfang et al. 2013; Wirth 2014). Emphasizing the importance of a formalized cooperative system in the choice of organizational form, Wirth (2014) pointed out four community-related institutional forces that are central to the design of a biogas plant: community spirit, cooperative tradition, locality, and responsibility. Walker et al. (2010) compared two cases of local RE projects and concluded that trust between local people and groups that take projects forward can help projects work. Bomberg and McEwen (2012) identified through case studies that a strong community identity and a search for local autonomy and community sustainability are relevant to community action concerning energy. Seyfang et al. (2013) conducted a web-based survey and concluded that the civil society basis of community energy groups and projects is fundamental to successful community projects.

To summarize the literature review regarding RE cooperatives, the most crucial factor for a local RE project to be successful may be community involvement in the project. RE cooperatives may be considered one of the forms that can realize this feature. Furthermore, community ownership of RE facilities is one aspect of community involvement. In light of this understanding, let us investigate why municipal RE companies attract increasing attention in Japan.

42.3.2 Literature Review on Public Service Motivation

To examine whether a municipal RE company is an organizational form connected with community involvement, let us look at the motivation of public employees. It has been pointed out that public employees are motivated by an intrinsic value for their work (Francois 2000; Wright 2001; Jensen and Stonecash 2005; Besley and Ghatak 2005; Bright 2009; Huang and Feeney 2016). Jensen and Stonecash (2005) argued that one reason why workers in the public sector may receive utility from their tasks is that identifying themselves with the mission of the public sector plays an important role in their work. Similarly, Besley and Ghatak (2005) stated that workers in a mission-oriented sector pursue the goals of the organization to which they belong since they perceive associated intrinsic benefits.

Public service motivation (PSM) is defined as "a general altruistic motivation to serve the interests of a community of people, a state, a nation or humankind" (Rainey and Steinbauer 1999). Francois (2000) argued that there is considerable survey-based evidence that PSM exists. Bright (2009) found through a questionnaire survey that an important relationship exists between PSM and the intrinsic nonmonetary work preferences of public employees. Furthermore, Huang and Feeney (2016) found, using data from national surveys of local government managers in the USA, that managers with higher PSM report greater levels of citizen participation in decision-making.

To summarize, public employees are typically motivated to serve their region and citizens. Hence, a municipal RE company, in which a municipality is deeply involved, may realize community involvement to a great degree.

42.3.3 Interviewing

To support the views deduced from the literature review, a municipal RE project will be examined as a case study. Company X is selected as a case, and its manager is interviewed mainly from an organizational point of view.

To begin with, the history of RE promotion in Municipality A is worth mentioning. The initiative to promote RE in that town was proposed by the mayor after the devastating earthquake and subsequent Fukushima nuclear disaster of 2011. Then, the municipality started a project regarding photovoltaic (PV) generation utilizing municipality-owned land in collaboration with a private company. Subsequently, two other PV generation projects were put into practice: the second project utilized state-owned land, and the third utilized private land; this private land was acquired later by the municipality. It should be noted that these three projects were set up through a public–private partnership (Engel et al. 2014, p. 2). In other words, the RE projects were run by a profit-making company at first, but they were monitored by a public authority.

The municipality was becoming more committed to RE promotion. To manage the three projects more efficiently and further RE development in the town, it established a foundation. This foundation later set up Company X to undertake RE projects in place of the foundation. In other words, the municipality invests in Company X indirectly through the foundation. Furthermore, a person who had been a public employee at Municipality A was appointed the manager of Company X. Additionally, Company X's office is in the municipality building. On balance, the municipality undertakes RE projects in the region.

The interview was carried out with the manager of Company X, who had been a public employee in charge of RE development in the town. The main findings obtained from the interview are as follows:

First, the motivation of Company X is to contribute to the increasing benefits of the region and its citizens. This shows a sharp contrast with that of a purely private company aiming to maximize profits. Although a municipality generally may have the same motivation, it would not undertake this kind of project directly. Rather, it would encourage private companies to do so. In fact, Municipality A had arranged a public–private partnership for the first RE project. The reason why a municipality does not directly undertake such a project may be that it falls into the realm of business beyond public service.

Second, the reason why Company X was set up is not to realize community involvement but to manage RE projects efficiently and further RE development. In other words, setting up the company, which was recommended by a consulting company, was done purely for managerial purposes. However, the company may have the trust of its citizens since it has a strong connection with Municipality A. It is noted that the second and third PV generation projects were launched through citizens' proposals. Company X can attain public acceptance due to the original rationale behind RE promotion in the town: the mayor aimed to facilitate energy transition in the face of the Fukushima nuclear disaster of 2011 and regenerate the town.

Lastly, the company may serve as a showcase for other municipalities that are planning to promote RE with the intention of regenerating the region. There are officials from many municipalities visiting the company to learn how to set up a municipal RE company and promote local RE development.

To summarize, the interview found that a municipal RE company has public service motivation and may work on behalf of the citizens. Therefore, a municipal RE company may be an effective organization for undertaking local RE projects.

42.4 Conclusions

This study investigated why municipal RE companies in an RE project attract increasing attention in Japan. A literature review on RE cooperatives and interviews with the manager of a municipal RE company were conducted to examine the effectiveness of municipal RE companies in undertaking RE projects. It is found that RE cooperatives and municipal RE companies share the feature of community involvement. The difference is that RE cooperatives are directly involved in the community whereas municipal RE companies are involved through public service motivation.

The establishment of RE cooperatives is closely related to the fact that cooperatives are historically familiar in such countries. On the other hand, there are many other countries like Japan where the organizational form of the cooperatives is less familiar. Furthermore, an RE cooperative in which many local residents participate is difficult to set up due in part to legal conditions in Japan (Yamamoto 2016). In such countries, municipal RE companies may have potential for undertaking local RE projects successfully.

Interestingly, many cooperatives work with municipalities. According to DGRV (2014), for some 60% of the energy cooperatives, a municipality is a member of the cooperative and/or actively engages in the cooperative committees. Schreuer and Weismeier-Sammerand (2010) also pointed out that the relationship between the municipality, in particular the mayor, and active citizens is of crucial importance. Moreover, Seyfang et al. (2013) observed that community energy groups often work in partnership with other organizations, most prominently with local authorities. These suggest the importance of the role a municipality plays in RE development in a region.

According to Tarhan (2015), local factors such as existing levels of trust, familiarity with cooperatives, and a history of cooperation affect success in RE cooperatives. Similarly, municipality involvement will be contextual. For example, some economies provide most goods and services publicly, while other economies provide these privately. Accordingly, the organizational form that is best suited for RE projects in an economy depends on its economic, social, environmental, historical, and cultural characteristics.

References

Besley, T., & Ghatak, M. (2005). Competition and incentives with motivated agents. *The American Economic Review, 95*(3), 616–636.

Bomberg, E., & McEwen, N. (2012). Mobilizing community energy. *Energy Policy, 51*, 435–444.

Bright, L. (2009). Why do public employees desire intrinsic nonmonetary opportunities? *Public Personnel Management, 38*(3), 15–37.

DGRV. (2014). *Energy cooperatives: Findings of survey conducted by the DGRV and its member associations*. Spring 2014. DGRV (German Cooperative and Raiffeisen Confederation), Berlin.

Engel, E., Fischer, R. D., & Galetovic, A. (2014). *The economics of public-private partnerships: A basic guide*. New York: Cambridge University Press.

Francois, P. (2000). 'Public service motivation' as an argument for government provision. *Journal of Public Economics, 78*, 275–299.

Huang, W.-L., & Feeney, M. K. (2016). Citizen participation in local government decision making: The role of manager motivation. *Review of Public Personnel Administration, 36*(2), 188–209.

Huybrechts, B., & Mertens, S. (2014). The relevance of the cooperative model in the field of renewable energy. *Annals of Public and Cooperative Economics, 85*(2), 193–212.

ICA. (2017). *What is a co-operative*. https://ica.coop/en/what-co-operative. Accessed in Jan 2017.

Jensen, P. H., & Stonecash, R. E. (2005). Incentives and the efficiency of public sector-outsourcing contracts. *Journal of Economic Surveys, 19*(5), 767–787.

Musall, F. D., & Kuik, O. (2011). Local acceptance of renewable energy—A case study from southeast Germany. *Energy Policy, 39*, 3252–3260.

Rainey, H. G., & Steinbauer, P. (1999). Galloping elephants: Developing elements of a theory of effective government organizations. *Journal of Public Administration Research and Theory, 9*, 1–32.

Schreuer, A., & Weismeier-Sammerand, D. (2010). *Energy cooperatives and local ownership in the field of renewable energy technologies: A literature review*. RiCC-research report 2010/4. Vienna University of Economics and Businesses, Berlin.

Seyfang, G., Park, J. J., & Smith, A. (2013). A thousand flowers blooming? An examination of community energy in UK. *Energy Policy, 61*, 977–989.

Tarhan, M. D. (2015). Renewable energy cooperatives: A review of demonstrated impacts and limitations. *Journal of Entrepreneurial and Organizational Diversity, 4*(1), 104–120.

Viardot, E. (2013). The role of cooperatives in overcoming the barriers to adoption of renewable energy. *Energy Policy, 63*, 756–764.

Walker, G. (2008). What are the barriers and incentives for community-owned means of energy production and use? *Energy Policy, 36*, 4401–4405.

Walker, G., & Devine-Wright, P. (2008). Community renewable energy: What should it mean? *Energy Policy, 36*, 497–500.

Walker, G., Devine-Wright, P., Hunter, S., High, H., & Evans, B. (2010). Trust and community: Exploring the meanings, contexts and dynamics of community renewable energy. *Energy Policy, 38*, 2655–2663.

Warren, C. R., & McFadyen, M. (2010). Does community ownership affect public attitudes to wind energy? A case study from south-west Scotland. *Land Use Policy, 27*, 204–213.

Wirth, S. (2014). Communities matter: Institutional preconditions for community renewable energy. *Energy Policy, 70*, 236–246.

Wright, B. E. (2001). Public-sector work motivation: A review of the current literature and a revised conceptual model. *Journal of Public Administration Research and Theory, 11*(4), 559–586.

Yamamoto, Y. (2016). The role of community energy in renewable energy use and development. *Renewable Energy and Environmental Sustainability, 1*(18), 1–4.

Yildiz, Ö. (2014). Financing renewable energy infrastructures via financial citizen participation—The case of Germany. *Renewable Energy, 68*, 677–685.

Yildiz, Ö., Rommel, J., Debor, S., Holstenkamp, L., Mey, F., Mükkerm, J. R., Radtke, J., & Rognli, J. (2015). Renewable energy cooperatives as gatekeepers or facilitators? Recent developments in Germany and a multidisciplinary research agenda. *Energy Research and Social Science, 6*, 59–73.

Chapter 43
Sustainable Supply Chain: Feedstock Logistics Issues of Palm Oil Biomass Industry in Malaysia

Puan Yatim, Sue Lin Ngan, and Hon Loong Lam

43.1 Introduction

A sustainable and reliable energy source is one of the main challenges that mankind will face over the coming decades, particularly because of the increasing trend in world energy use and the need to address climate change. The US Energy Information Administration recently projects that world energy consumption will grow by 48% between 2012 and 2040 (US EIA 2016). Most of this growth can be attributed to two main reasons: a growing world population and developing countries. As the population of the world continues to grow, the need for energy will increase with most of the energy consumed derived from nonrenewable sources and those resources are not infinite and over time they will deplete and cease to exist. Much of the global increase in energy demand occurs among developing countries such as China and India where strong economic growth coupled with expanding population lead the increase in world energy consumption. Further, since these countries are in the process of becoming industrialized, they may have not yet improved efficiency and sustainable use of energy, resulting in both more consumption and more waste. According to the US Energy Information Administration (EIA 2016), non-OECD countries demand for energy rises by 71% from 2012 to 2040, while in the OECD economies, total energy use increases by 18% during the same period. Therefore, effective actions for transition to a sustainable energy future and consumption of bio-based products may help decarbonize the global

P. Yatim (✉)
UKM-Graduate School of Business, Universiti Kebangsaan Malaysia, Bangi, Selangor Darul Ehsan, Malaysia
e-mail: puan@ukm.edu.my

S.L. Ngan • H.L. Lam
Centre of Excellence for Green Technology, Faculty of Engineering, University of Nottingham Malaysia Campus, Semenyih, Selangor Darul Ehsan, Malaysia

A. Sayigh (ed.), *Transition Towards 100% Renewable Energy*,
Innovative Renewable Energy, https://doi.org/10.1007/978-3-319-69844-1_43

energy mix and promote sustainable development as well as reduce the impact of climate change.

One of the policy agendas proposed by various governments, both in developed and developing economies, to address energy security and climate change concerns is bioeconomy policy strategies. The importance of the bioeconomy has grown considerably in policy agendas around the world with at least 45 countries now have bioeconomy initiatives while at least 8 including the European Union and the United States have holistic bioeconomy strategies (Viaggi 2016). The bioeconomy comprises those parts of the economy that use renewable resources to provide raw materials for existing and new industrial and energy uses (European Commission 2016). A strong bio-based industrial sector worldwide, particularly in developing countries, will significantly not only reduce the world's heavy reliance on fossil resources but also contribute to economic growth and create job opportunities as well as attract foreign investments. For instance, in European Union, the total turnover of the bio-based industries reaches €600 billion in 2013, while the total employment accounts for 13.8 million jobs (Piotrowski et al. 2016). In recent years, developing countries are also experiencing accelerated growth in their bioeconomy sector. For instance, India's bioeconomy surpasses USD 4 billion in 2013 while Brazil's sugarcane industry accounts for 2% of its gross domestic product (GDP) and provides 4.5 million jobs in 2012 (El-Chichakli et al. 2016). In Malaysia, bioeconomy contributes 13.4% of the total gross domestic product (GDP), a contribution equivalent to RM 106.7 billion (USD 34.7 billion) in 2010 (Malaysian Bioeconomy Development Corporation 2016).

At the heart of a robust and sustainable bioeconomy is the development of an integrated bio-based industries value chains which encompass three important elements, namely, biomass and waste supply chain, biorefineries, and final product applications. The focus of waste resources such as biomass as a component of the bioeconomy remains underdeveloped in many countries including in Malaysia, but the importance of "waste to wealth" in providing value-added products is increasingly recognized. Biomass feedstocks include dedicated energy crops (e.g., switchgrass and wheatgrass), agricultural crops (e.g.., corn and palm oil), agricultural crop residues (e.g., corn stover, rice straw, oil palm fronds), forest residues (e.g., woody materials), aquatic plants (e.g., seaweed and algae), and municipal wastes (e.g., industrial and commercial wastes). Mobilization of these biomass feedstocks with regard to logistics and competitive costs is a critical success factor to ensure a vibrant and sustainable bioeconomy. The biomass supply chain is facing a complex challenges such as feedstock supply risk, insufficient infrastructure, and fragmented supply chain. Biorefinering involves the process of converting biomass into value-added chemicals, plastics, and fuels. Although investment in research, development, demonstration, and deployment of technologies have increased in recent years, the biorefinery development and its activities are still evolving and exposed to several risks such as technical, commercial, financial, and technology. Refined processing methods and pathways coupled with advanced and efficient technologies for biomass conversion enable low value renewable resources to be transformed into value-added products. Biomass-derived final product applications include biofuels, biochemical and biomaterials, and feed and food ingredients. Although vast economic and environmental benefits can be derived from bio-based industries, the industries in general confront several challenges such as product development risk, delivery risk, and market access and acceptance.

Malaysia is the second largest producer of oil after neighboring Indonesia. The total oil palm planted area in Malaysia reached 5.64 million hectares in 2015 with crude palm oil production of 17.32 million tons in 2016 (MPOB 2017). The palm oil industry is expected to grow in the next decade largely due to business opportunities in upstream expansion, exploitation of existing downstream palm oil activities, as well as bioenergy production. With the expansion of the industry, the generation of large amount of oil palm biomass wastes is also expected to increase to about 100 million dry tons in 2020 (AIM 2013). Recognizing the economic importance of, particularly, oil palm biomass, the government introduced the National Biomass Strategy (NBS) in 2013, a strategic initiative to exploit biomass wastes for commercial opportunities.

Despite its economic potentials, the biomass industry in general remains relatively underdeveloped and underinvested. Many of the industry's activities are still confined to low value-added conversion of the biomass to downstream products due to a number of gaps existing along the industry's value chain, and one of these gaps is access to long-term supply of biomass and mobilization of the biomass for downstream utilization at competitive costs. A typical biomass industry value chain is comprised of several components or nodes, and their linkages are shown in Fig. 43.1.

The growth of the biomass industry depends largely on the management of supply chain risk through coordination or collaboration among supply chain actors

Fig. 43.1 Biomass industry value chain nodes

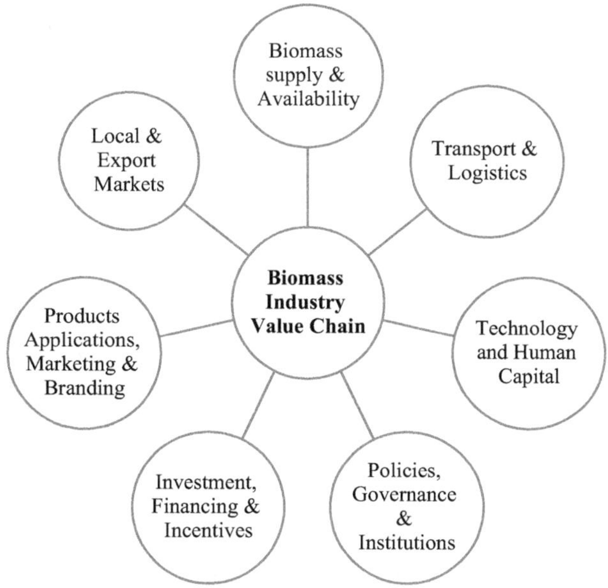

in the industry so as to ensure a guaranteed biomass feedstock and minimizes supply interruptions. As shown in Fig. 43.1, two of the important components of the industry's value chain are reliable supply of biomass feedstock and feedstock logistics. One of the barriers associated with low biomass utilization and mobilization often cited in the biomass supply chain literature is lack of reliable and sustainable supply of feedstock and feedstock logistics problems (Rentizelas et al. 2009). It is also obvious that the biomass supply chain literature has not paid to the issue of feedstock logistics the attention it deserves. Feedstock logistics plays a critical role in the biomass industry's supply chain as it links feedstock production and conversion. The economics of bioenergy highly relies on feedstock costs which are determined by several factors include feedstock types, location, physical and chemical properties, and logistics (Kurian et al. 2013). Feedstock logistics encompasses all the processes necessary to harvest or collect the biomass and mobilize it from the biomass locations such as forests, fields, and plantations to the throat of the conversion reactor at the biorefinery facility while simultaneously ensuring that the delivered feedstock meets the specifications of the biorefinery conversion process.

Numerous studies have been dedicated to analyze approaches associated with the different aspects of biomass supply chain, including approaches for initial network design and optimization of a specific supply chain for cost reduction and operation efficiency (Sharma et al. 2013; Tay et al. 2012; Lim and Lam 2016), biomass storage planning (Rentizelas et al. 2009; Darr and Shah 2012), and tactical and strategic decisions (Lin et al. 2014; De Meyer et al. 2015). Most of the existing research work on biomass supply chain, in general, concerns supply chain design and optimization using simulation and mathematical models. Our work aims at identifying issues and challenges faced by the biomass industry in Malaysia with the main focus on palm oil biomass feedstock logistics, a component of the biomass industry value chain. We also offer several recommendations to overcome feedstock logistics challenges confronting the industry and its value chain.

43.2 Current State of the Malaysian Biomass Industry

Over the last decade, government policies in promoting the renewable energy sector and in addressing waste disposal problems have encouraged active participation of plantation companies in the biomass industry. Joint efforts of public-private partnerships and linkages between industry participants and other stakeholders not only help achieving sustainable development goals but also foster innovation as well as introduce new technologies. Further, as sustainable development and sustainability have been mainstreamed into government policies and business practices, new markets for green products have emerged and increased investment deals. For instance, RM 2.07 billion (USD 0.51 billion) worth of renewable energy investments were approved during the third quarter of 2016, 33% higher than the whole of 2015 with many of these investment deals are related to manufacturing projects (Zakariah 2016). Nevertheless, many of these manufacturing activities are still confined to low value-added products, and abundance of biomass, particularly palm oil biomass, is still left unexploited. The industry still faces many issues and challenges that require institutional intervention and industry-wide participation to drive the industry toward the high value-added stage of its life cycle. Among the issues and challenges confronting the industry include limited participation by small- and medium-sized enterprises (SMEs) in biomass utilization to high-value products, low-value products from biomass with generic substitutes, gaps in the industry's value chain, lack of communication and information and technology sharing among industry participants, facilitators from government agencies, and foreign partners or investors.

43.3 Palm Oil Biomass Feedstock Logistics

The uptake of biomass industry depends on several factors including production costs and supply chain infrastructure. Several types of biomass sources are bulky, voluminous, and often seasonably available. Uncertainties in raw material supply and final product demand also complicate biomass supply chain management. Further, the ownership of the oil palm biomass feedstock is highly fragmented. MPOB (2016) reports that 61% of the total oil palm planted area is owned by private estates, followed by 16% of the planted area belongs to independent smallholders. The remaining planted areas are owned by state agencies (6%), FELDA (13%), FELCRA (3%), and RISDA (1%), respectively. Dispersed ownership of palm oil plantations coupled with remote locations of these plantations increase supply chain risk, making biomass projects more difficult to obtain financing. Unlike wind, solar, and hydro, biomass power generation requires a feedstock that must be produced, collected, transported, pretreated, and stored. The economics of renewable energy power generation and other bio-based industries heavily critically dependent upon the availability of a secure, long-term supply of biomass feedstock at a competitive cost.

In general, feedstock logistics are a part of biomass supply chain and involve four main processes of mobilizing biomass feedstock from the field to the throat of a biorefinery. It is a critical bridge between biomass productions and biomass utilization for producing high value-added products such as bioenergy, biomaterials, and biochemical. Logistics design and infrastructure, cost-benefit analysis, and sustainability of environment are important considerations in biomass feedstock logistics decisions. One approach to factor in these considerations in feedstock logistics designs or systems is by using optimization modeling so that feedstock logistics systems can be cost-effective and efficient and economically viable (Miao et al. 2012; Zhang et al. 2013).

Figure 43.2 illustrates palm oil biomass feedstock logistics. Depending on the types of palm oil wastes, some of these wastes are located in palm oil plantations or estates, while others are located at palm oil mills. For instance, fronds and trucks are located in the plantations areas, while empty fruit bunches (EFB), palm kernel

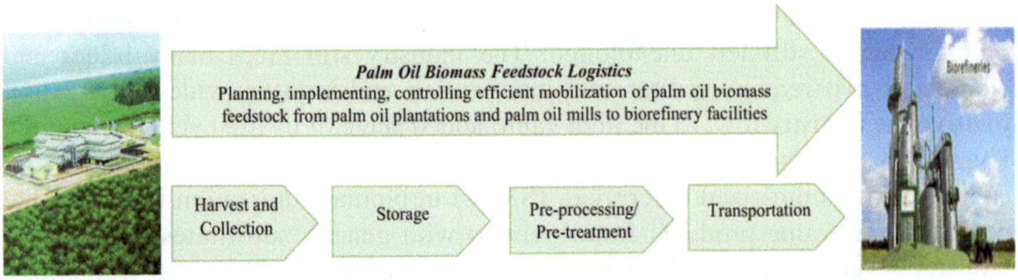

Fig. 43.2 Palm oil biomass feedstock logistics processes involving harvest and collection, storage, preprocessing, and transportation for delivery to biorefinery facilities

Biomass type		Site of production	Annual availability	
			Per ha (tons of dry biomass)	*National total (m tons of dry biomass)*
	Fronds	Plantation	9.6	46.4
	Trunks	Plantation	3.0	14.4
	EFB	Mill	1.4	6.7
	Shells (PKS)	Mill	0.8	4.1
	Fiber (MF)	Mill	1.4	6.9
	POME	Mill	12.2 (wet weight)	59.3 (wet weight)

Fig. 43.3 Annual availability of oil palm biomass by type (Source: National Biomass Strategy 2020: New wealth creation for Malaysia's biomass industry (Version 2.0, 2013))

shells (PKS), mesocarp fiber (MF), and palm oil mill effluent (POME) are located at mills (see Fig. 43.3). The following section discusses feedstock logistics and processes that are involved in the mobilization of palm oil biomass from the field to the throat of a biorefinery.

43.3.1 Harvest and Collection

The first stage of the palm oil biomass feedstock logistics process is harvesting and collecting palm oil wastes from plantations or mills to storage facilities. Factors such as size of plantations or estates, quantity of palm oil wastes, frequency of harvest and collection, locations of the wastes, availability of labor, and harvest and collection methods determine efficiency, effectiveness, and costs associated with the harvest and collection process. Fresh fruit bunches (FFBs) are harvested and collected in the plantations and subsequently transported to the mills for production of crude palm oil (CPO). Therefore, empty fruit bunches (EFBs) and palm kernel shells (PKS) and mesocarp fiber are available at the mills and can be transported to storage facilities within the mills or to other locations for downstream processing.

On the hand, palm fronds and trunks are not readily available, and these wastes can only be harvested and collected during plantation replanting and pruning. Currently, palm fronds are left in the plantations as top soil replacement and natural fertilizer. Methods of collection and mobilization of the fronds range from simple manual collection with wheelbarrow, buffalo carts, or motorized carts to advanced mechanization. The choice of collection methods also depends on terrain conditions in the plantation area (e.g., flat, undulating, or hilly terrain), labor constraints, and economies of scale. Oil palm trunks are available at the end of a plantation's life cycle every 25–30 years. Trunks are either felled or chipped and left in the plantation as fertilizer. A small percentage of trunks are processed and used in wood-based industry such as for flooring, plywood, and furniture. Out of 5.64 million hectares of planted area at the end of 2015, 86% was mature plantations, while 14% are immature trees (MPOB 2016). Over the next few years, approximately 240 million tons of palm oil trunks will become available during replanting with unevenly distributed supply of waste from palm oil trunks due to geographic locations and different maturities of the plantations (AIM 2013). Uncertainties of supply due to geographic and timing constraints pose challenges on collection methods and potential utilization of these wastes since many downstream industries rely on a consistent supply of biomass over time. As a result, downstream industries need to consider all direct and indirect costs associated with the harvest and collection process.

43.3.2 Storage

Sufficient storage for biomass is critical to accommodate seasonality of production and to ensure consistent supply to downstream industries. At the storage stage of feedstock logistics, decisions such as storage methods and storage costs are made so that quality and specifications of the biomass are not compromised. The type of storage required largely depends on the types of biomass and the properties of the biomass including the moisture content. For instance, high-moisture feedstock such as oil palm fronds which intended to be used wet is more suited for fermentation and anaerobic digestion conversion processes while low-moisture content feedstock is suitable for thermal process (McKendry 2002; Zahari et al. 2012). The wet storage system requires constant monitoring to avoid excessive degradation of feedstock, while dry storage system should take into account spontaneous combustion and excess decomposition problems (Badger and Fransham 2006). Further, the shrinkage, compositional and pretreatment impacts, and soluble sugar capture also need to be considered prior to deciding which storage system is appropriate to use (Kenney et al. 2013; Carpenter et al. 2014).

Cost of storage and constant supply of biomass feedstock are critical to ensure feasibility to developing a sustainable logistics infrastructure capable of supplying large volume of biomass to biorefinery facilities. The storage facilities can be built on the same location as the source of feedstock depending on space, quantities, and characteristics of the feedstock. The on-field storage method may cost less, but it exposes to high risk of biomass losses, and biomass moisture cannot be controlled which can lead to potential problems in the biorefinery at conversion stage (Rentizelas et al. 2009). In some cases, feedstock quantities are small and may not be economically viable to set up storage facilities on the same site as the source of feedstock. The size of the storage facility at a biorefinery plant also affects the transport arrangements (Allen et al. 1998). For relatively small capacity, storage facility (e.g., a few days' supply) will need regular restocking of feedstock than a plant with large storage capacity. Thus, in order to prevent supply shortage and interruption of conversion processes, reliable and flexible transport arrangements become important, particularly for a biorefinery plant with small storage facility.

43.3.3 Preprocessing/Pretreatment

Different biomass types undergo different forms of preprocessing or pretreatment processes which physically, chemically, or biologically transform biomass feedstock into a state or a form more suitable for transport or conversion to liquid fuels and fine chemicals. Some of the examples of preprocessing processes include densifying, on-site pyrolysis, grinding, drying, chemical treating, ensiling, fractionating, and blending (Tabil et al. 2011; Chin et al. 2013; Rodriguez et al. 2015; Oudenhoven et al. 2016). With regard to oil palm biomass feedstock, trunks and fronds can be chipped, dried, and/or pelletized, while empty fruit bunches and mesocarp fiber can be shredded, dried, and/or compacted. Palm kernel shells can be readily used or transported without further preprocessing due to their very low moisture content.

The choice of the preprocessing method and the pretreatment technology used for a particular biomass depends on its composition and the by-products produced. These factors significantly affect the costs associated with preprocessing and pretreatment methods (Kumar et al. 2009). For instance, due to high upfront investment cost needed for preprocessing of biomass such as shredding, drying, and pressing of biomass, most millers have been reluctant to utilize empty fruit bunches (EFBs) for renewable energy production, instead the EFBs are burned to produce fertilizers (Samiran et al. 2016; Shafawati and Siddiquee 2013). With preprocessing stage may take a large proportion of their total costs, downstream industries are likely to explore other cheaper preprocessing methods that do not require, for instance, biomass to be dried. Insufficient knowledge about properties of oil palm biomass with regard to its utilization as a bioenergy and biomaterial feedstock inhibit the design of cost-effective and efficient preprocessing technologies and equipment. One of the reasons for the lack of knowledge of biomass properties among downstream industries is because much of this knowledge is confined to research labs of local universities and research institutes and there exists inadequate two-way networking and research-industry collaboration (Hansen and Ockwell 2014).

43.3.4 Transportation

Biomass feedstock transportation is critical as it bridges biomass production, transformation, and conversion into a complete biorefinery system. Therefore, a reliable, flexible, and efficient transportation can significantly improve the cost-competitiveness and increase the viability of the biomass industry in Malaysia. The biomass types, forms, intended end use, supply and demand locations, and equipment and facility availability also affect the performance of the transportation system (Miao et al. 2012). Existing transport systems are largely fixed with well-established configurations and volumes that comply with road and transport laws as well as are constrained to several transport options such as truck, rail, or barge. Thus, the primary constraints on transport systems transporting and handling oil palm biomass feedstock include maximizing both truck capacity and efficiency by increasing the bulk density of feedstock and maximizing truck routes as well as minimizing wait times (Hess et al. 2007). Understanding these constraints help downstream industries design and manage efficient transport systems which can minimize costs associated with mobilizing biomass within the feedstock logistics systems. Several key technical and operational aspects closely linked to feedstock transportation also must be addressed such as handling operations and queuing systems required to unload feedstock from mode of transports (e.g., truck) to interim storage before conversion process (Hess et al. 2007; Miao et al. 2012). With regard to oil palm biomass feedstock, road transport is likely to be the preferred mode of moving biomass from plantations to biorefinery facilities due to scattered oil palm plantations in remote locations. Further, the dispersed ownership of the plantations complicated the feedstock logistics as different owners have their own preferences on the harvest and collection time. Most of the crude palm oil is currently transported from mills to port-based bulking and refinery installations. A similar structure is likely envisaged for the biomass supply chain in order to aggregate sufficiently large quantities of biomass in decentralized biomass collection and processing facilities. Since trucking is commonly used for biomass transportation, it often can be socially undesirable because it may damage roadways and increase traffic flow in rural areas. Innovative transportation solutions such as improved containers and lighter vehicles could reduce traffic and environmental impacts.

43.4 Conclusion

Although Malaysia is blessed with abundance of biomass from oil palm production, the growth of downstream activities in the oil palm industry is low (Khoo 2014). One of the factors that impede the growth of downstream industries is feedstock logistics. Feedstock logistics can be the highest cost component in the biomass supply change, ranging from 15% to 60% of the total cost of the production (AIM 2013). In general, the oil palm biomass supply chain infrastructure in Malaysia is almost nonexistent. The growth of the biomass industry depends heavily on the reliable, flexible, efficient, and cost-effective supply chain infrastructure as to ensure a guaranteed biomass feedstock supply and minimizes supply interruptions. Rapid growth in demand for bio-based products will require major improvements and configurations in supply chain infrastructure. Efficient feedstock logistics can be achieved through decentralized operations that facilitate local sourcing, storage, preprocessing, and transportation of biomass wastes from plantations to conversion facilities. Integrated feedstock logistics systems that are cost-effective and energy-efficient will require new ways of planning agricultural and energy policies, renewable energy infrastructure, and bio-based industry policy agenda. Developing integrated biomass logistics systems will require not only investment in technical and technological know-how but also research and innovation in novel technologies and efficient value chains of downstream industries. Further, sustainable socioeconomic and rural area development, policy, and institutional support can also help promote the growth of bio-based industries and meet challenges of oil palm biomass feedstock logistics.

The economic feasibility of the biomass industry remains highly uncertain without value-creating biomass supply chains. Without evidence of long-term and sustainable feedstock supplies, obtaining financing from traditional sources such as project finance and bank loans become difficult for bio-based industries. Financial institutions including insurance companies need to design suitable financing products that address the technological and financial barriers faced by industry players. In developed countries such as the United States, bond financing has already been used to support clean energy projects. Several types of tax-exempt and tax-subsidized bond programs have been created to support renewable energy projects in the country. The bio-based industries in Malaysia could form clusters with focal points in different downstream activities, and smaller biomass-based projects can be aggregated, and green bonds are issued backed by economic potentials of these projects.

Malaysian government has launched multiple policy initiatives to promote and support the biomass industry. These policy initiatives such as National Biomass Strategy 2013, Bioeconomy Transformation Program 2012, and Green Technology Financing Scheme 2009 aim at exploiting abundance of biomass for commercial opportunities by providing expert assistance and incentives in knowledge and skill capacity building and in development of technologies. These policies, however, lack emphasis and focus on building ecosystem, such as supply chain solutions, for robust bio-based industries.

One of the challenges faced by the oil palm industry is constant pressures from various stakeholders to demonstrate its commitment to sustainable agricultural practices. Environmentalists have long been argued that oil palm production poses threat to biodiversity in Malaysia and Indonesia as forests are being cleared and converted to oil palm plantations. Transportation sector represents a major source of GHGs emission in Malaysia. Since transportation system in Malaysia is still heavily dependent on fossil fuels, the GHGs emitted from the transportation activities of the feedstock logistics from the source point of feedstock to a biorefinery facility are likely to be harmful to the environment. Further, the life cycle of oil palm tree is about 30 years, and fruits can be harvested from year 5 onward. The consistent supply of the oil palm biomass depends on healthy soil condition. Thus, fertilizer is necessary to maintain healthy soil condition and oil palm tree productivity. However, excess use of fertilizer can degrade soil and may also lead to water pollution. Therefore, a well-planned supply chain management system with special considerations on its impacts on the environment is necessary in order to create a sustainable supply chain for biorefinery industry in Malaysia.

Currently, there is minimal infrastructure in place for biomass supply chain in Malaysian biomass industry. A key priority for project developers is to mitigate supply chain-related risks. Strategic partnerships with plantation owners and millers and vertical integration between upstream and downstream industries can help minimize logistics risks along the biomass supply chain. Partnerships with other industries such as paper or pulp industry which already has an established supply chain infrastructure can also be explored to achieve supply chain synergies such as cost savings.

References

AIM (Agensi Inovasi Malaysia). (2013). *National biomass strategy 2020*. www.nbs2020.gov.my/. Accessed 20.10.16.

Allen, J., Browne, M., Hunter, A., Boyd, J., & Palmer, H. (1998). Logistics management and costs of biomass fuel supply. *International Journal of Physical Distribution and Logistics Management, 28*(6), 463–477.

Badger, P. C., & Fransham, P. (2006). Use of mobile fast pyrolysis plants to densify biomass and reduce biomass handling costs – A preliminary assessment. *Biomass and Bioenergy, 30*(4), 321–325.

Carpenter, D., Westover, T. L., Czernik, S., & Jablonski, W. (2014). Biomass feedstocks for renewable fuel production: A review of the impacts of feedstock and pre-treatment on the yield and product distribution of fast pyrolysis bio-oils and vapors. *Green Chemistry, 16*, 384–406.

Chin, K. L., H'ng, P. S., Go, W. Z., Wong, W. Z., Lim, T. W., Maminski, M., & Luqman, A. C. (2013). Optimization of torrefaction conditions for high energy density solid biofuel from oil palm biomass and fast growing species available in Malaysia. *Industrial Crops and Products, 49*, 768–774.

Darr, M. J., & Shah, A. (2012). Biomass storage: An update on industrial solutions for baled biomass feedstocks. *Biofuels, 3*(3), 321–332.

De Meyer, A., Cattrysse, D., & Van Orshoven, J. (2015). A generic mathematical model to optimise strategic and tactical decisions in biomass-based supply chains (OPTIMASS). *European Journal of Operational Research, 245*(1), 247–264.

El-Chichakli, B., von Braun, J., Lang, C., Barben, D., & Philp, J. (2016). Five cornerstones of a global bioeconomy. *Nature, 535*, 221–223.

European Commission. (2016). *Bioeconomy*. www.ec.europa.eu/research/bioeconomy/. Accessed 15.08.16.

Hansen, U. E., & Ockwell, D. (2014). Learning and technological capability building in emerging economies: The case of the biomass power equipment industry in Malaysia. *Technovation, 34*, 617–630.

Hess, J. R., Wright, C. T., & Kenney, K. L. (2007). Cellulosic biomass feedstocks and logistics for ethanol production. *Biofuels, Bioproducts & Biorefining, 1*(3), 181–190.

Kenney, K. L., Smith, W. A., Gresham, G. L., & Westover, T. L. (2013). Understanding biomass feedstock variability. *Biofuels, 4*(1), 111–127.

Khoo, D. (2014). Palm oil industry needs to invest more in downstream activities. www.thestar.com.my/business/business-news/2014/06/21/call-to-venture-downstream-palm-oil-industry-needs-to-invest-more-in-downstream-activities-to-reduc/. Accessed 20.10.16.

Kumar, P., Barrett, D. M., Delwiche, M. J., & Stroeve, P. (2009). Methods for pre-treatment of lignocellulosic biomass for efficient hydrolysis and biofuel production. *Industrial & Engineering Chemistry Research, 48*(8), 3713–3729.

Kurian, J. K., Nair, G. P., Hussain, A., & Raghavan, G. S. V. (2013). Feedstocks, logistics and pre-treatment processes for sustainable lignocellulosic biorefineries: A comprehensive review. *Renewable and Sustainable Energy Reviews, 25*, 205–219.

Lim, C. H., & Lam, H. L. (2016). Biomass supply chain optimisation via novel Biomass Element Life Cycle Analysis (BELCA). *Applied Energy, 161*, 733–745.

Lin, T., Rodríguez, L. F., Yogendra, N., Shastri, Y. N., Hansen, A. C., & Ting, K. C. (2014). Integrated strategic and tactical biomass–biofuel supply chain optimization. *Bioresource Technology, 156*, 256–266.

Malaysian Bioeconomy Development Corporation. (2016). *Bioeconomy Malaysia*. www.bioeconomycorporation.my/bioeconomy-malaysia/. Accessed 02.08.16.

McKendry, P. (2002). Energy production from biomass (Part 1): Overview of biomass. *Bioresource Technology, 83*, 37–46.

Miao, Z., Shastri, Y., Grift, T. E., Hansen, A. C., & Ting, A. C. (2012). Lignocellulosic biomass feedstock transportation alternatives, logistics, equipment configurations, and modeling. *Biofuels, Bioproducts & Biorefining, 6*(3), 351–362.

MPOB. (2017). *Economics and industry development division*. Malaysia Palm Oil Board. http://bepi.mpob.gov.my/index.php/my/. Accessed 02.01.17.

Oudenhoven, S. R. G., van der Ham, A. G. J., van den Berg, H., Westerhof, R. J. M., & Kersten, S. R. A. (2016). Using pyrolytic acid leaching as a pre-treatment step in a biomass fast pyrolysis plant: Process design and economic evaluation. *Biomass and Bioenergy, 95*, 388–404.

Piotrowski, S., Carus, M., & Carrez, D. (2016). *European bioeconomy in figures, bio-based industries consortium*. Brussels. biconsortium.eu/sites/biconsortium.eu/files/downloads/20160302_Bioeconomy_in_figures.pdf. Accessed 30.04.16.

Rentizelas, A. A., Tolis, A. J., Tatsiopoulos, I. P., & I.P. (2009). Logistics issues of biomass: The storage problem and the multi-biomass supply chain. *Renewable and Sustainable Energy Reviews, 13*, 887–894.

Rodriguez, C., Alaswad, A., Mooney, J., Prescott, T., & Olabi, A. G. (2015). Pre-treatment techniques used for anaerobic digestion of algae. *Fuel Processing Technology, 138*, 765–779.

Samiran, N. A., Jaafar, M. N. M., Ng, J. N., Lam, S. S., & Chong, C. T. (2016). Progress in biomass gasification technique – With focus on Malaysian palm biomass for syngas production. *Renewable and Sustainable Energy Reviews, 62*, 1047–1062.

Shafawati, S. N., & Siddiquee, S. (2013). Composting of oil palm fibres and *Trichoderma* spp. as the biological control agent: A review. *International Biodeterioration & Biodegradation, 85*, 243–253.

Sharma, B., Ingalls, R. G., Jones, C. L., & Khanchi, A. (2013). Biomass supply chain design and analysis: Basis, overview, modeling, challenges, and future. *Renewable and Sustainable Energy Reviews, 24*, 608–627.

Tabil, L., Adapa, P., & Kashaninejad, M. (2011). In A. M. Santos Bernardes (Ed.), *Biofuel's engineering process technology* (pp. 411–438). Croatia: InTech.

Tay, D. H. S., Ng, D. K. S., & Tan, R. R. (2012). Robust optimization approach for synthesis of integrated biorefineries with supply and demand uncertainties. *Environmental Progress & Sustainable Energy, 32*(2), 384–389.

US Energy Information Administration (EIA). (2016). *Chapter 1: World energy demand and economic outlook*. www.eia.gov/outlooks/ieo/world.cfm. Accessed 20.12.16.

Viaggi, D. (2016). Towards an economics of the bioeconomy: Four years later. *Bio-based and Applied Economics, 5*(2), 101–112.

Zahari, M. A. K. M., Zakaria, M. R., Ariffin, H., Mokhtar, M. N., Salihon, J., Shirai, Y., & Hassan, M. A. (2012). Renewable sugars from oil palm frond juice as an alternative novel fermentation feedstock for value-added products. *Bioresource Technology, 110*, 556–561.

Zakariah, Z. (2016). *Boost for renewable energy sector*. www.nst.com.my/news/2016/11/193062/boost-renewable-energy-sector. Accessed 02.12.16.

Zhang, J., Osmani, I., Awudu, I., & Gonela, V. (2013). An integrated optimization model for switchgrass-based bioethanol supply chain. *Applied Energy, 102*, 1205–1217.

Chapter 44
Volume Segmentation in a Stratified Vertical EWH Tank Using Steady-State Element Cycles for Energy Balance

Yu-Chieh J. Yen, Ken J. Nixon, and Willie A. Cronje

44.1 Introduction

A domestic electric water heater (EWH) is a thermal energy store that converts electrical energy to thermal energy through a heating element. This interface between electrical energy and thermal energy is used to verify the calculated energy stored in the thermal system to a high level of accuracy. This kind of energy balance becomes prevalent in the presence of thermal stratification, where some portions of the energy store hold different temperatures. The usage of hot water from the EWH draws energy from the storage and causes the system to replenish the drawn thermal energy using electrical energy, which is often grid tied. This links to modelling loads for demand-side management and the development of "smart" devices in the context of "smart" grids. Baker (2008) presents a summary of existing and predicting future energy storage technologies.

Thermal stratification occurs in a storage tank because the density of water differs based on its associated temperature; some parts of the total volume may heat up or cool down at different rates depending on mass proximities to the heating element or tank walls. In a well-stratified tank, the hot layers and cold layers are separated in the tank with minimal mixing. Although thermal stratification is not the focus of this work, this is an acknowledgement that stratification is present in the tank and is observed through the use of multiple temperature sensors along the tank height. Characterising stratification in a tank is primarily classified in two major categories: methods based on the first law of thermodynamics (energy) and those based on the second law of thermodynamics (exergy). Haller et al. (2009) reviews the available literature on characterisation methods.

Y.-C.J. Yen (✉) • K.J. Nixon • W.A. Cronje
School of Electrical & Information Engineering, University of the Witwatersrand, Johannesburg, South Africa
e-mail: yu-chieh.yen@wits.ac.za

A. Sayigh (ed.), *Transition Towards 100% Renewable Energy*,
Innovative Renewable Energy, https://doi.org/10.1007/978-3-319-69844-1_44

In demand-side management, an EWH is assumed as a homogenous volume that holds one representative temperature to describe the thermal energy stored (Lu et al. 2005; Palensky et al. 2008). By using one temperature measurement, the states of the system are limited to steady state (around a hysteresis point) or draw state, which usually empties the entire energy store. Other models exist to describe the presence of thermal stratification that occurs primarily through draw events, such as the bi-nodal models (Kondoh et al. 2011). Since this work aims to validate a multi-nodal approach as presented by Fernández-Seara et al. (2007) and Campos Celador et al. (2011) through steady-state operation, other models and modes of operation are not discussed further in this paper.

The focus of this work is to provide a strategy for estimating the energy stored in a stratified vertical tank using a multi-nodal model to gain some degree of accuracy using an energy balance calculation. This is accomplished using steady-state heating element operations, which are the heating cycles caused by standing losses in the system. This work demonstrates the effects of different volume segmentation methods under different thermal insulation schemes to determine the best-fit model to use when extending to dynamic states. Furthermore, the basic methodology will be extended to horizontally oriented tanks.

44.2 Methodology

This study aims to validate a multi-nodal model of a cylindrical EWH by matching multiple PT-100 temperature measurements to volume segments of water inside the tank. The thermal states of the tank are matched to the states of the heating element during steady-state operation by comparing the calculated energy. The energy states of the tank depend on the volume segment sizing. This study shows how the volume segment sizing can affect the overall stored energy in the tank.

44.2.1 Multi-nodal Model for Energy in Stratified Tanks

The multi-nodal model presented by Fernández-Seara et al. (2007) provides an estimation of energy stored in a stratified EWH tank. To resolve different temperatures in a stratified tank, temperature measurements are made in fixed intervals along the height of the tank. Individual volumes are defined in the tank, which correlate to each temperature segment measured. In this analysis, it is assumed that the measured temperature is uniform in each associated volume segment. The way in which these volume segments are defined can greatly affect the calculated stored energy and distort any conclusions that may arise from incorrect segment sizing.

The estimated energy stored in the tank at a particular point in time is determined in Eq. 44.1, where N is the number of volume segments that correspond to temperature sensors.

$$E_t(t) = \sum_{j=1}^{N} E_j(t)$$
$$E_j(t) = V_j \rho C_p \left[T_j(t) - T_a(t)\right] \tag{44.1}$$

The total energy in the tank at time t is represented by E_t (kWh), which is the sum of energy in each volume segmented layer E_j (kWh), where j is the layer iterator and N is the total number of layers. Measured temperatures corresponding to each layer segment is represented by T_j (°C), which is compared to a fixed ambient temperature T_a (°C). When considering the thermal energy in the tank during the recharging state (from $t = 0$ to $t = t_{\text{on}}$), as shown in Fig. 44.1a, the thermal energy in tank can be summarised by Eq. 44.2, where T_a cancels out.

$$E_t(t_{\text{on}}) - E_t(0) = \sum_{j=1}^{N} V_j \rho C_p \left[T_j(t_{\text{on}}) - T_j(0)\right] \tag{44.2}$$

44.2.2 *Energy Balance per Element Cycle*

In the trends observed in Fig. 44.1a, the element operates to heat up the fixed volume of water to the maximum set-point temperature. The heating element switches off, and the tank losses heat at a fixed rate based on the tank insulation, which is known as standing losses. This represents thermal energy that is lost from the storage tank to the surrounding environment due to the properties of the tank materials, thermal-insulating materials protecting it from the ambient temperature. The temperature of the tank is controlled by a bimetallic strip thermostat, which operates around a hysteresis point. When the temperature of the water at the thermostat falls below a lower threshold, the element will operate to heat the water to an upper threshold. This is its steady-state operation.

The electrical energy used to heat up the water can be characterised with a degree of accuracy based on the time the element is operating and the current that is drawn, which is limited by the power rating of the heating element. Equation 44.3 provides the electrical energy in kilowatt-hours drawn from the heating element when it is operating, where E_e is the electrical energy drawn from the heating element, P is the power rating of the heating element and t_{on} is the amount of time the heating element operates in each steady-state cycle.

$$E_e = P t_{\text{on}} \tag{44.3}$$

The presence of standing losses in a system is an unintended consequence of storing water at temperatures above ambient temperature, which leads to the steady-state operation. However, these cycles can be used to analyse different states of a system within a controlled environment and can be a useful tool in performing

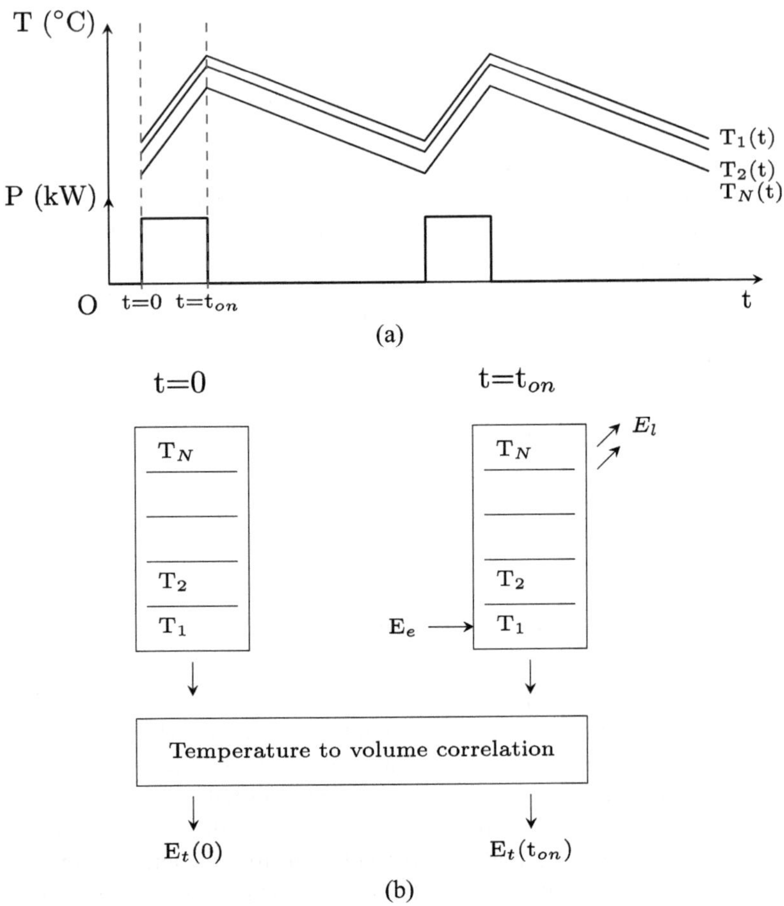

Fig. 44.1 Overview of energy balance for stratified vertical EWH cylinders showing the energy state just before the heating element operates and the energy state just after the element switches off. Schematic of temperature trends during standing loss process. For energy balance, $t = 0$ is defined as the time before the heating element operates, and $t = t_{on}$ is the amount of time the heating element draws electrical power to heat the system

an energy balance to check that estimated values or processes do not violate the laws of conservation of energy. This analysis makes use of only steady-state operation; no draw events are considered.

To conduct an energy balance analysis, the thermal state of the tank is evaluated just before the heating element operates at time $t = 0$ and just after the heating element operates at time $t = t_{on}$. The state of the tank is determined by multiple temperature measurements, $T_1(t)$, $T_2(t)$, ..., $T_N(t)$, in different positions in the tank as shown in Fig. 44.1a.

In each thermal state of the tank, measured temperatures need to be correlated with volumes that define water mass, as shown in Eq. 44.1. Depending on the volume definitions, an energy value will be determined for the tank in each thermal state. Energy balance is performed at time $t = t_{\text{on}}$ as shown in Eq. 44.4, the energy in the tank is evaluated before heating element operation, $E_t(0)$, and after heating element operation, $E_t(t_{\text{on}})$, and compared against the electrical energy, $E_e(t_{\text{on}})$, input into the system and its standing losses, $E_l(t_{on})$ over the time period t_{on}. This is represented in Fig. 44.1b.

$$E_t(t_{\text{on}}) = E_t(0) + E_e(t_{\text{on}}) - E_l(t_{\text{on}}) \tag{44.4}$$

Since there is a degree of uncertainty in the evaluation method of $E_t(0)$ and $E_t(t_{\text{on}})$, the left-hand side and right-hand side of Eq. 44.4 will not be exactly equal, and therefore there is an error value assigned to this method. This error represents energy that is unaccounted for in the energy balance and is denoted by ε and is defined by Eq. 44.5.

$$\epsilon = E_t(t_{\text{on}}) - E_t(0) - E_e(t_{\text{on}}) + E_l(t_{\text{on}}) \tag{44.5}$$

Since it is assumed that each volume layer holds a uniform temperature, each of these states evaluated is assumed to be stable.

44.2.3 Experimental Setup

The EWH under test is a 100-l vertically mounted cylindrical tank. It has a single heating element inside the tank rated at 2 kW. The tank is set with a preset thermostat temperature at the maximum setting. For a standing loss cycle, the heating element operates for approximately 17 ± 2 min, which results in approximately 0.6 kWh.

The tank is allowed to operate as normal, without any hot water draw off for several cycles. Each heating element cycle is defined by an "on" state followed by an "off" state, which results in the tank heating up from a local minimum temperature to a local maximum. A temperature string is inserted into a 100-l EWH tank in direct contact with the hot water. The string has ten PT100 sensors (Class A) placed 100 mm apart along the height of 1 m. In the scope of this study, only nine sensors are used. For extra data points, the tank is wrapped under different degrees of thermal insulation.

44.2.4 Resolving Volume Layers

In the presence of thermal stratification, it is a common practice to approximate volume layers as a 1D model in the tank to estimate the total stored energy. This

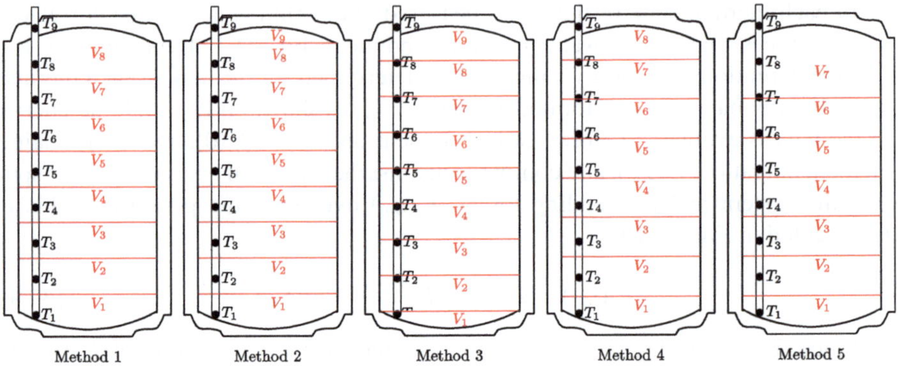

Fig. 44.2 Vertical volume division based on fixed temperature measurement positions for a stratified 100-l EWH cylinder

makes the assumption that the water holds a uniform temperature in each volume segment defined. In determining the volume segment, the main variable to account for is the height of the layer x_j, which is multiplied by the cross-sectional area of the tank base with radius r. This is shown in the following equation:

$$V_j = \pi r^2 x_j \tag{44.6}$$

There are five volume segmentation methods considered for this study, as shown in Fig. 44.2 and discussed in the subsections below. Because of the conservation of energy, it is expected that the most relevant method will convergence the unaccounted energy, ϵ, (from Eq. 44.5) to a value of zero. Since this is considered a stochastic process, the more data points acquired from each set of tests, the more a pattern will emerge. The volume segmentation in these methods assumes that the internal components of the EWH are negligible, such as the heating element, sacrificial anode, diffuser and temperature sensor string.

The tank is tested under five different insulation schemes to provide additional test cases; in each scenario ϵ should also converge to zero for the most relevant volume segmentation method – in particular with better insulation.

Method 1 This volume segmentation method is determined by the middle point between each sensor along the string. Volume V_1 includes a volume defined by $x_j = 50$ mm and the curvature of the bottom. Volumes V_2–V_7 are the same size defined by a layer height, $x_j = 100$ mm. Volume V_8 is also a large segment which is determined by temperature sensor T_8 with a volume defined by $x_j = 100$ mm and the curvature of the top (assumed to be equal to the bottom). Sensor T_9 is ignored in this scenario.

Method 2 This method is very similar to Method 1; however, it takes into account temperature sensor T_9 by adding volume V_9. Volume V_8 is equal to volumes V_2–V_7.

Method 3 This method defines the height of the volumes by the sensor position, i.e. the sensor position indicates the highest temperature in the volume. Volumes V_2–V_7 are the same size defined by a layer height, $x_j = 100$ mm. Volume V_1 is defined as smaller than previous methods and volume V_8 has a larger influence than previous methods.

Method 4 This method considers the tank divided into equal volumes based on eight temperature sensors. This methodology does not directly correlate volumes with sensor locations.

Method 5 This final method is similar to Method 4 for volumes V_2–V_6; volume V_7 is a combination of Method 4 V_7 and V_8, which ignores sensor T_8.

44.3 Results

The results of each test show the unaccounted energy (kWh), ϵ from Eq. 44.5, in each heating element cycle caused by standing losses in the tank. These ϵ values are summarised in the whisker-and-box plots presented in Fig. 44.3. The plots indicate for each dataset the upper and lower quartiles (represented by the upper and lower

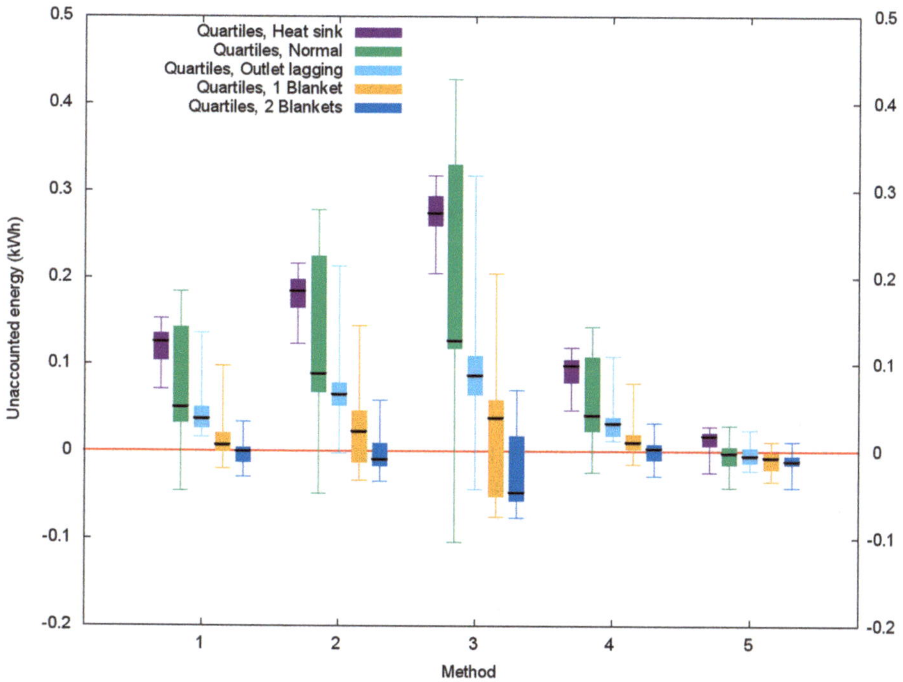

Fig. 44.3 Comparison of ϵ represented by quartiles for five volume segmentation methods using different thermal insulation schemes on a vertically mounted tank. Insulation types are as follows: (a) heat sink, (b) normal, (c) outlet lagging, (d) one blanket, (e) two blankets

bound of the box), the median value (represented by the line inside the box) and the maximum and minimum values (indicated by the upper and lower legs). The median values are presented in Table 44.1, as these are the values that are of interest in making conclusions.

As discussed, for the energy balance to hold, the value of ϵ must tend towards zero. This indicates that the left-hand side and right-hand side of Eq. 44.4 are equivalent and therefore there is no violation of the first law of thermodynamics. If ϵ holds a positive value, the energy in the tank from Eq. 44.1 is overestimated, and a negative value indicates an underestimation of energy.

From observing Fig. 44.3, it is clear that volume segmentation in Method 5 has the median values for unaccounted energy closest to zero in each thermal insulation scenario. Therefore Method 5 is the most appropriate volume segmentation method of the five presented. This method ignores temperature sensor T_8 and T_9 which is situated close to the outlet pipe and could be influenced by the pipe.

44.3.1 Accumulated Errors

$$\alpha = \sum_{m=1}^{M} \epsilon_m \tag{44.7}$$

Since the steady-state heating element events are taken from the same system with concurrent events, the sum of all unaccounted energy ϵ per insulation scheme, α, should also tend to zero for the most appropriate volume segmentation method, as shown in Eq. 44.7. The sample event is represented as m, and M is the total number of events in each insulation scheme.

44.3.2 Effect of the Estimation

By taking a percentage of the error, ϵ, against the average electrical energy, it is possible to determine what effect the error can have on the energy balance. Equation 44.8 uses the median error and the average electrical energy (taken as 0.6 kWh) to determine the effect of the error, δ, for each test set. It should be noted that $E_e << E_t$, so this error has a finer resolution than if it were compared to E_t.

$$\delta = \frac{\text{Median}(\epsilon)}{\text{Average}(E_e)} \times 100 \tag{44.8}$$

It has been determined that Method 5 is the most appropriate volume segmentation method, with percentage errors of -1.80 to 2.98 % as shown in Table 44.1.

Table 44.1 For each insulation scheme (a-e) test dataset and each volume segmentation method, accumulated errors and median values of unaccounted energy as a percentage of E_e, δ

		Accumulated errors, α (kWh)					Percentage error, δ (%)				
		Volume segmentation method					Volume segmentation method				
Insulation	Inst. standing loss (W)	1	2	3	4	5	1	2	3	4	5
(a) Heatsink	108.97	2.18	3.26	4.92	1.67	0.22	20.99	30.75	45.72	16.37	2.98
(b) Normal	109.03	1.73	3.09	4.69	1.34	−0.08	8.54	14.85	21.24	6.86	−0.29
(c) Outlet lagging	98.4	1.15	1.87	2.43	0.94	−0.09	6.26	10.85	14.57	5.42	−0.75
(d) 1 blanket	71.62	0.32	0.62	0.45	0.30	−0.25	1.17	3.83	6.36	1.78	−1.11
(e) 2 blankets	62.16	−0.02	0.00	−0.42	0.02	−0.22	0.11	−1.56	−7.88	0.54	−1.80

44.4 Conclusion

This paper details a methodology for accurately completing an energy balance for a thermally stratified EWH with multiple temperature measurements. This multi-nodal energy model will be extended for the dynamic state of the tank, where water is drawn from the tank and the mixing of volume segments is considered. Since the volume segmentation is specific to the geometry of stratified layers, this can also be extended in a similar manner in a horizontally oriented tank. The dynamic state of these mixing layers for the horizontal model will also need to be evaluated.

This analysis has concluded that Method 5 is the most appropriate volume segmentation method with low variability in samples, low accumulated errors and low percentage error to energy drawn from the heating element, E_e. This can now be extended to the dynamic state and take into account partial draw events and mixing characteristics using the volume segmentation method in a multi-nodal model. It remains to be seen what the introduction of mixing events will do to the assumption of uniform temperature in each volume layer.

Acknowledgements This work is based on the research supported in part by the Alstom Chair for Clean Energy Systems Technology (ACCEST), the National Research Foundation of South Africa (unique grant no: 98241) and the Department of Higher Education and Training (DHET). Further thanks extended to BlackDot Energy for the donation of thermal blankets used in this work.

References

Baker, J. (2008). New technology and possible advances in energy storage. *Energy Policy, 36*(12), 4368–4373.

Campos Celador, A., Odriozola, M., & Sala, J. (2011). Implications of the modelling of stratified hot water storage tanks in the simulation of CHP plants. *Energy Conversion and Management, 52*(8–9), 3018–3026.

Fernández-Seara, J., Uhía, F. J., & Sieres, J. (2007). Experimental analysis of a domestic electric hot water storage tank. Part I: Static mode of operation. *Applied Thermal Engineering, 27*(1), 137–144. https://doi.org/10.1016/j.applthermaleng.2006.05.004.

Haller, M. Y., Cruickshank, C. A., Streicher, W., Harrison, S. J., Andersen, E., & Furbo, S. (2009). Methods to determine stratification efficiency of thermal energy storage processes – Review and theoretical comparison. *Solar Energy, 83*(10), 1847–1860. https://doi.org/10.1016/j.solener.2009.06.019.

Kondoh, J., Lu, N., & Hammerstrom, D. J. (2011). An evaluation of the water heater load potential for providing regulation service. *IEEE Transactions on Power Systems, 26*(3), 1309–1316.

Lu, N., Chassin, D. P., & Widergren, S. E. (2005). Modeling uncertainties in aggregated thermostatically controlled loads using a state queueing model. *IEEE Transactions on Power Systems, 20*(2), 725–733. arXiv:0409038. https://doi.org/10.1109/TPWRS.2005.846072.

Palensky, P., Kupzog, F., Zaidi, A. A., & Zhou, K. (2008). Modeling domestic housing loads for demand response. In *Proceedings – 34th annual conference of the IEEE industrial electronics society, IECON 2008* (pp. 2742–2747). https://doi.org/10.1109/IECON.2008.4758392.

Chapter 45
Modular Pico-hydropower System for Remote Himalayan Villages

Alex Zahnd, Mark Stambaugh, Derek Jackson, Thomas Gross, Christoph Hugi, Rick Sturdivant, James Yeh, and Subodh Sharma

45.1 Introduction

Approximately 17% of the global population or 1.2 billion people (GSR 2016) still live without access to electricity. The majority of these people live in the Asia-Pacific region and sub-Saharan Africa. Rural electrification is the process of improving access to energy services by generating and bringing electrical power to rural communities. Families in the remote high-altitude villages in the Nepalese Himalayas work and struggle day to day under conditions of poor nutrition and sanitation, insufficient indoor lighting, lack of access to safe drinking water, and poor or absent education opportunities. RIDS-Nepal has worked to address these deficiencies by partnering with villages to increase their awareness, provide malnutrition and nonformal education programs, and help them install pit latrines, smokeless metal stoves, solar PV home systems for indoor lighting, safe drinking water systems, greenhouses, and solar driers; all part of RIDS-Nepal's developed

A. Zahnd (✉) • M. Stambaugh
RIDS-Switzerland, Lenzburg, Switzerland

RIDS-Nepal, Kathmandu, Nepal

RIDS-USA, San Diego, CA, USA
e-mail: azahnd@rids-nepal.org

D. Jackson
Aurora Power and Design, Inc., Boise, ID, USA

T. Gross • C. Hugi
FHNW Switzerland – Institute for Ecopreneurship, Muttenz, Switzerland

R. Sturdivant • J. Yeh
Azusa Pacific University, Azusa, CA, USA

S. Sharma
Kathmandu University, Dhulikhel, Nepal

A. Sayigh (ed.), *Transition Towards 100% Renewable Energy*,
Innovative Renewable Energy, https://doi.org/10.1007/978-3-319-69844-1_45

"Family of 4" and "Family of PLUS" holistic community development concepts. All these projects are important cornerstones for the villagers to escape the cycle of poverty and to improve their living conditions. However, improved access to locally generated energy services such as electricity is also needed to improve productivity and education through sufficient indoor lighting and the means to run motors, almost essential for economic development. Solar PV systems rely extensively on deep-cycle batteries that must be replaced occasionally, creating often environmental problems due to the difficulty of transporting the spent batteries and a lack of recycling facilities.

Many remote villages in the Himalayas are located near one of the over 6000 rivers and streams flowing from the mountains down to the Ganges river basin in India, providing a huge potential for hydropower and improved access to energy services. These services would contribute directly to the achievement of several sustainable development goals (SDGs). Apart from renewable energy sources (SDG7), locally produced electricity can reduce indoor air pollution due to the traditional burning of resin-soaked wood for light (SDG3), decrease deforestation, decrease greenhouse gas emissions, and empower women through more equitable education and lifelong learning by reducing the time they spend collecting wood and by providing sufficient light throughout the evening (SDGs4/15/13/5/10), improving access to clean hot water and sanitation (SDG6), improving food availability by heating greenhouses and drying food (SDG2), expanding local industries and skills (SDG9), and thus help reducing extreme poverty (SDG1).

The government of Nepal encourages villages without access to electricity to build a subsidized MHP system, often in the range of 15–50 kW. While this is a step into the right direction and the MHP system technology is in general mature and proven, it is RIDS-Nepal's experience that a majority of the MHP systems in the region within 3–12 months after they have been commissioned are either inoperable due to premature equipment breakage, inappropriate operation, and absent maintenance or they provide far less than the expected power output. Generally, no sustainable process is in place to pay for the used electricity, in order to build an operating and maintenance fund. The promised improvements of living conditions are shattered adding only negative publicity to the inappropriateness of MHP systems. A new, more bottom-up approach for rural electrification is therefore needed and proposed.

45.2 Methodology

45.2.1 Study Area

Moharigaun village (Lat: 29°20′06″/Long: 82°22′26″/Alt: 3′142 m.a.s.l.) is in the northeast of the Jumla district in the northwest of Nepal. Jumla ranks as the seventh least developed district of Nepal's 75 districts. RIDS-Nepal has been partnering with Moharigaun since 1998 in holistic community development projects, including building with each family a pit latrine (PL), smokeless metal stove (SMS) in their home, a solar PV home system (SHS), a high-altitude greenhouse (GH) for increased vegetable production (from previously 3 to now 10 months per year), and a solar drier (SD). A community-level drinking water system was also built.

45.2.2 Renewable Energy Resources

Renewable energy resources such as sunlight, wind, rain, tides, waves, biomass, and geothermal heat naturally replenish or are infinite on human timescales, compared to fossil fuel resources, such as coal, oil, and natural gas, which are finite or not able to be replenished. Renewable energy resources exist in all geographical areas, in contrast to nonrenewable energy resources, which are concentrated in a limited number of countries.

45.2.3 Solar PV Systems

Solar PV systems (Zahnd 2013) have become an often used renewable energy technology for first-time, small-scale electrifications of rural communities, partly because the costs of solar PV systems have come down significantly. However, the companion batteries are expensive, difficult to transport, and have limited lifetimes.

45.2.4 Micro-hydropower Systems

With Nepal's vast, economically feasible hydropower potential of 42,130 MW, it is obvious that this technology is a serious option. While MHP system technology is mature and used worldwide, in Nepal the experiences are mixed. The following practical parameters and issues have been noted since 1996 (Zahnd 2013) regarding the disappointing outcome of MHP systems in the remote, high-altitude areas:

- *Lack of understanding energy needs*: MHP developers are not aware of how and why users have so far generated their energy services, lack introduction of, and what the users expect of a MHP system.
- *Too rigid targets*: Adherence to the government's 100 watt/household policy, while less would be sufficient and would be a better fit for the village's ability to operate and maintain it.
- *Lack of cost awareness*: Users are not instructed that energy services have a cost.
- *Investment focus*: While the building of the MHP is highly subsidized, operation, maintenance and repair are not.

- *Inadequate quality*: Due to remoteness, high transport cost and limited budget, minimum and low quality equipment is included in the MHP design, often causing premature failure rates.
- *Oversized installations*: In villages with 30–100 households standard MHP with 10–50 kW power ratings are built, far larger than needed, resulting in dumping 60–80% of the electricity generated.
- *Waste of excess electricity*: The MHP dump load is a "dumb" rather and a "useful" dump load. The overcapacity heats river water at the turbine house, away from the village and thus of no use.
- *Lack of options beyond lighting*: MHPs operate only during the night for indoor lighting, and, thus no income generation is considered.
- *Inadequate transmission*: Transmission is done using bare wires suspended on dead trees rather than secure underground cabling.

Above issues raise serious concerns of applicability and sustainability of the present MHP systems, proposed and subsidized by the Nepal government in remote, high-altitude regions. In contrast, PHP systems are intended to generate <5 kW from nearby small streams/rivers. This presents a first step with small increments of changes from where the local community starts introducing electricity to cover basic needs like lighting, integrating the PHP concept quicker into to the local culture. With these aims in mind, RIDS-Nepal introduced a new, modular concept, enabling the local user community to expand the PHP system as the village grows in energy services demand and capabilities and economic vitality.

45.3 Results

45.3.1 Modular Pico-hydropower System Design

Modularity brings several advantages to a system including better fit of capacity to demand, an incremental step-and-repeat expansion as the village's energy demands and ability to finance them increase, use of proven and familiar technology, and improved reliability due to redundancy. Although the probability of a failure increases with the number of components in a system, part of a modular system can continue to function, albeit at reduced capacity, while a failed module is being repaired or replaced. This is particularly important for remote locations where transportation is difficult, time consuming, and expensive.

Each module of power generation (see Fig. 45.1) is composed of a set of components including a turbine, a generator, transmission cables, a charge controller, and a component to dump loads. Additional modules can be added as needed to generate more power. Modules of energy storage, composed of batteries, can also be added as needed to deliver more than the power capacity of the turbines during daily periods of higher demand such as early evening. Modules of AC power delivery, composed of inverters, can be added as needed to deliver more

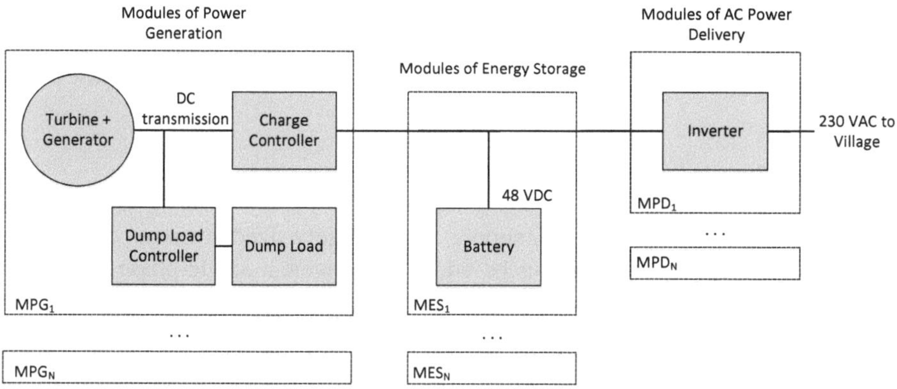

Fig. 45.1 High-level architecture

instantaneous AC power. The capacity of the transmission lines used in a PHP system is typically driven by mechanical and energy loss constraints versus amperage. This allows considerable modular expansion of the system without having to dig new trenches and run new cables. In the following subsections, the concrete setup of these modules is presented.

45.3.1.1 Modules of Power Generation

Each module of power generation will include one PowerSpout Turgo HP integrated turbine and permanent magnet generator using direct drive to eliminate the regular maintenance expense of belt replacement. A full-wave bridge rectifier converts the three-phase power from the generator to DC for transmission to the charge controllers and prioritized dump loads in the village. Use of DC allows the MPPT logic in the charge controllers to maximize the power output of each integrated turbine/generator, a benefit of not having to maintain 50 Hz frequency and 230 VAC. Charge controllers ensure the batteries are not overcharged which can severely degrade their life span. Studer VS-70 charge controllers are being considered to support 300VDC transmission. An external dump load controller is needed with the Studer VS-70. A custom-designed PWM dump load controller is being considered because so few 600VDC commercial versions exist. Battery side dump loads were rejected due to the high current needed to consume the power from four turbines, resulting in high cost and complexity of the SSR and resistor array needed to support prioritized dump loads.

45.3.1.2 Modules of Energy Storage

Many PHP sites have limited head or flow rates, so their instantaneous power capacity is limited. Excess electricity from 24/7 production is stored in batteries for delivery in excess demand periods. A small battery bank (high-quality 2 V tubular gel batteries) will be installed initially, and additional banks of batteries in parallel can be added as the village demand and economic capacity increases.

45.3.1.3 Modules of AC Power Delivery

In order to exceed the peak power supply, appropriately sized inverters, or multiple thereof, are needed. Inverters can be added to convert a single-phase system to three-phase to drive motors, or the inverters can be wired in parallel to increase the power generation capacity of a single phase. Three Studer XTM 4000-48 inverters are being considered, generating three-phase power.

45.3.2 Underground Infrastructure

Buried HDPE pipes will be used to deliver water from the inlet to the turbines. They will be thermally welded using a commercial HDPE welding machine. Buried armored cables will be used to eliminate the deforestation and maintenance issues inherent in the current practice of utilizing dead trees and bare aluminum wires for transmission and distribution. Armored cable consisting of four aluminum conductors ubiquitous for three-phase transmission, 650 V rating, is used.

45.3.3 Prioritized Dump Loads

MHP systems in Nepal dump excess energy by heating water at the turbines away from the village. The system in Moharigaun will use this excess energy for a prioritized set of services in the village, such as water heating for showers, heating and lighting of greenhouses, or heating biogas digesters for increased efficiency. A PWM controller routes excess power to the dump loads as the voltage generated by the turbines increases under conditions of low power demand. Prioritization is achieved by connecting all the dump load resistors in parallel, each enabled by their own SSR controlled by the PWM signal passed through a standard thermostat used to control the upper element in an electric hot water heater as shown in Fig. 45.2. Use of DC transmission also simplifies the circuitry needed for the dump loads, i.e., a single resistor is all that is needed instead of one per phase, and a single FET or IGBT based SSR can be used to switch power to it. This can add up when multiple dump loads are used.

Ubiquitous hot water heater elements are the dump loads. Typically starting with one turbine, this being a research project, there are four turbine/generators, each associated with their own transmission line and charge controller, and each prioritized dump load will consist of four SSRs and four hot water heater elements.

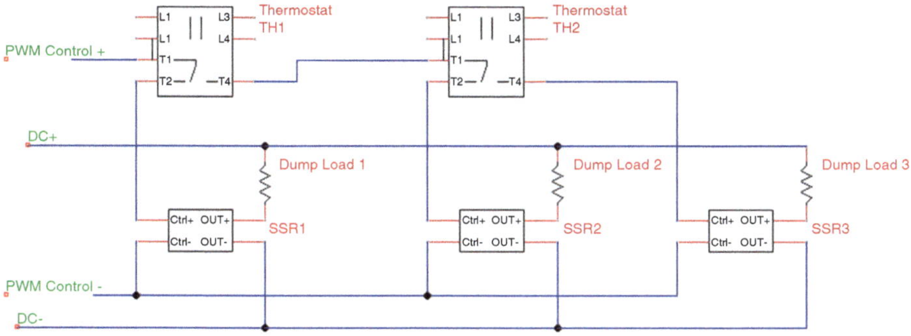

Fig. 45.2 Dump load prioritization technique

45.3.4 Electricity Payment System and Tariff Policies

Itron's ACE9000 SSP DIN-R prepayment system is planned for the PHP system. These meters are mounted on DIN rails in ten junction boxes throughout the village, paired with circuit breakers in the same box. A three-phase distribution spine is routed through the village to these junction boxes, and one-phase buried armored cables are routed in a star topology to each household. SPDs are installed in each junction box. Village operators of the system will collect money and credit the meter tokens. The ACE9000 pay-as-you-go system supports multiple tariff policies, including a monthly flat fee and a per-kWh fee.

45.3.5 Indoor and Community Lighting

Based on each family's own defined indoor lighting demands and financial capacity, up to six LED bulbs, each 3–5 watt, are considered for each home. Up to 15, streetlighting poles, mounted each with a 10 watt LED lamp at 5 m height, able to distribute the light evenly are considered to light up the village paths.

45.3.6 Remote Data Monitoring and Controlling

The operating parameters are monitored and controlled remotely through the purpose-built internet site for this pilot project so we can learn as much as possible about how to build and operate future systems in Northwestern Nepal.

45.3.6.1 Capturing Interval Consumption Data per Households

The ACE9000 meters also capture each household's historical interval power consumption data so we can better understand how the village adapts to electricity. This is needed to understand the generation and energy storage needs of the future systems deployed in similar villages in the area.

45.3.6.2 Capturing Operating Conditions

Sensors will measure turbine operating parameters including input water pressure, output voltage, current, interval energy production, and temperature. Operating parameters read from the charge controllers include input voltage, current, and interval energy delivered to the battery, heat sink temperature, battery voltage, and battery temperature. Operating parameters read from the dump load controller will include interval energy delivered to each dump load. Operating parameters read from the inverters include generated voltage, current, power factor, and heat sink temperature.

45.3.6.3 Modification of System Operational Parameters

Pursuant to the pilot's goals, the authors will be able to remotely set inverter and charge controller and inverter operating parameters.

45.3.7 Energy Theft

Although the ACE9000 meters will automatically disconnect a household if it is not recharged, there is a risk that an operator could recharge a meter for their neighbor without collecting money. Another form of energy theft involves connecting into the system ahead of the ACE9000 meters, though the use of buried armored cables makes this more difficult. Through the internet the pay-as-you-go system, accounting, and bank balance can be monitored. If needed, RIDS-Nepal staff can be dispatched to address any situation through education and reinforcement of why pay-as-you-go is important for long-term service provision.

45.3.8 Training and Follow-Up

The following trainings and follow-up courses are included in the 5-year planned project period. From each household at least two members are trained how to operate and maintain their new lighting system in regard to understanding the cost of electricity versus the benefits, how to recharge the repayment system, and how to safely and wisely use the available electricity. Up to five local users, chosen by the village community, will be trained and compensated to operate and maintain the PHP system, solving day-to-day problems such as cleaning intake, flushing pipe lines, greasing bearings, etc. and contacting RIDS-Nepal for more serious issues. Up to four of them are trained to run the prepayment system, able to disconnect users who have not paid, and make sure that the income is deposited in the bank. Follow-up training is given to all trainees annually and to new trainees.

45.4 Conclusions

The MHP approach used predominantly in Northwestern Nepal has proved to be inadequate, usually resulting in failed or poorly performing systems within a few months/years. The proposed pilot project described herein, with its modular PHP system approach, is believed to address most of the fundamental problems with the current MHP approach, while providing a way to start small and grow along with the served village's energy demands and the economic vitality it takes to operate and maintain the system. Remote monitoring will be used to validate this hypothesis, to help define the initial scale of future systems, to improve their design and operation, and to justify the prevalent use of this new modular PHP approach in the region.

References

GSR 2016. (2016). Renewable 2016 global status report, REN21.

RIDS-Nepal. (n.d.). *Rural integrated development services – Nepal*. www.rids-nepal.org

SDGs. (n.d.). *Sustainable development goals 2016–2030*. https://sustainabledevelopment.un.org/sdgs

Zahnd, A. (2013). *The role of renewable energy technology in holistic community development*. PhD Dissertation published by Springer Thesis Recognizing Outstanding Ph.D. Research, chapters 3–6.

Chapter 46
Impacts of Distributed Generators on the Protection System of Distribution Networks

Abdullah Zia, Mohammad T. Arif, and Amanullah M.T. Oo

46.1 Introduction

An increase in global energy demand due to factors like ever-growing reliance on electricity and the popularity of energy hungry devices has led to the realization that conventional means of energy generation have had and continue to have negative impacts on the sustainability of the planet we live in. Global warming, climate change, pollution, and the high level of expensive maintenance required for infrastructure used in the conventional grids have pushed engineers and scientists throughout the world to look for alternative means of harvesting energy which are now on the rise on a global level. Most common examples of such sources are renewable technologies such as solar, wind, hydropower, and geothermal. RE sources fall under the category of distributed energy resources (DERs) and are often also referred to as renewable embedded generation (EG). They often form the power supply in modular distributed generation (DG) systems where the source is located close to the load. The significance of RE sources has gone up in recent times, and it continues to grow at an exponential rate for a number of reasons. Some of the most important reasons includes (IEC 2012; Kennedy 2015)

- Sustainability and environmental concerns such as contribution in the carbon footprint reduction
- The possibility of making electricity accessible for new consumers of the developing and underdeveloped world
- Reduced losses in transmission and distribution network
- A process known as peak shaving which is responsible for reduction of energy consumption from the main grid during the periods of maximum demand

A. Zia (✉) • M.T. Arif • A.M.T. Oo
School of Engineering, Faculty of Science, Engineering and Built Environment, Deakin University, Waurn Ponds, VIC, Australia
e-mail: zab@deakin.edu.au; m.arif@deakin.edu.au; aman.m@deakin.edu.au

A. Sayigh (ed.), *Transition Towards 100% Renewable Energy*, Innovative Renewable Energy, https://doi.org/10.1007/978-3-319-69844-1_46

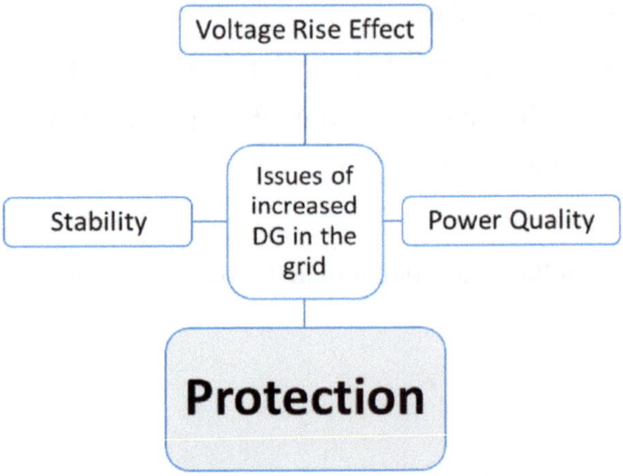

Fig. 46.1 Classification of problems caused by DG integration

- Increased reliability of the power network by supporting loads locally
- Voltage stability support

The renewable energy target (RET) scheme devised by the Australian Government expects that 23.5 % of the energy needs of Australia will be met by renewable sources by 2020 (Australian Government 2015). This transition to a more sustainable and planet-healthy mix of energy resources is not a very smooth one, as it implies major changes to the structure and functioning of the power network.

In order to succeed in the transition from largely conventional to a significantly alternative means of generation, a number of technical obstacles need to be resolved. As illustrated in Fig. 46.1, connecting renewables into the network brings with it a number of problems and is a lot more complicated than what the layman may perceive. The focus of this paper is the effect of this integration on the existing protection systems in place. Literature suggests that the problem at hand can be summarized into two points.

1. Conventional protection techniques that are presently in use by distribution side of the utilities are vulnerable to failure with the increased trends of small-scale generator and PV installations primarily due to the fact that the radial and passive nature of the network does not hold after RE integration.
2. The protection system needs to be modified/restructured/redesigned or completely replaced so it takes into consideration DER when they reach levels that can result in the malfunction of protection system.

The task at hand was to observe the aforementioned failures and then attempt to find solutions to combat these failures. In order to accomplish this, an understanding of conventional protection systems commonly used in DNs and understanding of possible problems is required.

46.2 Background

Overcurrent relays (OCRs) happen to be the most common protection devices used in radial distribution networks and they are known for their economic feasibility and effectiveness (Rezaei et al. 2015). For the sake of maximizing understanding, clarity, and convenience, the scope of this paper is limited to scenarios that involve simple radial systems that have DGs representing renewable energy integrated into them therefore deeming them nonradial upon connection. When DGs are added to the network, a reconfiguration of structure and components is required due to the change in nature of the network. Unwanted operation and failure to operate can both occur when conventional protection is being used despite the incorporation of DG into the network. Aside from the protection device coordination issue, a number of other impacts cause protection failure in DNs. They include recloser-associated problems such as prohibition of automatic reclosing and unsynchronized operation of reclosers. However, out of all of the impacts, "false tripping" and "protection blinding" are the most serious problems found in the literature.

False tripping is also known as sympathetic tripping or unnecessary tripping of a healthy feeder or part of the DN due to contribution of the DG that has been introduced to the network. In a regular, radial network, when a fault occurs at one of the feeders, the fault current flows exclusively from the main source toward the faulty feeder, due to which the respective overcurrent relay trips as a consequence of the pickup current value of the relay settings being less than the fault current value. It is important to know that in most cases, the OC relays are not concerned with the direction in which the power is flowing and hence will trip regardless of forward or backward flow of fault current. This feature of traditional OC relays poses a threat to the integrity of the protection system when DGs are integrated into the network. A schematic illustration of the phenomena is shown in Fig. 46.2. The probability of Relay 1 (R1) tripping due to a fault in a feeder adjacent to it is high in such a scenario because the nondirectional Relay 1 experiences reverse power flow and is unable to differentiate between forward and backward nature of the power flow. False tripping is an example of selectivity problem. As the condition illustrated in Fig. 46.1, Relay 1 (R1) would trip if the current contribution by the DG in the case of a fault exceeds the pickup value of R1. This issue cannot be solved by increasing the pickup value because that would lead to insensitivity to faults altogether for that feeder, and hence an alternate solution is required.

Protection system blinding, which is also known as protection underreach, takes place as a result of a decrease in current contribution to the fault location from the grid itself like in the case as shown in Fig. 46.3 where a DG contributes current to the fault thereby decreasing the current coming from the grid which is also the current that is detected by the relay. In such a scenario, the protection system will not be able to detect the fault properly and no measures will be taken to isolate the faulty segment of the system if the relay is a definite OCR. In case of an inverse time OCR, the fault will be detected, but due to the low value of current, the tripping action of the relay will be delayed. The minimum grid contribution compared to the

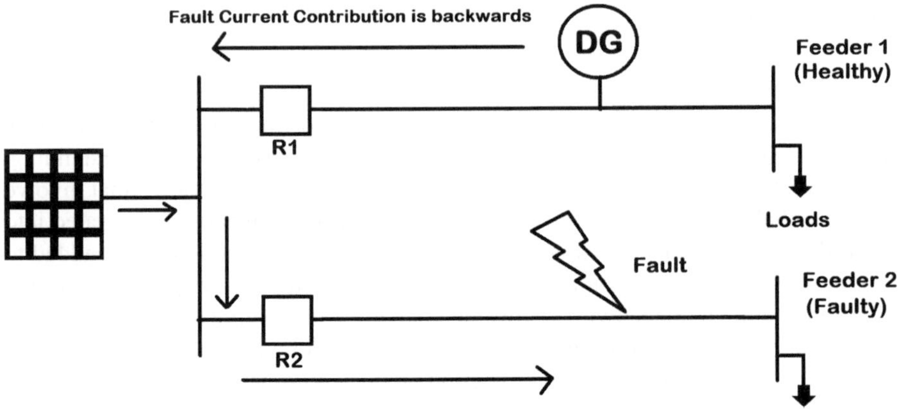

Fig. 46.2 Illustration of false tripping phenomena

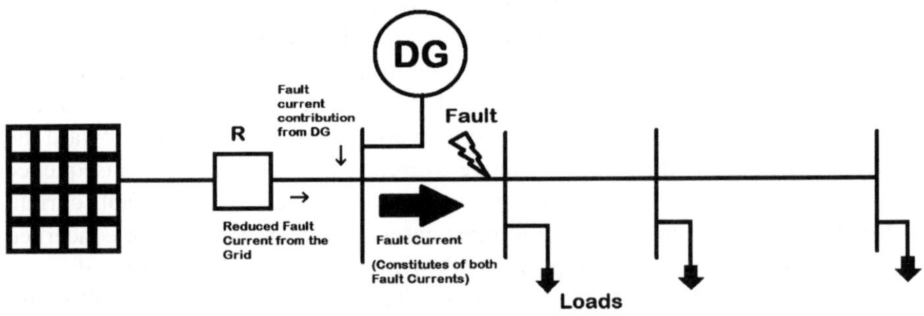

Fig. 46.3 Illustration of protection blinding phenomena

pickup current of the overcurrent protection is a determination factor to figure out if blinding of protection will occur. Protection blinding is a form of fault detection problem. The division of the current contribution depends on factors like the network configuration, grid impedance, and the size of the DG (Coster et al. 2011).

46.3 Methodology

In order to demonstrate the two aforementioned issues in a simulation environment, an inverse time OCR was simulated, by taking into consideration the simplified version of the inverse OCR characteristic Eq. (46.1), and a flowchart is devised for the proper functioning of the relay as shown in Fig. 46.4 inspired by Abdel-Salam et al. (2017) and Aman et al. (2011).

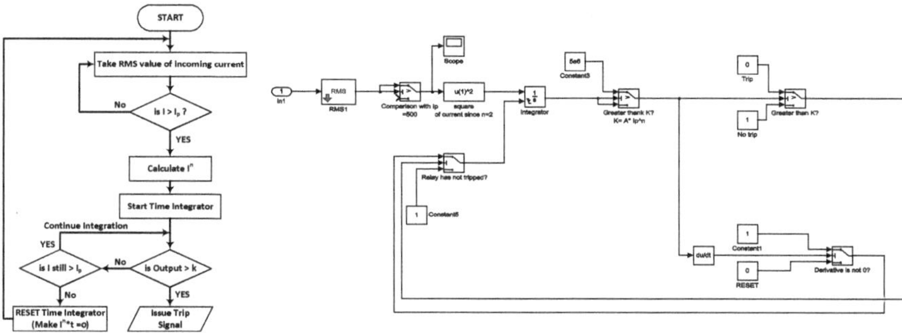

Fig. 46.4 Implementation of OCR model

$$t(I) = \left(\frac{AI_p^n}{I^n} + C \right) \times M \quad (46.1)$$

where

$t(I)$ is time taken for the relay to trip as a function of current flowing through the relay
I_p is the pickup current
I is the flowing current and
C, M, A, and n are constants that determine the desired functionality of the relay.

According to IEEE very inverse standard, the constants are defined as $A = 19.61$, $C = 0.491$, $M = 1$, $n = 2$ and AI_p^n is referred to k in the model.

Since three-phase systems are involved, three similar relays are constructed and their outputs are ANDed together so that the whole system trips in case of fault current experienced in any of the phases. Note that 1 is the default value of the input to the breakers, whereas the transition from 1 to 0 is termed a trip signal in the context of this simulation model.

Upon the testing and validating the operation of the OCR, it was utilized in simple simulated medium voltage (11 kV) DNs such as the conditions indicated by Figs. 46.2 and 46.3. The phenomena of false tripping and protection blinding were observed by simulating with and without integration of three-phase sources representing DGs. A solution for false tripping was then simulated by introducing a check on the direction of power flow. Considering the findings from literature, a technique was applied using incoming voltage and current waveforms as indicators for forward or reverse power flow with the help of switches, comparators, logic, and mathematical operators as indicated the implementation in Fig. 46.5.

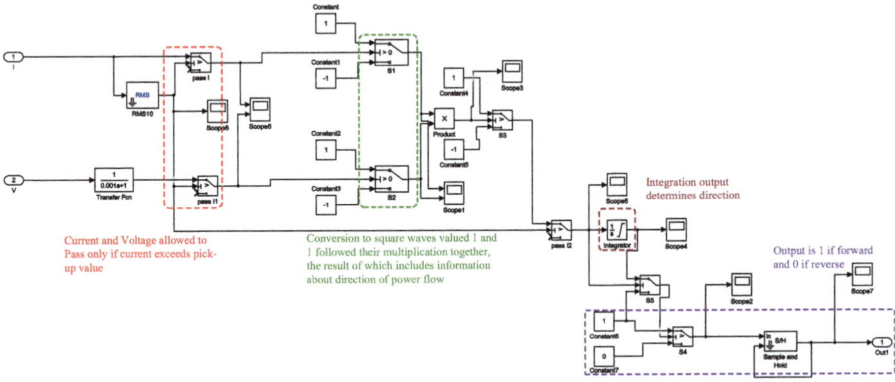

Fig. 46.5 Implementation of directional block in the OCR model

46.4 Results and Discussions

An 11 kV medium voltage DN model based on Fig. 46.2 was simulated and certain measurements were recorded for the false trip phenomena. Table 46.1 summarizes the results when a three-phase fault at the bottom feeder in Fig. 46.2 was introduced at 0.05 s. Upon adjusting the DG input, it was noticed that both the relays tripped but only the relay in the faulty feeder (R2) should have tripped. Thus false tripping was successfully demonstrated. The values obtained for the time taken for the relay to operate were double checked using Eq. 46.1 and proper functioning of the OCR was confirmed.

Similarly, considering the network structure as illustrated in Fig. 46.3, the issue of protection blinding was demonstrated in a simulation model and the value of the DG input was adjusted until a delay in tripping time was observed. This delay caused by DG integration is a form of protection blinding. Table 46.2 summarizes values taken down when a two-phase-to-ground fault was applied downstream the DG shortly after the simulation start time. It can be noticed that DG integration has caused a delay in tripping time due to a decrease in fault current passing through the relay (R).

A convenient way of addressing the false tripping in this particular simulation model, as discussed previously, involved implementing a directional block which stops the relay from operating in a case where reverse power flow is experienced. Hence upon integrating the directional element, it is observed that R2 from Fig. 46.2 trips after 2.6 s as expected. However, R1 rightly does not trip at all.

Table 46.1 Simulation parameters for false tripping model

Point of observation	Measurement
Pre-fault phase current for phase A (top healthy feeder)	31 A
Pickup current for each phase for R1	31*1.25 ≈ 40 A
Fault level after fault introduction for R1	250 A
Time taken to trip for R1	0.9 s
Pre-fault phase current for phase A (bottom faulty feeder)	480 A
Pickup current for each phase for R2	480*1.25–580 A
Fault level after fault introduction for R2	1,772 A
Time taken to trip for R2	2.6 s

Table 46.2 Simulation parameters for protection blinding model

Point of observation	Measurement
Pre-fault phase current before DG introduction	300 A
Pickup current	300 *1.25 ≈ 375 A
Fault current value (phase B) without DG	540 A
Fault current value (phase B) with DG	525 A
Relay tripping time without DG	9.83 s
Relay tripping time with DG	10.83 s

46.5 Conclusion

To summarize, protection systems play a vital role in the reliable functioning of a DN, and for conventional OC protection systems, DG integration poses a threat and can lead to scenarios like the ones demonstrated in this paper. In reality, the DN involved is a lot more complex compared to the samples shown in this context. However, the examples mentioned can help in understanding the problem of a complicated nature using simple simulation models. The two common problems associated with DG integration in the context of electrical systems protection false tripping and protection blinding were demonstrated, and a solution of the false tripping problem was demonstrated as well. The models can be improved by using detailed models for DGs and integrating actual load profiles instead of constant arbitrary ones.

Acknowledgment Authors would like to acknowledge the help of Mr. Mohsen Khalaf (m.khalaf@uwaterloo.ca) from ECE Department, University of Waterloo, Waterloo, ON, Canada.

References

Abdel-Salam, M., Kamel, R., Sayed, K., & Khalaf, M. (2017). Design and implementation of a multifunction DSP-based-numerical relay. *Electric Power Systems Research, 143*, 32–43.

Aman, M. M., Khan, M. Q. A., & Qazi, S. A. (2011). Digital directional and non-directional overcurrent relays (modeling and performance analysis). *NED University Journal of Research, VIII* (2).

Australian Government. (2015). *The renewable energy target (RET) scheme*. Department of the Environment and Energy, Australian Government. Available: https://www.environment.gov.au/climate-change/renewable-energy-target-scheme. Access date: 25 May 2016.

Coster, E. J., Myrzik, J. M. A., Kruimer, B., & Kling, W. L. (2011). Integration issues of distributed generation in distribution grids. *Proceedings of the IEEE, 99*(1), 28–39.

IEC. (2012). Grid integration of large-capacity renewable energy sources and use of large-capacity electrical energy storage. White Paper.

Kennedy, J. (2015). *Distribution protection in a modern grid embedded with renewable resources*. Doctor of Philosophy thesis, School of Electrical, Computer and Telecommunications Engineering, University of Wollongong, http://ro.uow.edu.au/theses/4367

Rezaei, N., Othman, M. L., Wahab, N. I. A., & Hizam, H. (2015). Coordination of overcurrent relays protection systems for wind power plants. In *2014 I.E. international Conference Power & Energy (PECON)* (pp. 394–399), Kuching 2014.

Chapter 47
Rapid Decarbonisation of Australian Housing in Warm Temperate Climatic Regions for 2050

John J. Shiel, Behdad Moghtaderi, Richard Aynsley, Adrian Page, and John M. Clarke

47.1 Introduction

The number of natural disasters are increasing at an exponential rate, with a threefold increase from 1980 to 2014 (MunichRe et al. 2015), and global temperatures are accelerating at an unprecedented rate, 20 times faster than at the end of the last ice age (Nuccitelli 2016). As carbon is added to the troposphere, temperatures will continue to rise, leading to more natural disasters.

To keep its 2015 Paris Conference of Parties (COP21) commitments, implying a zero-carbon economy by 2050 (ASBEC 2016), Australia needs to rapidly lower its residential carbon emissions. This is because it has one of the world's worst-performing building stocks per person, and residential emissions make up 13% of total building emissions (WRI 2005; ASBEC 2008).

However, this is a difficult problem due to the large number of stakeholders, high retrofit costs, and issues such as split incentives for landlords and tenants, where those who pay for energy efficiency upgrades do not enjoy the benefits (WBCSD

J.J. Shiel (✉)
The Priority Research Centre for Frontier Energy Technologies and Utilisation, University of Newcastle, University Drive, Callaghan, NSW, Australia

EnviroSustain, PO Box 265, Jesmond, NSW, Australia
e-mail: jshiel@westnet.com.au

B. Moghtaderi • A. Page
The Priority Research Centre for Frontier Energy Technologies and Utilisation, University of Newcastle, University Drive, Callaghan, NSW, Australia

R. Aynsley
Building Energetics Pty Ltd, PO Box 219, Pomona, QLD, Australia

J.M. Clarke
CSIRO Climate Science Centre, PB1 Aspendale, VIC, Australia

A. Sayigh (ed.), *Transition Towards 100% Renewable Energy*,
Innovative Renewable Energy, https://doi.org/10.1007/978-3-319-69844-1_47

2009; ASBEC 2016). These upgrades will be more likely to be adopted if they are convenient and affordable (Bond et al. 2011).

So this study aimed to find affordable retrofits for 2050 to lower heating and cooling carbon emissions of existing Australian housing in a warm temperate climate. This climate is where around 50% of the population lives (BZE et al. 2013), and housing makes up 80% of the residential stock (ABS 2012; DEWHA 2008).

47.2 Methodology

Three actual Australian houses were modelled with the Australian Nationwide House Energy Rating Scheme (NatHERS) tool AccuRate for the warm temperate climate of Adelaide, with affordable envelope retrofits. We estimated the single retrofit carbon savings and DIY costs and ranked them by simple payback period (SPP). We combined the most cost-effective single retrofits for the three payback periods, small, medium, and large, and calculated their carbon savings and total DIY capital costs. We also presented a method to retrofit a house into a resilient zero-carbon state.

47.2.1 Climate Change for 2050

The climate change method developed annual hourly AccuRate weather input files to represent Adelaide in 2050 (Shiel 2017) for the climate scenarios by:

- Establishing the change in climate parameters for 2050 from a general circulation model (GCM) also known as a global climate model
- Employing the change factor method (CFM) to find the regional climate

The Australian Climate Futures (ACF) online tool (Clarke et al. 2011; Whetton et al. 2012) was used to project changes in the temperature and humidity climate parameters from 1995 to 2050, for two scenarios of climate change. These were an Extreme Climate Change (RCP8.5) scenario which is reported here, where RCP is a Representative Concentration Pathway (van Vuuren et al. 2011) and a Scarce Resource scenario (RCP4.5) with resource depletion affecting energy supply and consumption of materials including for construction (Barrett 2014; Hall 2010; Steffen et al. 2015).

The ACF tool evaluates how many GCMs agree across two spectrums of weather parameters that are relevant to the application, in this case buildings, e.g. temperature and rainfall, and also takes account of model skill (Moise et al. 2015). The HadGEM2-ES GCM was found to best represent for each scenario the 2050 climate futures with the most number of GCM's, called the 'maximum consensus' case (Shiel 2017).

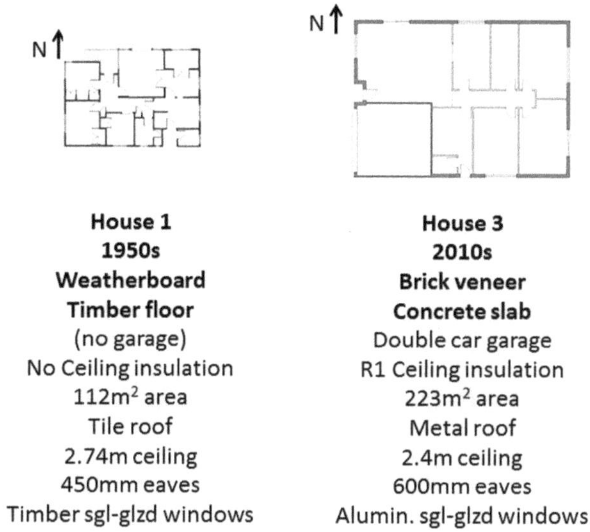

Fig. 47.1 Actual house plans representing the 1950s and 2010s eras in Australia (Source: Shiel)

The change factor method (CFM) chosen was Belcher's 'morphing' technique (Belcher et al. 2005), which is popular with building researchers (Chen et al. 2012; Hacker et al. 2008), and the NatHERS Adelaide 1990 Reference Meteorological Year (RMY) climate file was used as the baseline.

47.2.2 Houses Modelled

While three actual house plans were modelled (Shiel 2017), only two are reported here due to space constraints: the 1950s weatherboard lightweight wall and timber floor house and the 2010s brick veneer wall with concrete slab on ground, as shown in Fig. 47.1.

47.2.3 Retrofit Modelling Approach

There were 76 retrofit designs considered for each house, reduced to around 50 for DIY retrofits, including novel ones such as aquariums and wine for added thermal mass and a PV panel parasol roof.

The main categories of retrofits were partial air conditioning of the house (partitioned), conduction controls with perimeter insulation, radiation controls with window resizing and treatments including vegetation, convection controls with sealing, extra thermal inertia with more thermal mass, and additional air movement controls from vegetation wing walls, louvre windows, and ceiling fans.

The carbon emissions and cost savings per year for each retrofit over the base case were estimated using a typical split-system air conditioner for South Australia (SA) (ABS 2014), the state's electricity carbon intensity factor of 0.64 $kgCO_{2-e}$/kWh (DEE 2016), and an average electricity rate of $0.35 per kWh.

One of the novel retrofits was an optimum, healthy level of ten air changes per hour at 50 Pascals pressure ($10ACH_{50}$) infiltration (Ambrose et al. 2013; Lstiburek 2013). This level was estimated using an unsealed exhaust fan approach from the Moreland Energy Foundation (MEFL 2010), for which energy recovery ventilation (ERV) systems were not required.

47.2.4 *Retrofit Combinations to Suit Occupant Categories*

Retrofit combinations with small, medium, and large payback periods of less than 3 years, 3–10 years, and 10–20 years, respectively, were designed to suit various occupant categories:

- Tenants, owners with a large mortgage or short term residency, or those less wealthy to carrying out some retrofits
- Owners with a mortgage, landlords who may wish to keep expenses low, or those wanting to do more retrofits than above
- Occupants with more disposable income, with long-term occupancy, or desiring deeper retrofits.

47.3 Results

The Extreme Climate Change scenario for two houses and DIY costings are reported here, with contractor costs in (Shiel 2017) and the Scarce Resource scenario results are reported in (Shiel et al. 2017).

47.3.1 *Climate Change*

The projected surface temperature increase from the HadGEM2-ES GCM from 1995 to 2050 for the Extreme Climate Change scenario is 1.8 Kelvin, with only a 0.9% increase in humidity.

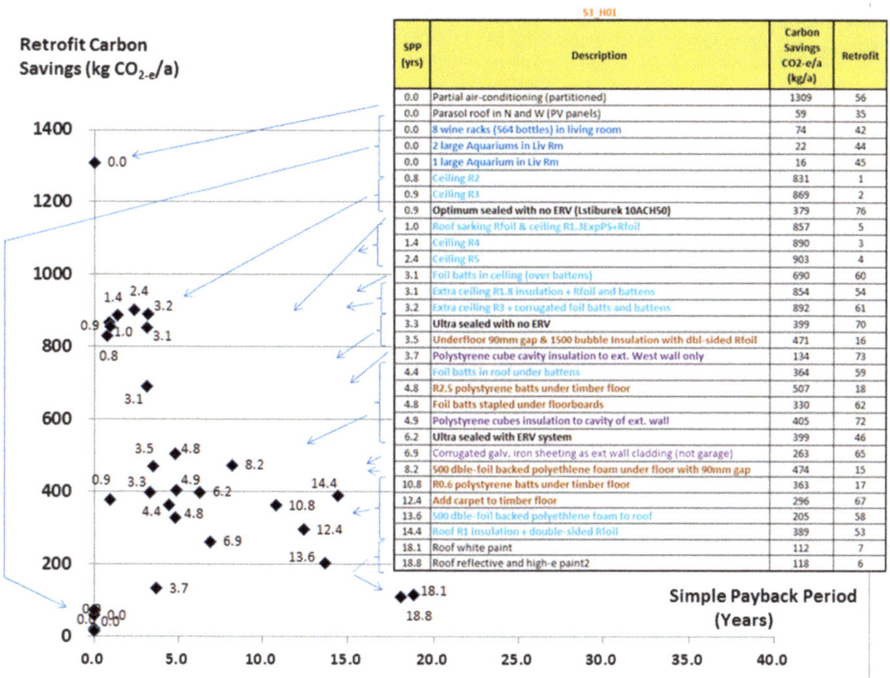

SPP (yrs)	Description	Carbon Savings CO2-e/a (kg/a)	Retrofit
0.0	Partial air-conditioning (partitioned)	1309	56
0.0	Parasol roof in N and W (PV panels)	59	35
0.0	8 wine racks (564 bottles) in living room	74	42
0.0	2 large Aquariums in Liv Rm	22	44
0.0	1 large Aquarium in Liv Rm	16	45
0.8	Ceiling R2	831	1
0.9	Ceiling R3	869	2
0.9	Optimum sealed with no ERV (Lstiburek 10ACH50)	379	76
1.0	Roof sarking Rfoil & ceiling R1.3ExpPS+Rfoil	857	5
1.4	Ceiling R4	890	3
2.4	Ceiling R5	903	4
3.1	Foil batts in ceiling (over battens)	690	60
3.1	Extra ceiling R1.8 insulation + Rfoil and battens	854	54
3.2	Extra ceiling R3 + corrugated foil batts and battens	892	61
3.3	Ultra sealed with no ERV	399	70
3.5	Underfloor 90mm gap & 1500 bubble Insulation with dbl-sided Rfoil	471	16
3.7	Polystyrene cube cavity insulation to ext. West wall only	134	73
4.4	Foil batts in roof under battens	364	59
4.8	R2.5 polystyrene batts under timber floor	507	18
4.8	Foil batts stapled under floorboards	330	62
4.9	Polystyrene cubes insulation to cavity of ext. wall	405	72
6.2	Ultra sealed with ERV system	399	46
6.9	Corrugated galv. iron sheeting as ext wall cladding (not garage)	263	65
8.2	500 dble-foil backed polyethlene foam under floor with 90mm gap	474	15
10.8	R0.6 polystyrene batts under timber floor	363	17
12.4	Add carpet to timber floor	296	67
13.6	500 dble-foil backed polyethlene foam to roof	205	58
14.4	Roof R1 insulation + double-sided Rfoil	389	53
18.1	Roof white paint	112	7
18.8	Roof reflective and high-e paint2	118	6

Source: Shiel

Fig. 47.2 The carbon savings by payback period for individual DIY retrofits of a 1950s weatherboard timber floor house for the Adelaide 2050 Extreme Climate Change scenario

47.3.2 Single Retrofits

There are graphs of the carbon savings of individual retrofits plotted against the DIY SPPs with labels identifying each SPP, and they include a table listing the retrofits ranked by SPP and colour-coded according to Table 47.1:

- In Fig. 47.2 for the 1950s weatherboard walls with a timber floor (house number 1)
- In Fig. 47.3 for the 2010 brick veneer walls with concrete floor (house number 3).

The most cost-effective single retrofits were partial house conditioning to reduce the conditioned volume (which was not included in the combinations of retrofits); ceiling and roof insulation; gap sealing to an optimum level; and external wall cavity insulation; and then retrofits that suited the house construction. These included under-floor insulation, vegetation, ceiling fans and low-e films to the windows for the weatherboard and timber-floored house, and corrugated galvanised iron sheeting to the external wall for the brick veneer house and internal brick walls if there was a concrete floor for added thermal mass (see Table 47.1 and Figs. 47.2 and 47.3).

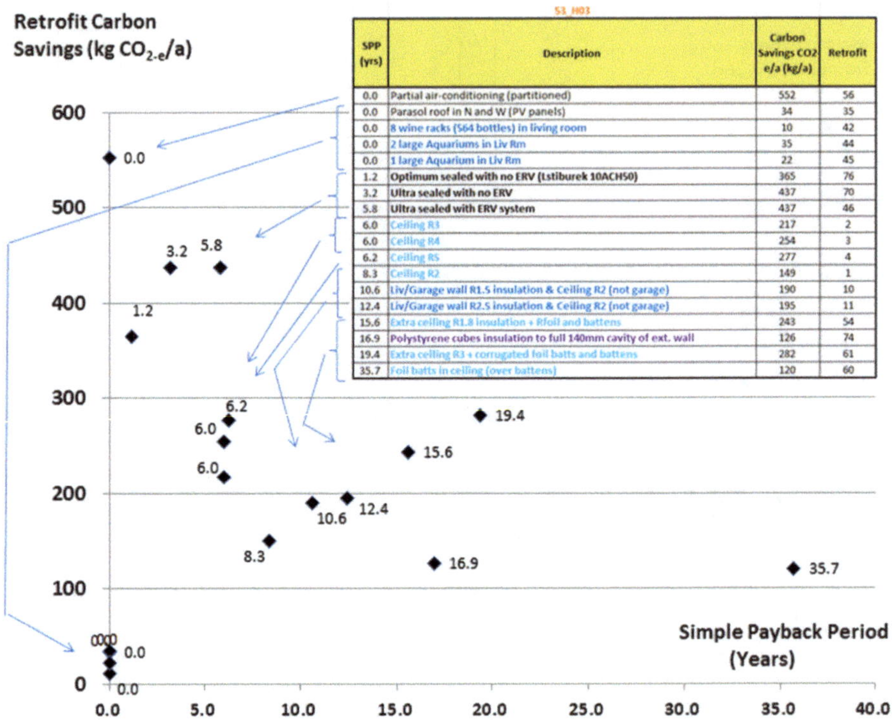

SPP (yrs)	Description	Carbon Savings CO2 e/a (kg/a)	Retrofit
0.0	Partial air-conditioning (partitioned)	552	56
0.0	Parasol roof in N and W (PV panels)	34	35
0.0	8 wine racks (564 bottles) in living room	10	42
0.0	2 large Aquariums in Liv Rm	35	44
0.0	1 large Aquarium in Liv Rm	22	45
1.2	Optimum sealed with no ERV (Lstiburek 10ACH50)	365	76
3.2	Ultra sealed with no ERV	437	70
5.8	Ultra sealed with ERV system	437	46
6.0	Ceiling R3	217	2
6.0	Ceiling R4	254	3
6.2	Ceiling R5	277	4
8.3	Ceiling R2	149	1
10.6	Liv/Garage wall R1.5 insulation & Ceiling R2 (not garage)	190	10
12.4	Liv/Garage wall R2.5 insulation & Ceiling R2 (not garage)	195	11
15.6	Extra ceiling R1.8 insulation + Rfoil and battens	243	54
16.9	Polystyrene cubes insulation to full 140mm cavity of ext. wall	126	74
19.4	Extra ceiling R3 + corrugated foil batts and battens	282	61
35.7	Foil batts in ceiling (over battens)	120	60

Source: Shiel

Fig. 47.3 The carbon savings by payback period for individual DIY retrofits of a 2010s brick veneer concrete floor house for the Adelaide 2050 Extreme Climate Change scenario

Table 47.1 The retrofits of the large payback period combination. The retrofits are colour coded according to their type and location: **thermal mass**; **floor**; **sealing**; insulation; ext. walls; **ext. wall cavity insulation**; **vegetation**; black for others.

House Type	Large Payback Period Retrofit Combination (for Deep Retrofit Households)
H01 – 1950s Weatherboard, timber floor	35 - Parasol roof in N & W roof; **42 - 8 wine racks (564 bottles); 44 & 45 - 3 large Aquariums;** 3 - Ceiling R4 Insulation; 16 - Underfloor 90mm gap & 1500 bubble Insulation with dbl-sided Rfoil; 59 - Foil batts in roof under battens; 62 - Foil batts stapled under floor; **72 - Polystyrene cubes insulation to cavity of ext. wall; 55 - Green ivy on North & West walls; 76 – Optimum sealed & no ERV (Lstiburek 10 ACH_{50});** 67 - Carpet to timber floor; 53 - Roof R1 insulation; **26 - 5m high deciduous trees to North windows;** 49-1.4m dia Ceiling fans; 32 - Low-e film applied to windows
H03 – 2010 Brick veneer (brick cladding on a timber stud wall), concrete floor	35 - Parasol roof in N & W roof; **42 - 8 wine racks (564 bottles); 44 & 45 - 3 large Aquariums; 76 – Optimum sealed & no ERV (Lstiburek $10ACH_{50}$)**; 3 - Ceiling R4 Insulation; **10 – Add Liv/Garage brick wall with R1.5**; **74 - Polystyrene cubes to full 140mm cavity of ext. wall;** 65 – Corrugated galv. iron sheeting as ext wall; 59 - Foil batts in roof under battens

Source: Shiel

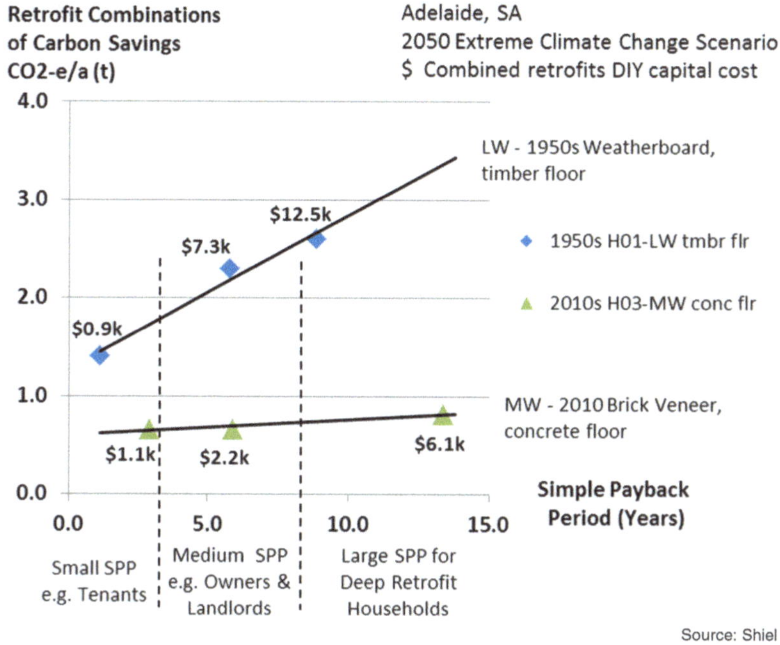

Fig. 47.4 Combined retrofit carbon savings by payback period with DIY costings for a 1950s weatherboard wall and timber floor house, and a 2010s brick veneer wall with concrete slab on ground house

47.3.3 Retrofit Combinations

The retrofits were combined in the order of their cost-effectiveness for each component such as ceiling and wall. These are listed in Table 47.1 for the large payback period retrofit combination, which may suit deeper retrofit households for each house type.

Figure 47.4 shows the carbon savings of combined retrofits for the three payback periods for the two house types, with typical occupant categories separated by dashed lines and a dollar label on each point for the DIY cost of that retrofit combination.

47.3.4 Resilient Zero-Carbon Retrofit Method

Figure 47.5 shows a method to retrofit a house to become a resilient zero-carbon one, with a PV system, battery, and additional inverter for operating essential services during power outages.

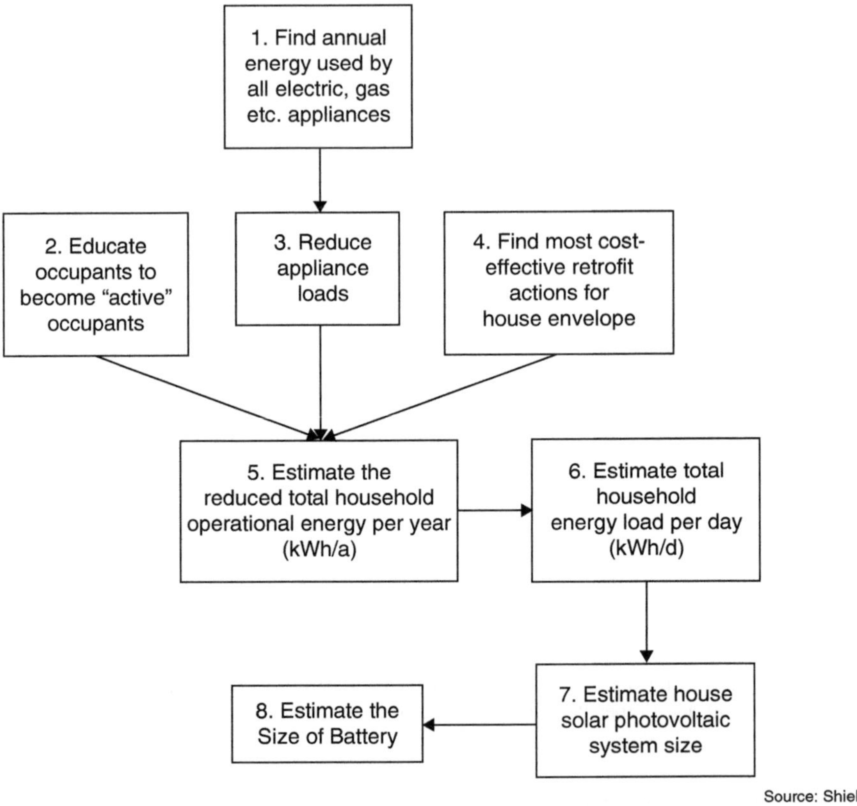

Fig. 47.5 A resilient operational zero-carbon retrofit method, which includes a battery system which can have its own inverter for essential services

47.4 Discussion and Conclusion

47.4.1 Discussion

47.4.1.1 Climate Change

The Extreme Climate Change scenario with RCP8.5 is the current temperature trajectory (Peters et al. 2013), and HadGEM2-ES shows if this trend continues, there will be a 1.8 K rise in temperature from 1995 to 2050 for Adelaide.

47.4.1.2 Single Retrofits

Some of the novel retrofits, e.g. partial conditioning, vegetation, and liquids for thermal mass, did not comply with the NatHERS star rating protocols, since the goal was to find affordable retrofits that reduced energy and carbon use, rather than increase the star rating.

An optimum level of sealing of $10ACH_{50}$ was trialled, together with ultra-sealed levels with and without an ERV system, but an ERV system requires power that may not always be available in a warming world (World Bank 2012; Murdoch 2011; Kempton 2015; Mohr et al. 2015). House designs that have well-sealed envelopes below $10ACH_{50}$ are rewarded by NatHERS, even though this can impair the indoor air quality. It is noted that Australia has no regulations for minimum ventilation levels (Aynsley and Shiel 2017).

Thermal mass is very beneficial for moderating temperatures without air conditioning in temperate or subtropical climates where there are large diurnal temperature swings (Barrios et al. 2011). So thermal mass retrofits were trialled, even those with zero cost (assumed because they are hobbies) such as liquids in aquariums and in wine racks.

47.4.1.3 Rapid Decarbonisation with Retrofit Combinations

Table 47.1 shows the list of retrofits that make up the large SPP combination for each house type, at a DIY cost of $12,500 and $6,100 for the 1950s and 2010s house types, respectively. The table includes single retrofits with carbon savings too low to be effective for the tables in the figures they have low or zero DIY cost. Partial house conditioning retrofits were ignored due to their different conditioned area, making ranking more difficult.

By interpolating the trends of carbon savings in Fig. 47.4, around 2 tonnes of CO_2-e/a could be saved for the 1950s house for a combined retrofit costing around $6,000. This contrasts with only around 1 tonne of CO_2-e/a which could be saved for the 1950s house for a combined retrofit costing around the same amount. Further, the slope of the 2010s house line is very shallow which means that it takes longer to payback the cost of the retrofit combinations than for the 1950s house. This is because the older house has very poor thermal performance, and the most affordable retrofits were very effective, whereas the 2010s house already had a very effective cost-effective ceiling insulation retrofit.

Assuming the trends in Fig. 47.4 hold for other retrofits of these types of houses, decarbonisation of the existing stock could be achieved by educating households with weatherboard and brick veneer houses about the typical estimated costs, payback periods, and carbon-saving impacts of selected retrofit combinations.

Rapid decarbonisation could then be achieved if the large renovation network was involved, if tenants adopted the small simple payback period combination of retrofits, and if governments promoted schemes with rebates to lower the capital retrofit costs further, or implemented a mandatory residential disclosure scheme, where perfomance ratings are carried out on leasing or selling residences. There needs to be government intervention because these schemes would greatly assist the climate change goals of various government levels. For tenants, if they negotiated a 3 year lease without a rental increase and invested around $1,000 in weatherboard or brick veneer house retrofit combinations, there would be a small net gain over 3 years. They would then enjoy the social, environmental, and economic benefits that accrue from retrofitting older dwellings especially for those who may have difficulties paying large energy bills, such as the elderly, or those suffering energy deprivation (Shalekoff 2017). This approach is one way to help overcome the split incentive issue.

47.4.1.4 Resilient Zero-Carbon Retrofit Method

A resilient zero-carbon retrofit method is shown in Fig. 47.5, based on retrofitting a house after an energy audit and improving appliance energy efficiency and occupant behaviour. A solar photovoltaic (PV) system and battery can be sized from estimated energy consumption, and resilience can be added if an additional inverter is added to the battery to maintain essential services on grid outages.

47.4.1.5 The Reliability of the Retrofit Approach

Since actual houses that have been built were modelled, the AccuRate models are realistic and they are also representative of two of the three main types of wall construction in Australia (DEWHA 2008). DIY costings were used because of the significant renovation movement in Australia. The combination of retrofits also provides a robust approach by adding retrofits that are ranked most cost-effective for each component, e.g. for the external walls of the 1950s house, both polystyrene cube insulation and corrugated galvanised iron sheeting were both used.

While the simple payback period is a reliable cost-benefit metric to provide an estimate of the payback period for short terms, various factors are usually not taken into account, such as the future energy price; inflation; or costs of retrofit replacement, repair, or refinishing. However, the longevity of the retrofits was considered, and replacement cost was included in this analysis.

These results are from AccuRate simulations and so adjustments are needed for carbon and cost savings and payback periods, once actual conditioned volumes and occupancy levels are compared to those assumed in NatHERS.

47.5 Conclusion

Affordable carbon-reducing single retrofits were found for modern brick veneer Australian housing with a concrete floor, with more retrofits found for weatherboard older style housing with timber floors, due to their poorer thermal performance. Two key retrofits for a termperate climate, and for climate change, were an optimal level of thermal mass to stabilise internal temperatures without the need for air conditioning, and a healthy minimum infiltration level of $10ACH_{50}$ rather than relying on an ERV system due to possible power supply issues in the future.

This research suggests that combinations of retrofits with simple payback periods from small to large, can be found to upgrade the thermal performance of the existing stock of weatherboard and brick veneer houses to suit the budgets of tenants, owners and landlords. Tenants and landlords who can agree a 3 year lease without rent increase, and where the tenant can implement the combination of retrofits with a payback period of less than 3 years, could help to overcome the split incentives problem.

Government retrofit subsidies and rebate schemes should be considered for household categories such as such as tenants, the elderly and those suffering energy deprivation to assist both the federal government to meet its greenhouse gas target and the state government to reduce peak power loads to ensure reliability of supply for their region.

Finally, the research suggests that weatherboard and brick veneer housing in Australia can be rapidly decarbonised if the deep retrofit combinations with a large payback period for each house type are implemented.

Acknowledgements The authors would like to thank CSIRO's Dr. Dong Chen for NatHERS assistance and discussions, Architect Graham Hunt for assistance with AccuRate modelling, and four seasons and other suppliers for assistance with retrofit costings.

References

ABS. (2012). *Year book Australia*. Canberra: Australian Bureau of Statistics, Commonwealth of Australia, Canberra A.C.T.

ABS. (2014). *ABS 4602.0.55.001 – Environmental issues: Energy use and conservation – Sources of energy used by households*. Canberra: Australian Bureau of Statistics, Canberra, ACT.

Ambrose, M., James, M., Law, A., Osman, P., & White, S. (2013). *CoA-DI: The evaluation of the 5-star energy efficiency standard for residential buildings*. Canberra: Commonwealth of Australia, Department of Industry & CSIRO, CSIRO.

ASBEC. (2008). *The second plank – Building a low carbon economy with energy efficient buildings*. Flemington: The Australian Sustainable Built Environment Council (ASBEC).

ASBEC. (2016). *Low carbon, high performance – How buildings can make a major contribution to Australia's emissions and productivity goals*. Surry Hills: Australian Sustainable Built Environment Council.

Aynsley, R., & Shiel, J. (2017). Ventilation strategies for a warming world. *Architectural Science Review, 60*, 249–254. https://doi.org/10.1080/00038628.2017.1300764.

Barrett, B. (2014). *Renewed economic growth or financial crash? – Our world*. http://ourworld.unu.edu/en/2014-renewed-economic-growth-or-financial-crash

Barrios, G., Huelsz, G., Rechtman, R., & Rojas, J. (2011). Wall/roof thermal performance differences between air-conditioned and non air-conditioned rooms. *Energy and Buildings, 43*, 219–223. https://doi.org/10.1016/j.enbuild.2010.09.015.

Belcher, S., Hacker, J., & Powell, D. (2005). Constructing design weather data for future climates. *Building Services Engineering Research and Technology, 26*, 49–61.

Bond, S., Pacifici, C., Newman, P. (2011). Sustainability in housing: Perceptions of real estate agents, building professionals & householders. In *17th annual pacific rim real estate society conference, Gold Coast*.

BZE, H. T., Keech, R., Eales, D., & McConnell, D. (2013). *Zero carbon Australia buildings plan*. Melbourne: Beyond Zero Emissions.

Chen, D., Wang, X., & Ren, Z. (2012). Selection of climatic variables and time scales for future weather preparation in building heating and cooling energy predictions. *Energy and Buildings, 51*, 223–233. https://doi.org/10.1016/j.enbuild.2012.05.017.

Clarke, J., Whetton, P., & Hennessy, K. (2011). Providing application-specific climate projections datasets: CSIROs climate futures framework. In: *Proceedings of MODSIM, international congress on modelling and simulation, modelling and simulation society of Australia and New Zealand* (pp. 12–16). Perth.

DEE. (2016). *National greenhouse accounts factors*. http://www.environment.gov.au/climate-change/greenhouse-gas-measurement/publications/national-greenhouse-accounts-factors-aug-2016

DEWHA. (2008). *Energy use in the Australian residential sector 1986–2020*. Canberra: Department of the Environment, Water, Heritage and the Arts.

Hacker, J., De Saulles, J., Minson, A., & Holmes, M. (2008). Embodied and operational carbon dioxide emissions from housing: A case study on the effects of thermal mass and climate change. *Energy and Buildings, 40*, 375–384.

Hall, M. R. (Ed.). (2010). *Materials for energy efficiency and thermal comfort in buildings*. Boca Raton: CRC Press.

Kempton, H. (2015). *Fault found in Basslink cable 100km offshore*. http://www.themercury.com.au/news/tasmania/fault-found-in-basslink-cable-100km-offshore/news-story/38f74e0516c2745dcfa4bcc4808bde4a

Lstiburek, J. (2013). Deal with the manure and then don't suck. *ASHRAE Journal, 55*, 40–46.

MEFL. (2010). On-ground assessment of the energy efficiency potential of Victorian homes – Report on pilot study by Moreland Energy Foundation Ltd. Sustainability Victoria.

Mohr, S. H., Wang, J., Ellem, G., Ward, J., & Giurco, D. (2015). Projection of world fossil fuels by country. *Fuel, 141*, 120–135. https://doi.org/10.1016/j.fuel.2014.10.030.

Moise, A., Wilson, L., Grose, M., Whetton, P., Watterson, I., Bhend, J., Bathols, J., Hanson, L., Erwin, T., Bedin, T., Heady, C., & Rafter, T. (2015). Evaluation of CMIP3 and CMIP5 models over the Australian region to inform confidence in projections. *Australian Meteorological and Oceanographic Journal, 65*, 19–53.

MunichRe, Cooper, S., Hedde, C., Rauch, E., & Hartwig, R. (2015). What's going on with the weather? Webinar. Germany. Retrieved from https://www.munichre.com/site/mram/get/documents_E-1959049670/mram/assetpool.munichreamerica.wrap/PDF/2014/MunichRe_III_NatCatWebinar_01072015w.pdf

Murdoch, L. (2011). *Qld floods disaster "worst in history"*. http://www.smh.com.au/environment/weather/qld-floods-disaster-worst-in-history-20110116-19sja.html

Nuccitelli, D. (2016). *Earth is warming 50× faster than when it comes out of an ice age*. https://www.theguardian.com/environment/climate-consensus-97-per-cent/2016/feb/24/earth-is-warming-is-50x-faster-than-when-it-comes-out-of-an-ice-age

Peters, G. P., Andrew, R. M., Boden, T., Canadell, J. G., Ciais, P., Le Quéré, C., Marland, G., Raupach, M. R., & Wilson, C. (2013). The challenge to keep global warming below 2 °C. *Nature Climate Change, 3*, 4–6. https://doi.org/10.1038/nclimate1783.

Shalekoff, A. (2017). When energy efficiency is a healthy pursuit. In *Proceedings of the improving residential energy efficiency international conference*, 2017. Wollongong.

Shiel, J. J. (2017). *Low-carbon and affordable retrofits of Australian housing for climate change and scarce resource scenarios*. (PhD dissertation – forthcoming).

Shiel, J. J., Moghtaderi, B., Aynsley, R. M., Page, A., & Clarke, J. M. (2017). Decarbonising the Australian housing stock for climate change. In *Proc. of Australasian building simulation conference 2017*. Melbourne: AIRAH.

Steffen, W., Broadgate, W., Deutsch, L., Gaffney, O., & Ludwig, C. (2015). The trajectory of the anthropocene: The great acceleration. *Anthropological Review, 2*, 81–98. https://doi.org/10.1177/2053019614564785.

van Vuuren, D. P., Edmonds, J., Kainuma, M., Riahi, K., Thomson, A., Hibbard, K., Hurtt, G. C., Kram, T., Krey, V., Lamarque, J.-F., Masui, T., Meinshausen, M., Nakicenovic, N., Smith, S. J., & Rose, S. K. (2011). The representative concentration pathways: An overview. *Climatic Change, 109*, 5–31. https://doi.org/10.1007/s10584-011-0148-z.

WBCSD. (2009). *Energy efficiency in buildings: Transforming the market*. Geneva: World Business Council for Sustainable Development.

Whetton, P., Hennessy, K., Clarke, J., McInnes, K., & Kent, D. (2012). Use of representative climate futures in impact and adaptation assessment. *Climatic Change, 115*, 433–442. https://doi.org/10.1007/s10584-012-0471-z.

World Bank. (2012). *Turn down the heat: Why a 4 degree celsius warmer world must be avoided.* Washington, DC: The World Bank.

WRI. (2005). *Navigating the numbers – Greenhouse gas data and international climate policy*. Washington, DC: World Resources Institute.

Chapter 48
Miscibility Gap Alloys: A New Thermal Energy Storage Solution

Erich Kisi, Heber Sugo, Dylan Cuskelly, Thomas Fiedler, Anthony Rawson, Alex Post, James Bradley, Mark Copus, and Samuel Reed

48.1 Introduction

A majority of renewable energy sources are intermittent, and therefore energy storage is a key factor in the continuous supply of electricity from renewable sources. Almost all energy generation and usage involve heat and most energy dissipation occurs during energy conversion from one form to another. Consequently, it is more efficient for thermal systems to store energy in the form of heat whenever possible. Hence the focus of this work is on heat storage for effective electricity generation, process heat delivery, waste heat recovery and thermal management within energy systems.

Thermal storage materials have been reviewed many times in the recent literature (Rawson 2016; Xu et al. 2015; Rathod and Bannerjee 2013; Laing et al. 2012; Khare et al. 2012; Kenisarin 2010). Three primary storage types, solid sensible heat storage, liquid sensible heat storage and latent heat storage, have emerged, and some typical characteristics are compared broadly in Table 48.1. Solid sensible heat storage materials are common engineering materials such as rocks, concrete and brickwork which have the advantage of extremely low cost and high availability. Their disadvantages include very low thermal conductivity, lower energy density and limitations on maximum operating temperature. Liquid sensible heat storage materials such as thermal oils and molten salts are also typically characterised by low thermal conductivity, low to moderate energy density and strict upper and lower temperature limits. At an operational level, sensible heat storage materials also incur system penalties as heat is delivered over a wide temperature range. In

E. Kisi (✉) • H. Sugo • D. Cuskelly • T. Fiedler • A. Rawson
A. Post • J. Bradley • M. Copus • S. Reed
Structure of Advanced Materials Group, School of Engineering, University of Newcastle, Callaghan, NSW, Australia
e-mail: erich.kisi@newcastle.edu.au

A. Sayigh (ed.), *Transition Towards 100% Renewable Energy*,
Innovative Renewable Energy, https://doi.org/10.1007/978-3-319-69844-1_48

Table 48.1 Generalised characteristics of some current and proposed thermal storage material types (Rawson 2016; Kuravi et al. 2013)

Type	Maximum temperature (°C)	Thermal conductivity (W/mK)	Energy density (MJ/L)[a]
Solid sensible heat			
Concrete, rocks, sand	400–450	1–1.5	0.18–0.22
Graphite	2000	20–100	0.15
Liquid sensible heat			
Thermal oils	300–345	0.1	0.18–0.22
$NaNO_3$-KNO_3 salts	290–565	0.57	0.27
PCM			
Waxes	60–150	0.3	0.50
Alkali metal salts	400–1000	0.5–2	0.8–3

[a]All values based upon 100 °C of sensible heat capacity

solid storage materials, this means that the heat absorption or delivery is always on a heat flow transient and the energy flow rate (power) can vary by up to an order of magnitude from the top to the bottom of the operating temperature range. Latent heat in phase change materials (PCM) does not suffer from this effect as heat is absorbed/delivered over a narrow temperature band around the phase change. Commonly investigated phase change materials have high energy density; however this is generally coupled with very low thermal conductivity, a large freezing contraction and often corrosive nature making implementation troublesome (Khan et al. 2016).

Low thermal conductivity in the majority of sensible heat materials has necessitated the development of complex engineering solutions to move heat around the system using high flow rates of working fluids and heat exchange augmentation. In some cases, particularly molten salts (Kuravi et al. 2013), the working fluid is itself a liquid sensible heat storage material. Workable engineering solutions to storage using these materials have been implemented; however, there is a significant whole-of-life cost trade-off between the use of cheap storage materials and high cost, maintenance intensive heat transfer and pumping infrastructure. In addition, due to the large pumping requirement and the need to keep the salt molten, there is a large parasitic energy consumption recently found to be 12% in summer and as high as 24% in winter (Relebohele and Dinter 2015). Due to their adoption in the concentrated solar thermal power (CSP) industry, molten KNO_3-$NaNO_3$ salt mixtures are the benchmark against which all other thermal storage materials are judged.

Motivated by a desire to simultaneously overcome problems of low thermal conductivity, modest energy density and complex plant design, a different kind of thermal storage material has been constructed from an intimate mixture of two metals which are largely immiscible. By using powder metallurgical techniques to manipulate the microstructure, it is possible to manufacture a miscibility gap alloy (MGA) which contains discrete, fully encapsulated, particles of a lower melting point metal (Tm_1) trapped with a dense matrix of a higher melting point metal

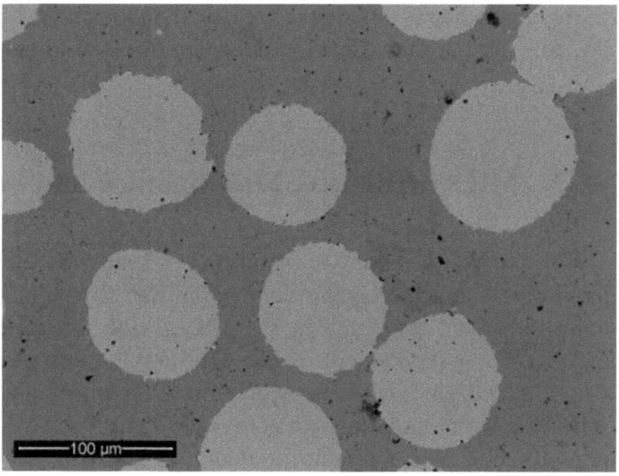

Fig. 48.1 An ideal MGA microstructure of Cu in Fe. The Cu inclusion particles do not form continuous paths and therefore remain trapped when molten

(Tm_2) (Sugo et al. 2013). An ideal MGA microstructure is shown in Fig. 48.1. If the operating temperature range is set to say $Tm_1 \pm 50$ °C, heat may be stored as a combination of the latent heat of fusion of the lower melting point metal and 100 °C of sensible heat storage.

MGA materials have many advantages for practical thermal storage systems including

1. Externally the material remains and behaves as a solid meaning: (a) the storage unit can be modular 'blocks' shaped for convenience with integrated heat transfer tubing to convey the working fluid; (b) no movement (convection, pumping, etc.) of the storage material is required, greatly reducing infrastructure, maintenance costs and plant footprint.
2. Thermal conductivity is 50–200 times greater than the majority of installed thermal storage materials. Consequently (a) heat transfer within a storage block is very rapid leading to relatively uniform temperature distribution; (b) the storage block can be directly heated from heat sources such as a concentrated solar power (CSP) receiver; and (c) heat transfer infrastructure costs can be further reduced as large tube spacings (0.5–2 m) can be used to transfer heat into working fluids such as steam to operate a turbine/generator.
3. Energy density is in the moderate to high range which means (a) further reduction in the plant footprint with associated cost savings and (b) reduced socio-environmental impact.
4. There are good opportunities for reuse and recycle: (a) modular MGA thermal storage blocks can be reconfigured if they become redundant in one application, and (b) MGA, being composed of immiscible metals, can be readily separated by melting and recycled at end of life.
5. Thermodynamically stable MGA are expected to function for decades with little maintenance.

This paper presents recent technical advances in MGA manufacture and scale-up, new alloy system developments and utilisation concepts.

48.2 Microstructure, Manufacture and Scale-Up

The key to MGA storage materials is to circumvent the microstructure which forms naturally when a melt comprising two immiscible metals (or semimetals) is allowed to solidify. As illustrated in Fig. 48.2a), the resulting solid would have the low melting point component as the continuous matrix with no mechanical stability when heated past Tm_1. To avoid this microstructure, manufacturing in the solid state by powder metallurgy was invoked to produce the microstructure illustrated in Fig. 48.2b, a real example of which is shown in Fig. 48.1 (Sugo et al. 2013). Optimisation of the manufacturing conditions for several MGA systems has been undertaken. This has comprised determining the effect of the relative particle sizes of the two components (matrix and inclusions), optimum pressing and sintering temperatures and in systems with a graphite matrix, the effectiveness of various binders. Several general conclusions can be drawn from this exercise.

Mixing and Die Loading The ability of the matrix phase to fully encapsulate the low melting point included phase that is determined first by simple geometric effects. The primary aim is to keep particles of the low melting point phase separated during powder handling and die loading and thereby avoid percolation of the molten-included phase which would facilitate its flow out of the solid matrix. Consider the process of loading the powders into dies. For equal-sized spheres of the matrix and inclusion components in a dense random packing arrangement (i.e. prior to pressing), the percolation threshold, above which it is more likely than not that continuous paths of the included phase will exist in any large ensemble of particles, is only 28.95 vol.% (Rintoul and Torquato 1997). If the microstructure of the loose powder is above the percolation threshold, it will remain so after compaction in the die as there is no physical mechanism by which the included phase can separate during pressing. As the matrix particle size is reduced relative to the inclusion size, the percolation threshold increases because the matrix particles

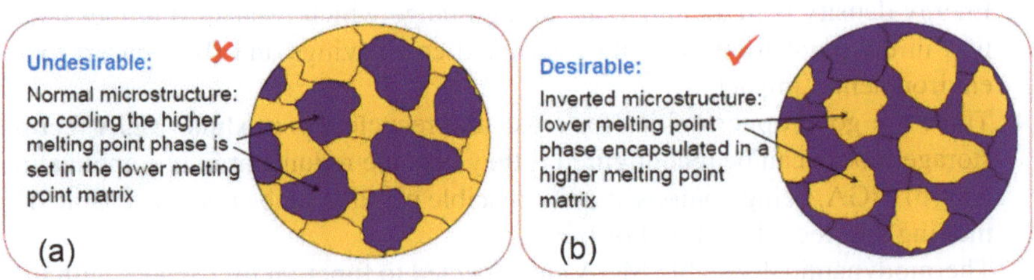

Fig. 48.2 Illustration of (**a**) the 'natural' microstructure of immiscible metals cooled from the melt and (**b**) the inverted microstructure produced using powder metallurgy

act increasingly fluidlike by filling the interstices. The upper limit of 63.4% is given by the packing density in a dense random packing of spheres (Song et al. 2008). Therefore our first general conclusion relating to manufacture is that MGA microstructures approaching this absolute limit can be manufactured when the size ratio of included particles to matrix particles is precisely tailored.

Pressing The formation of microstructures with very high volume fractions of discrete low melting point inclusion particles is assisted by plasticity in the matrix particles making them more fluidlike. Therefore, a secondary optimisation of microstructure involves the control of plastic flow and work hardening during pressing so that no included phase drains out upon heating. During early experiments (Sugo et al. 2013), the pressing pressure was arbitrarily 300 MPa. Further experiments have shown that for each system studied, there is an optimum pressing pressure which depends on the die dimensions, yield strength and work hardening exponent of the matrix material.

Sintering In many cases, the melting temperature of the included phase (T_{m1}) is higher than the sintering temperature of the matrix phase. This is the case, for example, in the Fe-Cu MGA where Fe sinters quite well at $\leq$1000 °C, whereas Cu melts at 1085 °C. The well-formed microstructure in Fig. 48.1 is the result of this situation, and it may be seen that the spheres of Cu have changed little from their initial state. In systems with a lower Tm_1, e.g. Al-Sn, the situation can be different. The Al matrix does not sinter well below the rather low melting point of Sn (232 °C) after low-pressure cold pressing (Sugo et al. 2013). Since that early work, a process of 'warm' pressing was developed whereby dies and powders are heated to 200 °C during pressing. Due to the low yield strength and low work hardening exponent of Al, an alternative is to increase the pressing pressure to 450 MPa so that plasticity in the Al is the major mechanism for encapsulation of the Sn droplets. The resulting microstructure is shown in Fig. 48.3 in a large (0.5 L) Al-Sn MGA storage block. It should be noted that Al-Sn is a model system for microstructure and manufacturing developments, although it is not likely to become commercially competitive due to the high cost of Sn and related, more economical systems (e.g. Al-Cd), respond well to the same developments.

Scale-Up For ultimate use in large-scale installations, much larger MGA blocks need to be manufactured. The storage elements in Fig. 48.3 represent an important stage in the scale-up of MGA storage technology. The 0.5 L blocks shown can store 0.26 MJ of heat in the target range of 183–283 °C. To improve the space-filling properties of stacks of modular elements, experiments have been conducted using a square cross-section die to produce flat storage tiles. A 0.75 L tile in the C-Zn system, capable of storing 0.49 MJ of heat, is shown in Fig. 48.4. Tiles such as these can be stacked into storage units in the sub-m^3 size range for small-scale use of MGA storage and also act as intermediate-sized test beds for much larger-scale implementation. Additional details are given elsewhere in these proceedings (Copus et al. 2017).

Fig. 48.3 Exterior and interior of large Al-Sn MGA samples showing well dispersed Sn particles in an Al matrix. Dark regions show Sn particles lost during cutting for examination

Fig. 48.4 Large C-Zn MGA storage block

48.3 Alloy Developments

Since our initial investigations of the Al-Sn and Fe-Cu systems (Sugo et al. 2013), a great many other systems have been investigated. There are a number of ways MGA systems may be categorised: according to the matrix material (Al, Fe, C, etc.), the number of components (binary, ternary, quaternary, etc.), the types of

Table 48.2 Properties of some MGA thermal storage materials (Rawson 2016)[b]

Type	Melting temperature of active phase (°C)	Thermal conductivity (W/mK)	Energy density (MJ/L)[a]
Sn 50% – Al	230	120	0.43
Zn 50% – C	420	70	0.65
Mg 50% – Fe	650	100	0.58
Cu 50% – Fe	1085	200	1.2
Si 50% – SiC	1410	75	2.5

[a]All values include 100 °C of sensible heat capacity

[b]A convenient graphical method of displaying the thermal properties of MGA was pioneered in earlier work and may be consulted for further comparison with other storage materials (Rawson et al. 2014)

components (all-metal, metal-semimetal, metal-nonmetal, etc.) or features relating to applications (energy density, thermal conductivity, etc.). Here we have opted for the latter and simply categorise according to the projected temperature range of operation as this is the primary driver in selecting an MGA system. The thermal properties of a range of the systems studied are given in Table 48.2 according to this criterion. The newest systems are presented elsewhere in these proceedings (Reed et al. 2017).

We envisage that low temperature systems (with phase change <300 °C) will be useful for space heating, the provision of process heat, HVAC systems and for waste heat recovery. As the crystal binding energy (per atom) scales approximately with the melting temperature of a solid, the energy density of these systems is the lowest among the MGA currently under investigation. Medium temperature systems (with phase change 300–500 °C) have higher energy density and are expected to be useful in a variety of process heat applications as well as trough concentrator-type concentrated solar thermal power (CSP) plants. Of these, the C-Zn system shows the most promise. High-temperature systems (with phase change 500–750 °C) have high energy density and are suitable for the production of superheated steam in power tower or dish CSP applications to produce steam at 500–600 °C for conventional high-efficiency subcritical Rankine cycle turbine generators as used in fossil fuel power stations. Very high-temperature systems (>750 °C) have the highest energy density and will be able to service advanced supercritical Rankine or Brayton cycles which have drawn considerable research interest. It should be noted that there are very few other thermal storage options in this temperature range. It should also be noted that maintaining the temperature of the storage material well above the temperature of the end use (steam, etc.) is not a disadvantage as judicious use of steam flow control and desuperheaters can easily adjust the steam conditions prior to entry into the turbine whilst benefitting from the very high energy density of the higher-temperature systems. Therefore, the very high-temperature systems can be adapted to conventional Rankine cycle systems.

48.4 Utilisation Concepts

The overriding concept for MGA implementation, illustrated in Fig. 48.5, is to have modular MGA blocks which can be easily assembled into larger units before being surrounded by an insulating envelope. The metallic or graphitic matrix of MGA modules can be readily shaped during manufacture or machined post-manufacture to have integral heat transfer tubing. Thermal storage blocks constructed in this way can be used in a number of ways. The configuration shown in Fig. 48.5 represents the situation where during daylight hours, a heat source such as a solar trough concentrator field provides heat to the storage block via a heat transfer fluid shown in red. As the opportunity for direct electricity generation declines at the end of the day, heat is extracted from the block, most likely by passing water into the block (shown in blue) and producing steam which can be used directly in a turbine/ generator.

There are a number of variations of this solar boiler concept. For higher-temperature sources such as solar power towers, there is an opportunity to couple the solar receiver directly to the thermal storage block. In this case, the storage block is also the boiler for 'normal' or direct operation during daylight, and there is only one heat transfer circuit.

A 1.5 kW unit to demonstrate these concepts is under construction at the University of Newcastle. The unit will use C-Zn storage modules of the type shown in Fig. 48.4, assembled into a 600 × 600 × 450 mm block (162 L). This storage block will be capable of storing and supplying 64 MJ of energy (equivalent to approximately 100 lead-acid automotive batteries). The aim is to use the stored heat to run the 1.5 kW steam impulse turbine shown in Fig. 48.6 from the stored heat. Initially, a small electric boiler will supply saturated steam at 0.5 MPa, 152 °C, and the MGA block will act only as a superheater, bringing the steam temperature into 200–220 °C range for optimum operation of the small turbine. In later trials, the electric boiler will be removed and the storage unit used to directly

Fig. 48.5 MGA modular storage block concept

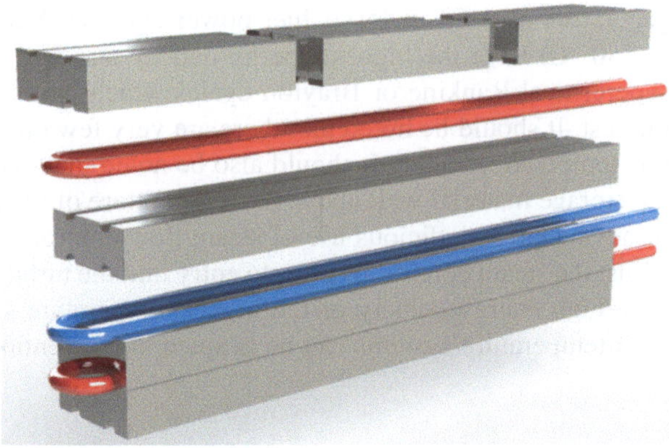

Fig. 48.6 1.5 kW turbine to be used in the demonstration unit

convert the condenser outlet stream (water at 40 °C) into superheated steam at 200–220 °C.

Even at the relatively low efficiency of the small turbine (~10%), the electricity generated would be sufficient to supply the average consumption of one household. The entire steam system has been assembled and is undergoing testing and certification. Manufacture of sufficient C-Zn thermal storage modules is underway with full testing on the unit expected in the near future.

Additional development work within the group at Newcastle includes a high-efficiency 2 kW Tesla turbine which is under development and thermal modelling of MGA storage blocks under various operational conditions. Scientific studies using in situ neutron diffraction and neutron imaging experiments to define the energy content in partially charged MGA samples are also under way (Sugo et al. 2017).

48.5 Conclusions

Miscibility gap alloys have high energy density, very high thermal conductivity, a range of operating temperature available by alloy selection and a number of plant complexity and operational advantages that make them very promising thermal storage materials. Their cost structure, though different from conventional TES materials, provides a competitive alternative (Post et al. 2017). A large number of systems have shown suitability for a variety of operating conditions. Scale-up is underway and a demonstration unit is in the final stages of construction.

References

Copus, M., Reed, S., Kisi, E. H., Sugo, H. O., & Bradley, J. (2017). Scaling up miscibility gap alloy thermal storage materials. In *Transition Towards 100% Renewable Energy: Selected papers from the World Renewable Energy Congress (WREC XVI)*, 2017, ISBN: 978-3-319-69843-4.

Kenisarin, M. M. (2010). High-temperature phase change materials for thermal energy storage. *Renewable and Sustainable Energy Reviews, 14*, 955–970.

Khan, Z., Khan, Z., & Ghafoor, A. (2016). A review of performance enhancement of PCM based latent heat storage system within the context of materials, thermal stability and compatibility. *Energy Conversion and Management, 115*, 132–158.

Khare, S., Dell'Amico, M., Knight, C., & McGarry, S. (2012). Selection of materials for high temperature latent heat energy storage. *Solar Energy Materials & Solar Cells, 107*, 20–27.

Kuravi, S., Trahan, J., Goswami, D. Y., Rahman, M., & Stefanakos, E. K. (2013). Thermal energy storage technologies and systems for concentrating solar power. *Progress in Energy and Combustion Science, 39*, 285–319.

Laing, D., Bahl, C., Bauer, T., Fiss, M., Breidenbach, N., & Hempel, M. (2012). High-temperature solid media thermal energy storage for solar thermal plants. *Proceedings of the IEEE, 100*(2), 516–524.

Post, A., Rawson, A. J., Sugo, H. O., Cuskelly, D. T., Bradley, J., & Kisi, E. H. (2017). Price estimation for miscibility gap alloy thermal storage systems. *Renewable Energy and Environmental Sustainability 2*, 32.

Rathod, M. K., & Bannerjee, J. (2013). Thermal stability of phase change materials used in latent heat energy storage systems: A review. *Renewable and Sustainable Energy Reviews, 18*, 246–258.

Rawson, A. J., Sugo, H. O., & Kisi, E. H. (2014). Characterising thermal properties of miscibility gap alloys for thermal storage applications. In *Proceedings 52nd conference of the Australian Solar Council, Melbourne*.

Rawson, A. J. (2016). *Modelling and application of advanced thermal storage materials*. PhD thesis, University of Newcastle, Newcastle.

Reed, S., Sugo, H., & Kisi, E. H. (2017). New highly thermally conductive thermal storage media. In *Transition Towards 100% Renewable Energy: Selected papers from the World Renewable Energy Congress (WREC XVI)*, 2017, ISBN: 978-3-319-69843-4.

Relebohele, J. R., & Dinter, F. (2015). Evaluation of parasitic consumption for a CSP plant. In *Proceedings solar PACES 2015*, AIP conf. proceedings 1734, document 070027 (pp. 1–8).

Rintoul, M. D., & Torquato, S. (1997). Precise determination of the critical threshold and exponents in a three – Dimensional continuum percolation model. *Journal of Physics A: Mathematical and General, 30*, L585–L592.

Song, C., Wang, P., & Makse, H. A. (2008). A phase diagram for jammed matter. *Nature, 453*, 629–632.

Sugo, H. O., Kisi, E. H., & Cuskelly, D. T. (2013). Miscibility gap alloys with inverse microstructures and high thermal conductivity for high energy density thermal storage applications. *Applied Thermal Engineering, 51*, 1345–1351.

Sugo, H. O., Kisi, E. H., Bradley, J., Fiedler, T., & Luzin, V. (2017). In situ Neutron diffraction studies of operating MGA thermal storage materials. *Renewable Energy Environmental Sustainability 2*, 34.

Xu, B., Li, P. W., & Chan, C. (2015). Application of phase change materials for thermal energy storage in concentrated solar thermal power plants: A review to recent developments. *Applied Energy, 160*, 286–307.

Index

A. Sayigh (ed.), *Transition Towards 100% Renewable Energy*,
Innovative Renewable Energy, https://doi.org/10.1007/978-3-319-69844-1

Y

Z

MIX
Papier aus verantwortungsvollen Quellen
Paper from responsible sources
FSC® C105338

If you have any concerns about our products,
you can contact us on
ProductSafety@springernature.com

In case Publisher is established outside the EU,
the EU authorized representative is:
Springer Nature Customer Service Center GmbH
Europaplatz 3, 69115 Heidelberg, Germany

Printed by Libri Plureos GmbH
in Hamburg, Germany